AF361709

Virus-Like Particle Vaccines

Virus-Like Particle Vaccines

Editors

Martin F Bachmann
Monique Vogel

MDPI • Basel • Beijing • Wuhan • Barcelona • Belgrade • Manchester • Tokyo • Cluj • Tianjin

Editors
Martin F Bachmann
University Hospital
Switzerland

Monique Vogel
University Hospital
Switzerland

Editorial Office
MDPI
St. Alban-Anlage 66
4052 Basel, Switzerland

This is a reprint of articles from the Special Issue published online in the open access journal *Viruses* (ISSN 1999-4915) (available at: https://www.mdpi.com/journal/viruses/special_issues/VLP_vaccines).

For citation purposes, cite each article independently as indicated on the article page online and as indicated below:

LastName, A.A.; LastName, B.B.; LastName, C.C. Article Title. *Journal Name* **Year**, *Article Number*, Page Range.

ISBN 978-3-03943-074-1 (Hbk)
ISBN 978-3-03943-075-8 (PDF)

Contents

About the Editors

Martin F. Bachmann (martin.bachmann@dbmr.unibe.ch) conducted his PhD at the Institute for Experimental Immunology at ETH in Zürich (1995). He has served as a postdoc at the Department of Immunology, Toronto, Canada (1995–1997); member at the Basel Institute for Immunology, Basel (1997–2000), Chief Scientific Officer at Cytos Biotechnology AG, Schlieren-Zürich (2000–2012) ; Visiting Professor of Immunology at the University of Zürich (2012–present); and Associate Professor of Immunology at the University of Oxford (2013–present). Since 2015, he has served as Head of Immunology at the Clinic of Rheumatology, Immunology and Allergology and Professor of Immunology at the University of Bern.

Monique Vogel (monique.vogel@dbmr.unibe.ch) completed her PhD at the Institute of General Microbiology, University of Bern (1989). She has served as Research Assistant and Senior Scientist at the Institute of Immunology, University of Bern (1989–2015); Consulting Immunologist for the transplantation Diagnostic Laboratory at the Institute of Immunology (2004–2015); Head of the Allergy Laboratory, INO, Inselspital, Bern (2011–2015); and, since 2015, as Deputy Head of Immunology at the Clinic of Rheumatology, Immunology and Allergology.

 *viruses*

Editorial

Special Issue "Virus-Like Particle Vaccines"

Monique Vogel [1,2] **and Martin F. Bachmann** [1,2,3,*]

1   University Hospital for Rheumatology, Immunology, and Allergology, University of Bern, 3010 Bern, Switzerland; monique.vogel@dbmr.unibe.ch
2   Department of BioMedical Research, University of Bern, 3010 Bern, Switzerland
3   Nuffield Department of Medicine, Centre for Cellular and Molecular Physiology (CCMP), The Jenner Institute, University of Oxford, Oxford OX3 7BN, UK
*   Correspondence: martin.bachmann@me.com

Received: 3 August 2020; Accepted: 4 August 2020; Published: 10 August 2020

Virus-like particles (VLPs) have become a key tool for vaccine developers and manufacturers. They can be broadly used to develop prophylactic as well as therapeutic vaccines in a vast number of indications for human as well as companion animals and animals for food production. An additional use of VLPs is to tune the type and duration of immune responses. In this Special Issue "Virus-like Particles Vaccines", essentially all of these aspects and applications are discussed and various aspects of VLP vaccinology are highlighted, including VLP characterization. The manuscript by Irene Gonzales-Dominguez et al. is an interesting example, where six different biophysical methods were assessed and compared for the characterization of HIV-1-based VLPs produced in mammalian and insect cell platforms [1]. An important role for VLPs in recent development has been their use as platforms to display antigens. In this context, Ina Balke and Andris Zeltins made an interesting contribution with respect to plant virus-derived VLPs as display platforms [2] as well as Kara-Lee Aves et al. describing the very popular Tag/Catcher system to display antigens on VLPs [3]. As expected, most of the manuscripts focused on the development of prophylactic vaccines in humans. Many VLPs are still developed against classical pathogens, such as influenza virus, norovirus, hepatitis B or E, human cytomegalovirus and human papilloma virus. Peter Pushko and Irina Tretyakova give an interesting outlook for the development of VLP-based vaccines against H7N9 influenza [4] while Arturo Cérbulo-Vazquez et al. present reports on medical outcomes in women that became pregnant after immunization with a VLP-based vaccine against Influenza H1N1 during the 2009 pandemic [5]. Maria Malm et al. make the interesting observation that simultaneous immunization with a multivalent norovirus VLP-based vaccine induces better immune responses than sequential vaccination, reminding readers of the old concept of original antigenic sin [6]. Joan Kha-Tu Ho et al. describe the classical use of HBsAg as vaccine against hepatitis B as well as a novel display platform used e.g., in the malaria vaccine RTS,S [7]. Yike Li et al. describe a novel and interesting VLP-based vaccine against Hepatitis E, currently registered in China [8]. Human Cytomegalovirus has been a long-standing vaccine target with little success. Michela Perotti and Laurent Perez describe an interesting novel VLP-based vaccine designed by structural approaches to combat this virus [9]. Virally sexually transmitted diseases (STDs) are often resistant to current therapeutic treatments. Human papillomavirus (HPV) is the most common sexually transmitted infection and some HPV types are the main causes of cervical cancers. Rashi Yadav et al. present an interesting review on a single VLP-based L2 vaccine which elicit a strong protective immune response against many different types of HPV types [10]. VLPs may not only be used to immunize human prophylactically but also animals. An interesting example for a new animal vaccine candidate is described by Fangfang Wu et al., who describe a VLP-based vaccine against Sudan Virus which is immunogenic in mice and horses [11]. An essential factor for all prophylactic vaccines is their ability to induce long-lived antibody responses, a problem discussed by Bryce Chackerian and David Peabody [12]. Therapeutic vaccines are a new and important emerging topic, covered by vaccines against cancer, atopic dermatitis and cat allergy. Jerri Caldeira et al. give a general introduction to the use of VLPs for the treatment of cancer [13]. John Foerster and Aleksandra Moleda present the concept of displaying cytokines on the surface of VLPs in order to induce anti-cytokine antibodies for the treatment of chronic disease. They use IL-13 as an example [14]. Franziska Thoms et al. finally present the concept of immunizing cats against Fel d 1, the major cat allergen in humans. This reduces Fel d 1 levels in cats and here they demonstrate that this improves the interaction of the allergic cat owner with his cat, as the two can spend more quality time together due to reduced allergic symptoms [15].

Conflicts of Interest: M.F.B. is involved with the development of several VLP-based vaccines. He is a founder and shareholder of Saiba AG, Evax AG, Hypopet AG, DeepVax GmbH and HealVax GmbH. M.V. declares no conflict of interest.

References

1. Gonzalez-Dominguez, I.; Puente-Massaguer, E.; Cervera, L.; Godia, F. Quality Assessment of Virus-Like Particles at Single Particle Level: A Comparative Study. *Viruses* **2020**, *12*, 223. [CrossRef] [PubMed]
2. Balke, I.; Zeltins, A. Recent Advances in the Use of Plant Virus-Like Particles as Vaccines. *Viruses* **2020**, *12*, 270. [CrossRef] [PubMed]
3. Aves, K.L.; Goksoyr, L.; Sander, A.F. Advantages and Prospects of Tag/Catcher Mediated Antigen Display on Capsid-Like Particle-Based Vaccines. *Viruses* **2020**, *12*, 185. [CrossRef] [PubMed]
4. Pushko, P.; Tretyakova, I. Influenza Virus Like Particles (Vlps): Opportunities for H7n9 Vaccine Development. *Viruses* **2020**, *12*, 518. [CrossRef] [PubMed]
5. Cérbulo-Vázquez, A.; Arriaga-Pizano, L.; Cruz-Cureño, G.; Boscó-Gárate, I.; Ferat-Osorio, E.; Pastelin-Palacios, R.; Figueroa-Damian, R.; Castro-Eguiluz, D.; Mancilla-Ramirez, J.; Isibasi, A.; et al. Medical Outcomes in Women Who Became Pregnant after Vaccination with a Virus-Like Particle Experimental Vaccine against Influenza a (H1n1) 2009 Virus Tested During 2009 Pandemic Outbreak. *Viruses* **2019**, *11*, 868. [CrossRef] [PubMed]
6. Malm, M.; Vesikari, T.; Blazevic, V. Simultaneous Immunization with Multivalent Norovirus Vlps Induces Better Protective Immune Responses to Norovirus Than Sequential Immunization. *Viruses* **2019**, *11*, 1018. [CrossRef] [PubMed]
7. Ho, J.K.; Jeevan-Raj, B.; Netter, H.J. Hepatitis B Virus (Hbv) Subviral Particles as Protective Vaccines and Vaccine Platforms. *Viruses* **2020**, *12*, 126. [CrossRef] [PubMed]
8. Li, Y.; Huang, X.; Zhang, Z.; Li, S.; Zhang, J.; Xia, N.; Zhao, Q. Prophylactic Hepatitis E Vaccines: Antigenic Analysis and Serological Evaluation. *Viruses* **2020**, *12*, 109. [CrossRef] [PubMed]
9. Perotti, M.; Perez, L. Virus-Like Particles and Nanoparticles for Vaccine Development against Hcmv. *Viruses* **2020**, *12*, 35. [CrossRef] [PubMed]
10. Yadav, R.; Zhai, L.; Tumban, E. Virus-Like Particle-Based L2 Vaccines against Hpvs: Where Are We Today? *Viruses* **2020**, *12*, 18. [CrossRef] [PubMed]
11. Wu, F.; Zhang, S.; Zhang, Y.; Mo, R.; Yan, F.; Wang, H.; Wong, G.; Chi, H.; Wang, T.; Feng, N.; et al. A Chimeric Sudan Virus-Like Particle Vaccine Candidate Produced by a Recombinant Baculovirus System Induces Specific Immune Responses in Mice and Horses. *Viruses* **2020**, *12*, 64. [CrossRef] [PubMed]
12. Chackerian, B.; Peabody, D.S. Factors That Govern the Induction of Long-Lived Antibody Responses. *Viruses* **2020**, *12*, 74. [CrossRef] [PubMed]
13. Caldeira, J.C.; Perrine, M.; Pericle, F.; Cavallo, F. Virus-Like Particles as an Immunogenic Platform for Cancer Vaccines. *Viruses* **2020**, *12*, 448. [CrossRef] [PubMed]
14. Foerster, J.; Moleda, A. Virus-Like Particle-Mediated Vaccination against Interleukin-13 May Harbour General Anti-Allergic Potential Beyond Atopic Dermatitis. *Viruses* **2020**, *12*, 438. [CrossRef] [PubMed]
15. Thoms, F.; Haas, S.; Erhart, A.; Nett, C.S.; Rufenacht, S.; Graf, N.; Strods, A.; Patil, G.; Leenadevi, T.; Fontaine, M.C.; et al. Immunization of Cats against Fel D 1 Results in Reduced Allergic Symptoms of Owners. *Viruses* **2020**, *12*, 228. [CrossRef] [PubMed]

Article

Quality Assessment of Virus-Like Particles at Single Particle Level: A Comparative Study

Irene González-Domínguez *,†, **Eduard Puente-Massaguer** *,†, **Laura Cervera and Francesc Gòdia**

Departament d'Enginyeria Química Biològica i Ambiental, Universitat Autònoma de Barcelona, Cerdanyola del Vallès, 08193 Barcelona, Spain; laura.cervera@uab.cat (L.C.); francesc.godia@uab.cat (F.G.)
* Correspondence: irene.gonzalez@uab.cat (I.G.-D.); eduard.puente.massaguer@gmail.com (E.P.-M.);
  Tel.: +34-93-58-13302 (I.G.-D.)
† These authors contributed equally to the work.

Received: 22 December 2019; Accepted: 11 February 2020; Published: 17 February 2020

Abstract: Virus-like particles (VLPs) have emerged as a powerful scaffold for antigen presentation and delivery strategies. Compared to single protein-based therapeutics, quality assessment requires a higher degree of refinement due to the structure of VLPs and their similar properties to extracellular vesicles (EVs). Advances in the field of nanotechnology with single particle and high-resolution analysis techniques provide appealing approaches to VLP characterization. In this study, six different biophysical methods have been assessed for the characterization of HIV-1-based VLPs produced in mammalian and insect cell platforms. Sample preparation and equipment set-up were optimized for the six strategies evaluated. Electron Microscopy (EM) disclosed the presence of several types of EVs within VLP preparations and cryogenic transmission electron microscopy (cryo-TEM) resulted in the best technique to resolve the VLP ultrastructure. The use of super-resolution fluorescence microscopy (SRFM), nanoparticle tracking analysis (NTA) and flow virometry enabled the high throughput quantification of VLPs. Interestingly, differences in the determination of nanoparticle concentration were observed between techniques. Moreover, NTA and flow virometry allowed the quantification of both EVs and VLPs within the same experiment while analyzing particle size distribution (PSD), simultaneously. These results provide new insights into the use of different analytical tools to monitor the production of nanoparticle-based biologicals and their associated contaminants.

Keywords: VLP; viral quantification; NTA; flow virometry; SRFM; cryo-TEM; SEM

---

1. Introduction

Virus-like particles (VLPs) are considered a promising platform in the field of vaccine development. Nowadays, there are several licensed VLP-based vaccines, such as Cervarix®, Gardasil®, Hecolin® or Porcilis PCV® and more than 100 candidates are undergoing clinical trials [1]. Their success as immunogens lies on their ability to mimic native viruses without containing a viral genome. Their highly organized and repetitive antigen structure has shown effective cellular and humoral immune responses [2]. Furthermore, advances in the field of bioengineering have widened their possible applications; VLP technology accepts several modifications including encapsulation, chemical conjugation or genetic engineering. By doing so, VLPs can be pseudotyped or used either as DNA or drug nanocarriers [1,3].

VLP quality assessment is of major importance since both, the physicochemical and biological properties, are responsible of their clinical efficacy. The preservation of their structural integrity during all the stages of vaccine manufacturing, storage and administration is critical to ensure their success [4]. The study of particle size distribution (PSD) or particle concentration are some of the critical quality attributes (CQA) that could be monitored in this regard [5]. Overall, the specific detection and quantification of VLPs entails several difficulties, especially for enveloped VLPs, which are composed of a protein capsid surrounded by the host-cell lipid membrane. VLPs must be distinguished from other similar nanovesicle structures; extracellular vesicles (EVs), including exosomes and microvesicles [6], adventitious viruses, or baculoviruses (BV) in insect cell systems [7], are important process-related impurities. In this sense, traditional quantification methods such as $TCID_{50}$ or PCR have a limited applicability due to the non-infective nature of VLPs.

Comprehensive studies on VLP-based vaccine candidates have been conducted by multiple approaches, including biochemical, biological and biophysical methods [3]. Biochemical protein gels, biological enzyme-linked immunosorbent assay (ELISA) or immunoblot are normally used [8–11]. Nonetheless, these assays cannot distinguish assembled from non-assembled structures [12]. Among biophysical methods, analytical ultracentrifugation, dynamic light scattering (DLS) and transmission electron microscopy (TEM) are the reference methods used to assess VLP physical properties [3]. Recently, technical progress in the field of microscopy, as well as the application of nanotechnology to virology, have given rise to several single nanoparticle analytical technologies. These techniques represent the most advanced methods to evaluate VLP size, polydispersity, purity and even nanoparticle composition simultaneously [3,12].

Among them, electron microscopy (EM) has traditionally been the preferred technique since resolution at the nanometric or even atomic level is achieved [13]. Within EM methods, transmission (TEM), scanning (SEM) and cryogenic (cryo-TEM) methodologies are frequently used. TEM is the gold standard technique for the characterization of virus-like structures as reported in a myriad of studies [3,14]. This methodology requires a contrast medium for sample visualization, typically a heavy metal solution containing a cationic or anionic salt, being negative staining the most extended strategy [15]. In TEM-Negative staining, a thin layer of biological material is covered by a dried non-crystalline amorphous layer of a heavy metal salt, typically uranyl acetate. Differential electron scattering between the biological material and the surrounding staining layer enables the visualization of the specimen. The application of SEM in the characterization of different materials has been demonstrated in several works [16]. However, few studies address its use as a tool for VLP characterization. The addition of Alcian Blue solution to the grid before sample deposition results in the activation of the grid with a net positive charge, with successful results reported for the visualization of other negatively charged specimens such as nucleic acids [17]. Since viral structures and EVs are known to have an overall negative charge at physiological pH [18,19], this strategy could improve their adsorption and reduce nanoparticle loss during the sample preparation process. Cryo-TEM has also gained increasing interest as a tool for nanoparticle visualization over the last years [13]. Essentially, this technique enables the visualization of viruses and VLPs in their native conformation at nanometric and even atomic scale [3], and the addition of a contrast solution is not required. A key point of this technique is the rapid freezing process which reduces sample damage. Therefore, the selection of an adequate grid and support film is of upmost importance since the correct formation of a thin ice film is pivotal for an adequate sample visualization. Perforated carbon films are generally the preferred option since they allow the biological material to be imaged in the ice generated between holes in the carbon support film [20,21].

The study of viral vectors and nanoparticles by confocal microscopy has been traditionally restricted by the Abbe diffraction limit. However, the appearance of super-resolution fluorescence microscopy (SRFM) enabling to surpass this constraint has opened a breadth of opportunities to apply confocal microscopy to the nanoscale [22]. Despite SRFM has been mainly used to study cellular processes, its application to appraise individual viral structures is becoming more popular [23]. In previous works, a method for VLP quantification by HyVolution2 SRFM has been described by González-Domínguez and co-workers [24,25]. This method combines sub-Airy confocal microscopy with mathematical deconvolution, which has been described to achieve resolutions up to 140 nm [26]. Finally, light scatter-based technologies, such as nanoparticle tracking analysis (NTA) and flow virometry are also gaining attention for viral particle and EV quantification [5,12,27,28]. NTA is a method to characterize and quantify nanoparticles in solution that relates the rate of Brownian motion to nanoparticle size. Its use in the assessment of nanoparticles has been reported for viruses, VLPs and other nanoparticles [5,29–31]. This technique is theoretically able to detect nanoparticles with a size comprised between 30 and 1000 nm, but the nanoparticle concentration has to be maintained around 10^8 particles/mL and 20–60 particles/frame [29]. The latter indicates that the range of possible nanoparticle concentrations is narrow, and it is often required to dilute the sample to meet this criterion, which is generally based on trial-and-error. Flow virometry has recently emerged as a technique to specifically detect viruses similarly to conventional cell-based flow cytometry [27]. Labeling studies at single particle level, particle quantification or virus sorting are some of the applications that can be performed with this technology [27]. Considering the high difference in volume between a cell and a nanoparticle, which can be one million-fold [32], the acquisition settings need to be adjusted to detect the scattered or fluorescence signal from nanoparticles. Still, a significant loss of scattered light that fall in the range of the background noise of the instrument and different sensitivities between equipments are a general concern [33]. To address this issue, the implementation of the violet (405 nm) side scatter (V-SSC) has been reported to improve the sensitivity but also the resolution of the technique [34]. Owing to the specific features of each analytical method, characterization results such as particle concentration obtained by different techniques are often difficult to compare.

The aim of this work is the characterization of VLPs using several advanced nanoparticle analytical methods, and to discuss the technological limitations that may affect their use, including sample preparation or equipment set-up. The study of PSD, particle ultrastructural analysis, VLP quantification and differentiation from other nanoparticle subpopulations has been performed. The VLPs analyzed in this work are HIV-1 Gag VLPs, which are a promising platform for the development of a vaccine candidate against HIV, but also as a scaffold for chimeric or multivalent vaccine development [2,35]. Upon expression in the host cell, the Gag polyprotein travels to the cell membrane and after an oligomerization process, HIV-1 Gag VLPs are released from the cell through a budding process [36,37]. Thus, the final nanoparticles are enveloped by the host cell lipid membrane [38], with sizes comprised between 100–200 nm. In previous works, the Gag polyprotein has been fused to GFP to track the VLP production process [36]. By doing so, VLPs could be easily quantified and distinguished from other contaminant particles. Besides, product characteristics are known to be affected by the expression system selected for VLP production [39]. Here, the two most relevant systems for the generation of HIV-1 Gag-based VLPs have been used [40]. VLPs obtained by transient gene expression in HEK 293 cells and baculovirus (BV) infection in Sf9 cells have been characterized in parallel. This study provides relevant data on the use of different analytical methods to evaluate the production of VLPs and their associated contaminants in animal cell-based bioprocesses.

2. Materials and Methods

2.1. HEK 293 Mammalian Cell Line, Culture Conditions and Transient Transfection

The mammalian cell line used in this work is a serum-free suspension-adapted HEK 293 cell line (HEK 293SF-3F6 from NRC, Montreal, QC, Canada) kindly provided by Dr. Amine Kamen from McGill University (Montreal, QC, Canada). Cells were cultured as previously described [36]. HIV-1 Gag-eGFP VLP production was achieved by transient transfection. Briefly, the pGag-eGFP plasmid encoding for the Gag-eGFP polyprotein, is diluted with FreeStyle 293 medium (Invitrogen, Carlsbad, CA, USA) and vortexed for 10 s, then polyethylenimine (PEI) is added at 1:2 DNA:PEI ratio (*w/w*) and vortexed three times, the mixture is incubated for 15 min at RT and is added to the cell culture, where a medium exchange has been already performed. HIV-1 Gag-eGFP VLPs were harvested at 72 h post transfection (hpt) by centrifugation at 1000× *g* during 15 min. Supernatants were stored at 4 °C until analysis. Non-transfected negative controls reproducing cell growth conditions were also produced, for comparison.

2.2. Sf9 Insect Cell Line, Culture Conditions and Baculovirus Infection

The suspension-adapted lepidopteran insect cell line used in this work is the *Spodoptera frugiperda* (Sf9, cat. num. 71104, Merck, Darmstadt, Germany) gently provided by Dr. Berrow (Institute of Biomedical Research, Barcelona, Spain). Sf9 cells were cultured in Sf900III medium (Thermo Fisher Scientific, Grand Island, NY, USA) in 125-mL disposable polycarbonate Erlenmeyer flasks [25]. Cell cultures were shaken at 130 rpm using an orbital shaker (Stuart, Stone, United Kingdom) and maintained at 27 °C. HIV-1 Gag-eGFP VLP production was achieved through infection with the recombinant baculovirus *Autographa californica* multiple nucleopolyhedrovirus (*Ac*MNPV) (BD Biosciences, San José, CA, USA) enconding for the Gag-eGFP protein. Shortly, Sf9 cells were grown to $2–3 \times 10^6$ cells/mL and were infected at a multiplicity of infection (MOI) of 1. HIV-1 Gag-eGFP VLPs were harvested at 40 or 72 h post infection (hpi) by centrifugation at 1000× g during 15 min, and supernatants were kept at 4 °C until analysis. Non-infected negative controls reproducing cell growth conditions were also produced, for comparison.

Biophysical analyses were performed on HIV-1 Gag-eGFP VLP supernatants obtained as previously described. FreeStyle and Sf900III cell culture media and conditioned cell culture media obtained from HEK 293 and Sf9 non-transfected/infected conditions were also assessed in each analysis as negative controls.

2.3. Electron Microscopy (EM)

EM analyses were performed at Servei de Microscòpia at Universitat Autònoma de Barcelona (UAB, Barcelona, Spain). Particle size distribution (PSD) analyses were performed with ImageJ Fiji (ImageJ, NIH, WI, USA) and SigmaPlot 12.0 (Systat Software Inc., San Jose, CA, USA).

2.3.1. Transmission Electron Microscopy (TEM)-Negative Staining

Prior to negative staining, HIV-1 Gag-eGFP VLPs were concentrated by double sucrose cushion 25–45% (w:v) at 31.000 rpm and centrifuged for 2.5 h at 4 °C in a Beckman Optima L100XP centrifuge using a SW32 Ti rotor (Beckman Coulter, Brea, CA, USA). The 25–45% interphase was recovered and stored at 4 °C until analysis [41]. TEM micrographs were analyzed after air-dried negative staining. The protocol used is summarized in Figure 1A. Briefly, VLP samples were deposited onto carbon-coated copper or Holey carbon 200 mesh grids (Micro to Nano, Wateringweg, the Netherlands). Grids were glow discharged in a PELCO easiGlow glow discharge unit (PELCO, Fresno, CA, USA). Thereafter, 8 μL of sample were loaded onto the grid and incubated at RT for 1 min. Excess sample was carefully drained off the grid with the aid of filter paper. Samples were negatively stained with 8 μL of 2% w:v uranyl acetate by incubation at RT for 1 min. Excess stain was drained off as previously indicated and grids were dried at RT until analysis. TEM examinations were performed with a Jeol JEM-1400 (JEOL USA, Pleasanton, CA, USA) transmission electron microscope equipped with an ES1000W Erlangshen charge-coupled device camera (CCD) (Model No. 785; Gatan, Pleasanton, CA, USA).

2.3.2. Scanning Electron Microscopy (SEM)-Alcian Blue Staining

HIV-1 Gag-eGFP VLP visualization by SEM was performed by staining VLP-containing supernatants with Alcian Blue solution, adapted from Gállego [17]. The protocol used is shown in Figure 2A. Briefly, Alcian Blue solution 1% w:v in 3% v:v acetic acid (Sigma Aldrich, St Louis, MO, USA) was diluted in ultrapure water achieving a final concentration of 1 μg/mL. Carbon-coated copper or Holey carbon 200 mesh grids were placed in this solution for 5 min, then the excess of Alcian Blue solution was removed by washing the grid in ultrapure water. Grids were dried with filter paper. Thereafter, 8 μL of the sample were placed on the grid and incubated at RT for 5 min. Excess sample was carefully drained off the grid with the aid of filter paper. Negative staining was applied to the samples when required, using the protocol described in Section 2.3.1. SEM micrographs were performed in a FE-SEM Merlin scanning electron microscope (Zeiss, Jena, Germany). Micrographs were taken with the lens mode and secondary electron detector, with an electron high tension (EHT) comprised between 1–2 eV and 2.9–7.5 mm of working distance using a protocol adapted from González-Domínguez et al. [31].

2.3.3. Cryogenic Transmission Electron Microscopy (cryo-TEM)

Cryo-TEM analyses of HIV-1 Gag-eGFP VLPs were conducted from harvested supernatants. In Figure 3A, the sample preparation procedure for VLP visualization is presented. 2–3 μL of sample were blotted onto 200 or 400 mesh Holey carbon grids (Micro to Nano, Wateringweg, the Netherlands) previously glow discharged in a PELCO easiGlow glow discharge unit. Samples were subsequently plunged into liquid ethane at −180 °C using a Leica EM GP cryo workstation and observed in a Jeol JEM-2011 TEM electron microscope operating at 200 kV. During imaging, samples were maintained at −173 °C, and pictures were taken using a CCD multiscan camera (Model No. 895, Gatan).

2.4. Super-Resolution Fluorescence Microscopy (SRFM)

SRFM was performed with a TCS SP8 confocal microscope equipped with Huygens deconvolution suite embedded via a direct interface with LAS X software and GPU arrays (Leica Microsystems, Wetzlar, Germany) at Servei d'Anatomia Patològica from Hospital Sant Joan de Déu (Esplugues de Llobregat, Barcelona, Spain), as previously described [24]. A summary of the protocol is depicted in Figure 4A. Briefly, harvested VLPs were directly loaded onto the microscope slide and adsorbed to the surface of the cover glass after an incubation time of 30 min at RT (Figure 4A). HIV-1 Gag-eGFP VLP preparations were analyzed with 100 X magnification (zoom 5), a line average of 3 and 496 x 496 pixels with HC PL APO CS2 100 X/1.40 OIL objective with the HyVolution2 mode (Leica). Five fields of 13 sections per each biological triplicate were studied in harvested HIV-1 Gag-eGFP VLPs. Deconvolution was performed with the SVI Huygens Professional program and the best resolution strategy (Scientific Volume Imaging B.V., Hilversum, the Netherlands). HIV-1 Gag-eGFP VLP concentration was calculated based on the division of particle number by 3D image volume as previously described [24]. Briefly, direct quantification was performed on deposited HIV-1 Gag-eGFP VLP samples with $23 \times 23 \times 3$ μm in *xyz* from a total loaded volume of 50 μL distributed in 24×60 mm under the cover glass and a total height of 34 μm. Assuming complete sample deposition, minimum concentration of Gag-eGFP VLPs was also calculated. PSD analyses were performed using SigmaPlot 12.0 software.

2.5. Nanoparticle Tracking Analysis (NTA)

HIV-1 Gag-eGFP VLPs and total particle content were analyzed by NTA. A NanoSight® NS300 device (Malvern Panalytical, Malvern, United Kingdom) equipped with a blue filter module (488 nm) and a neutral filter was used to quantify GFP-fluorescent nanoparticles and total particle by light diffraction, respectively. The measurements were performed at Service of Preparation and Characterization of Soft Materials located at Institut de Ciència de Materials de Barcelona (ICMAB, CSIC, Campus UAB). The workflow used for HIV-1 Gag-eGFP VLP quantification is summarized in Figure 5A. Prior to injection into the device chamber, each sample was diluted to obtain 1 mL sample with a concentration of around 10^8 particles/mL. Sample injection into the chamber was continuously added using a pump to improve the robustness of the measurement and minimize the photobleaching effect due to fluorescence depletion over time (Figure 5B) [42]. The videos recorded were then analyzed with the NTA 3.2 software (Malvern Panalytical) by tracking the individual particle movement, where camera level and detection threshold were adjusted manually for each sample (Table 1). Three independent analyses were carried out and videos of 60 s were recorded at RT, with particles identified and tracked by their Brownian motion. HIV-1 Gag-eGFP VLP concentrations were calculated as the total fluorescent particles and the concentration of EVs was calculated as the difference between light scattering particles and fluorescent particles. PSD analyses were performed with NTA 3.2 software and SigmaPlot 12.0 software.

Table 1. NTA analysis settings.

	VLPs		Total Nanoparticles	
2-5	Camera Level	Threshold	Camera Level	Threshold
1-5	16	4	10	4
HEK 293 supernatants				
Sf9 supernatants	16	3	8	3
HEK 293 conditioned medium	-	-	10	4
Sf9 conditioned medium	-	-	14	3
FreeStyle culture medium	-	-	13	5
Sf900III culture medium	-	-	13	4

2.6. Flow Virometry

Flow virometry experiments were performed with a CytoFLEX LX (Beckman Coulter) with violet side scatter (V-SSC) 405 nm filter configuration at Servei de Cultius Cel·lulars, Producció d'Anticossos i Citometria (UAB, Barcelona, Spain). The different steps required in the analysis are described in Figure 6A. Measurements from different experiments were standardized using a mixture of Megamix-Plus Side Scatter and Forward Scatter (FSC) fluorescent polystyrene beads (100, 160, 200, 240, 300, 500 and 900 nm; Biocytex, Marseille, France). The threshold of height trigger signal in Violet Side Scatter (V-SSC) was manually adjusted to 1200 and laser gains were set as 72, 9 and 106 for FSC, V-SSC and B525-FITC, respectively. Samples were diluted with 1 X Dulbecco's phosphate-buffered saline (DPBS, Thermo Fisher Scientific) until the abort rate value was below the 2%. Three-hundred thousand events were analyzed per sample at a flow rate of 10 µL/min. V-SSC vs B525-FITC density plots were used to gate the different EVs and VLPs populations. Gating was manually adjusted for each channel. Results were analyzed with the CytExpert v.2.3 software (Beckman Coulter). Nanoparticle concentrations were calculated with Equation (1):

$$Particle\ concentration\left(\frac{Events}{mL}\right) = (Events)\cdot\frac{\mu L}{mL}\cdot Dilution \tag{1}$$

Particle size diameters based on the median violet side scatter were calculated by Mie correlation [43] using FCM$_{PASS}$ software [33]. Megamix-Plus SSC and FSC fluorescent polystyrene beads were used as the reference material with a refractive index (RI) of 1.633 with a 405 nm illumination wavelength [43]. The following RI were used for Mie correlation: 1.374 (vesicle cytosol), 1.394 (vesicle upper cytosol), 1.354 (vesicle lower cytosol), 1.474 (vesicle membrane) and 1.345 (vesicle surrounding medium) considering 405 nm as the illumination wavelength. The vesicle membrane thickness was defined as 10 nm [43]. The half angle parameter was set as circular aperture geometry. The maximum diameter was fixed as 3000 nm, the triggering threshold as 60 arbitrary units (a.u.) and the log decade number parameter comprised between 0–6. Thereafter, HIV-1 Gag-eGFP VLP and EV mean sizes were exported with the FlowJo v.10 software (BD Biosciences).

3. Results

3.1. Transmission Electron Microscopy (TEM)-Negative Staining

Virus-like particle (VLP) characterization by TEM with uranyl acetate as the negative staining solution is shown in Figure 1. VLPs were concentrated by ultracentrifugation according to previous studies [36]. VLPs produced in HEK 293 (Figure 1B,D) and Sf9 cells (Figure 1C,E) were examined in order to identify differences arising from the expression of these nanoparticles in each specific cell platform. Uranium strongly reacts with phosphate and amino groups and typically stains proteins, nucleic acids and lipid membranes [44]. Thus, VLPs were observed as spherical electrodense structures surrounded by a bright corona (white arrows) that might correspond to structured Gag-eGFP monomers surrounded by the lipid membrane [36,45]. As expected, the presence of baculoviruses (BV) (black arrows) was detected along with VLPs (white arrows) in infected Sf9 samples. EVs were observed as less electrodense nanoparticles (dashed grey arrows) and that could also be identified in conditioned medium samples (Supplementary materials S1) [46]. Interestingly, differences between VLPs and EVs were less evident in Sf9 micrographs compared to HEK 293 ones. A mean population diameter of 165 ± 54 nm (n = 71) and 146 ± 33 nm (n = 69) was quantified in HEK 293 and Sf9 samples, respectively (Figure 1F,G). Moreover, TEM characterizations revealed that only the 52 ± 6% and 55 ± 14% of the nanoparticles analyzed were VLPs in HEK 293 and Sf9 concentrated samples. The use of negative staining for sample visualization hindered a more detailed characterization of the structure of the different specimens due to the presence of artifacts and background noise in all samples (Figure 1B,E) [1].

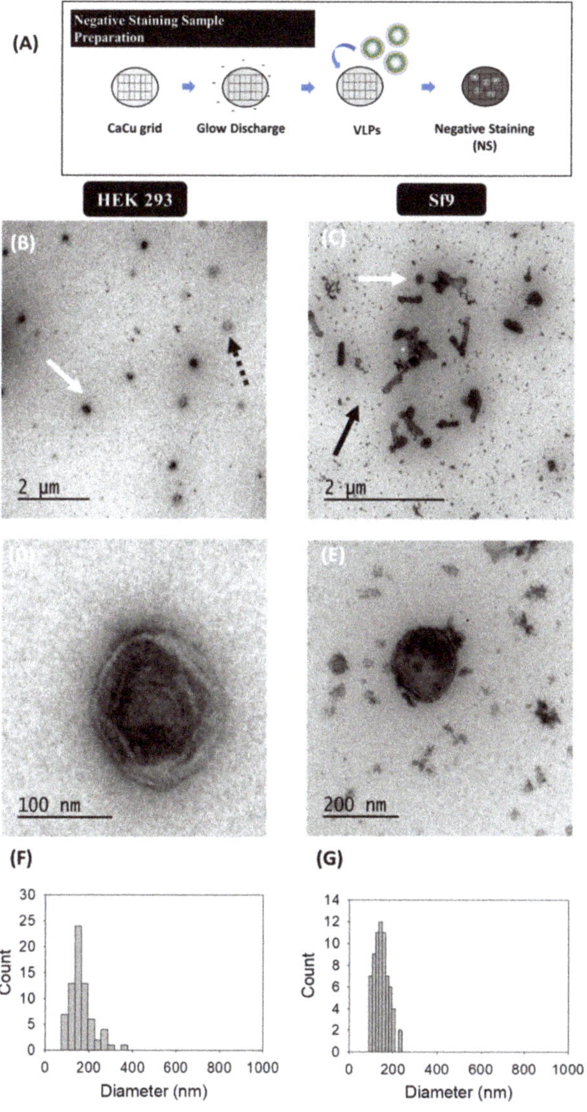

Figure 1. Transmission electron microscopy with negative staining analysis of HIV-1 Gag-eGFP VLPs produced in HEK 293 and Sf9 cells harvested at 72 hpt and 72 hpi, respectively. (**A**) Sample preparation

protocol; (**B–E**) samples corresponding to the 25–45% ultracentrifugation fraction from HEK 293 supernatants (**B,D**) and Sf9 supernatants (**C,E**). All samples were stained with uranyl acetate in Ca-Cu grids. (**F,G**) PSD analysis of HEK 293 and Sf9 concentrated samples, respectively. White and grey dashed arrows indicate the presence of VLPs and EVs in HEK 293 and Sf9 productions, respectively, and black arrows indicate the presence of BVs in Sf9 samples. Negative controls were assessed using the same conditions as in VLP samples (Supplementary materials S1).

3.2. Scanning Electron Microscopy (SEM)-Alcian Blue Staining

SEM analyses were directly performed on harvested VLP supernatants from transfected HEK 293 (Figure 2B,D,F) and BV infected Sf9 cells (Figure 2E,G). Initial analyses using standard sample preparation protocols resulted in the loss of a high portion of the nanoparticles prior to visualization [47]. Thus, the combination of SEM with Alcian Blue solution was investigated (Figure 2) and compared to uranyl acetate staining as the reference methodology. The comparison between negative staining, Alcian Blue grid pre-treatment followed by negative staining, and only Alcian Blue grid pre-treatment is shown in Figure 2B–E, respectively.

Samples stained with uranyl acetate presented an important background regardless the grid treatment or not with Alcian Blue solution (Figure 2B,C and Figure S1). Indeed, large and irregular salt stacks deposited on the sample grid were detected probably related to the interaction of uranium with phosphate salts and amino acids from cell culture media, with similar results as observed in TEM micrographs (Figure 1). A decrease in the load of background signal was achieved by Alcian Blue grid pre-treatment without negative staining (white arrows) and nanoparticles could be individually resolved as 3D sphere-like structures. Despite the improvement in nanoparticle resolution achieved by the Alcian Blue grid pre-treatment, VLPs and EVs could not be distinguished in these analyses. Thus, the calculation of particle size distribution (PSD) was performed considering all nanoparticles as a single population, resulting in 296 ± 88 nm ($n = 94$) for HEK 293 and 162 ± 60 nm ($n = 57$) for Sf9 supernatants. The presence of EVs in conditioned supernatants could also be observed by SEM-Alcian Blue staining (Supplementary Materials S1), as well as the typical rod-shaped capsids of BVs (black arrows) in infected Sf9 supernatants (Figure 2E).

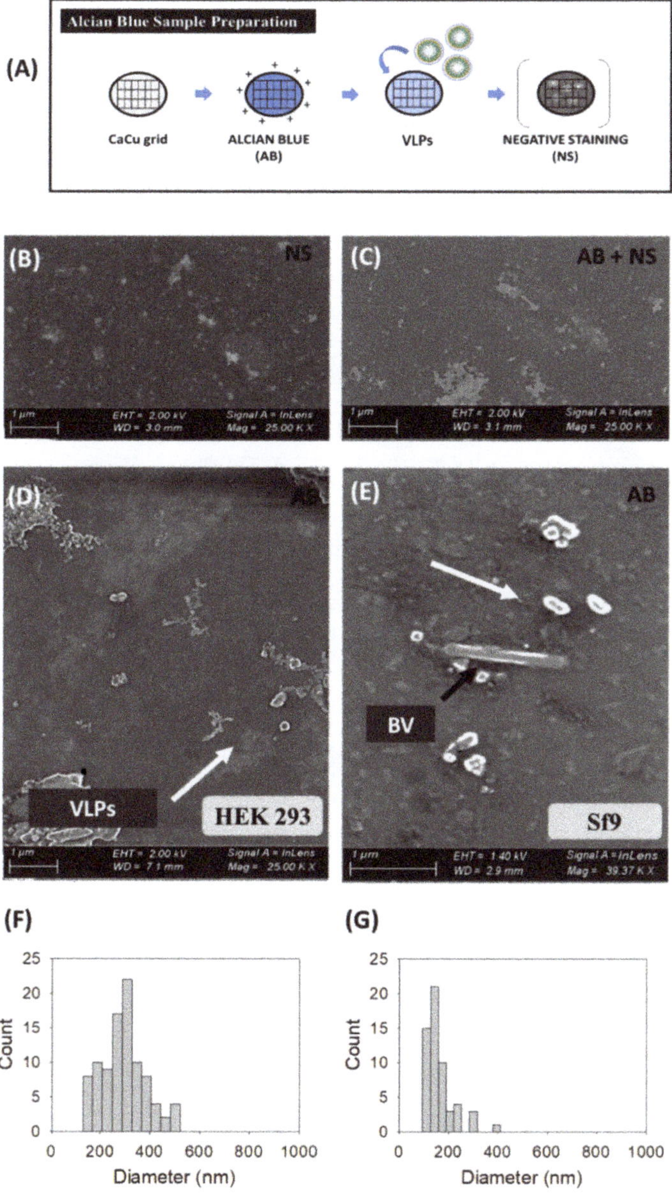

Figure 2. Scanning electron microscopy analysis of HIV-1 Gag-eGFP VLPs produced in HEK 293 and Sf9 cell lines harvested at 72 hpt and 72 hpi, respectively. (**A**) Sample preparation protocol; (**B–E**) Comparison of different sample preparation methods: (**B**) negative staining (NS) (**C**), combining Alcian Blue solution and negative staining (AB + NS) on HEK 293 supernatants; (**D**) Supernatant from transfected HEK 293 cells and (**E**) supernatant from BV infected Sf9 cells treated with Alcian Blue (AB) for 1 min in a Holey carbon 200 mesh grid. (**F–G**) PSD analysis of HEK 293 and Sf9 supernatants, respectively. White arrows indicate the presence of nanoparticles in HEK 293 and Sf9 supernatants and black arrows indicate the presence of BV. Negative controls were analyzed using the same conditions as in VLP samples (Supplementary materials S1).

3.3. Cryogenic Transmission Electron Microscopy (cryo-TEM)

Ultrastructural analysis of VLP, EV and BV populations was conducted in harvested supernatants by cryo-TEM (Figure 3B–F). Gag-eGFP VLPs were observed as electrodense nanoparticles surrounded by a lipid envelope with a granular-like heterogeneous internal structure (Figure 3B,C, white arrows). Gag-eGFP VLPs produced in both platforms displayed an average size of 202 ± 68 nm ($n = 59$) for HEK 293 and 146 ± 42 nm ($n = 188$) for Sf9 cells (Figure 3G,H, respectively). In parallel, different EV subpopulations could be detected, including exosomes (30–100 nm), microvesicles (50–2000 nm) and multivesicular bodies (MVB) (Figure 3D–F) [6]. Similar nanoparticle populations were detected in conditioned medium samples from both cell lines (Supplementary materials S1), indicating a basal expression of EVs in these cell lines. As for infected Sf9 supernatants, a large concentration of BVs was detected (black arrows), encompassing different BV phenotypes: the occlusion-derived BV (ODV), a relaxed form of the BV (rBV) (Figure 3E) and the typical infective BV (budded virus) (Figure 3F). ODVs presented several rod-shaped nucleocapsids arranged in parallel disposition inside vesicular bodies (light blue arrow), while a spiral-like nucleocapsid organization could be clearly distinguished in rBV (mid blue arrow) [7]. Budded BVs displayed ovoid-like structures containing one nucleocapsid, with the DNA highly compacted in supercoiled structures (black arrow). Moreover, the apical spikes and the BV lateral pocket, which is the space between the lipid bilayer and the nucleocapsid, were resolved using this technique (Figure 3G) [7]. Other BV forms could be identified in infected Sf9 samples and consisted of BV containing vesicles or protein structures besides the nucleocapsid (cBV, dark blue arrow, Figure 3E). Differences in Gag-eGFP VLPs (green) and Gag VLPs (brown) were also evaluated by cryo-TEM (Figure 3I,J). Gag VLPs evidenced a higher internal degree of ordered Gag arrangement in comparison to Gag-eGFP VLPs, similar to that of immature HIV-1 virions [38], while Gag-eGFP VLPs did not attain such level of structural organization.

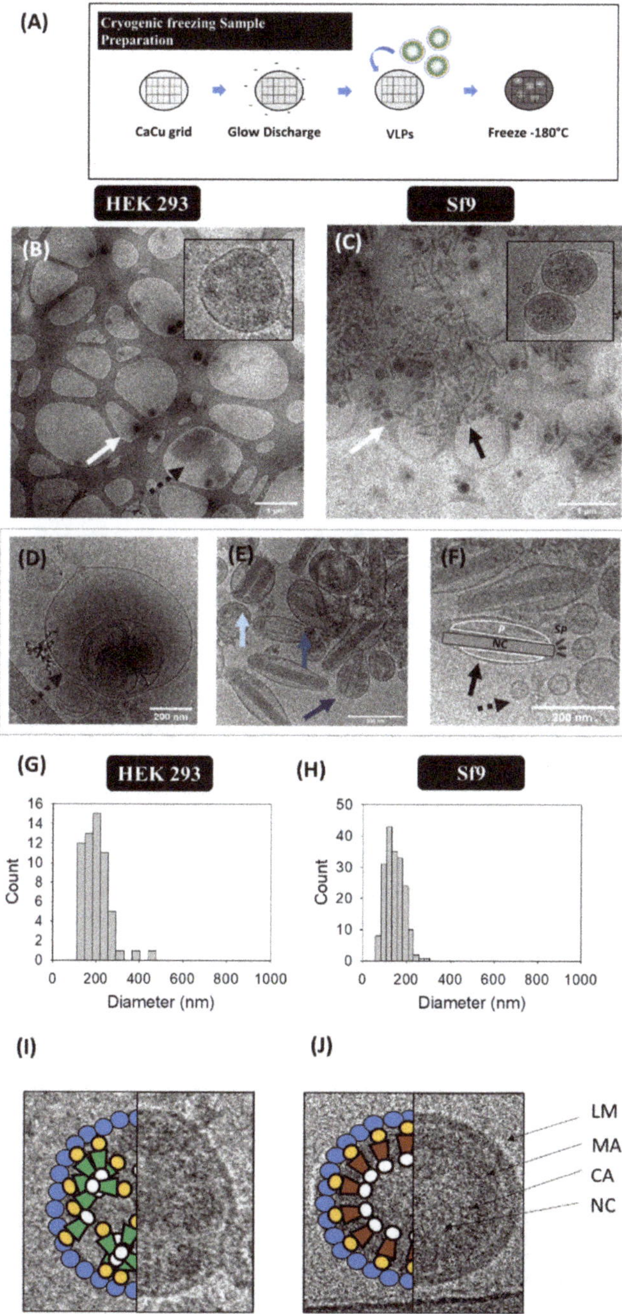

Figure 3. Cryo-TEM analysis of HIV-1 Gag-eGFP VLPs produced in HEK 293 and Sf9 cells harvested at 72 hpt and 72 hpi, respectively. (**A**) Sample preparation protocol. Supernatant from HEK 293 (**B**) and Sf9 cells (**C**). Both samples were prepared and visualized in Holey carbon grids. (**D**–**F**) Morphological

characterization of contaminant particles including multivesicular bodies (MVB) (**D**), occlusion-derived BV (ODV, light blue), a relaxed-form of the BV (rBV, mid blue) or BV-containing vesicles or protein structures (cBV, dark blue) (**E**), and analysis of the infective BV (budded virus) structure: nucleocapsid (NC), lateral pocket side (P) and apical spikes (SP) (**F**). (**G–H**) PSD analysis of HEK 293 and Sf9 supernatants, respectively. White arrows indicate the presence of VLPs, dashed grey arrows point EVs and infective BVs are shown in black arrows. (**I,J**) Ultrastructural organization of an HIV-1 Gag-eGFP VLP (**I**) and an HIV-1 Gag VLP (**J**) produced in Sf9 cells by BV infection. LM: lipid membrane, MA: matrix; CA: capsid, NC: nucleocapsid. Negative controls were assessed using the same conditions as in VLP samples (Supplementary materials S1).

3.4. Super-Resolution Fluorescence Microscopy (SRFM)

HyVolution2 SRFM nanoparticle analyses were conducted in HEK 293 and Sf9 supernatants by a simple preparation strategy similar to common cell visualization methods used in confocal microscopy. Images were acquired at 100 X magnification as depicted in Figure 4B, where VLPs correspond to green dots. VLPs presented an elongated structure in the Z plane while a smaller diameter was observed in the XY axis due to the convolution effect of light. A mean VLP diameter of 268 ± 77 nm ($n = 371$) was obtained in HEK 293 samples, whereas an average of 261 ± 104 nm ($n = 794$) was measured for Sf9 supernatants (Figure 4C,D). VLP concentrations were comprised in the range of 10^9–10^{10} VLPs/mL in both platforms, representing a 1.7-fold difference for Sf9 over HEK 293-derived VLPs (Table 2). The same acquisition strategy was performed in conditioned medium samples from both cell lines, and no detection of GFP signal was observed (Supplementary materials S1).

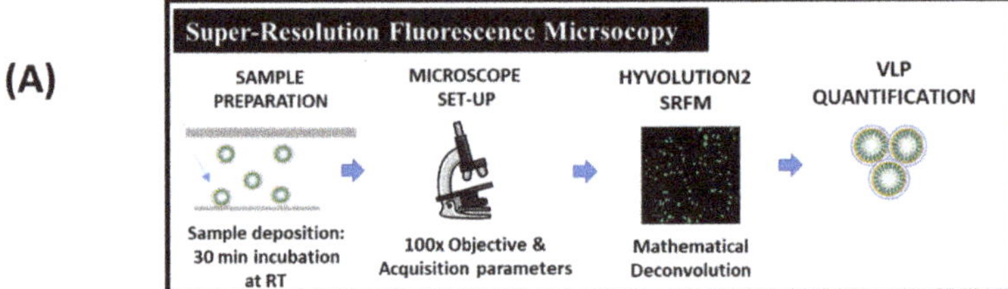

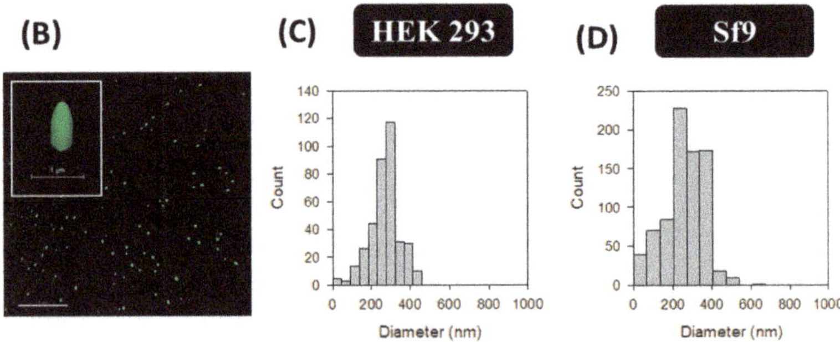

Figure 4. Super-resolution fluorescence microscopy analysis of HIV-1 Gag-eGFP VLPs produced in HEK 293 and Sf9 cells harvested at 72 hpt and 40 hpi, respectively. (**A**) Sample preparation and analysis workflow and (**B**) Gag-eGFP VLP imaging. (**C,D**) PSD analysis of HEK 293 and Sf9 supernatants directly quantified after their addition to the microscope slide, respectively. Negative controls were analyzed in the same conditions as in VLP samples (Supplementary materials S1).

Table 2. Quantification of VLP and EV concentrations in transfected HEK 293 and BV infected Sf9 supernatants using different analytical methods.

Technique	VLPs (10^9/mL)			EVs (10^9/mL)	
	FV	NTA	SRFM	FV	NTA
HEK 293	0.2 ± 0.0	5.0 ± 0.5 *	$4.0\text{--}43.0$ *	0.2 ± 0.0	16.6 ± 2.9 *
Sf9	0.2 ± 0.0	8.6 ± 2.1 *	$6.6\text{--}74.5$ *	0.3 ± 0.0	9.7 ± 1.6 *
Fold difference (Sf9:HEK 293)	1.5	1.7	1.7	1.2	0.6

* From González-Domínguez et al. [24].

3.5. Nanoparticle Tracking Analysis (NTA)

The total amount of nanoparticles produced in each platform was evaluated by NTA with the light scattering mode and VLPs were analyzed with the fluorescent filter module (Figure 5A). The PSD and concentration of the different nanoparticle populations was conducted by the combination of both modules. An average diameter of 143 ± 39 nm for VLPs and 161 ± 66 nm considering all nanoparticles as a single population was measured in HEK 293 samples. As for Sf9 supernatants, an average diameter of 213 ± 95 nm was measured for VLPs and 194 ± 75 nm for total nanoparticles. Comparison of VLP concentrations in both platforms resulted in a 1.7-fold increase in Sf9 compared to HEK 293 samples (Figure 5C,D). On the contrary, a higher EV content was found in HEK 293 samples (Table 2). Analysis of original FreeStyle and Sf900III cell culture media by light scattering NTA resulted in a concentration of $0.6 \pm 0.1 \times 10^9$ and $24.4 \pm 1.1 \times 10^9$ diffracting particles/mL, respectively, with a mean size diameter of 174 ± 39 nm and 92.7 ± 74.6 nm. Thus, a difference of two orders of magnitude in particle concentration was obtained in Sf900III over FreeStyle medium. Evaluation of conditioned FreeStyle and Sf900III media was also assessed by NTA and yielded a concentration of $17.6 \pm 0.9 \times 10^9$ and $27.3 \pm 1.5 \times 10^9$ diffracting particles/ml, respectively, with a mean diameter of 174 ± 65 nm and 101 ± 47 nm for HEK 293 and Sf9 samples, respectively.

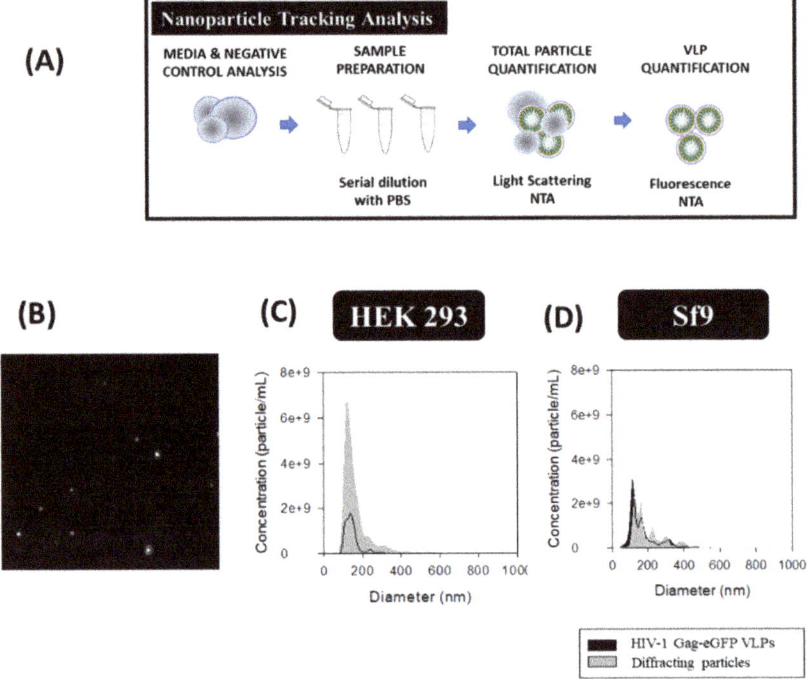

Figure 5. Nanoparticle tracking analysis of HIV-1 Gag-eGFP VLPs produced in HEK 293 and Sf9 cells harvested at 72 hpt and 40 hpi, respectively. (**A**) Sample preparation and analysis workflow. (**B**) Image of nanoparticles tracked by NTA. (**C,D**) PSD analysis of HEK 293 and Sf9 supernatants, respectively. HEK 293 and Sf9 supernatants were diluted with filtered DPBS to adjust their concentration prior to analysis. Negative controls were analyzed in the same conditions as in VLP samples (Supplementary materials S1).

3.6. Flow Virometry

The production of HIV-1 Gag-eGFP VLPs and other nanoparticle populations was simultaneously analyzed by flow virometry. Equipment calibration for VLP analysis was performed with commercial beads of known diameter (100, 160, 200, 240, 300, 500 and 900 nm) in order to calculate the mean nanoparticle diameter (Figure 6B, B.1 and B.2). Mie scatter modelling allowed to convert V-SSC intensities of VLPs and EVs to their corresponding size with FCM$_{PASS}$ software developed by Welsh and co-authors (Figures 3 and 6 and Supplementary materials S2) [33]. By doing so, the detection of nanoparticles with diameters down to 100 nm could be achieved.

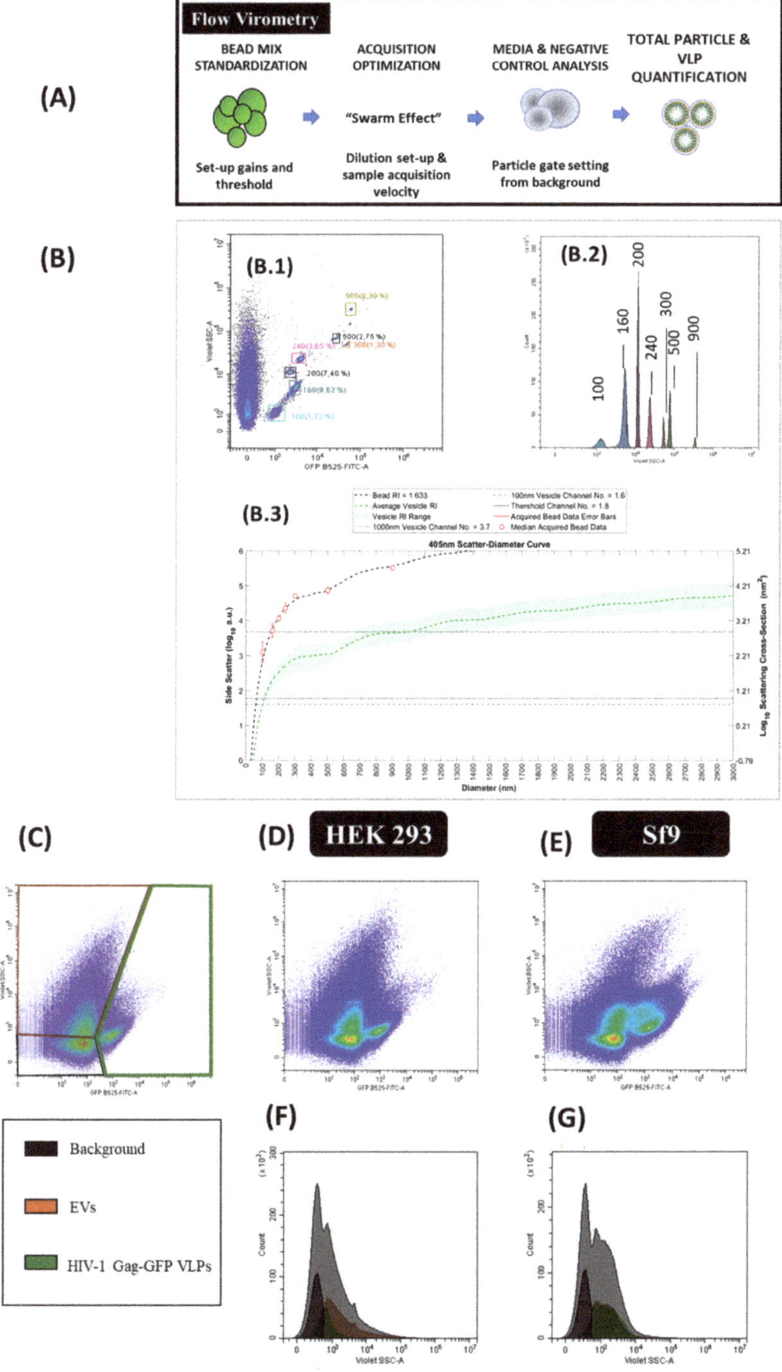

Figure 6. Flow virometry analysis of HIV-1 Gag-eGFP VLPs produced in HEK 293 and Sf9 cells harvested at 72 hpt and 40 hpi, respectively. (**A**) Sample preparation and analysis workflow and (**B**) equipment calibration using beads with a known diameter. Megamix-Plus fluorescent beads of 100,

160, 200, 240, 300, 500 and 900 nm were used. (B.1) GFP B525-FITC-A vs. V-SSC density plots of the beads, (B.2) GFP B525-FITC-A histogram of the beads and (B.3) Mie correlation of the beads with FCM$_{PASS}$ software [33]. (**C**) Gating of the different nanoparticle populations in a density plot and defined as EVs, HIV-1 Gag-eGFP VLPs and background signal of the equipment. (**D–G**) Density plots and histograms of nanoparticles produced in HEK 293 (**D,F**) and Sf9 supernatants (**E,G**), respectively. HEK 293 and Sf9 supernatants were diluted with filtered DPBS prior to analysis. Negative controls were analyzed using the same conditions as in VLP samples (Supplementary materials S1).

After equipment set-up, different nanoparticle populations could be assessed in one single analysis. Three main populations were detected in transfected HEK 293 and BV infected Sf9 samples (Figure 6G). Two of these populations corresponded to EVs, classified as small and large EVs, and the third one was related to VLPs (fluorescent particles). EVs displayed a high level of heterogeneity in the V-SSC, with values ranging from 10^2 to 10^5 a.u. Interestingly, a second subpopulation of VLPs with a higher V-SSC intensity was detected in BV infected Sf9 samples, which could be probably associated to the aggregation of VLPs or the interaction of VLPs with other cellular compounds released to the medium. This second VLP subpopulation displayed a more pronounced right-skewed V-SSC distribution compared to the more homogeneous VLP population observed in HEK 293 samples (Figure 6F). In terms of quantification, VLP concentrations in the range of 10^8 particles/mL were measured with a 1.5-fold increase in VLP concentration in Sf9 over HEK 293 supernatants (Table 2). As regards EVs, higher levels of these nanoparticles were quantified in conditioned media from Sf9 over HEK 293 cells, respectively (Supplementary materials S1). Mean size analysis of VLPs and EVs by Mie correlation resulted in 114 ± 26 nm for VLPs and 115 ± 26 nm for EVs in HEK 293 supernatants, respectively, while a mean diameter of 117 ± 22 nm and 115 ± 24 nm was measured for VLPs and EVs in Sf9 supernatants (Figure 6D–G).

4. Discussion

4.1. Sample Preparation and Equipment Set-up

Among the different techniques used, the cost, technical requirements and the time needed for sample analysis are practical issues that have to be considered to select an adequate analytical technique for nanoparticle characterization (Table 3) [12]. Regarding sample preparation, purified VLPs were loaded and stained by the addition of uranyl acetate in TEM-Negative staining. Altogether, sample preparation time required less than 10 min per sample, without considering the previous ultracentrifugation step, which required one day of experimentation. Compared to the TEM-Negative staining protocol (Figure 1A), SEM-Alcian Blue grids were not treated by glow discharge before the addition of Alcian Blue solution. The VLP-containing supernatant was directly loaded, representing an overall process duration of 15 min. As for cryo-TEM, sample preparation required an approximate time of 10 min; however, the equipment set-up required a longer time of pre-conditioning compared to TEM and SEM due to the low temperature conditions employed. In SRFM, acquisition of 3D VLP images took around 10 min per field and data analysis with Imaris software required 15–20 min to process the different channels and construct the final image. As for NTA, the time needed to conduct a complete analysis lasted 1 h due to sample preparation and software analysis (Figure 5A), while in flow virometry the different nanoparticle populations could be analyzed in one single run (Table 3).

Table 3. Comparison of the different analytical technologies assessed in this work.

Parameter	TEM	SEM	Cryo-TEM	HyVolution2 SRFM	NTA	Flow Virometry
Expertise	+++	+++	+++++	+++	++	++
Time of measurement	++++	++++	+++++	+++	++	+
Resolution	Nanometric (<10 nm) [14]	Nanometric (<10 nm) [48]	Atomic [13]	Nanometric (140 nm) [26]	Nanometric (30 nm) [29]	Nanometric (100–200 nm) [5]
Pros	Widely used, ultrastructural analysis	Surface analysis of nanoparticle populations	Ultrastructural analysis in its native form	Direct visualization, compatible with fluorescence	Easy handling, compatible with fluorescence	Simultaneous analysis of light scattering and fluorescence labelling
Cons	Staining required, working under vacuum conditions	Working under vacuum conditions	Time consuming, working at cryogenic temperatures	Data analysis, convolution effect of light	Screening effect, variability due to acquisition settings	Sensitivity, inter-equipment variability

In most cases, sample preparation and equipment set-up are based on trial-and-error, which can introduce a certain degree of variability in the characterization of nanoparticles. In the case of TEM, the low electron density present in biological samples as well as the need to work under vacuum conditions require the addition of contrast agents for a better resolution of the specimens under evaluation. Uranium strongly reacts with phosphate and amino groups [44], thus conferring a higher level of electron density to biological samples and VLPs are observed as round structures surrounded by a bright corona. Despite the simplicity of negative staining, the presence of artifacts and background due to the composition of the contrast medium are detected (Figure 1B–D), as reported in the literature [3,15]. Depending on the sample origin, this can have a major impact, as observed in Sf9 cell micrographs in comparison to HEK 293 samples, where a higher level of background was identified and EVs could not be distinguished from VLPs (Figure 1C). This background is probably caused by the larger load of particles present in insect cell culture media [49]. In this sense, new commercially available chemically defined insect cell culture media could alleviate this problem [50]. Different approaches have been investigated to overcome the drawbacks of negative staining [1]. SEM-Alcian Blue has been studied in this work as a possible alternative. On the one hand, TEM has been traditionally used for viral intraparticle characterization, while SEM is generally applied for a broader screening of biological populations [48,51]. Novel preparation methods, such as ionic liquid infiltration [51] or virus quantification by Prep/g [52] have been reported as strategies to assess bacterial and viral preparations with SEM in the recent years. In the same line, the use of Alcian Blue as a solution for grid pre-treatment was shown to increase the adsorption of nanoparticles in this work, thus allowing to improve the detection of these particles avoiding negative staining (Figure 2). Nevertheless, the differentiation of VLPs from EVs was not possible due their similar morphology, but this new label-free method has a potential applicability in combination with X-ray spectroscopy [31], SRFM [53], or "wet" SEM [51] to deepen into nanoparticle characterization. Alternatively, the low contrast present in biological samples can also be overcome in EM by using cryo-TEM, which does not require sample staining (Figure 3). However, an adjustment of the cryogenic freezing protocol for each sample depending on its physicochemical properties is required [21]. Overall, all EM techniques demand a high expertise by the user and often entail long sample analysis times (Table 3). Therefore, the development of more automatic and high throughput complementary methods is required to process nanoparticle samples in a faster manner.

SRFM, NTA and flow virometry can be implemented to this purpose since they allow mass quantification of nanoparticles. Nonetheless, sample standardization and technical expertise is still critical for an adequate nanoparticle assessment (Table 3). A microscope equipped with advanced capture and image processing modules, and the optimization of acquisition conditions together with different imaging software are required in SRFM. Additional process automatization should be developed to widen its application since the time required for sample analysis is still high and operator-dependent. In NTA, relevant parameters to be considered during nanoparticle analysis are the camera level and the detection threshold [5]. These settings need to be manually adjusted in each sample for an adequate tracking of the nanoparticles recorded and their automatic quantification. NTA measurements depend on the refractive index (RI) of each particle and the nanoparticle containing solution. Thus, different specific settings were applied to measure HEK 293 and Sf9 supernatants (Table 1). Another important feature affecting the final output with NTA is size heterogeneity of nanoparticle populations in the sample, which hinder the quantification of the small subpopulations due to the screening effect of higher particles, as reported by van der Pol et al. [5]. Regarding flow virometry, differences in the measurements due to inter-equipment variability have been pointed as one of the major challenges of this technique. Nevertheless, the use of commercial beads as standard nanoparticles and the implementation of Mie correlations have contributed to solve this problem [33,52,54]. Similarly, tunable resistive pulse sensing (TRPS), Atomic Force Microscopy (AFM) and field-flow fractionation coupled to multiangle light scattering (FFF-MALS) are single particle analysis methods that can be implemented for nanoparticle characterization [12]. However, none of these techniques enable to differentiate VLPs from EVs without additional sample treatments.

4.2. Ultrastructural Analysis

Single particle evaluation conducted by SRFM, NTA and flow virometry is an interesting option for characterization analyses since specific labeling can be applied to study different nanoparticle populations [27,55]. The use of the fluorescently tagged Gag polyprotein enabled the differentiation of VLPs from the other co-produced nanoparticle populations in this work. These analytical methods have also been successfully used in combination with immunolabeling to quantitatively asses viral populations [27]. A step further has been recently achieved in SRFM with the possibility to detect nucleic acids and the lipid membrane in VLPs [25]. The commercially available ViroCyt®flow cytometer has been similarly used to quantify different viral isolates, including the Ebola virus [56]. Still, the detection of VLPs with ViroCyt® has not been reported [57]. Nevertheless, none of these techniques achieve the levels of nanoparticle resolution of EM methods (Table 3).

Among the different EM methods evaluated, cryo-TEM showed a higher benefit compared to TEM and SEM since the native conformation of the different nanoparticles could be assessed. The ultrastructural analysis of the different VLP, EV and BV subpopulations could be described in detail, with remarkable differences detected between Gag-eGFP and Gag VLPs. The Gag polyprotein is known to travel to the vicinity of the plasma membrane, aggregate with other Gag monomers, and bud to the extracellular space as an immature HIV-1 particle (Figure 3J) [38,58]. Interestingly, Gag-eGFP VLPs do not achieve the expected organized structure as Gag VLPs probably due to the eGFP fusion that alters the native budding process [59]. These differences could not be detected by TEM due to the interference of negative staining with the samples [11,36,60]. To our knowledge, this is the first time were Gag-eGFP VLP intraparticle organization is observed. Therefore, it is shown that fusion proteins can introduce morphological alterations in the structure of the nanoparticles produced and cryo-TEM is an interesting method to identify them.

4.3. Particle Size Distribution

The PSD of VLPs has been studied with the six methods evaluated in this work. Typically, the VLP mean diameter is comprised between 100 and 200 nm, as observed by TEM, cryo-TEM, NTA and flow virometry [61]. Non-symmetric right-skewed distributions were observed for VLPs measured with the different techniques. Similar PSDs of HIV-1 Gag VLPs produced in CAP-T and CHO cells were shown by Gutierrez-Granados et al. and Steppert et al., respectively [37,62]. The presence of larger population of VLPs, especially when measuring Sf9 samples by NTA and flow virometry, could be related to the aggregation of VLPs or their interaction with other cellular compounds released to the medium. SEM and SRFM presented higher PSD in comparison to the rest of analytical technologies. SEM analysis of HEK 293 samples resulted in a PSD of 296 ± 88 nm, which could indicate that several EVs were quantified instead of VLPs since microvesicles can reach up to 2000 nm [6]. HyVolution2 SRFM yielded the highest VLP PSD values in both cell platforms (>250 nm), which has also been observed by Xiao and co-workers when analyzing Cy5-labeled adeno-associated viruses [63], and could be attributed to the convolution effect of light. In this case, the individual fluorochrome intensity does not necessarily correlate with the real particle size and might result in particle size diameter overestimation [64]. The use of more powerful SRFM techniques, such as stimulated emission depletion (STED) or stochastic optical reconstruction microscopy (STORM) could reduce this effect [65]. However, STED and STORM require specific fluorophores that tolerate the high laser intensities applied and also a higher level of expertise and equipment infrastructure in comparison to HyVolution2 SRFM.

The PSD of EVs was assessed by NTA and flow virometry (Figures 5 and 6). In both cases, EVs presented an average diameter comprised between 100 and 200 nm, similar to that of VLPs. Nonetheless, a right-skewed distribution with vesicles diameters from 45 to 500 nm was observed, especially in Sf9 samples. These results correlate with the existence of several types of EVs, encompassing exosomes and microvesicles, and BVs as visualized in cryo-TEM micrographs (Figure 3). Of note, the PSD of rod-shaped baculoviruses could not be accurately characterized by any of these methods since they assume that all nanoparticles are spherical. Van der Pol and coworkers also observed the size heterogeneity of EVs, with diameters ranging between 70 and 800 nm. They report a minimum detection limit for EVs of 70–90 nm with NTA and 150–190 nm by flow virometry [5]. Lower EV detection limits of 45 nm with NTA and 100 nm by flow virometry were reported in this work. The sample composition and also the implementation of the V-SSC instead of the 488 nm SSC in flow virometry could contribute to explain these differences [34]. Still, a higher level of accuracy is possible by using National Institute of Standards and Technology (NIST)-traceable beads in the Mie correlation with flow virometry.

4.4. Particle Concentration

SRFM, NTA and flow virometry were compared for the quantification of nanoparticles in HEK 293 and Sf9 samples (Figures 4–6 and Table 2). Compared to SRFM and NTA, flow virometry yielded a lower concentration of nanoparticles being around 30-fold lower for VLPs and >60-fold lower for EVs (Table 2). Similar differences in quantification by flow virometry have also been reported by van der Pol and co-authors in the analysis of urinary vesicles [5]. Differences in nanoparticle quantification between techniques could be related to the swarm effect, which is based on the fact that more than one particle passes through the detector simultaneously. However, the analysis of the same sample dilutions by flow virometry and NTA resulted in a linear correlation between both methods (VLPs, R^2 > 0.99; total particles R^2 > 0.91), indicating that the swarm effect could not be the main reason behind these results (Supplementary materials S3). On the other hand, it is also possible that the detection of small particles fell in the range of the background signal of the cytometer, thus contributing to reduce the final titers. Even though the average particle size of EVs and VLPs was above the minimum detection particle diameter of flow virometry, additional refinement of the Mie correlation could still be required to adequately explain these data.

Analysis of the different nanoparticle populations in each platform revealed that at least 50% of the total particles produced did not correspond to VLPs (Table 2). The possibility to discriminate but also simultaneously quantify VLPs and the rest of specimens by NTA and flow virometry represents a remarkable advantage for nanoparticle-based bioprocesses. However, further refinement of light scattered-based methodologies is required since a remarkable background signal is detected in medium solutions devoid of particles (Supplementary materials S1).

5. Conclusions

The selection of an adequate analytical method is essential in VLP characterization processes. Among the six methodologies studied in this work, cryo-TEM was shown as the best method to resolve nanoparticle structures while maintaining its native conformation. This technique allowed for a detailed characterization of the different EV and BV subpopulations co-produced with VLPs in HEK293 and Sf9 productions. Alternatively, the high-throughput analysis of VLPs and their differentiation from other contaminant particles was achieved by flow virometry, SRFM and NTA. Among them, flow virometry showed to be the fastest method for PSD analysis while also allowing to simultaneously quantify different nanoparticle subpopulations. Nonetheless, further improvements of these methods are required since different quantification results are observed between techniques.

Supplementary Materials: The following are available online at http://www.mdpi.com/1999-4915/12/2/223/s1, Figure S1: Negative control analyses: cell culture and conditioned media controls, Figure S2: Mie correlation for EV analysis with FCM$_{PASS}$ software, Figure S3: Nanoparticle correlation between flow virometry and NTA.

Author Contributions: Conceptualization, I.G.-D. and E.P.-M.; methodology, I.G.-D. and E.P.-M.; formal analysis, I.G.-D. and E.P.-M.; investigation, I.G.-D. and E.P.-M.; writing—original draft preparation, I.G.-D. and E.P.-M.; review, editing and supervision, L.C. and F.G. All authors have read and agreed to the published version of the manuscript.

Funding: Irene González-Domínguez (FPU16/02555) and Eduard Puente-Massaguer (FPU15/03577) are recipients of an FPU grant from Ministerio de Educación, Cultura y Deporte of Spain. The research group is recognized as 2017 SGR 898 by Generalitat de Catalunya.

Acknowledgments: The help of Manuela Costa (Servei de Cultius Cel·lulars, Producció d'Anticossos i Citometria, UAB) and Ángel Calvache and Jorge Fomaro (Beckman Coulter) for facilitating access to the Cytoflex LX equipment and the Megamix-Plus beads is very appreciated. The authors would like to acknowledge the help of Marti de Cabo and Emma Rossinyol (Servei de Microscòpia, UAB, Barcelona, Spain) for the assistance with EM methodologies, Mònica Roldan for the support with SRFM (Servei d'Anatomia Patològica, Hospital Sant Joan de Déu, Barcelona, Spain) and Joshua Welsh (NIH, Bethesda, MD, USA) for the help with FCM$_{PASS}$ Software.

Abbreviations

AB	Alcian Blue
AFM	Atomic Force Microscopy
BV	Baculovirus
CQA	Critical Quality Attributes
Cryo-TEM	Cryogenic Transmission Electron Microscopy
DLS	Dynamic Light Scattering
EM	Electron Microscopy
EV	Extracellular Vesicle
cBV	filled baculovirus with vesicle-like or protein content
LM	Lipid Membrane
MVB	Multivesicular Body
NS	Negative Staining
NTA	Nanoparticle Tracking Analysis
ODV	Occlusion-Derived Baculovirus
PSD	Particle Size Distribution
RI	Refractive Index
rBV	relaxed-form of baculovirus
RT	Room Temperature
SEM	Scanning Electron Microscopy
SRFM	Super-Resolution Fluorescence Microscopy
TEM	Transmission Electron Microscopy
TRPS	Tunable Resistive Pulse Sensing
V-SSC	Violet Side Scatter
VLP	Virus-Like Particle

References

1. Charlton Hume, H.K.; Vidigal, J.; Carrondo, M.J.T.; Middelberg, A.P.J.; Roldão, A.; Lua, L.H.L. Synthetic biology for bioengineering virus-like particle vaccines. *Biotechnol. Bioeng.* **2019**, *116*, 919–935. [CrossRef]
2. Cervera, L.; Gòdia, F.; Tarrés-Freixas, F.; Aguilar-Gurrieri, C.; Carrillo, J.; Blanco, J.; Gutiérrez-Granados, S. Production of HIV-1-based virus-like particles for vaccination: Achievements and limits. *Appl. Microbiol. Biotechnol.* **2019**, *103*, 7367–7384. [CrossRef] [PubMed]
3. Lua, L.H.L.; Connors, N.K.; Sainsbury, F.; Chuan, Y.P.; Wibowo, N.; Middelberg, A.P.J. Bioengineering virus-like particles as vaccines. *Biotechnol. Bioeng.* **2014**, *111*, 425–440. [CrossRef] [PubMed]
4. Jain, N.K.; Sahni, N.; Kumru, O.S.; Joshi, S.B.; Volkin, D.B.; Russell Middaugh, C. Formulation and stabilization of recombinant protein based virus-like particle vaccines. *Adv. Drug Deliv. Rev.* **2015**, *93*, 42–55. [CrossRef] [PubMed]
5. Van der Pol, E.; Coumans, F.a.W.; Grootemaat, a.E.; Gardiner, C.; Sargent, I.L.; Harrison, P.; Sturk, A.; van Leeuwen, T.G.; Nieuwland, R. Particle size distribution of exosomes and microvesicles determined by transmission electron microscopy, flow cytometry, nanoparticle tracking analysis, and resistive pulse sensing. *J. Thromb. Haemost.* **2014**, *12*, 1182–1192. [CrossRef] [PubMed]
6. Akers, J.C.; Gonda, D.; Kim, R.; Carter, B.S.; Chen, C.C. Biogenesis of extracellular vesicles (EV): Exosomes, microvesicles, retrovirus-like vesicles, and apoptotic bodies. *J. Neurooncol.* **2013**, *113*, 1–11. [CrossRef] [PubMed]
7. Wang, Q.; Bosch, B.-J.; Vlak, J.M.; van Oers, M.M.; Rottier, P.J.; van Lent, J.W.M. Budded baculovirus particle structure revisited. *J. Invertebr. Pathol.* **2016**, *134*, 15–22. [CrossRef]
8. Transfiguracion, J.; Manceur, A.P.; Petiot, E.; Thompson, C.M.; Kamen, A.A. Particle quantification of influenza viruses by high performance liquid chromatography. *Vaccine* **2015**, *33*, 78–84. [CrossRef]
9. Steppert, P.; Burgstaller, D.; Klausberger, M.; Tover, A.; Berger, E.; Jungbauer, A. Quantification and characterization of virus-like particles by size-exclusion chromatography and nanoparticle tracking analysis. *J. Chromatogr. A* **2017**, *1487*, 89–99. [CrossRef]
10. Gutiérrez-Granados, S.; Cervera, L.; Gòdia, F.; Carrillo, J.; Segura, M.M. Development and validation of a quantitation assay for fluorescently tagged HIV-1 virus-like particles. *J. Virol. Methods* **2013**, *193*, 85–95. [CrossRef]
11. Venereo-Sanchez, A.; Gilbert, R.; Simoneau, M.; Caron, A.; Chahal, P.; Chen, W.; Ansorge, S.; Li, X.; Henry, O.; Kamen, A. Hemagglutinin and neuraminidase containing virus-like particles produced in HEK-293 suspension culture: An effective influenza vaccine candidate. *Vaccine* **2016**, *34*, 3371–3380. [CrossRef] [PubMed]
12. Heider, S.; Metzner, C. Quantitative real-time single particle analysis of virions. *Virology* **2014**, *462–463*, 199–206. [CrossRef] [PubMed]
13. Jiang, W.; Tang, L. Atomic cryo-EM structures of viruses. *Curr. Opin. Struct. Biol.* **2017**, *46*, 122–129. [CrossRef]
14. Popov, V.L.; Tesh, R.B.; Weaver, S.C.; Vasilakis, N.; Popov, V.L.; Tesh, R.B.; Weaver, S.C.; Vasilakis, N. Electron Microscopy in Discovery of Novel and Emerging Viruses from the Collection of the World Reference Center for Emerging Viruses and Arboviruses (WRCEVA). *Viruses* **2019**, *11*, 477. [CrossRef]
15. De Carlo, S.; Harris, J.R. Negative staining and cryo-negative staining of macromolecules and viruses for TEM. *Micron* **2011**, *42*, 117–131. [CrossRef]
16. Kalsoom, A.; Khan, S.A.; Khan, S.B.; Asiri, A.M. Chapter 4. Scanning Electron Microscopy: Principle and Applications in Nanomaterials Characterization. In *Handbook of Materials Characterization*; Sharma, S.K., Ed.; Springer International Publishing: Cham, Switzerland, 2018; pp. 113–146. ISBN 978-3-319-92954-5.
17. Gállego Ossul, I. *Estructura y propiedades de las placas de cromatina de los cromosomas metafásicos: Estudio mediante técnicas de microscopía TEM, AFM y Espectroscopia de Fuerza Atómica*; Universitat Autònoma de Barcelona: Bellaterra, Spain, 2010.
18. Michen, B.; Graule, T. Isoelectric points of viruses. *J. Appl. Microbiol.* **2010**, *109*, 388–397. [CrossRef]
19. Deregibus, M.C.; Figliolini, F.; D'antico, S.; Manzini, P.M.; Pasquino, C.; De Lena, M.; Tetta, C.; Brizzi, M.F.; Camussi, G. Charge-based precipitation of extracellular vesicles. *Int. J. Mol. Med.* **2016**, *38*, 1359–1366. [CrossRef]

20. Murata, K.; Wolf, M. Cryo-electron microscopy for structural analysis of dynamic biological macromolecules. *Biochim. Biophys. Acta Gen. Subj.* **2018**, *1862*, 324–334. [CrossRef]

21. Thompson, R.F.; Walker, M.; Siebert, C.A.; Muench, S.P.; Ranson, N.A. An introduction to sample preparation and imaging by cryo-electron microscopy for structural biology. *Methods* **2016**, *100*, 3–15. [CrossRef]

22. Hanne, J.; Zila, V.; Heilemann, M.; Müller, B.; Kräusslich, H.-G. Super-resolved insights into human immunodeficiency virus biology. *FEBS Lett.* **2016**, *590*, 1858–1876. [CrossRef] [PubMed]

23. Godin, A.G.; Lounis, B.; Cognet, L. Super-resolution Microscopy Approaches for Live Cell Imaging. *Biophysj* **2014**, *107*, 1777–1784. [CrossRef] [PubMed]

24. González-Domínguez, I.; Puente-Massaguer, E.; Cervera, L.; Gòdia, F. Quantification of the HIV-1 virus-like particle production process by super-resolution imaging: From VLP budding to nanoparticle analysis. **2020**. Under review.

25. Puente-Massaguer, E.; Lecina, M.; Gòdia, F. Application of advanced quantification techniques in nanoparticle-based vaccine development with the Sf9 cell baculovirus expression system. *Vaccine* **2020**, *38*, 1849–1859. [CrossRef] [PubMed]

26. Borlinghaus, R.T.; Kappel, C. HyVolution—The smart path to confocal super-resolution. *Nat. Methods* **2016**, *13*, i. [CrossRef]

27. Zamora, J.L.R.; Aguilar, H.C. Flow virometry as a tool to study viruses. *Methods* **2017**, *134–135*, 87–97. [CrossRef]

28. Maguire, C.M.; Rösslein, M.; Wick, P.; Prina-Mello, A. Characterisation of particles in solution—A perspective on light scattering and comparative technologies. *Sci. Technol. Adv. Mater.* **2018**, *19*, 732–745. [CrossRef]

29. Kramberger, P.; Ciringer, M.; Štrancar, A.; Peterka, M. Evaluation of nanoparticle tracking analysis for total virus particle determination. *Virol. J.* **2012**, *9*, 265. [CrossRef]

30. Aguilar, P.P.; González-Domínguez, I.; Schneider, T.A.; Gòdia, F.; Cervera, L.; Jungbauer, A. At-line multi-angle light scattering detector for faster process development in enveloped virus-like particle purification. *J. Sep. Sci.* **2019**, *42*, 2640–2648. [CrossRef]

31. González-Domínguez, I.; Grimaldi, N.; Cervera, L.; Ventosa, N.; Gòdia, F. Impact of physicochemical properties of DNA/PEI complexes on transient transfection of mammalian cells. *N. Biotechnol.* **2019**, *49*, 88–97. [CrossRef]

32. Nolan, J.P.; Duggan, E. Analysis of Individual Extracellular Vesicles by Flow Cytometry. In *Flow Cytometry Protocols, Methods in Molecular Biology*; Hawley, T.S., Hawley, R.G., Eds.; Springer Science+Business Media LLC: Berlin, Germany, 2018; Volume 1678, pp. 79–92. ISBN 978-1-4939-7346-0.

33. Welsh, J.A.; Horak, P.; Wilkinson, J.S.; Ford, V.J.; Jones, J.C.; Smith, D.; Holloway, J.A.; Englyst, N.A. FCM PASS Software Aids Extracellular Vesicle Light Scatter Standardization. *Cytom. Part A* **2019**. [CrossRef]

34. McVey, M.J.; Spring, C.M.; Kuebler, W.M. Improved resolution in extracellular vesicle populations using 405 instead of 488 nm side scatter. *J. Extracell. Vesicles* **2018**, *7*, 1–11. [CrossRef] [PubMed]

35. Charlton Hume, H.K.; Lua, L.H.L. Platform technologies for modern vaccine manufacturing. *Vaccine* **2017**, *35*, 4480–4485. [CrossRef]

36. Cervera, L.; Gutiérrez-Granados, S.; Martínez, M.; Blanco, J.; Gòdia, F.; Segura, M.M. Generation of HIV-1 Gag VLPs by transient transfection of HEK 293 suspension cell cultures using an optimized animal-derived component free medium. *J. Biotechnol.* **2013**, *166*, 152–165. [CrossRef] [PubMed]

37. Gutiérrez-Granados, S.; Cervera, L.; de L.M. Segura, M.; Wölfel, J.; Gòdia, F. Optimized production of HIV-1 virus-like particles by transient transfection in CAP-T cells. *Appl. Microbiol. Biotechnol.* **2016**, *100*, 3935–3947. [CrossRef] [PubMed]

38. Göttlinger, H.G. HIV-1 Gag: A Molecular Machine Driving Viral Particle Assembly and Release. In *HIV Sequence Compendium 2011*; Kuiken, C., Foley, B., Leitner, T., Apetrei, C., Hahn, B., Mizrachi, I., Mullins, J., Rambaut, A., Wolinsky, S., Korber, B., Eds.; Los Alamos National Laboratory, Theoretical Biology and Biophysics: Los Alamos, NM, USA, 2011; pp. 2–28, LA-UR-11-11440.

39. Thompson, C.M.; Petiot, E.; Mullick, A.; Aucoin, M.G.; Henry, O.; Kamen, A.A. Critical assessment of influenza VLP production in Sf9 and HEK293 expression systems. *BMC Biotechnol.* **2015**, *15*, 1–12. [CrossRef] [PubMed]

40. Genzel, Y. Designing cell lines for viral vaccine production: Where do we stand? *Biotechnol. J.* **2015**, *10*, 728–740. [CrossRef]

41. Puente-Massaguer, E.; Lecina, M.; Gòdia, F. Integrating nanoparticle quantification and statistical design of experiments for efficient HIV-1 virus-like particle production in High Five cells. *Appl. Microbiol. Biotechnol.* **2020**, *104*, 1569–1582. [CrossRef]

42. De Rond, L.; Coumans, F.A.W.; Nieuwland, R.; van Leeuwen, T.G.; van der Pol, E. Deriving Extracellular Vesicle Size From Scatter Intensities Measured by Flow Cytometry. *Curr. Protoc. Cytom.* **2018**, *86*, 1–14. [CrossRef]

43. Malvern Instruments Limited. Determining Fluorescence Limit of Detection with Nanoparticle Tracking Analysis (NTA). 2015. Available online: https://www.malvernpanalytical.com/en/learn/knowledge-center/application-notes/AN150507FluorescenceLOD (accessed on 17 February 2020).

44. Ellis, E.A. Staining Sectioned Biological Specimens for Transmission Electron Microscopy: Conventional and En Bloc Stains. In *Electron Microscopy: Methods and Protocols, Methods in Molecular Biology*; Kuo, J., Ed.; Humana Press: Totowa, NJ, USA, 2014; pp. 57–72. ISBN 978-1-62703-776-1.

45. Cervera, L.; Fuenmayor, J.; González-Domínguez, I.; Gutiérrez-Granados, S.; Segura, M.M.; Gòdia, F. Selection and optimization of transfection enhancer additives for increased virus-like particle production in HEK293 suspension cell cultures. *Appl. Microbiol. Biotechnol.* **2015**, *99*, 9935–9949. [CrossRef]

46. González-Domínguez, I.; Gutiérrez-Granados, S.; Cervera, L.; Gòdia, F.; Domingo, N. Identification of HIV-1–Based Virus-like Particles by Multifrequency Atomic Force Microscopy. *Biophys. J.* **2016**, *111*, 1173–1179. [CrossRef]

47. Puente-Massaguer, E.; González-Domínguez, I.; Universitat Autònoma de Barcelona, Barcelona, Spain. Preliminary experiments on HIV-1 Gag-eGFP VLP characterization by Scanning Electron Microscopy. Personal communication, 2020.

48. Gencer, D.; Bayramoglu, Z.; Nalcacioglu, R.; Kleespies, R.G.; Demirbag, Z.; Demir, I. Characterisation of three Alphabaculovirus isolates from the gypsy moth, Lymantria dispar dispar (Lepidoptera: Erebidae), in Turkey. *Biocontrol Sci. Technol.* **2018**, *28*, 107–121. [CrossRef]

49. Palomares, L.A.; Realpe, M.; Ramírez, O.T. An Overview of Cell Culture Engineering for the Insect Cell-Baculovirus Expression Vector System (BEVS). In *Methods in Molecular Biology*; Humana Press: Totowa, NJ, USA, 2015; Volume 5, pp. 501–519. ISBN 978-3-319-10319-8.

50. Thermo Fisher Scientific. ExpiSfTM Expression System. 2020. Available online: https://assets.thermofisher.com/TFSAssets/LSG/manuals/MAN0017532_ExpiSfExpressionSystem_UG.pdf (accessed on 17 February 2020).

51. Golding, C.G.; Lamboo, L.L.; Beniac, D.R.; Booth, T.F. The scanning electron microscope in microbiology and diagnosis of infectious disease. *Sci. Rep.* **2016**, *6*, 26516. [CrossRef] [PubMed]

52. Van Der Pol, E.; Hoekstra, A.G.; Sturk, A.; Otto, C.; Van Leeuwen, T.G.; Nieuwland, R. Optical and non-optical methods for detection and characterization of microparticles and exosomes. *J. Thromb. Haemost.* **2010**, *8*, 2596–2607. [CrossRef] [PubMed]

53. Pedersen, M.; Jamali, S.; Saha, I.; Daum, R.; Bendjennat, M.; Saffarian, S. Correlative iPALM and SEM resolves virus cavity and Gag lattice defects in HIV virions. *Eur. Biophys. J.* **2019**, *48*, 15–23. [CrossRef] [PubMed]

54. Welsh, J.A.; Holloway, J.A.; Wilkinson, J.S.; Englyst, N.A. Extracellular Vesicle Flow Cytometry Analysis and Standardization. *Front. Cell Dev. Biol.* **2017**, *5*, 1–7. [CrossRef]

55. Yang, Z.; Sharma, A.; Qi, J.; Peng, X.; Lee, D.Y.; Hu, R.; Lin, D.; Qu, J.; Kim, J.S. Super-resolution fluorescent materials: An insight into design and bioimaging applications. *Chem. Soc. Rev.* **2016**, *45*, 4651–4667. [CrossRef]

56. Rossi, C.; Kearney, B.; Olschner, S.; Williams, P.; Robinson, C.; Heinrich, M.; Zovanyi, A.; Ingram, M.; Norwood, D.; Schoepp, R. Evaluation of ViroCyt® Virus Counter for Rapid Filovirus Quantitation. *Viruses* **2015**, *7*, 857–872. [CrossRef]

57. Blancett, C.D.; Fetterer, D.P.; Koistinen, K.A.; Morazzani, E.M.; Monninger, M.K.; Piper, A.E.; Kuehl, K.A.; Kearney, B.J.; Norris, S.L.; Rossi, C.A.; et al. Accurate virus quantitation using a Scanning Transmission Electron Microscopy (STEM) detector in a scanning electron microscope. *J. Virol. Methods* **2017**, *248*, 136–144. [CrossRef]

58. Strauss, J.D.; Hammonds, J.E.; Yi, H.; Ding, L.; Spearman, P.; Wright, E.R. Three-Dimensional Structural Characterization of HIV-1 Tethered to Human Cells. *J. Virol.* **2016**, *90*, 1507–1521. [CrossRef]

59. Pornillos, O.; Higginson, D.S.; Stray, K.M.; Fisher, R.D.; Garrus, J.E.; Payne, M.; He, G.P.; Wang, H.E.; Morham, S.G.; Sundquist, W.I. HIV Gag mimics the Tsg101-recruiting activity of the human Hrs protein. *J. Cell Biol.* **2003**, *162*, 425–434. [CrossRef]

60. Chen, Y.; Wu, B.; Musier-Forsyth, K.; Mansky, L.M.; Mueller, J.D. Fluorescence fluctuation spectroscopy on viral-like particles reveals variable Gag stoichiometry. *Biophys. J.* **2009**, *96*, 1961–1969. [CrossRef] [PubMed]

61. Inamdar, K.; Floderer, C.; Favard, C.; Muriaux, D. Monitoring HIV-1 Assembly in Living Cells: Insights from Dynamic and Single Molecule Microscopy. *Viruses* **2019**, *11*, 72. [CrossRef] [PubMed]

62. Steppert, P.; Burgstaller, D.; Klausberger, M.; Berger, E.; Aguilar, P.P.; Schneider, T.A.; Kramberger, P.; Tover, A.; Nöbauer, K.; Razzazi-Fazeli, E.; et al. Purification of HIV-1 gag virus-like particles and separation of other extracellular particles. *J. Chromatogr. A* **2016**, *1455*, 93–101. [CrossRef] [PubMed]

63. Xiao, P.J.; Li, C.; Neumann, A.; Samulski, R.J. Quantitative 3D tracing of gene-delivery viral vectors in human cells and animal tissues. *Mol. Ther.* **2012**, *20*, 317–328. [CrossRef]

64. Sarder, P.; Nehorai, A. Deconvolution methods for 3-D fluorescence microscopy images. *IEEE Signal Process. Mag.* **2006**, *23*, 32–45. [CrossRef]

65. Schermelleh, L.; Ferrand, A.; Huser, T.; Eggeling, C.; Sauer, M.; Biehlmaier, O.; Drummen, G.P.C. Super-resolution microscopy demystified. *Nat. Cell Biol.* **2019**, *21*, 72–84. [CrossRef]

Article

Immunization of Cats against Fel d 1 Results in Reduced Allergic Symptoms of Owners

Franziska Thoms [1,2], Stefanie Haas [1,2], Aline Erhart [3], Claudia S. Nett [4], Silvia Rüfenacht [5], Nicole Graf [6], Arnis Strods [7], Gauravraj Patil [7], Thonur Leenadevi [7], Michael C. Fontaine [7], Lindsey A. Toon [7], Gary T. Jennings [1,2], Gabriela Senti [8], Thomas M. Kündig [9] and Martin F. Bachmann [10,11,*]

[1]   Department of Dermatology, Zurich University Hospital, Wagistrasse 12, 8952 Schlieren/Zurich, Switzerland; franziskazabel@hotmail.com (F.T.); haasstefanie206@hotmail.com (S.H.); Gary.jennings@usz.ch (G.T.J.)

[2]   HypoPet AG, Moussonstrasse 2, 8091 Zurich, Switzerland

[3]   Clinical Trials Center Zurich, University Hospital Zurich, Moussonstrasse 2, 8044 Zurich, Switzerland; alineerhart@hotmail.com

[4]   vetderm.ch, Ennetseeklink für Kleintiere, Rothusstrasse 2, 6331 Hünenberg, Switzerland; cnett@vetderm.ch

[5]   dermaVet, Tierklinik Aarau West AG, Muhenstrasse 56, 5036 Oberentfelden, Switzerland; s.ruefenacht@dermavet.ch

[6]   Graf Biostatistics, Amelenweg 5, 8400 Winterthur, Switzerland; graf@biostatistics.ch

[7]   Benchmark Animal Health, Benchmark Holdings Plc, 8 Smithy Wood Dr, Sheffield S35 1QN, UK; arnis.strods@bmkvaccines.com (A.S.); gauravraj.patil@bmkvaccines.com (G.P.); thonur.leenadevi@bmkvaccines.com (T.L.); michael.fontaine@bmkanimalhealth.com (M.C.F.); lindsey.toon@bmkanimalhealth.com (L.A.T.)

[8]   Director Research and Education, University Hospital Zurich, Rämistrasse 100, 8091 Zurich, Switzerland; Gabriela.senti@usz.ch

[9]   Department of Dermatology, University Hospital Zurich, Gloriastrasse 31, 8091 Zurich, Switzerland; Thomas.kuendig@usz.ch

[10] Department of Immunology, Inselspital, University of Bern, Salihaus 2, 3007 Bern, Switzerland

[11] Jenner Institute, University of Oxford, Old Road Campus, Roosevelt Drive, Oxford OX3 7BN, UK

*   Correspondence: martin.bachmann@me.com

Received: 28 January 2020; Accepted: 26 February 2020; Published: 6 March 2020

Abstract: An innovative approach was tested to treat cat allergy in humans by vaccinating cats with Fel-CuMV (HypoCat™), a vaccine against the major cat allergen Fel d 1 based on virus-like particles derived from cucumber mosaic virus (CuMV-VLPs). Upon vaccination, cats develop neutralizing antibodies against the allergen Fel d 1, which reduces the level of reactive allergen, thus lowering the symptoms or even preventing allergic reactions in humans. The combined methodological field study included ten cat-allergic participants who lived together with their cats ($n = 13$), that were immunized with Fel-CuMV. The aim was to determine methods for measuring a change in allergic symptoms. A home-based provocation test (petting time and organ specific symptom score (OSSS)) and a general weekly (or monthly) symptom score (G(W)SS) were used to assess changes in allergic symptoms. The petting time until a pre-defined level of allergic symptoms was reached increased already early after vaccination of the cats and was apparent over the course of the study. In addition, the OSSS after provocation and G(W)SS recorded a persistent reduction in symptoms over the study period and could serve for long-term assessment. Hence, the immunization of cats with HypoCat™ (Fel-CuMV) may have a positive impact on the cat allergy of the owner, and changes could be assessed by the provocation test as well as G(W)SS.

Keywords: cat allergy; vaccination; Fel d 1; HypoCat™; virus-like particle

---

1. Introduction

Cats are among the most popular and common pets worldwide and are a significant source of indoor allergens [1]. Hence, allergies to cats are widespread, with a prevalence of 10%–30% in the Western population [2]. A total number of 10 *Feline domesticus* (Fel d) allergens, that are recognized by human IgEs, have been identified [3–9]. Fel d 1, an uteroglobin-like protein, is considered to be the major cat allergen. In fact, 94% of patients allergic to cats have Fel d 1-specific IgE [10]. Fel d 1 belongs to the family of secretoglobins with homologies to uteroglobin. Its function is unknown but it has been postulated to play a potential role in skin protection and pelt conditioning or have an involvement in the transport of steroids, hormones and pheromones [11,12]. Fel d 1 is produced in sebaceous, salivary, lacrimal, and anal glands and is present in the saliva, tears, skin and fur [13–16]. It is shed from the cat to the environment through airborne dander and if inhaled by humans may result in sensitization and induction of cat allergy [17].

The immune response against innocuous cat allergens is characterized as type I and IV hypersensitivity, involving Th2 cells shaping the environment for production of IgE antibodies by B and plasma cells and recruitment of additional inflammatory cells [18–20]. Affected patients suffer from mild symptoms, e.g., sneezing, itchiness of skin and eyes, to severe symptoms ranging from conjunctivitis, rhinitis to asthma, which, upon direct exposure to cats, can lead to life-threatening conditions. There are several recommendations for dealing with cat allergy [21]. Allergic people are advised to avoid allergen exposure by removal of all potential allergen-containing or contaminated objects in the households, e.g., pillows, blankets, carpets, rugs. Environmental cleaning and the use of air humidifiers and HEPA filters can also contribute to the relief of symptoms. Another approach is to remove the cat. However, the bond between owners and their cats is often so strong that they are more likely to accept the risk to their health, which they may not even be fully aware of, than give up their pet [22].

Cat allergic subjects usually treat their allergic symptoms with antihistamines and corticosteroids. Another possibility is allergen specific immunotherapy (AIT), which is the only disease-modifying option, but carries the risk of inducing serious side effects and may take years. In fact, AIT can require 30–80 injections over a duration of three to five years with a low chance of success. New approaches explore different routes of administration (e.g., epicutaneous, sublingual, intralymphatic), different formulations of allergens with adjuvants (e.g., MPL, MCT), the introduction of mutations into the protein sequence which delete T cell or IgE epitopes, and finally the use of short peptides instead of full-length allergens [23–31]. The challenge for the development of new therapies is exemplified by the recent failure of a phase III clinical study testing a peptide-based vaccine to treat cat allergy [32]. Orengo et al. are developing a monoclonal IgG antibody therapy targeting Fel d 1 in humans that aims to increase the allergen-specific IgG/IgE ratio and relieve symptoms and showed good clinical impact [33].

An alternative approach to the problem of cat allergy, and one that does not involve separation of the cat from its owner, is to lower Fel d 1 on the animal itself. One recently described method is the addition of anti-Fel d 1 IgY harvested from chicken eggs to cat food. A reduction in active Fel d 1 in saliva and fur has been reported but whether this will result in clinically significant reductions in allergy still needs to be addressed [34,35].

Another approach to lowering allergenic Fel d 1 levels on the cat is active immunization with the aim of inducing anti-Fel d 1 antibodies in the animal itself. Towards this end, a feline vaccine targeting Fel d 1 in cats to treat cat allergy in humans is being developed [36]. The vaccine is based on a recombinantly expressed Fel d 1 protein covalently linked to a virus-like particle (VLP) derived from the Cucumber mosaic virus (CuMV) [37]. The VLP consists of the CuMV coat protein, without any viral genetic information, which serves as a carrier and induces, due to its particulate and repetitive structure, strong and sustained antibody responses, even against self molecules like the Fel d 1 protein in cats [38,39]. To date, vaccination with Fel-CuMV (HypoCat^{TM}) has been tested in 70 cats and was well tolerated without short- or long-term (two years) side effects. Furthermore, vaccination induced strong neutralizing anti-Fel d 1 IgG responses lowering levels of reactive allergen in tear extracts of study cats tested with human basophils from cat allergic subjects [36].

In the current manuscript, we report the results of a first field trial with ten cat allergic participants living together with their cats, and the cats were vaccinated with Fel-CuMV. The aim of this exploratory methodology study was 1) to determine a suitable method for measuring a change in the allergic symptoms of the owner and 2) quantify changes in the interactions between the cat and the owner. These methods may then be used in a larger trials in the future. Three parameters were monitored over a duration of almost two years. A home-based provocation test determined the petting time, defined as the time during which the owner was able to interact with the cat until a certain level of symptoms using a visual analogue scale (VAS score of 5) was reached and their organ-specific symptoms score (OSSS) after petting their cats. In addition, a general weekly or monthly symptom score (GWSS and GSS, respectively) assessed overall changes in allergic symptoms in human subjects without interaction with the cat. The provocation test was assessed as a parameter of the acute allergic reaction, comparable to a hospital-based provocation test (e.g., nasal or conjunctival test), whereas the G(W)SS served as readout of chronic symptoms. As observed in previous studies [36], vaccination with Fel-CuMV was well-tolerated by cats and induced strong serum antibody responses. Changes in symptoms of cat owners and interaction times with their cats were observed after vaccination of the pets. Thus, both tests, the provocation test and the G(W)SS, are suitable to detect changes in the symptoms of cat allergic patients. In particular, the petting time may serve as an early efficacy read-out following vaccination of the cats, whereas the OSSS and G(W)SS seem to be more appropriate for long-term monitoring.

2. Materials and Methods

2.1. Study Population

The study participants were recruited from March till July 2017 and the study commenced in April 2017. Upon completion in February 2018, an extension study with seven of the study participants was conducted from May 2018 until April 2019. Males or females aged 18-65 with a history of cat allergy that were cohabitating with cat(s) were eligible for study inclusion. An understanding of the nature, meaning and scope of the study and signing of an informed consent were also required. Participants were also required to test positive for skin prick tests performed with histamine dihdrochloride and cat allergen extract. In order to confirm the cat allergy to their own cat, participants were also tested positive by a screening scratch test to fur of their own cat.

Participants were not enrolled if they suffered from immunosuppression or anemias, leukemia or other hematological diseases. Additional exclusion criteria were: pregnancy, breast feeding or intent to become pregnant during the course of the study; a positive skin prick test with the negative control; a known history of anaphylactic reactions to pet allergens; the use of Beta-blockers, neuroleptic drugs and tricyclic anti-depressants.

Medications like ACE-inhibitors and Beta2-agonists as well as anti-histamines and corticosteroids could influence the study results and were therefore prohibited within 3 days prior to the application of allergen extracts or the provocation test.

Main study: Ten cat allergic participants ranging from 21 to 51 years old, including eight women and two men, were screened and enrolled in the study at the University Hospital Zurich, Switzerland (Table 1). A total of 13 cats (two participants had two or three cats, respectively) were enrolled in the animal part of the study.

Extension study: Seven participants, including six women and one man, aged between 22 and 52 years old continued in the extension study. A total of nine cats (one participant had three cats) were enrolled in the animal part of the study.

2.2. Study Design

The study was a single-center, open-label, non-placebo controlled, combined, methodological field trial to determine (a) method(s) for measuring cat allergy symptoms in participants that have immunized their cats with Fel-CuMV.

Human part: *Main study*—Screening visits took place at the Clinical Trials Center, University Hospital Zurich (USZ), CH in collaboration with the Department of Dermatology (USZ). After the informed consent was given, the participants underwent two tests to confirm their cat allergy. A skin-prick test was performed using a standard cat allergen and a scratch-test was performed with fur from the own cat. Home-based provocation tests were performed in study weeks 1–3 before vaccination of the cat(s), and in weeks 4, 8, 12, 16, 20 and 24 after vaccination of the cat(s). The provocation test assessed two parameters, the petting time and organ-specific symptoms score (OSSS). The test was only valid if participants had not taken medication containing anti-histamines less than 3 days before the test, otherwise they had to reschedule the test accordingly. If the rescheduling did not happen, the test for that timepoint was invalid and not considered in the statistical analysis. Once a week, the participants filled in a questionnaire retrospectively recording their organ-specific symptoms of the previous week (i.e., general weekly symptom score GWSS). A close-out phone call at the end of the study in week 29 was done to follow up on the well-being and health of the study participant.

Extension study - Seven of ten participants of the main study signed the informed consent form and were enrolled in the extension study. They performed a provocation test before the booster injection of their cat(s) in study week 1 (or week 54 in the combined schedule of main and extension study including the intervening time) followed by four additional tests in study weeks 5, 9, 25 and 45 (weeks 58, 62, 78 and 98, respectively). Again, a provocation test was only valid if participants had not taken medication containing anti-histamines less than 3 days before the test. In the event of a breach, the procedure was followed as described in the main study. In addition, participants recorded their general organ-specific symptoms without provocation monthly (i.e., general specific symptoms GSS). The study finished with a close-out phone call by the clinical study team in study week 45 (week 98, respectively).

Animal part: All interventions for the cats were performed by study veterinarians at the "Kleintierpraxis Schwäntenmos" in Zumikon, Zurich, "Ennetseeklinik für Kleintiere AG" in Hünenberg, Zug and "Tierklinik Aarau West AG" in Aarau, Aargau.

Main study—Thirteen cats were enrolled and received three subcutaneous injections of 100 µg Fel-CuMV formulated in 1 mL in study weeks 4, 7, and 10. Sera were collected in study weeks 4, 10, and 27 and analyzed for specific antibody responses.

Extension study—Nine cats received a subcutaneous booster injection of 100 µg Fel-CuMV formulated in 1 mL in week 3 (or week 56 in the combined study schedule). Sera were collected before the boost (week 56), 6 weeks after the boost (week 62) and at the end of the extension cat study (week 78).

2.3. Ethics Approvals

The protocol, participant information and consent form, as well as other study-specific documents, were submitted to the Zurich-based properly constituted cantonal ethic committee (KEK) in agreement with local legal requirements for formal study approval. The decision of the KEK concerning the conduct of the study had been made in writing to the Sponsor-Investigator before commencement of this study. The KEK decided the trial could be conducted without its approval, as the law for human research was judged to be not applicable. From an ethical perspective, the extension study was considered the same and was conducted without particular approval of the KEK. However, both studies were conducted in accordance with ICH/GCP Guidelines and registered with CLINICALTRIALS.gov as NCT03089788.

The Veterinary Offices of the participating cantons Zurich, Zug and Aargau approved the animal study ZH245/16 (19th March 2016). All cat owners signed informed consent for their cats. All interventions and examinations performed by the study veterinarians were in accordance to the Swiss Animal Welfare Ordinance and Animal Welfare Act on Animal Experimentation (2005, TSchG; 2008, TschV). The animal study was designed as an open-label, non-placebo controlled, multi-center, tolerability and immunogenicity study.

2.4. Study Objective

The objective of the study was to determine if the methods tested herein (provocation test and G(W)SS) were suitable to measure changes in allergy symptoms and interaction of the participants with their cats following immunization with Fel-CuMV (HypoCat^{TM}). The change in symptoms was assessed at baseline versus week 24 at the end of the main study and over the course of the entire study (main and extension study). The selected method(s) could be used in future clinical trials.

2.5. Provocation Test

In an earlier study "ZU_Hyposcore-001" (NCT02399579. https://clinicaltrials.gov/ct2/show/results/ NCT02399579) the validity of the provocation test (i.e., HypoScore), a new self-assessed, home-based score specific for cat allergy, was evaluated in cat-allergic participants without vaccination of their cats. In the current study, the provocation test was used to observe changes in allergic symptoms upon vaccination of cats with Fel-CuMV. The test assessed two parameters: the petting time and organ-specific symptoms score (OSSS). The test was performed by petting the cat in order to measure the time until the participant reached a defined symptom strength level (self-assessed, 5 on a VAS ranging from 0–10). If the participant did not reach a symptom level of 5, the petting was stopped at 45 min. When symptom severity of 5 was not reached but provocation time was <45 min (e.g., did not tolerate further petting), for data analysis the time of the last provocation test was carried forward (LOCF). The petting times were measured at baseline on three occasions (week 1–3) before immunization of the cat and over the course of main and extension study. After the provocation test, the participants filled in an OSSS questionnaire on a 4-point scale (0 = no symptoms up to 3 = severe symptoms) regarding their symptoms of eyes, nose, bronchia, lung and palate. Values of the OSSS could be between 0 and 30. The baseline of the HypoScore was measured weekly on three occasions before vaccination of the cats and was compared to the end (week 24) and over the course of the study (weeks 8, 12, 16, 20 and 24 in the main study and weeks 54, 58, 62, 78 and 98 in the extension study) after vaccination of the cats in weeks 4, 7, 10 and booster injection in week 56.

2.6. General Weekly or Monthly Specific Symptoms (GWSS or GSS)

Changes in the general organ specific symptoms on a 4-point scale (0 = no symptoms up to 3 = severe symptoms) of eyes, nose, bronchia, lung, and palate were assessed weekly in the main study (GWSS) or monthly in the extension study (GSS) before and after vaccination of cats with Fel-CuMV. The test was performed without provocation (i.e., direct interaction with their cat). Values of the G(W)SS could be between 0 and 30. The baseline of the general symptoms was determined on three occasions (week 1–3) before vaccination and on 22 occasions during the main study (week 4–25) and 12 occasions during the extension study (week 54-98) after vaccination of the cats in study weeks 4, 7, 10 and booster injection in week 56.

2.7. Cat Population

The cats which participated in the study were privately owned and included 13 cats of six breeds including; British Shorthair, European Shorthair, Russian Blue, Ragdoll, Abyssinian, and Egyptian Mau of both sexes (7 ♀; 6 ♂). The age ranged from 4 to 13 years and body weights of 2.5 to 7.4 kg. Thirteen animals were enrolled in the main and nine cats in the extension study. Where several cats lived in the same household, all cats were immunized, but only one cat was designated as the study cat. The cat owner performed the provocation tests and answered all questionnaires regarding the single study cat.

2.8. Vaccination of Cats

The vaccine Fel-CuMV and the vaccine production was described previously [36]. A vaccine dose contained 100 μg Fel-CuMV (HypoCat^{TM}) formulated in 1 mL aqueous buffer solution. Subcutaneous injections were applied in study weeks 4, 7 and 10 (main study) and 56 (extension study). The vaccine used in this study was produced by Benchmark Vaccines Limited, UK using a GMP-like production process. Prior to each vaccination, the cats were thoroughly examined by one of the participating veterinarians for their general health status, including a routine physical exam with measurement of the body weight and temperature, checking the site of injection, pulse and breathing rate, abdomen palpation and general appearance. During the course of the study, the owners regularly checked the health of their animal. Twenty-four hours after every injection the owners were called by the study personnel to enquire after the well-being of the cat.

2.9. Antibody Responses in Cats

Blood for serum antibody measurements was collected from the vena jugularis or cephalica using serum tubes (Sarstedt, Germany). Sampling was done in the study weeks 4, 7, 10, 27, 56, 62, and 78. After clotting (30 min) and centrifugation (1000 $\times g$, 5 min) of blood samples, sera were transferred to labeled polypropylene tubes (Eppendorf tubes, 1.5 mL, Germany) and stored at $\leq -15\,^{\circ}$C until thawed for antibody analyses.

A validated ELISA (developed by HypoPet AG) was performed to determine anti-Fel d 1 IgG in cat sera previously described [36]. ELISA titers are given as the reciprocals of the dilutions needed to achieve 50% of the optical density of the maximal signal measured at saturation (OD50). The geometric mean titers were calculated from the individual titers of the cats from each group.

2.10. Data and Statistical Analyses

The analysis for the human part included data from 9 out of 10 participants and was performed according to the "per protocol" (PP) analysis. Analyses were conducted in this way because one participant had urticaria, accompanied by pruritus and wheal formation, on two occasions during the study. The participant was unable to distinguish urticaria symptoms from the symptoms of cat allergy and was thus excluded from analysis. The analyses of cat data included all 13 cats, whereas the correlation of antibody responses in cats with measurements of human allergic symptoms included data from 9 out of 10 participants with their respective study cats ($n = 9$).

Statistical analyses were performed with Excel, Graph Pad Prism and R. Data were summarized by descriptive statistics: mean, standard deviation, median, first and third quartile, minimum, and maximum. This study was classified as a pilot study because there was no pre-existing information available concerning the expected change in symptoms of allergic cat owners before and after immunization of their cats (the effect size). The changes in the symptoms of the main study were evaluated using an exact Wilcoxon matched-paired signed rank test. Hypotheses were tested at a two-sided significance level of $\alpha = 0.05$. For comparison between baseline and course of study, the mean of all values was computed.

3. Results

3.1. Study Design and Population

This trial was an open-label, non-placebo controlled, single-center, exploratory study involving both cat-allergic humans and their cats (Figure 1). The aim of this methodology field trial was to assess different tests to measure allergic symptoms and to monitor possible changes in allergic symptoms in the study participants after their cats were immunized with the cat vaccine Fel-CuMV.

The study included 10 human participants (Table 1), ranging from 21 to 51 years old, and 13 cats (one participant had two cats and another three cats), ranging from 4 to 13 years old, of both sexes and six different breeds. All ten participants who were screened were enrolled in the study. There were no drop-outs and 10 of 10 participants completed the study. One participant suffered from recurrent chronic urticaria of unknown origin, but exhibited no symptoms at the beginning of the study. However, the urticaria occurred in study week 2 and again in week 16. On each occasion the urticaria lasted for 4–5 weeks until recovery. The participant was not able to distinguish symptoms of urticaria from those of cat allergy and was excluded from the analysis. Data are thus presented as "per protocol" (PP) analysis.

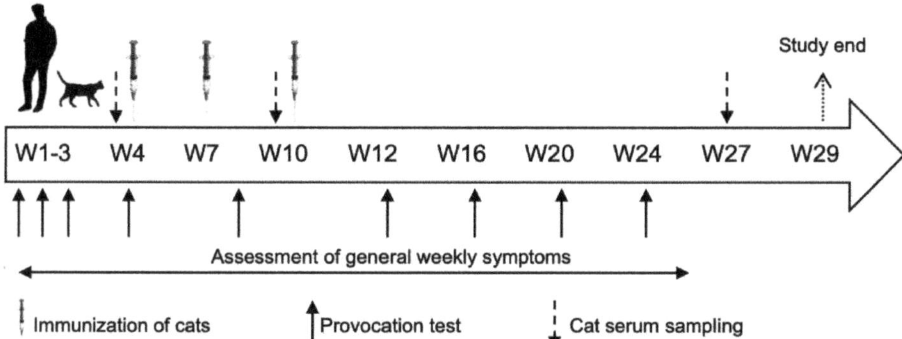

Figure 1. Study design of the combined human and animal trial. Prior to immunization of their cats, participants performed a baseline assessment of their allergic symptoms using the HypoScore (OSSS and petting time) and GWSS. In the course of the study, they performed the provocation test every four weeks and recorded their general symptoms weekly without provocation using the GWSS. Cats received three injections of 100 μg Fel-CuMV (HypoCat™) at intervals of three weeks (in study weeks 4, 7 and 10) subcutaneously. Cat sera were collected at baseline, week 10, and week 27.

3.2. Adverse Events

Since there was no treatment of the participants, the risk of inducing study-related adverse events was considered low. Nevertheless, some adverse events were recorded and included a cat bite during a veterinary visit, chronic urticaria, erythema, common cold, and pruritus (Table 2). The participant bitten by their cat during an immunization procedure at the veterinary practice was referred to their doctor and received prophylactic antibiotic treatment. They fully recovered from the wound after 1 week. Another participant #10 (excluded from analysis) suffered from an urticaria and presented with strong pruritus and wheal formation. The symptoms were treated with topical urea, macrogol-6-laurylether and fexofenadine-hydrochloride.

There were no severe adverse events observed in the cat population. Several minor clinical signs were noted upon immunization with Fel-CuMV and were judged to be normal vaccination reactions in cats. They disappeared within 72 h of administration. All reactions during and after injection of Fel-CuMV were less frequent on the second and third dosing occasions. There were no treatment related changes in body weight or temperature observed over the study period. Cat owners reported no change in the behavior or appearance (e.g., fur condition) of the cats.

Table 1. Study population.

Participant#	Age	Sex	Age When Allergy Was Observed the First Time	Affected Organs				Baseline VAS Score (0–10) Without Direct Interaction with Cat	Maximal VAS score (0–10) After direct interaction with cat	Continued in Follow up Study
				Eyes	Nose	Lungs/Bronchia	Palate			
1	50	female	10	x			x	2.5	8.8	x
2	29	female	8	x	x	x		3.5	9.2	x
3	21	male	21	x	x	x		3.9	7.2	x
4	43	female	28	x		x		1.1	7	x
5	24	female	22	x	x		x	3.7	7.2	x
6	45	male	23	x		x		4.2	9	-
7	40	female	10	x	x			1.3	5.6	x
8	28	female	12	x	x			0.3	7.7	x
9	51	female	41	x	x	x		1.4	8.5	-
10	37	female	23		x	x		1.9	4.9	-
Mean/Aver.	36.8 ±10.8	8 ♀; 2 ♂	19.0 ± 11.3	9/10	7/10	6/10	2/10	2.4 ± 1.4	7.5 ± 1.4	7/10

Table 2. Adverse events.

Adverse Event	Severity	Participant #	Total Number	Percentage of All AEs	Related to Intervention
Erythema and pruritus	mild	9	3	17.6%	Possible
Tooth infection	moderate	9	1	5.9%	No
Common cold	mild	2, 4, 6, 8	6	35.2%	No
Nausea with vomiting	moderate	10	1	5.9%	No
Nettle rash	moderate	10	1	5.9%	No
Cat allergy reactions	mild	24	1	5.9%	No
Swollen lips	mild	10	1	5.9%	No
Gastroenteritis	moderate	8	1	5.9%	No
Cat bite (hand)	severe	2	1	5.9%	Possible
Chronic urticaria (strong pruritus and wheal formation)	moderate	10	1	5.9%	Possible

3.3. Assessment of the OSSS after Provocation

Seven of nine participants showed lower OSSS at week 24 compared to baseline (Figure 2A) demonstrating a change in the OSSS after provocation. The change from mean 11.7 at baseline to mean 7.3 at week 24 was not significant ($p = 0.098$). The change in the OSSS over the course of the study was also assessed (Figure 2B–D). There was a significant reduction in the mean OSSS from week 8 to 24 (mean OSSS ranging from 6.4–9.1) compared to baseline (mean OSSS 11.7) (Figure 2B). The reduction in the OSSS in seven of nine participants over the study period compared to baseline was statistically significant ($p = 0.03$) (Figure 2C,D), demonstrating a sustained reduction in organ-specific symptoms after vaccination of the cat observed from week 8 to week 24.

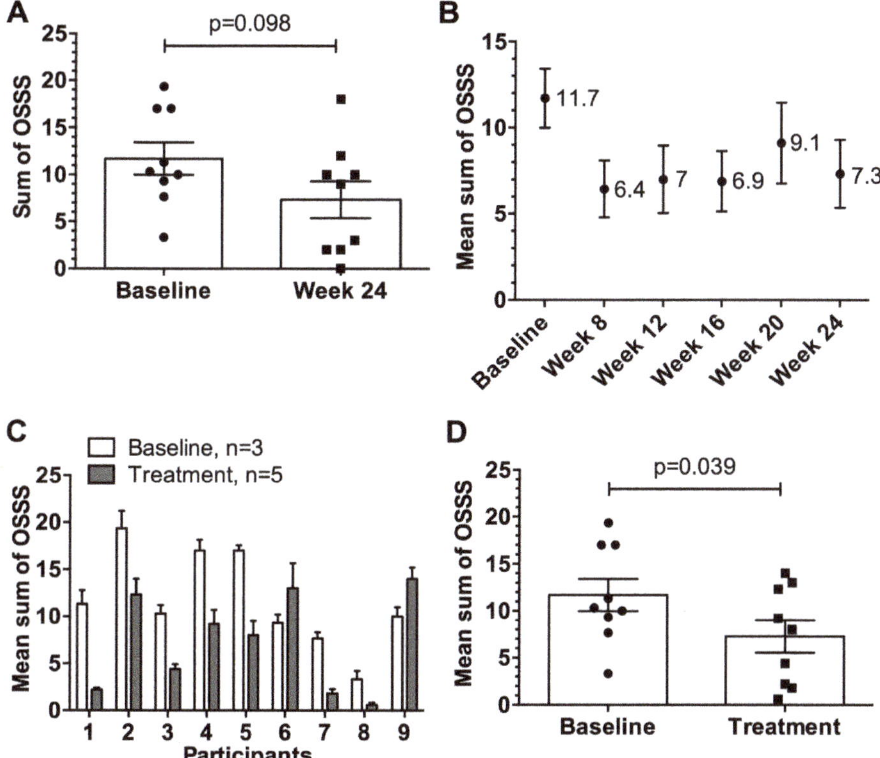

Figure 2. Organ specific symptom score (OSSS) after provocation. (**A**) Individual OSSS with SEM of participants ($n = 9$) at baseline versus week 24 at the end of the study. (**B**) Mean OSSS with SEM of participants ($n = 9$) at baseline and over the course of the study. (**C**) Individual mean OSSS with SEM comparing baseline (weeks 1–3, $n = 3$) vs. treatment period (weeks 8, 12, 16, 20 and 24, $n = 5$). (**D**) Individual mean OSSS ($n = 9$) with SEM comparing baseline ($n = 3$, weeks 1–3) vs. treatment period ($n = 5$, weeks 8, 12, 16, 20 and 24). Statistical significances were obtained by an exact Wilcoxon matched-paired signed rank test. Possible OSSS values from 0–30.

3.4. Petting Time

There were eight out of nine possible successes, meaning that the time of interaction with their cats to reach a defined level of allergic symptoms for eight participants was greater at week 24 than the mean baseline of weeks 1–3 (Figure 3A). The change from mean 16.9 min (1016 s) at baseline to mean

27.7 min (1659 s) was statistically significant ($p = 0.02$). Of note, three participants had reached the maximum interaction time of 45 min by the end of the study.

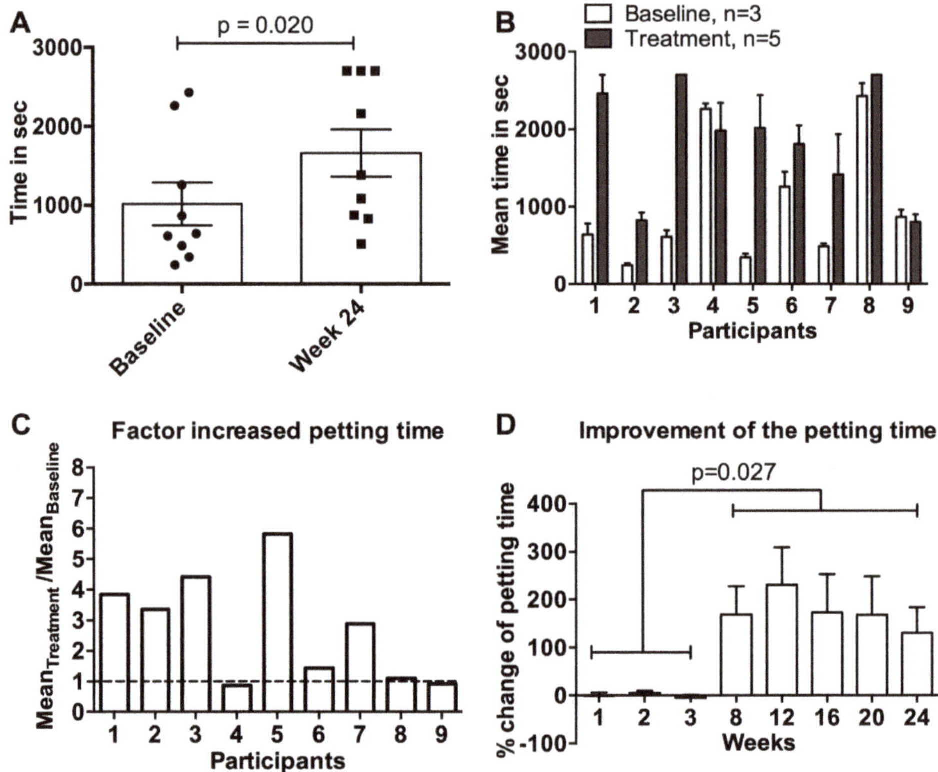

Figure 3. Time of petting as part of the provocation test. (**A**) Mean petting time with SEM in seconds of nine participants comparing baseline ($n = 3$) versus week 24 at the end of the study. (**B**) Individual mean petting time with SEM in seconds at baseline ($n = 3$, weeks 1–3) compared to treatment period ($n = 5$, weeks 8, 12, 16, 20 and 24). (**C**) Factor of changed petting time for each participant shown as ratio of the mean timepoints during treatment ($n = 5$, weeks 8, 12, 16, 20 and 24) to mean baseline time ($n = 3$, weeks 1–3). (**D**) Change in % of the mean petting time with SEM of nine participants during the treatment period in weeks 8, 12, 16, 20 and 24 compared to mean baseline in weeks 1–3. Statistical significances were obtained by an exact Wilcoxon matched-paired signed rank test.

The improved petting time was apparent over the entire course of the study after vaccination of the cats in weeks 4, 7 and 10 (Figure 3B–D). On a per participant basis, the petting time was increased in seven of nine participants up to 6-fold (Figure 3B,C), which corresponded to an average improvement of 100%–230% across weeks 8, 12, 16, 20, and 24 compared to baseline (Figure 3D).

3.5. General Weekly Symptom Score

There were eight of possible nine successes, meaning that the GWSS of eight participants were lower at week 24 compared to baseline (Figure 4A). The mean GWSS at baseline compared to week 24 was significantly reduced from 7.2 to 4.4 ($p = 0.023$).

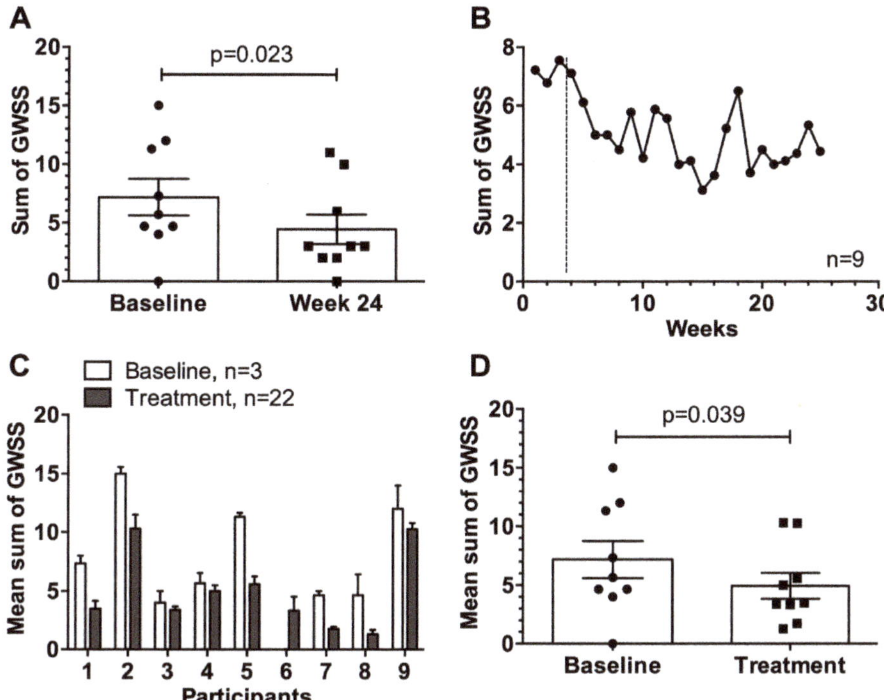

Figure 4. General weekly symptom score GWSS. (**A**) Mean GWSS with SEM of participants ($n = 9$) at baseline ($n = 3$, mean of weeks 1–3) versus week 24 at the end of the study. (**B**) Mean GWSS with SEM of participants ($n = 9$) at baseline ($n = 3$, weeks 1–3) and over the course of the study ($n = 22$, weeks 4–25). (**C**) Individual mean GWSS with SEM at baseline ($n = 3$, weeks 1–3) compared to treatment period ($n = 22$, weeks 4–25). (**D**) Mean GWSS with SEM of nine participants comparing baseline ($n = 3$, weeks 1–3) vs. treatment period ($n = 22$, weeks 4–25). Statistical significances were obtained by an exact Wilcoxon matched-paired signed rank test. Possible GWSS values from 0–30.

The average GWSS of all participants over the entire study period was also assessed (Figure 4B–D). Towards the end of the study, the mean GWSS was consistently lower than the baseline values (Figure 4B). Comparing the average baseline with the average treatment values of the GWSS, a change in the general allergic symptoms ($p = 0.039$) was observed in eight of nine participants, demonstrating a persistent improvement in symptoms upon immunization of the cats (Figure 4C,D).

3.6. Induction of Anti-Fel d 1 IgG Antibody Responses in Cats upon Vaccination with Fel-CuMV

Subcutaneous injection of Fel-CuMV vaccine was considered to be well-tolerated and safe, as only mild and transient side reactions were noted. Cat sera were assessed for Fel d 1-specific IgG antibody responses throughout the course of the study (Figure 5). Participant #8 had three cats, identified as cats 8.1, 8.2, and 8.3. There are no serological data available for cat 8.3 because it rejected the sampling procedure. There was a significant increase in anti-Fel d 1 IgG in all vaccinated cats ($p < 0.001$) detectable after two immunizations at week 10. By the end of the study at week 27, the anti-Fel d 1-specific antibody response in sera ($p = 0.001$) was still significantly increased compared to baseline.

The antibody responses of the study cats were compared with the OSSS and petting time of the provocation test and the GWSS of the participants at the corresponding timepoints (Figure 5C–E). The parameters for assessing the allergic symptoms of the participants correlated well with the antibody responses in cats. Petting time showed a direct correlation with the anti-Fel d 1 IgG titers, whereas the OSSS and the GWSS measurements were inversely correlated with the anti-Fel d 1 IgG titers.

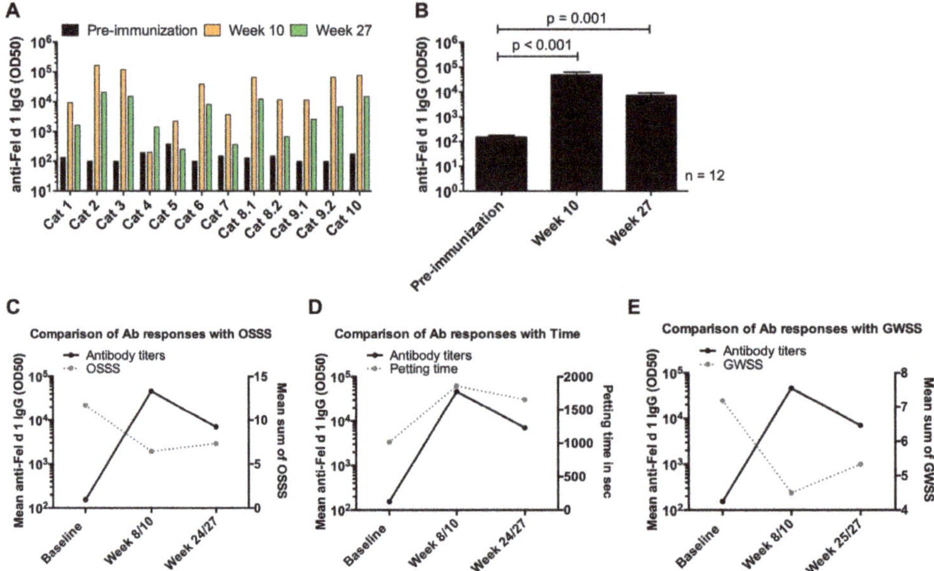

Figure 5. Antibody responses in cat sera. (**A**) Anti-Fel d 1 IgG antibody titers in cats before and after immunization, n = 12. (**B**) Mean anti-Fel d 1 IgG antibody titers with SEM, n = 12. (**C**) Anti-Fel d 1 IgG antibody titers in cats at baseline (before immunization), week 10 and 27 vs. OSSS at corresponding timepoints baseline (before immunization), week 8 and 24. (**D**) Anti-Fel d 1 IgG antibody titers in cats at baseline (before immunization), week 10 and 27 vs. petting time at corresponding timepoints baseline (before immunization), week 8 and 24. (**E**) Anti-Fel d 1 IgG antibody titers in cats at baseline (before immunization), week 10 and 27 vs. GWSS at corresponding timepoints baseline (before immunization), week 8 and 25. (C-E) Data included from participants (n = 9) with corresponding study cats (n = 9). Statistical significances were obtained by an exact Wilcoxon matched-paired signed rank test.

3.7. Data of the Extension Study

After completion of the main study, an extension study was conducted with seven of the original participants (Figure 6A). Reasons for the discontinuation of three participants were the exclusion of the urticaria patient, the cat of one participant died due to a study-unrelated icterus, and personal preferences of one participant.

The analysis presented in Figure 6 includes the data obtained from the seven participants and their respective cats (seven study cats plus two cats; one participant had three cats), who participated in both studies.

As described above, there was a reduction in the OSSS from week 8 to 24 (mean OSSS ranging from 4.3 to 5.6) compared to baseline (mean OSSS 12.3). The mean OSSS increased to 9.3 in the intervening period between both studies (i.e., from week 24 to 54). After the boost immunization in week 56, there was a small decrease in mean OSSS observed in study weeks 58 to 98 ranging from 6.0 to 7.4. Although the reduction in the OSSS over both study periods compared to baseline was not statistically significant, a trend of reduced OSSS was observed over the entire study period of almost 2 years (94 weeks) after vaccination of the cats.

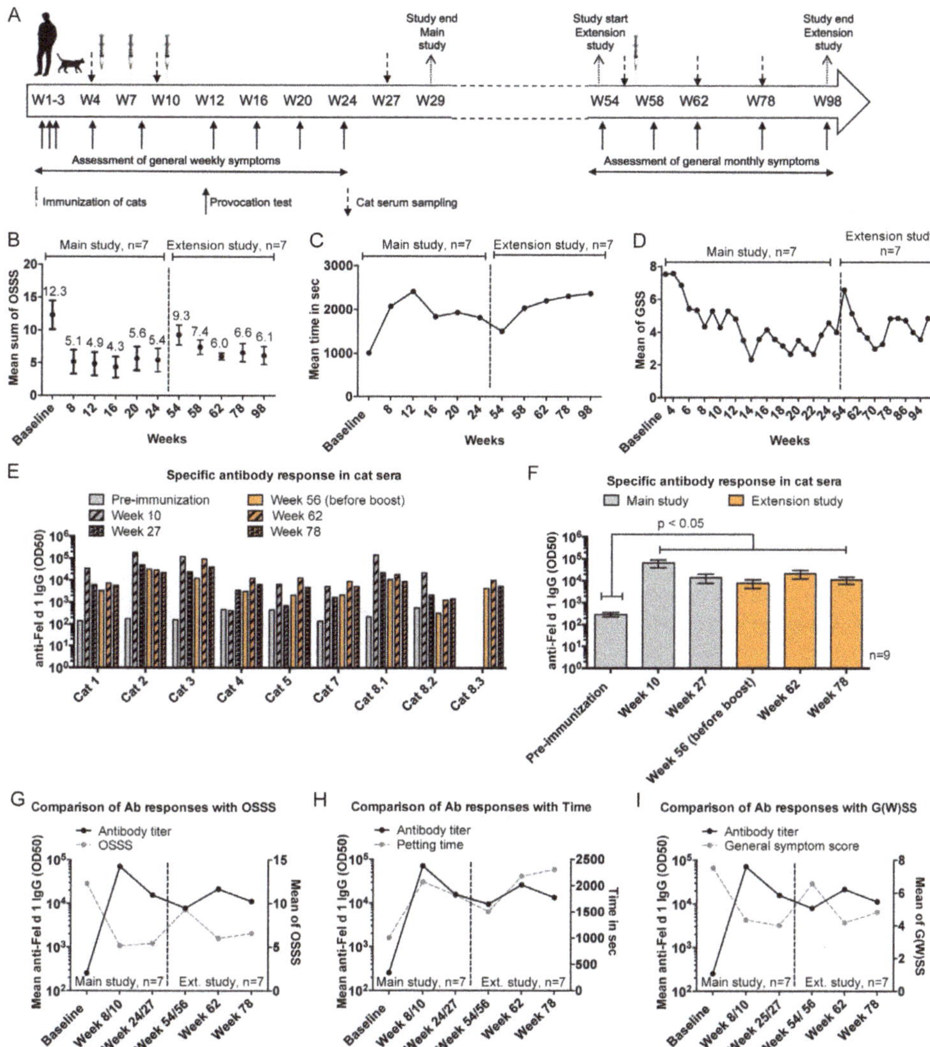

Figure 6. Extension study: Changes in allergic symptoms of participants after booster injection of cats. (**A**) Study design of the main and extension study including seven participants and nine cats. In addition to the main study, the participants performed a provocation test before the boost (week 56) followed by four additional provocation tests in study weeks 58, 62, 78, and 98. Participants also recorded their general organ specific symptoms (GSS) without provocation at monthly intervals during the extension study. The cats received a boost immunization of 100 µg Fel-CuMV subcutaneously in week 56. In addition to the collected serum samples of the main study, cat sera were also collected in study weeks 56, 62, and 78. (**B**) Mean OSSS with SEM over the course of the main and extension study of seven participants. (**C**) Mean petting time over the course of the main and extension study of seven participants. (**D**) Mean general symptoms (G(W)SS) over the course of the main (assessed weekly) and extension (assessed monthly) study of seven participants. (**E**) Measurement of anti-Fel d 1 IgG in cat sera of nine cats participating in the main and extension study. (**F**) Mean anti-Fel d 1 IgG antibody titers with SEM in cat sera. (**G**) Mean anti-Fel d 1 IgG antibody titers in cats vs. mean OSSS over the course of the main (baseline (before immunization), week 8/10 and 24/27) and extension study (week 54/56 (before boost), week 62 and 78). (**H**) Mean anti-Fel d 1 IgG antibody titers in cats vs. mean petting time

over the course of the main (baseline (before immunization), week 8/10 and 24/27) and extension study (week 54/56 (before boost), week 62 and 78). (**I**) Mean anti-Fel d 1 IgG antibody titers in cats vs. mean G(W)SS over the course of the main (baseline (before immunization), week 8/10 and 24/27) and extension study (week 54/56 (before boost), week 62 and 78). (**G-I**) Data included from participants ($n = 7$) with corresponding study cats ($n = 7$). Statistical significances were obtained by an exact Wilcoxon matched-paired signed rank test.

The GSS followed a consistent trend of lower symptoms from week 5 to 25 after the initial cat immunizations in weeks 4, 7 and 10 (Figure 6D). The mean GSS at baseline (weeks 1–3) of 7.5 improved to a mean score of 6.8, which was already detectable in week 5, and further improved to 2.3 in study week 14. By the end of the main study in week 25, the mean GSS of 4 was still lower compared to baseline (i.e., GSS 7.5). Before cats received the boost injection in week 56, the mean GSS increased to 6.5 but was still lower compared to baseline (i.e., GSS 7.5 of weeks 1–3). Upon booster injection, allergic symptoms improved again, reaching the lowest mean GSS of 3 in week 70. Clinical improvement was still evident by the end of the study in week 98 (i.e., GSS 4.8). Notably, the participants had reduced GSS at any timepoint after the first vaccination of their cats in study week 4, as the mean GSS over both study periods was always lower compared to baseline (weeks 1–3).

Cat sera were assayed for Fel d 1-specific IgG antibodies throughout both study periods (Figure 6E,F). The mean Fel d 1-specific IgG titers increased ~300-fold measured after two injections in week 10 and were still on a high level at the end of the main study at week 27 (Figure 6F). Specific antibodies declined between weeks 27 and 56 (before the boost) compared to the peak response detected in week 10. Following administration of a booster injection in week 56, the specific IgG levels increased again in eight of nine cats when measured by week 62, 6 weeks after the boost. Almost 4 months later, by the end of the study in week 78, the anti-Fel d 1 IgG responses were still significantly higher compared to baseline and were slightly above the antibody level measured before boost (week 56).

The antibody responses were inversely related to the OSSS and G(W)SS over the period of the main and extension studies (Figure 6G,I). In contrast, antibody responses directly correlated to the petting time of the provocation test (Figure 6H), indicating that specific antibodies reduced symptoms and increased petting time. Specifically, after the induction of anti-Fel d 1 antibodies detected in week 10, the OSSS decreased from a mean score of 12.5 at baseline to 5.1 and 5.4, measured in week 8 and 24, respectively. During the intervening phase of the main and extension study, the antibody titers declined and the mean OSSS increased to 9.3, detected in week 54. After the boost injection which caused an increase in specific antibodies, the mean OSSS decreased to a mean score of 6.0 and 6.6 in study weeks 62 and 78, respectively. Similar observations were made for G(W)SS.

The antibody titers also showed a correlation with petting time. Relative to baseline (week 1–3), mean petting time and anti-Fel d 1 IgG, determined in week 8 and 10, increased the mean petting time by a factor of 2, demonstrating that owners were found to be able to interact with their cats twice as long before they developed an allergic symptom strength of 5 according to VAS. The subsequent decrease in specific antibodies was associated with a decrease in petting time observed in weeks 24 and 54, although still above baseline. Upon the increase in specific antibodies after the booster injection, petting time increased again by a factor of 2.3 compared to baseline.

4. Discussion

Allergy is the most common chronic disease in Europe and U.S and often manifests with chronic conditions including rhinitis, conjunctivitis and asthma, which is associated with individual morbidity and high socio-economic costs. In fact, more than 150 million Europeans are affected by chronic allergic diseases (EAACI, 2016). The danger of developing chronic asthma is particularly high, especially in allergies, which are caused by air-borne allergens such as hay fever and cat allergy. One in five patients with allergies lives with the constant fear of getting an asthmatic or anaphylactic shock at any time or even dying as a result of a severe allergic reaction upon allergen encounter. The World Health Organization estimates that approximately 300 million individuals currently suffer from asthma worldwide and expect an increasing incidence to 400 million by 2025 [21,40]. This development is worrisome, threatening health and economies alike, and demands action of the global community for the development of new therapies, medications and diagnostic tools to address this major challenge [7,41]. Cat allergy contributes to a significant proportion of the allergic disease burden.

There are several new therapeutic developments to treat cat allergy that are currently pursued by various research groups and clinicians. Our novel approach of vaccinating the cat against Fel d 1, offers a cost-effective therapy without the risk of inducing severe side effects in cat-allergic patients, as they often occur during AITs, and without adversely affecting the cat. The vaccine targets Fel d 1, the major cat allergen for humans. By immunizing cats, the reactive allergen level can be lowered, thereby alleviating the symptoms of cat allergic patients. Here, we present clinical data of a first combined human and animal field trial conducted as an open-label, exploratory methodological study. The aim of the study was to assess different methods of measuring the allergic symptoms and determine if those methods could detect changes in allergic symptoms after vaccination of cats with Fel-CuMV (HypoCat™).

The collection of clinical research data must be reproducible, valid and traceable using standardized procedures [42,43]. However, the selection and combination of various tests and scoring systems for monitoring allergic symptoms and their changes usually have a significant impact on the outcome of clinical trials and the development of innovative medications. Therefore, it is advisable to select as few questionnaires, tests and scoring systems as possible to capture the changes that best reflect the effect of the treatment [44,45]. To this end, we developed a new method to record allergic symptoms by a self-assessed, home-based symptom score, the provocation test. The validity of the test was evaluated in a previously conducted human clinical trial involving non-immunized cats (NCT02399579). The test consists of two elements: an assessment of organ-specific symptom score (OSSS) after petting—a standardized score—and measurement of the time of petting the cat until a defined level of allergy symptom was reached (level of 5 on a VAS). Both parameters, the OSSS and the time, are analyzed separately. In addition to the provocation test, another self-assessed, home-based standard method was investigated in this study: a general symptom score (G(W)SS) assessing the allergic symptoms without provocation.

The mean OSSS as part of the provocation test at week 24 was reduced in seven of nine participants compared to baseline. Moreover, the change in the OSSS was apparent throughout the course of the main study (weeks 8, 12, 16, 20 and 24 vs. baseline) and showed a variable yet sustained reduction in seven of nine participants. It should be noted that a reduction in the OSSS was not necessarily expected to be large, since the cat owners were required to record their symptoms after petting the cat until they had reached a symptom strength of 5 on the VAS. Hence, upon cessation of the provocation test (i.e., petting), the OSSS would have been similar on each occasion. A reason why a substantial decrease in OSSS in the main study was achieved may have been due to the fact that three participants, at week 24, were able to pet their cat for the maximum time of 45 min. In this circumstance, the VAS score of 5 was not reached, and thus the OSSS was lower. Regarding the extension study, several participants at several occasions could pet their cat to the maximum petting time of 45 min and did not reach the symptom strength of 5 on a VAS. Thus, the allergic symptoms were not as pronounced, resulting in lower OSSS.

The second parameter of the provocation test was the petting time. Relative to baseline, an increased time of petting was observed in eight of the nine participants at week 24 and over the entire period of the main study in seven of the nine. The increase in petting time was even more pronounced in the extension study. Several participants could pet their cat to the maximum time of 45 min on several occasions. The average petting time increased by a factor of two upon immunization of the cats and showed a sustained improvement over the period of the main and extension studies, demonstrating that the participants could interact longer with their cats.

A change in general allergic symptoms without provocation assessed by the G(W)SS was noted upon vaccination of the cats. Relative to baseline, a reduction in the GWSS at week 24 and over the period of the main study was observed in eight of nine participants. Moreover, the GSS measured throughout the course of the extension study showed a variable yet sustained reduction in seven of seven participants. Of note, the improvement in general allergic symptoms of the participants recorded without direct interaction with their cats at all timepoints after immunization of the cats over the period of the main and extension studies indicates that the participants generally felt better and suffered less from allergic symptoms.

Subcutaneous injection of three doses of Fel-CuMV vaccine followed by a single boost injection a year later was considered to be well tolerated in several breeds of privately owned adult cats. No serious adverse event occurred during the study. The clinical signs and reactions upon vaccination with Fel-CuMV were mild and reversible. No related effects on body weight, food consumption, behavior and appearance were observed. Moreover, the intended immunological response in cats immunized with Fel-CuMV, i.e., induction of anti-Fel d 1 IgG antibodies, was achieved. Anti-Fel d 1 antibodies were measured in sera at week 10, three weeks after the second immunization. Previous studies in cats ($n > 60$) with Fel-CuMV have shown that about 50% of immunized animals achieved peak titers 2–3 weeks after the second immunization [36]. This would correspond to week 10 of the current study. These findings support the observation of the improved symptoms assessed by the provocation test and G(W)SS of the participants from week 8 and 6, respectively onwards, as Fel d 1-specific antibodies were present and had the potential to neutralize Fel d 1 as previously shown [36]. The improvement of symptoms had lessened by the end of the main study, which may be related to the kinetics of the antibody response, which showed a decline from week 10 to week 27. Upon administration of a booster injection, the symptoms and petting time improved again.

However, the results and conclusions of this open label, exploratory methodology study with a small sample size ($n = 9$ in main and $n = 7$ in extension study) must be taken with caution. There was no pre-existing information available regarding the expected change in symptoms of the participants before and after immunization of their cats. As a consequence, no sample size calculation could be done beforehand. Nevertheless, it is interesting to note that several parameters measured in the study were suggestive that targeting the major cat allergen Fel d 1 by immunization of cats with Fel-CuMV had a positive impact on the allergic sensation of the participant. In particular, the time that participants were able to interact with their cats before a particular level of allergic symptoms was reached increased significantly.

Improvements of the organ specific symptoms with (OSSS) and without provocation (G(W)SS) over the course of the study were also noted. Notably, it was observed that the improvement in these measures had lessened at the end of the main study but increased again after the booster injection. This observation of the kinetics is valuable for planning and conducting further studies. Therefore, all measured parameters (i.e., provocation test and GSS) are suitable to assess changes in allergic symptoms of the cat owners.

Finally, from a veterinary perspective, targeting Fel d 1, the major cat allergen in humans, by active vaccination has, to date, been well-tolerated. The general health status of the cats was not compromised by the induction of auto-antibodies against Fel d 1. However, allergic symptoms of the cat allergic owners were alleviated. As a result of the owner being less burdened by their allergy, the quality of life of their cat may be improved. The ability of allergic cat owners to better tolerate and increase the duration of their interactions with their pet can benefit the animal through better training and socialization and awareness of the animals' overall health. Moreover, the likelihood of abandonment and subsequent euthanasia in animal shelters could be decreased.

Author Contributions: Conceptualization, F.T., G.T.J., S.H., G.S., T.M.K. and M.F.B.; methodology, F.T., S.H., G.T.J., G.S., T.M.K. and M.F.B.; software, F.T., S.H., N.G.; validation, F.T., S.H., G.T.J., T.M.K. and M.B.F.; formal analysis, F.T., S.H., A.E., N.G., G.T.J. and M.F.B.; investigation, G.S., T.M.K. and M.F.B.; resources, F.T., G.T.J., S.H., G.S., T.M.K., A.S., G.P., T.L., M.C.F., L.A.T., C.S.N., S.R., and M.F.B.; data curation, F.T., S.H., N.G., G.T.J. and M.F.B.; writing—original draft preparation, F.T., G.T.J., S.H., G.S., C.S.N., S.R., N.G., T.M.K. and M.F.B; writing—review and editing, A.S., G.P., T.L., M.C.F. and L.A.T.; visualization, F.T. and S.H.; supervision, N.G., G.T.J., T.M.K. and M.F.B.; project administration, F.T., G.T.J.; funding acquisition, M.C.F., L.A.T., G.S., T.M.K. and M.F.B. All authors have read and agreed to the published version of the manuscript.

Funding: The work in this manuscript was funded by Benchmark Animal Health, UK and supported by the Swiss Commission for Technology and Innovation, (CTI project 16015.2 PFLS-LS to M.F.B., T.M.K. and G.S.).

Acknowledgments: We thank Antonia Gabriel-Fettelschoss, Monique Vogel and Pål Johansen for scientific discussion or careful reading of the manuscript.

Conflicts of Interest: Authors F.T., G.T.J., S.H., G.S., T.M.K. and M.F.B. have a financial relationship with HypoPet AG involving employment, stock ownership or payments for research activities. Authors A.S., G.P., T.L., M.C.F. and L.A.T. have a financial relationship with Benchmark Animal Health involving employment, or stock ownership. F.T., G.T.J., G.S., T.M.K., and M.F.B. own a patent for "Compositions against cat allergy". A.E., C.N., S.R., and N.G. have no conflict of interest.

Abbreviations

Ab	Antibody
AE	Adverse Event
CSR	Clinical Study Report
eCRF	Electronic Case Report Form
ELISA	Enzyme-linked immunosorbent assay
Fel d 1	Felis domesticus 1 allergen
Fel-CuMV	Fel d 1 - Cucumber mosaic virus-like particle conjugate vaccine
GCP	Good Clinical Practice
GSS	General montly Symptom Score
GWSS	General Weekly Symptom Score
ICH	International Conference on Harmonization
IEC	Independent Ethics Committee
ITT	Intention to Treat
KEK	Cantonal Ethics Committee Zürich
LOCF	Last Observation Carried Forward
LPLV	Last Patient Last Visit
OSSS	Organ Specific Symptom Score
PP	Per Protocol
SAE	Serious Adverse Event
VAS	Visual Analogue Scale

References

1. Platts-Mills, T.A.; Carter, M.C. Asthma and indoor exposure to allergens. *N. Engl. J. Med.* **1997**, *336*, 1382–1384. [CrossRef] [PubMed]
2. Rance, F. Animal dander allergy in children. *Arch. Pediatr.* **2006**, *13*, 584–586. [PubMed]
3. Anderson, M.C.; Baer, H. Allergenically active components of cat allergen extracts. *J. Immunol.* **1981**, *127*, 972–975. [PubMed]
4. Lowenstein, H.; Lind, P.; Weeke, B. Identification and clinical significance of allergenic molecules of cat origin. Part of the DAS 76 Study. *Allergy* **1985**, *40*, 430–441. [CrossRef] [PubMed]
5. Smith, W.; O'Neil, S.E.; Hales, B.J.; Chai, T.L.; Hazell, L.A.; Tanyaratsrisakul, S.; Piboonpocanum, S.; Thomas, W.R. Two newly identified cat allergens: The von Ebner gland protein Fel d 7 and the latherin-like protein Fel d 8. *Int. Arch. Allergy Immunol.* **2011**, *156*, 159–170. [CrossRef]
6. Smith, W.; Butler, A.J.; Hazell, L.A.; Chapman, M.D.; Pomes, A.; Nickels, D.G.; Thomas, W.R. Fel d 4, a cat lipocalin allergen. *Clin. Exp. Allergy* **2004**, *34*, 1732–1738. [CrossRef]
7. Konradsen, J.R.; Fujisawa, T.; van Hage, M.; Hedlin, G.; Hilger, C.; Kleine-Tebbe, J.; Matsui, E.C.; Roberts, G.; Ronmark, E.; Platts-Mills, T.A. Allergy to furry animals: New insights, diagnostic approaches, and challenges. *J. Allergy Clin. Immunol.* **2015**, *135*, 616–625. [CrossRef]
8. Ohman, J.L.; Lowell, F.C.; Bloch, K.J. Allergens of mammalian origin: Characterization of allergen extracted from cat pelts. *J. Allergy Clin. Immunol.* **1973**, *52*, 231–241. [CrossRef]

9.   Bonnet, B.; Messaoudi, K.; Jacomet, F.; Michaud, E.; Fauquert, J.L.; Caillaud, D.; Evrard, B. An update on molecular cat allergens: Fel d 1 and what else? Chapter 1: Fel d 1, the major cat allergen. *Allergy Asthma Clin. Immunol.* **2018**, *14*, 14. [CrossRef]

10.   Ukleja-Sokolowska, N.; Gawronska-Ukleja, E.; Zbikowska-Gotz, M.; Socha, E.; Lis, K.; Sokolowski, L.; Kuzminski, A.; Bartuzi, Z. Analysis of feline and canine allergen components in patients sensitized to pets. *Allergy Asthma Clin. Immunol.* **2016**, *12*, 61. [CrossRef]

11.   Kaiser, L.; Gronlund, H.; Sandalova, T.; Ljunggren, H.G.; van Hage-Hamsten, M.; Achour, A.; Schneider, G. The crystal structure of the major cat allergen Fel d 1, a member of the secretoglobin family. *J. Biol. Chem.* **2003**, *278*, 37730–37735. [CrossRef] [PubMed]

12.   Karn, R.C. The mouse salivary androgen-binding protein (ABP) alpha subunit closely resembles chain 1 of the cat allergen Fel dI. *Biochem. Genet.* **1994**, *32*, 271–277. [CrossRef] [PubMed]

13.   Charpin, C.; Mata, P.; Charpin, D.; Lavaut, M.N.; Allasia, C.; Vervloet, D. Fel d I allergen distribution in cat fur and skin. *J. Allergy Clin. Immunol.* **1991**, *88*, 77–82. [CrossRef]

14.   Dabrowski, A.J.; Van der Brempt, X.; Soler, M.; Seguret, N.; Lucciani, P.; Charpin, D.; Vervloet, D. Cat skin as an important source of Fel d I allergen. *J. Allergy Clin. Immunol.* **1990**, *86*, 462–465. [CrossRef]

15.   De Andrade, A.D.; Birnbaum, J.; Magalon, C.; Magnol, J.P.; Lanteaume, A.; Charpin, D.; Vervloet, D. Fel d I levels in cat anal glands. *Clin. Exp. Allergy* **1996**, *26*, 178–180. [CrossRef]

16.   van Milligen, F.J.; Vroom, T.M.; Aalberse, R.C. Presence of Felis domesticus allergen I in the cat's salivary and lacrimal glands. *Int. Arch. Allergy Appl. Immunol.* **1990**, *92*, 375–378. [CrossRef]

17.   Custovic, A.; Simpson, B.M.; Simpson, A.; Hallam, C.L.; Marolia, H.; Walsh, D.; Campbell, J.; Woodcock, A.; National Asthma Campaign Manchester Asthma and Allergy Study Group. Current mite, cat, and dog allergen exposure, pet ownership, and sensitization to inhalant allergens in adults. *J. Allergy Clin. Immunol.* **2003**, *111*, 402–407. [CrossRef]

18.   Platts-Mills, T.A. The role of immunoglobulin E in allergy and asthma. *Am. J. Respir. Crit. Care Med.* **2001**, *164*, S1–S5. [CrossRef]

19.   Platts-Mills, T.; Vaughan, J.; Squillace, S.; Woodfolk, J.; Sporik, R. Sensitisation, asthma, and a modified Th2 response in children exposed to cat allergen. A population-based cross-sectional study. *Lancet* **2001**, *357*, 752–756. [CrossRef]

20.   Anderson, G.P.; Coyle, A.J. TH2 and 'TH2-like' cells in allergy and asthma: Pharmacological perspectives. *Trends Pharmacol. Sci.* **1994**, *15*, 324–332. [CrossRef]

21.   Pawankar, R.; Canonica, G.W.; Holgate, S.T.; Lockey, R.F.; Blaiss, M.S. *WAO The White Book on Allergy*; World Allergy Organization: Milwaukee, WI, USA, 2013.

22.   Turner, D.C. A review of over three decades of research on cat-human and human-cat interactions and relationships. *Behav. Processes* **2017**, *141*, 297–304. [CrossRef] [PubMed]

23.   Leuthard, D.S.; Duda, A.; Freiberger, S.N.; Weiss, S.; Dommann, I.; Fenini, G.; Contassot, E.; Kramer, M.F.; Skinner, M.A.; Kundig, T.M.; et al. Microcrystalline Tyrosine and Aluminum as Adjuvants in Allergen-Specific Immunotherapy Protect from IgE-Mediated Reactivity in Mouse Models and Act Independently of Inflammasome and TLR Signaling. *J. Immunol.* **2018**, *200*, 3151–3159. [CrossRef]

24.   Zubeldia, J.M.; Ferrer, M.; Davila, I.; Justicia, J.L. Adjuvants in Allergen-Specific Immunotherapy: Modulating and Enhancing the Immune Response. *J. Investig. Allergol. Clin. Immunol.* **2019**, *29*, 103–111. [CrossRef] [PubMed]

25.   Klimek, L.; Willers, J.; Hammann-Haenni, A.; Pfaar, O.; Stocker, H.; Mueller, P.; Renner, W.A.; Bachmann, M.F. Assessment of clinical efficacy of CYT003-QbG10 in patients with allergic rhinoconjunctivitis. A phase IIb study. *Clin. Exp. Allergy* **2011**, *41*, 1305–1312. [CrossRef]

26.   Senti, G.; Crameri, R.; Kuster, D.; Johansen, P.; Martinez-Gomez, J.M.; Graf, N.; Steiner, M.; Hothorn, L.A.; Gronlund, H.; Tivig, C.; et al. Intralymphatic immunotherapy for cat allergy induces tolerance after only 3 injections. *J. Allergy Clin. Immunol.* **2012**, *129*, 1290–1296. [CrossRef] [PubMed]

27.   Senti, G.; Graf, N.; Haug, S.; Ruedi, N.; von Moos, S.; Sonderegger, T.; Johansen, P.; Kundig, T.M. Epicutaneous allergen administration as a novel method of allergen-specific immunotherapy. *J. Allergy Clin. Immunol.* **2009**, *124*, 997–1002. [CrossRef] [PubMed]

28.   Senti, G.; Johansen, P.; Haug, S.; Bull, C.; Gottschaller, C.; Muller, P.; Pfister, T.; Maurer, P.; Bachmann, M.F.; Graf, N.; et al. Use of A-type CpG oligodeoxynucleotides as an adjuvant in allergen-specific immunotherapy in humans. A phase I/IIa clinical trial. *Clin. Exp. Allergy* **2009**, *39*, 562–570. [CrossRef]

29. Senti, G.; von Moos, S.; Tay, F.; Graf, N.; Sonderegger, T.; Johansen, P.; Kundig, T.M. Epicutaneous allergen-specific immunotherapy ameliorates grass pollen-induced rhinoconjunctivitis. A double-blind, placebo-controlled dose escalation study. *J. Allergy Clin. Immunol.* **2012**, *129*, 128–135. [CrossRef]
30. Basomba, A.; Tabar, A.I.; de Rojas, D.H.; Garcia, B.E.; Alamar, R.; Olaguibel, J.M.; del Prado, J.M.; Martin, S.; Rico, P. Allergen vaccination with a liposome-encapsulated extract of Dermatophagoides pteronyssinus. A randomized, double-blind, placebo-controlled trial in asthmatic patients. *J. Allergy Clin. Immunol.* **2002**, *109*, 943–948. [CrossRef]
31. Fonseca, D.E.; Kline, J.N. Use of CpG oligonucleotides in treatment of asthma and allergic disease. *Adv. Drug Deliv. Rev.* **2009**, *61*, 256–262. [CrossRef]
32. Patel, D.; Couroux, P.; Hickey, P.; Salapatek, A.M.; Laidler, P.; Larche, M.; Hafner, R.P. Fel d 1-derived peptide antigen desensitization shows a persistent treatment effect 1 year after the start of dosing. A randomized, placebo-controlled study. *J. Allergy Clin. Immunol.* **2013**, *131*, 103–109.e7. [CrossRef] [PubMed]
33. Orengo, J.M.; Radin, A.R.; Kamat, V.; Badithe, A.; Ben, L.H.; Bennett, B.L.; Zhong, S.; Birchard, D.; Limnander, A.; Rafique, A.; et al. Treating cat allergy with monoclonal IgG antibodies that bind allergen and prevent IgE engagement. *Nat. Commun.* **2018**, *9*, 1421. [CrossRef] [PubMed]
34. Satyaraj, E.; Li, Q.; Sun, P.; Sherrill, S. Anti-Fel d1 immunoglobulin Y antibody-containing egg ingredient lowers allergen levels in cat saliva. *J. Feline Med. Surg.* **2019**, 1098612X19861218. [CrossRef] [PubMed]
35. Satyaraj, E.; Gardner, C.; Filipi, I.; Cramer, K.; Sherrill, S. Reduction of active Fel d1 from cats using an antiFel d1 egg IgY antibody. *Immun. Inflamm. Dis.* **2019**, *7*, 68–73. [CrossRef] [PubMed]
36. Thoms, F.; Jennings, G.T.; Maudrich, M.; Vogel, M.; Haas, S.; Zeltins, A.; Hofmann-Lehmann, R.; Riond, B.; Grossmann, J.; Hunziker, P.; et al. Immunization of cats to induce neutralizing antibodies against Fel d 1, the major feline allergen in human subjects. *J. Allergy Clin. Immunol.* **2019**, *144*, 193–203. [CrossRef] [PubMed]
37. Zeltins, A.; West, J.; Zabel, F.; El Turabi, A.; Balke, I.; Haas, S.; Maudrich, M.; Storni, F.; Engeroff, P.; Jennings, G.T.; et al. Incorporation of tetanus-epitope into virus-like particles achieves vaccine responses even in older recipients in models of psoriasis, Alzheimer's and cat allergy. *NPJ Vaccines* **2017**, *2*, 30. [CrossRef]
38. Bachmann, M.F.; Rohrer, U.H.; Kundig, T.M.; Burki, K.; Hengartner, H.; Zinkernagel, R.M. The influence of antigen organization on B cell responsiveness. *Science* **1993**, *262*, 1448–1451. [CrossRef]
39. Bachmann, M.F.; Jennings, G.T. Vaccine delivery. A matter of size, geometry, kinetics and molecular patterns. *Nat. Rev. Immunol.* **2010**, *10*, 787–796. [CrossRef]
40. Pawankar, R. Allergic diseases and asthma. A global public health concern and a call to action. *World Allergy Organ. J.* **2014**, *7*, 1–3. [CrossRef]
41. Chan, S.K.; Leung, D.Y.M. Dog and Cat Allergies: Current State of Diagnostic Approaches and Challenges. *Allergy Asthma Immunol. Res.* **2018**, *10*, 97–105. [CrossRef]
42. Pfaar, O.; Demoly, P.; Gerth van Wijk, R.; Bonini, S.; Bousquet, J.; Canonica, G.W.; Durham, S.R.; Jacobsen, L.; Malling, H.J.; Mosges, R.; et al. Recommendations for the standardization of clinical outcomes used in allergen immunotherapy trials for allergic rhinoconjunctivitis: An EAACI Position Paper. *Allergy* **2014**, *69*, 854–867. [CrossRef] [PubMed]
43. Anonymous. *Guideline on the Clinical Development of Products for Specific Immunotherapy for the Treatment of Allergic Diseases*; European Medecine Agency: London, UK, 2008.
44. Pfaar, O.; Alvaro, M.; Cardona, V.; Hamelmann, E.; Mosges, R.; Kleine-Tebbe, J. Clinical trials in allergen immunotherapy: Current concepts and future needs. *Allergy* **2018**, *73*, 1775–1783. [CrossRef] [PubMed]
45. Macchia, D.; Melioli, G.; Pravettoni, V.; Nucera, E.; Piantanida, M.; Caminati, M.; Campochiaro, C.; Yacoub, M.-R.; Schiavino, D.; Paganelli, R.; et al. Guidelines for the use and interpretation of diagnostic methods in adult food allergy. *Clin. Mol. Allergy* **2015**, *13*, 27. [CrossRef] [PubMed]

Article

A Chimeric Sudan Virus-Like Particle Vaccine Candidate Produced by a Recombinant Baculovirus System Induces Specific Immune Responses in Mice and Horses

Fangfang Wu [1,†], Shengnan Zhang [1,2,†], Ying Zhang [1,2], Ruo Mo [1,3], Feihu Yan [1,4,5], Hualei Wang [1,6], Gary Wong [7,8], Hang Chi [1,4,5], Tiecheng Wang [1,4,5], Na Feng [1,4,5], Yuwei Gao [1,4,5], Xianzhu Xia [1,4,5], Yongkun Zhao [1,4,5,*] and Songtao Yang [1,4,5,*]

[1] Institute of Military Veterinary Medicine, Academy of Military Medical Sciences, Changchun 130122, China; wufang_0617@163.com (F.W.); Zhang_Shengnan1992@163.com (S.Z.); zhangy99321@163.com (Y.Z.); moruo1996@163.com (R.M.); yanfh1990@gmail.com (F.Y.); whl831125@163.com (H.W.); ch_amms@163.com (H.C.); wgcha@163.com (T.W.); fengna0308@126.com (N.F.); yuwei0901@outlook.com (Y.G.); xiaxzh@cae.cn (X.X.)

[2] College of Wildlife Resources, Northeast Forestry University, Harbin 150040, China

[3] Animal Science and Technology College, Jilin Agricultural University, Changchun 130118, China

[4] Key Laboratory of Jilin Province for Zoonosis Prevention and Control, Changchun 130000, China

[5] Jiangsu Co-innovation Center for Prevention and Control of Important Animal Infectious Disease and Zoonoses, Yangzhou 225009, China

[6] College of Veterinary Medicine, Jilin University, Changchun 130062, China

[7] Institute Pasteur of Shanghai, Chinese Academy of Science, Shanghai 20031, China; garyckwong@hotmail.com

[8] Special Pathogens Program, Public Health Agency of Canada, Winnipeg, MB R3E3R2, Canada

[*] Correspondence: zhaoyongkun1976@126.com (Y.Z.); yst62041@163.com (S.Y.)

[†] These authors contributed equally to this work.

Received: 19 October 2019; Accepted: 1 January 2020; Published: 3 January 2020

Abstract: Ebola virus infections lead to severe hemorrhagic fevers in humans and nonhuman primates; and human fatality rates are as high as 67%–90%. Since the Ebola virus was discovered in 1976, the only available treatments have been medical support or the emergency administration of experimental drugs. The absence of licensed vaccines and drugs against the Ebola virus impedes the prevention of viral infection. In this study, we generated recombinant baculoviruses (rBV) expressing the Sudan virus (SUDV) matrix structural protein (VP40) (rBV-VP40-VP40) or the SUDV glycoprotein (GP) (rBV-GP-GP), and SUDV virus-like particles (VLPs) were produced by co-infection of Sf9 cells with rBV-SUDV-VP40 and rBV-SUDV-GP. The expression of SUDV VP40 and GP in SUDV VLPs was demonstrated by IFA and Western blot analysis. Electron microscopy results demonstrated that SUDV VLPs had a filamentous morphology. The immunogenicity of SUDV VLPs produced in insect cells was evaluated by the immunization of mice. The analysis of antibody responses showed that mice vaccinated with SUDV VLPs and the adjuvant Montanide ISA 201 produced SUDV GP-specific IgG antibodies. Sera from SUDV VLP-immunized mice were able to block infection by SUDV GP pseudotyped HIV, indicating that a neutralizing antibody against the SUDV GP protein was produced. Furthermore, the activation of B cells in the group immunized with VLPs mixed with Montanide ISA 201 was significant one week after the primary immunization. Vaccination with the SUDV VLPs markedly increased the frequency of antigen-specific cells secreting type 1 and type 2 cytokines. To study the therapeutic effects of SUDV antibodies, horses were immunized with SUDV VLPs emulsified in Freund's complete adjuvant or Freund's incomplete adjuvant. The results showed that horses could produce SUDV GP-specific antibodies and neutralizing antibodies. These results showed that SUDV VLPs demonstrate excellent immunogenicity and represent a promising approach for vaccine development against SUDV infection. Further, these horse anti-SUDV purified immunoglobulins lay a foundation for SUDV therapeutic drug research.

Keywords: Sudan virus; virus-like particle; vaccine; mice; horse; purified IgG

---

1. Introduction

Ebola virus infections lead to severe hemorrhagic fevers in humans and nonhuman primates, with human fatality rates of up to 90% [1]. The first outbreak of Ebola virus disease occurred in the Republic of Zaire (now the Democratic Republic of the Congo, DRC) and southern Sudan in 1976. From these independent outbreaks, two distinct viruses were identified, Zaire Ebolavirus (EBOV) [2] and Sudan virus (SUDV) [3], which are members of the genus Ebolavirus and have been the cause of sporadic outbreaks in humans throughout the years [1]. Species of Ebolavirus include EBOV, SUDV, Tai Forest Ebolavirus (TAFV), Bundibugyo Ebolavirus (BDBV), Reston Ebolavirus (RESTV) [4]. Both EBOV and SUDV are pathogenic to humans and nonhuman primates, causing severe hemorrhagic fever with high mortality rates of 67–90% [5,6]. Past EBOV outbreaks have been sporadic in nature and confined to central Africa, and thus far, the biggest outbreak on record is the 2013–2016 epidemic in western Africa, with 28,464 cases and 11,323 fatalities [7]. Most of the cases were in Guinea, Liberia, and Sierra Leone, but some of them were imported into Europe and the United States (https://app.who.int/ebola/current-situation/ebola-situation-report-30-march-2016). Currently, the latest devastating outbreak of the Ebola virus disease in the DRC is ongoing. As of today, a total of 3250 cases were reported, including 3133 confirmed and 117 probable cases, of which there have been 2174 fatalities with a mortality rate of 67% (WHO situation report https://www.who.int/csr/don/24-october-2019-ebola-drc/en/).

The Ebola virus is a non-segmented negative-strand RNA virus belonging to the family Filoviridae and the genus *Ebolavirus* [8]. The genome of filoviruses consists of a single-strand, negative-sense RNA genome of approximately 19 kb length, encoding the following genes in the following orientation: 3′-nucleoprotein (NP)-polymerase cofactor (VP35)-matrix protein (VP40)-glycoprotein (GP)-soluble GP (sGP)-small soluble GP (ssGP)-transcription activator (VP30)-minor matrix protein (VP24)-RNA dependent RNA polymerase (L)-5′ [9]. GP, the critical target antigen that is expressed on the surface of mature virions, is responsible for mediating cell attachment and viral entry [10]. Several vaccine platforms have been reported for EVD vaccine research, including vesicular stomatitis virus (VSV) [11,12], DNA replication-defective adenovirus vectors (Adv) [13,14], human parainfluenza virus type 3 [15], rabies virus [16,17], cytomegalovirus [18], Venezuelan equine encephalitis virus (VEEV) replicons [19] and virus-like particles (VLPs) [20]. There are eight vaccine candidates currently in human clinical trials. The two most promising preclinical vaccine candidates, VSV-EBOV [21] and chAd-EBOV, are in phase 3 clinical trials. No effective treatment for EVD is commercially available; however, China and Russia were the first to license EBOV vaccines in 2018 [22]. At the same time, passive immunotherapy with sera of animal origin has been used for over 120 years to treat bacterial and viral infections and drug intoxications. Currently, there are many therapeutic antibody drugs for treating Ebola virus disease, such as ZMappTM [23] and immunoglobulin F(ab′) 2 fragment [24].

VLPs represent a promising vaccine platform for a diverse array of viruses that include influenza virus, rotoviruses, noroviruses, HIV, hepatitis B virus, parvoviruses, rift valley fever virus, human papillomavirus and filoviruses [25–27]. VLPs are assembled by one or several proteins, with the distinct advantage of being noninfectious because they lack the viral genome required for replication. VLPs are highly ordered compounds similar to an actual live virus in terms of structure and size. The granular structure of VLPs is beneficial for antigen presentation and cell uptake, which can stimulate powerful innate and adaptive immune responses [28]. VLPs have the advantages of rapid production in large quantities and can generate robust innate, humoral and cellular immunity in animals and humans [29]. Furthermore, pre-existing immunity associated with live carrier vaccines is not hindered by VLP-based immunizations. Previous findings showed that SUDV VLPs could be readily assembled by the co-expression of insect cells with baculoviruses expressing GP, NP, and VP40 [30].

There are currently no approved specialized drugs or vaccines to protect against SUDV disease outbreaks, and thus there is an urgent need for the development of an efficacious, safe and economically viable vaccine or therapeutic antibody to control SUDV infections. Here, we report that production of SUDV VLPs has been accomplished in insect cells by the co-infection with recombinant baculoviruses rBV-GP-GP and rBV-VP40-VP40, and evaluate the ability of SUDV-VLPs to induce SUDV-specific humoral and cellular immune responses in vaccinated mice. Further, horses were immunized with SUDV VLPs, and horse serum was purified to prepare purified immunoglobulins and the purified immunoglobulins had neutralizing activity.

2. Materials and Methods

2.1. Cells and Animals

The 293T and Huh7 cells (ATCC) were cultured in Dulbecco's modified Eagle's medium (DMEM; Corning-Costar, Coring, NY, USA) supplemented with 10% fetal bovine serum and penicillin-streptomycin (FBS; Gibco, Grand Island, NY, USA). These cells were maintained at 37 °C. The Spodoptera frugiperda (Sf9) cells (Invitrogen, San Diego, CA, USA) were cultured in suspension in SF-900 II serum-free medium (Invitrogen, San Diego, CA, USA) supplemented with 10% FBS and penicillin-streptomycin. These cells were maintained at 27 °C.

Female BALB/c mice, 6–8 weeks old, were purchased from Changchun Institute of Biological Products (Changchun, China). All experiments involving mice adhered to the principles of the Welfare and Ethics Committee of the Military Veterinary Research Institute at the Academy of Military Sciences. The mice were provided ad libitum access to sterilized water and food throughout the study and were vaccinated under BSL-2 conditions.

Healthy male horses 2–6 years old and 400–500 kg were provided by Red Hill Military horse farm (Changchun, China). Horse studies were conducted with prior approval from the Animal Welfare and Ethics committee of the Institute of Military Veterinary, Academy of Military Medical Sciences (permit number SCXK-2012-017), according to Horse Quarantine and Immunization Protocols for Equine Serum Production.

2.2. Construction of Recombinant Baculoviruses

The cloning and construction of the recombinant bacmids, pFastBacDual-GP-GP, and pFastBacDual-VP40-VP40, were carried out. Briefly, we synthesized the full-length GP and VP40 genes (Supplementary material) according to nucleotide frequency (SUDV GP genbank accession no. AY729654.1, AY344234.1, KR063670.1, KU182912.1, NC006432.1, KC545392.1, KC545391.1, KC545390.1, KC545389.1, JN638998.1, KC589025.1, EU338380.1, U23069.1, KC242783.2, KT878488.1, FJ968794.1) (SUDV VP40 genbank accession no. KC545390.1, KC545389.1, KC545391.1, KC545392.1, JN638998.1, NC006432.1, AY729654.1, KU182912.1, KR063670.1, KC589025.1, KT878488.1, KC242783.2, FJ968794.1, KT750754.1, EU338380.1). Two identical full-length GP genes inserted into the pFastBacDual vector (Invitrogen, San Diego, CA, USA) by using KpnI, SmaI, EcoRI and XbaI, the GP genes were under the control of P10 and polyhedron promoters, generating the recombinant plasmid pFastBacDual-GP-GP (Supplementary Figure S1A). The recombinant plasmid pFastBacDual-VP40-VP40 (Supplementary Figure S1B) was constructed using the same strategy. Recombinant plasmids were used to transform *E. coli* DH10Bac competent cells to generate recombinant bacmids. Sf9 cells were transfected with recombinant bacmids using Cellfectin® II Reagent (Thermo Scientific, Carlsbad, CA, USA) according to the manufacturer's instructions. Transfected cells were incubated at 27 °C for 4 d and harvested, the recombinant baculoviruses rBV-GP-GP and rBV-VP40-VP40 by collecting the tinfected-Sf9 cells and supernatant.

2.3. Expression and Purification of SUDV GP and VP40 Antigens and Preparation of Polyclonal Antibodies

The SUDV GP (158~368 aa) and VP40 proteins were generated by a prokaryotic expression system. Briefly, the synthesized genes encoding the SUDV GP (158–368 aa) and VP40 proteins were separately inserted into the prokaryotic expression vector PET 30a (+) by using BamHI and NotI, in-frame with the 6×His-tag on the C- and N-terminus to construct PET 30a (+) -GP-his-C/N and PET 30a (+) -VP40-his-C/N. After the recombinant plasmids were confirmed by restriction digest, the two recombinant plasmids were transformed into *E. coli* BL21(DE3) (Transgen Biotech, Beijing, China) and the transformants selected on Luria-Bertani (LB) agar plates with 100 μg/mL kanamycin. A single clone was picked and inoculated into 4 mL of LB medium with 100 μg/mL kanamycin at 37 °C for overnight growth. Expression was induced with the addition of 0.4 mM IPTG when the optical density at 600 nm (OD_{600}) reached 0.6. The proteins were purified using a HisPurTM Ni-NTA Spin Purification kit (Thermo Scientific, Carlsbad, CA, USA) according to the manufacturer's instructions. To visualize the expression of the purified proteins, samples were resolved on a 12% SDS-polyacrylamide gel (SDS-PAGE) and stained with Coomassie blue, and protein concentrations were determined using a bicinchoninic acid (BCA) assay (Thermo Scientific, Carlsbad, CA, USA) followed by analysis at an absorbance of 570 nm; bovine serum albumin (BSA; Sigma-Aldrich, St. Louis, MO, USA) was used as the protein standard.

Polyclonal antisera against SUDV GP (158~368 aa) or VP40 were prepared by immunizing BALB/c mice with 10 μg purified GP or VP40 recombinant proteins twice at 2-week intervals, and harvesting the mouse serum, which were mouse polyclonal antisera against SUDV GP (158~368 aa) or VP40.

2.4. Immunofluorescence Testing of the Recombinant Baculoviruses

The Sf9 cells were infected with the fourth generation recombinant baculovirus rBac-GP-GP or rBac-VP40-VP40 for approximately 48 h, infected Sf9 cells were fixed with 4% paraformaldehyde for 15 min at the room temperature. Infected-Sf9 cells were washed with PBS and blocked with PBS containing 5% BSA (Sigma-Aldrich, St. Louis, MO, USA). Then, the cells were incubated with a 1:100 dilution of a mouse polyclonal antisera against SUDV GP or a mouse polyclonal antisera against SUDV VP40 (generated in our lab) for 1 h at 37 °C, and then washed with PBS and incubated with FITC-conjugated goat anti-mouse IgG (Bioss antibodies, Beijing, China) containing 1% diluted Evans blue for 1 h at 37 °C; then the infected Sf9 cells were washed with PBST (containing 0.05% Tween-20) three times and were observed with a fluorescence microscope. The control cells were incubated with the two primary antisera at the same time. Baculovirus titers were determined using a BacPAKTM Baculovirus Rapid Titer kit (TaKaRa, Dalian, China).

2.5. VLP Preparation

To generate SUDV VLPs, Sf9 cells were co-infected with recombinant baculoviruses rBV-GP-GP and rBV-VP40-VP40 at different ratios of 1:1, 1:2, 1:2.5, 1:3, 2:1, 2.5:1, or 3:1. The SUDV VLPs were harvested at 4-d post-infection. To purify the VLPs, culture supernatants were harvested and spun at 2000× *g*. The crude VLPs were then concentrated by ultracentrifugation at 30,000× *g* for 1 h, and the pellets were resuspended in PBS before purification with a 10–30–50% discontinuous sucrose gradient. Bands between 30–50% sucrose were collected and resuspended in endotoxin-free PBS, and the VLPs concentrations were determined using a BCA assay (Thermo Scientific, Carlsbad, CA, USA) followed by analysis at an absorbance of 570 nm. The VLP preparations were not tested for endotoxin after production.

2.6. Western Blotting and Transmission Electron Microscopy (TEM) Analysis of VLPs

Purified SUDV VLPs were processed and examined by Western blotting and transmission electron microscopy. For Western blotting, aliquots containing 10 μg of total protein were diluted with reducing buffer and denatured by heating at 95 °C for 10 min. Proteins were separated in 12% acrylamide gels, before they were transferred onto polyvinylidene fluoride (PVDF) membranes (Merck Millipore, Darmstadt, Germany) under denaturing conditions. For protein detection, two polyclonal antisera were used: mouse anti-SUDV GP polyclonal antisera and mouse anti-SUDV VP40 polyclonal antisera were mixed at a dilution of 1:1500 as a primary antibody for blotting; a goat anti-mouse IgG HRP-conjugated antibody (Bioss antibodies, Beijing, China) was used at a dilution of 1:5000 as a secondary antibody.

Negative staining transmission electron microscopy (TEM) was used to analyze the shape and size of purified SUDV VLPs. In short, 30 µL of sucrose gradient-purified SUDV VLPs were fixed for 15 min on carbon-coated formvar grids, grids were washed with 30 µL PBS and then treated with 1% phosphotungstate acid for 5 min. Grids were left to air dry and observed by using a HITACHI H-7650 transmission electron microscope.

2.7. Animal Immunizations

A total of two batches of BALB/c mice (6 weeks old, female) were purchased from the Changchun Yisi Laboratory Animal Technology Co., Ltd. (Changchun, China) and immunized. In batch one, 24 mice were randomized into four groups (n = 6 per group) and were vaccinated intramuscularly. Mice in group one were vaccinated with PBS, mice in group two were vaccinated with Montanide ISA 201 VG (ISA 201) adjuvant (Seppic, Paris, France), mice in group three were vaccinated with 20 µg of SUDV VLPs-only, and mice in group four were vaccinated with 20 µg of SUDV VLPs mixed with an equal volume of Montanide ISA 201 VG (ISA 201) adjuvant, and all groups were vaccinated twice at 3-week intervals (Figure 5A). The mouse sera were collected at two-weeks after every immunization. One week after the booster immunization, splenocytes from 3 mice of each group were isolated. In batch two, nine mice were randomly distributed into three groups (n = 3 per group) (PBS group, ISA 201 adjuvant, 20 µg of SUDV VLPs mixed with an equal volume of ISA 201 adjuvant) and were vaccinated intramuscularly. One week after the primary immunization, the inguinal lymph nodes were collected from 3 mice.

Two healthy male horses (numbered #392 and #18), 2–6 years old, 400–500 kg in weight and without detectable antibodies against SUDV detected by indirect ELISA, were supplied by Red Hill Military horse farm. The horses were multipoint injected subcutaneously in the rear area with 1.0, 2.0, 3.0, 3.0, or 4.0 mg of purified SUDV-VLPs for a total of 5 times, primary immunization mixed with an equal volume of Freund's incomplete adjuvant (Thermo Scientific, Rockford, IL, USA) and booster immunization mixed with an equal volume of Freund's complete adjuvant, maximum immune volume does not exceed 4 mL, with boosting at two week intervals (Figure 5B). The horse sera were collected before each immunization and stored at −20 °C for further studies.

2.8. Detection of SUDV GP-Specific Antibody by ELISA

The mice and horse serum samples were collected two weeks after each immunization. Levels of SUDV GP-specific antibodies were detected by indirect ELISA. Briefly, 100 µL of purified -prokaryotic expressed SUDV GP was coated on ELISA plates at a concentration of 10 µg/mL overnight at 4 °C, and then the plate was blocked with 2% BSA for 2 h at 37 °C. The serum was added (100 µL/well), and it was 2-fold serially diluted in 2% BSA; plates were subsequently incubated for 1.5 h at 37 °C, and 100 µL of HRP-labeled goat anti-mouse IgG or HRP-labeled goat anti-horse IgG (Bioss antibodies, Beijing, China), diluted 1:10,000 in 2% BSA was added to each well. After one hour of incubation at 37 °C, 100 µL of 3,3′,3,5′-tetramethylbenzidine (TMB) (Sigma-Aldrich, St. Louis, MO, USA) substrate solution was added to each well and then was stopped with the addition of 50 µL of 0.5 M H_2SO_4. Optical density (OD) values were measured at a wavelength of 450 nm (OD_{450}). After each incubation step, ELISA plates were washed five times with PBST. Mean OD values were considered positive if the OD values were more than two times the value of the negative control.

2.9. Detection of SUDV Neutralizing Antibodies by a Pseudotyped Virus

A pseudotyped virus neutralization assay was performed to test the neutralzing antibodies. Recombinant lentiviral vectors that express SUDV glycoprotein and carry a luciferase reporter gene were produced as described previously [31], with some modifications. Briefly, 239T cells were seeded in 60-mm culture dishes (Corning, NY, USA), and 24 h later, the 80% subconfluent cells were co-transfected with 3 µg of pcDNA 4.0-GP (the glycoprotein expression vector) and 3 µg of pNL4-3.Luc. RE vector by Lipofectamine^{TM} 3000 Transfection Reagent (Thermo Scientific, Carlsbad, CA, USA). After 2 days, the supernatant containing the pseudotyped viruses was harvested, and the titer was determined in Huh7 cells as previously described [32]. For testing of neutralizing antibodies, 2-fold serially diluted serum samples were mixed with 100 $TCID_{50}$ of pseudotyped viruses for 0.5 h at 37 °C and were then added to Huh7 cells. After 4 h, the inoculum was removed and replaced with fresh media. Then, cells were lysed at 48 h with cell lysis buffer, that was followed by the addition of 100 µL of luciferase substrate (Promega, Madison, WI, USA) to determine luciferase activity. The luciferase activity of the samples was measured with the Infinite M200 Microplate Spectrophotometer (Tecan, Mannedorf, Switzerland). The percent inhibition rate was calculated as previous showed [1]. The experiments were independently repeated three times.

2.10. Cell-Mediated Immune Responses in Mice

One week after the booster immunization in batch one, three mice from each group were randomly selected and euthanized. Their spleens were harvested into a tissue culture dish, and each spleen was roughly minced and pressed through a 5-mL syringe. The cells were filtered through a 40-µm filter (BD Falcon 40-µm strainer) and centrifuged at 2000 rpm for 10 min at the room temperature. The cells were processed by resuspension in a red blood cell lysis buffer and centrifuged at 2000 rpm for 10 min at twice the room temperature. The splenocytes were harvested and washed with RPMI 1640 medium (Gibco, San Diego, CA, USA) containing 10% FBS (Gibco, San Diego, CA, USA). Then, the splenocytes were cultured in RPMI 1640 medium containing 10% FBS and stimulated with or without the purified SUDV-GP antigen (10 µg/mL). Following incubation at 37 °C in 5% CO_2 for 48 h, the frequencies of splenocytes producing IFN-γ or IL-4 were measured using mouse ELISpot kits (Mabtech AB, Stockholm, Sweden) according to the manufacturer's instructions. Spot-forming cells (SFCs) were enumerated by an automated ELISpot reader (AID ELISPOT reader-iSpot, Germany).

The levels of cytokines in the supernatant of stimulated splenocytes were detected by commercial ELISA. Splenocytes were stimulated as described above and were then incubated for 48 h. The cell culture supernatant was collected by centrifugation at 3000 rpm for 15 min, and the manufacturer directions were followed to detect IL-2, IL-4, IL-10, IFN-γ, or TNF-α by ELISA kits (Mabtech AB, Stockholm, Sweden).

2.11. SUDV VLPs Induce Activation of B cells

Inguinal lymph node samples were harvested from batch two vaccinated mice 7 days after the primary immunization. Lymphoid cells were harvested into a tissue culture dish, and were roughly minced and pressed through a 5 mL syringe. The cells were filtered through a 40-µm filter (BD Falcon 40-µm strainer). Then, prepared in PBS with 2% FBS and were stained with equal volumes of 1:250 dilutions of anti-CD19-APC and anti-CD40-FITC antibodies (BD Biosciences, Franklin, VA, USA) for 30 min at 4 °C; the labeled B cells were then washed twice with PBS with 2% FBS and analyzed using a FACSAria TM Cell Sorter (BD Biosciences, Franklin, VA, USA).

2.12. Horse Immunoglobulin Purification

For equine antisera, the blood was taken from the jugular vein of #392 immunized horses, and the sera were collected. The horse serum was diluted 8-fold with PBS and then added to a saturated ammonium sulfate solution until the ammonium sulfate concentration was 50%. The solution was allowed to stand at 4 °C for 3 h, and it was centrifuged at 5000 rpm for 20 min. After removing the supernatant, the precipitate was dissolved in PBS, and saturated ammonium sulfate was added until the concentration of ammonium sulfate was 33%. The solution was allowed to stand at 4 °C for 3 h, and centrifuged at 5000 rpm for 20 min. The precipitate was dissolved in PBS and dialyzed against PBS at 4 °C for 18 h to remove the ammonium salt.

2.13. Laboratory Facility and Ethics Statement

The treatment of all mice was in accordance with the Welfare and Ethical guidance of Laboratory Animals of China (GB 14925-2001). The agreement was approved by the Animal Welfare and Ethics Committee of the Institute of Veterinary Medicine of the Military Academy of Sciences (Laboratory Animal Care and Use Committee Authorization, permit number JSY-DW-2018-02).

3. Results

3.1. Expression of SUDV GP and VP40 Proteins

The sequences encoding the GP (158–368 aa) and VP40 proteins of SUDV were cloned into the prokaryotic expression vector PET 30a (+), resulting in the plasmid PET 30a (+) -GP-his-C/N and PET 30a (+) -VP40-his-C/N (C and N terminal 6×His-tag). Correct insertion of the sequences in the recombinant plasmid was confirmed by restriction digest mapping analysis and DNA sequencing. The recombinant plasmids PET 30a (+) -GP-his-C/N and PET 30a (+) -VP40-his-C/N were transformed into *E. coli* BL21(DE3) cells. Recombinant proteins SUDV GP and VP40 were experessed in cells after added 0.4 mM IPTG. The proteins were purified by 6×His-tag affinity chromatography and detected by SDS-PAGE analysis (Figure 1). The purpose of expressing SUDV GP and VP40 proteins is to prepare anti-GP polyclonal antisera and anti-VP40 polyclonal antisera, and these two polyclonal antisera were successfully made through immunized mice with these two proteins twice respectively.

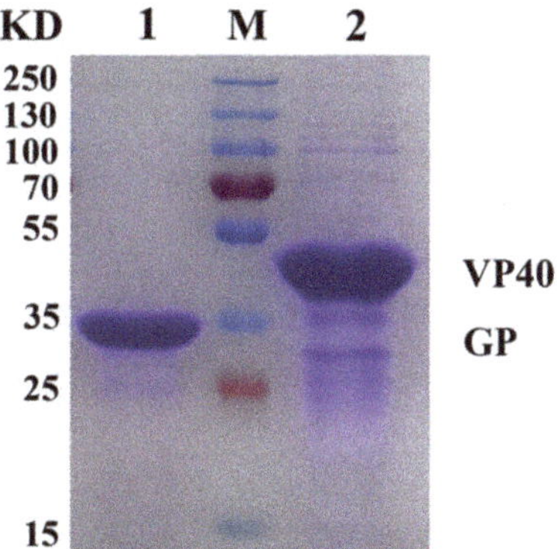

Figure 1. SDS-PAGE analysis of purified SUDV GP and VP40 proteins. Prokaryotic-expressed SUDV GP (158–368 aa) and VP40 proteins were purified by 6×His-tag affinity chromatography and detected by Coomassie-stained gel. The Lane 1 is the purified SUDV GP protein (32 kDa) and lane 2 is the purified SUDV VP40 protein (40 kDa), M is the protein molecular marker.

3.2. Verification of Baculovirus-Expressed SUDV GP and VP40 Proteins

To identify the successful rescue of the recombinant baculovirus and effectively expressed the SUDV GP and VP40 proteins, the fourth generation of recombinant baculovirus were detected by IFA. Sf9 cells infected with recombinant baculovirus rBac-GP-GP or rBac-VP40-VP40 were incubated with mouse anti-GP polyclonal or anti-VP40 polyclonal antisera respectively, and then incubated with FITC-conjugated goat anti-mouse IgG. The immunofluorescence results showed that Sf9 cells infected with recombinant baculovirus rBac-GP-GP or rBac-VP40-VP40 were shown in specific green (Figure 2D,E) and non-infected Sf9 cells (Figure 2F) did not show, which means that SUDV GP and VP40 proteins were successfully expressed in infected Sf9 cells.

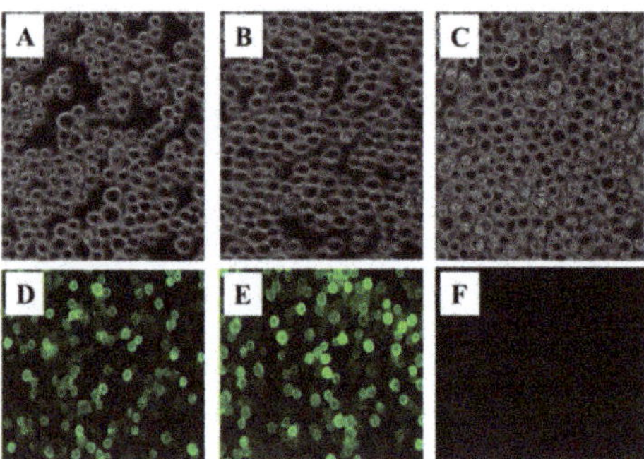

Figure 2. Immunofluorescence assay confirms the expression of SUDV GP and SUDV VP40 proteins in Sf9 cells (magnification of microscopy images, 200×). Recombinant baculovirus infected Sf9 cells became larger and rounder (**A,B**). Anti-SUDV GP and anti-SUDV VP40 polyclonal antisera were used to detect respectively the expression of SUDV GP and VP40. A specific green fluorescent signal around the GP-expressing (**D**) and VP40-expressing (**E**) cells indicates that these two proteins are expressed. The mock cells (**C**) were incubated the two polyclonal antisera showed negative staining (**F**).

3.3. Production and Characterization of SUDV VLPs

SUDV VLPs were generated by co-infecting with recombinant baculoviruses rBV-GP-GP and rBV-VP40-VP40 in Sf9 cells. The SUDV VLPs were harvested from the culture supernatant and purified as described in the Material and Methods. To further characterize the structure of the SUDV VLPs, we examined purified material by transmission electron microscopy. A negative staining study revealed that the SUDV VLPs were found to mimic the naive virus in structure and size; they were approximately 80 nm in diameter and 800–1500 nm in length (Figure 3A).

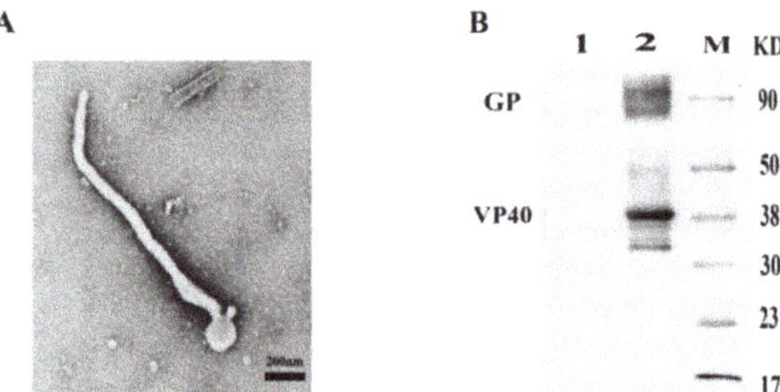

Figure 3. Characterization of the SUDV VLPs. Representative electron microscopy image of the SUDV VLPs; scale bar = 200 nm (**A**). Western blot analyses of SUDV GP and SUDV VP40 in purified SUDV VLPs by incubating with mouse anti-SUDV GP polyclonal antisera and mouse anti-SUDV VP40 polyclonal antisera at the same time. Lane 1 is the uninfected-Sf9 cells and lane 2 is the purified SUDV VLPs, M is the protein molecular marker (**B**).

The identity of the protein component of the purified SUDV VLPs was assessed by Western blot analysis with mouse anti-SUDV GP polyclonal antisera and mouse anti-SUDV VP40 polyclonal antisera. Both VP40 and GP proteins are present in the VLPs (Figure 3B). These results demonstrate that SUDV GP and VP40 efficiently assembled SUDV VLPs and released by insect cells.

3.4. SUDV VLPs Elicit Antibody Responses in Vaccinated Mice

To detect a SUDV-VLP-induced humoral immune response in vaccinated BALB/c mice, indirect ELISA and virus neutralization assay were performed to evaluate the production of SUDV GP-specific antibodies and neutralizing antibodies. At the fifth week of the immunization, the sera from mice treated with 20 µg SUDV VLPs mixed with an ISA 201 adjuvant group had high titers of SUDV specific IgG antibodies with endpoint titers up to 1:81,920 (Figure 4C), virus-neutralizing antibody titers averaged 1:320 and individual mice neutralizing antibody titers were up to 1:640 (Figure 4D), showing significant divergence when compared to the PBS group and ISA 201 adjuvant group. But the SUDV specific IgG antibodies and neutralizing antibodies were not detected in the VLPs-only group mice. These results demonstrate that SUDV VLPs at 20 µg antigen dose cannot stimulate the production of antibodies in mice, but this antigen dose mixed with ISA201 adjuvant can stimulate mice to produce specific IgG antibodies and neutralizing antibodies.

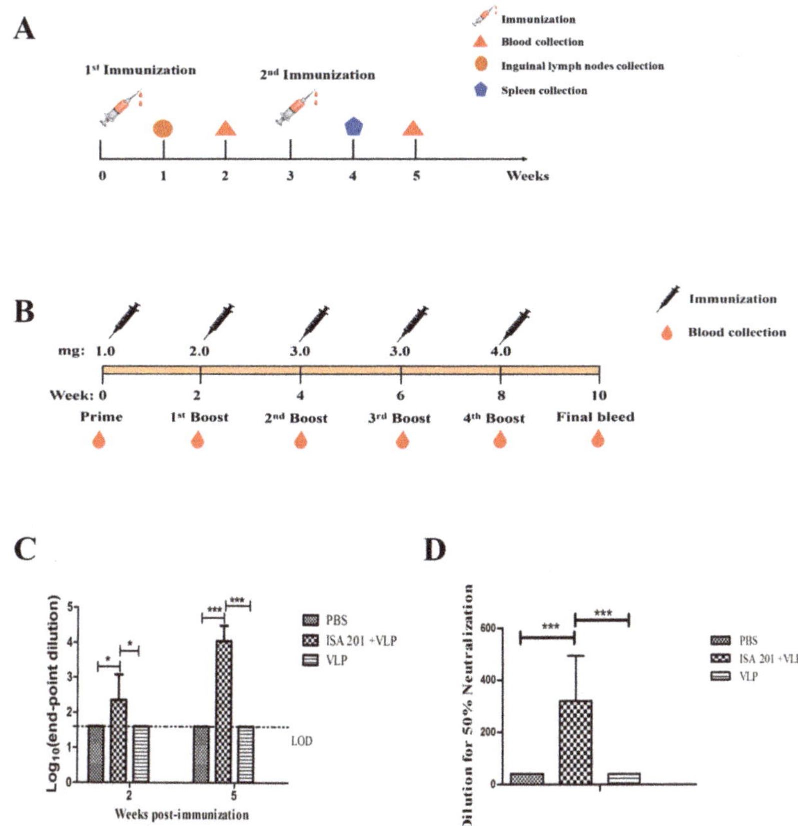

Figure 4. Mice and horse immunization procedure, analysis of SUDV GP-specific antibody, and neutralizing antibodies in vaccinated mice. The immunizations of mice (**A**) or horses (**B**). Analysis of SUDV-VLP-induced specific IgG antibody response by indirect ELISA at two weeks after every immunization of vaccinated mice (**C**). SUDV neutralizing antibody titers are detected in immunized mice at two weeks after booster immunization (**D**). Limit of detection (LOD) means the minimum concentration or content that can be detected under the determined experimental conditions. Error bars represent the standard deviation. The *p*-values were determined according to a Tukey's multiple comparison test (* $p < 0.05$, *** $p < 0.001$).

3.5. Antigen-Specific Cellular Immune Responses in Vaccinated Mice

ELISpot assays were used to evaluate the antigen-specific cellular immune response in vaccinated mice. Since 20 µg VLPs-only did not stimulate the production of specific antibodies and neutralizing antibodies, the VLPs-only group was removed in the cellular immune response experiment. The results from the IL-4 ELISpot assay are shown in Figure 5A. The SFCs produced from the splenocytes of mice immunized with 20 µg SUDV VLPs mixed with an ISA 201 adjuvant were significantly higher compared with mice immunized with PBS or ISA 201. The immunization with 20 µg SUDV VLPs mixed with an ISA 201 adjuvant also resulted in increased IFN-γ responses, with SFCs production that were significantly higher than that following immunization with PBS or ISA 201 (Figure 5B). These results demonstrate that 20 µg SUDV VLPs mixed with an ISA 201 adjuvant enhances IL-4 and IFN-γ responses in mice.

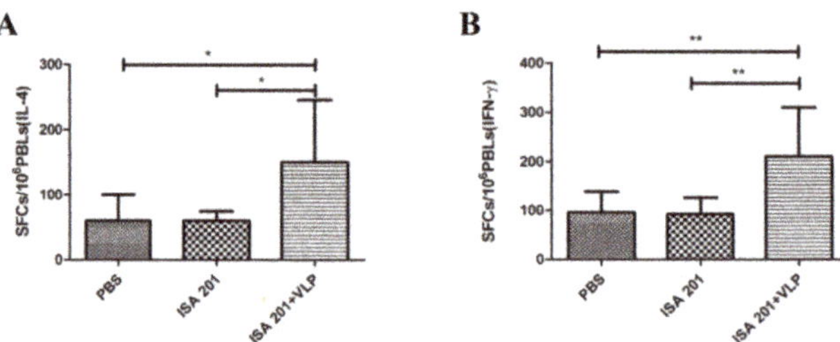

Figure 5. IFN-γ and IL-4 secretion by proliferating splenic induced by a purified baculovirus-expressed SUDV GP protein. Splenocytes from immunized mice and were stimulated with purified -prokaryotic expressed SUDV GP (10 µg/mL) for 48 h, and the level of IL-4 (**A**) or IFN-γ (**B**) were quantitated using an ELISpot assay. The data represent the mean ± standard deviation (SD) of SFCs per million cells. Statistical analysis between the two groups was analyzed by using Tukey's multiple comparison test ($* p < 0.05$, $** p < 0.01$).

3.6. SUDV VLPs-Enhanced Splenocytes Cytokine Secretion

To further investigate the antigen-specific cellular immune responses induced by SUDV VLPs, cytokines secreted by splenocytes were measured by commercial ELISA kits. Splenocytes were harvested from mice one week after the booster immunization and stimulated with Prokaryotic-expressed SUDV GP (158~368 aa). Cytokines secreted into the supernatant by splenocytes were assessed. Type 1 cytokines such as IL-2 (Figure 6A), IFN-γ (Figure 6B), and TNF-α (Figure 6C) were significantly higher in the group immunized with 20 µg SUDV VLPs mixed with an ISA 201 adjuvant than they were in the PBS and ISA 201 adjuvant group. Type 2 cytokines, such as IL-4 (Figure 6D) and IL-10 (Figure 6E), showed the same trend. These results indicate that 20 µg SUDV VLPs mixed with an ISA 201 adjuvant potently enhance cellular immune response in mice.

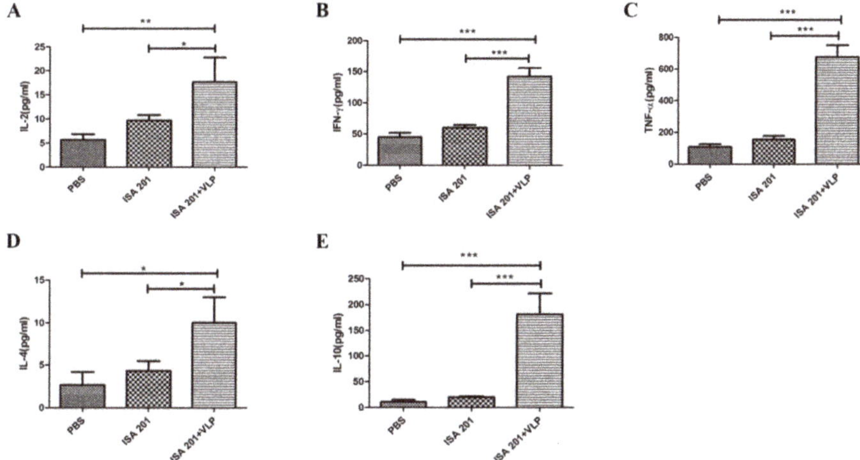

Figure 6. Detection of cytokine secretion in splenocytes. Splenocytes were collected from mice one week after following the booster immunization and were stimulated with purified-prokaryotic expressed SUDV GP protein for 48 h. The level of IL-2 (**A**), IFN-γ (**B**), TNF-α (**C**), IL-4 (**D**), and IL-10 (**E**) in the supernatant were measured with commercial ELISA kits ($n = 3$). Data are shown as the means ± SDs and were analyzed by using Tukey's multiple comparison test (* $p < 0.05$, ** $p < 0.01$, *** $p < 0.001$).

3.7. Enhancing Effects of SUDV VLP on B Cell Activation

To study whether the vaccine can induce B cell activation, inguinal lymph nodes were collected from the mice vaccinated with PBS, ISA 201 adjuvant and 20 μg SUDV VLPs mixed with ISA 201 adjuvant. The result showed (Figure 7) that the percentage of activated B cells (CD19+CD40+) was significantly higher in the 20 μg SUDV VLPs mixed with ISA 201 adjuvant group than in the PBS group and ISA 201 adjuvant group.

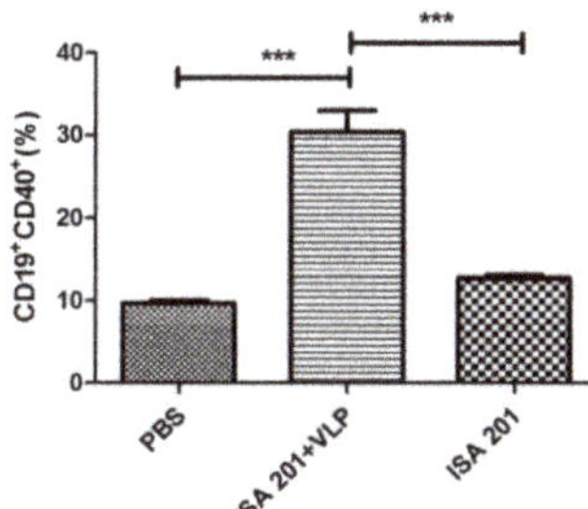

Figure 7. Activation of B cell in BALB/c mice. At seven days after the primary immunization, inguinal lymph nodes were collected from the mice treated with PBS, ISA 201 adjuvant and 20 μg SUDV VLPs mixed with ISA 201 adjuvant, and B cell activation was analyzed by staining with anti-CD19-APC and anti-CD40-FITC antibodies. Data are shown as the means ± SDs and were analyzed by using Tukey's multiple comparison test (*** $p < 0.001$).

3.8. Antibody Response in Vaccinated Horses

The humoral immune response in horses was assessed at 2, 4, 6, and 8 weeks after immunization (Figure 4B). Serum samples were collected to measure the SUDV GP-specific antibodies and neutralizing antibody. The level of SUDV GP-specific IgG antibody was detected from the second week (Figure 8A). After the fifth immunization, the titer of SUDV GP-specific IgG antibody of horse #392 was 1:40,960, and for horse #18 was 1:8192. The results revealed that SUDV VLPs mixed with Freund's adjuvant could stimulate horses to produce humoral immune response. Regarding the research on SUDV therapeutic antibodies, we selected #392 horse serum for further study. The #392 serum was extracted and purified to obtain purified IgG; #392 horse serum and #392 purified IgG were simultaneously subjected to virus neutralization. These findings showed that the neutralizing antibody titer of #392 horse serum was 1:10,240 after the fifth immunization, correspondingly, the neutralizing antibody titer of purified IgG was 1:5120 (Figure 8B).

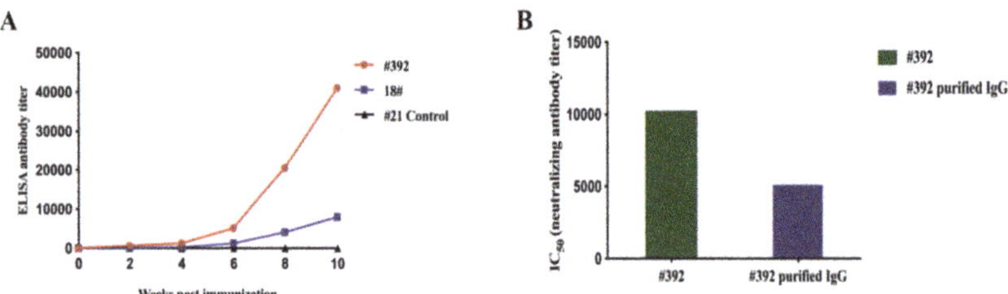

Figure 8. The detection of humoral immune responses in horses after immunized SUDV VLPs mixed with Freund's adjuvant. (**A**) Measurement of SUDV GP-specific IgG antibody by indirect ELISA and (**B**) detection of neutralizing activity by using pseudotyped viruses.

4. Discussion

The mechanism of natural immunity against EBOV remains unclear. The efficacy of different types of vaccines may vary depending on the vaccine platform: antibody is the major immune correlate factor of the rVSV-ZEBOV vaccine [33]. Meanwhile, CD8 T cell responses have been attributed to protection by the Ch-Ad5 vaccine [34]. Previous VLP vaccine studies showed it is efficacious against lethal Ebola challenge, however, with different adjuvants to save antigen and enhance vaccine-induced immune responses [35,36].

The impact of SUDV outbreaks in recent years, and the potential for viral spread to non-endemic regions or countries, makes the development of safe and efficacious vaccines urgent.

Previous findings revealed that SUDV VLPs could be readily assembled by the co-expression of baculoviruses expressing GP, NP and VP40 in insect cells [30]. Here, we demonstrated that co-infection of rBV-GP-GP and rBV-VP40-VP40 recombinant baculoviruses can result in the successful assembly of VLPs in insect cells, the morphology of SUDV VLPs is similar to a native virus, and the immunogenicity of VLPs was tested with an ISA 201 adjuvant as a candidate vaccine in mice.

In our study, we generated recombinant baculaviruses rBV-GP-GP and rBV-VP40-VP40 using the pFastBacDual vector to increase the protein production utilizing the dual promoter. To determine the optimal proportion of two recombinant baculaviruses rBV-GP-GP and rBV-VP40-VP40, we also tried to co-infect Sf9 cells with recombinant baculaviruses rBV-GP-GP and rBV-VP40-VP40 at ratios of 1:1, 1:2, 1:2.5, 1:3, 2:1, 2.5:1, and 3:1. At a ratio of 2.5:1, the morphology of VLPs was similar to the native virus, as assessed by TEM (Figure 3A). The Western blot result (Figure 3B) showed that SUDV GP and VP40 efficiently assembled SUDV VLPs.

The immunization results depicted the failure of the group vaccinated with 20 µg of SUDV VLPs-only to stimulate the production of humoral immune responses in mice, while the group vaccinated with 20 µg of VLPs mixed with an ISA 201 adjuvant could stimulate the production of humoral immunity in mice. Presumably, immunization with 20 µg of SUDV VLPs-only is an insufficiently low dose of an immunogen that cannot stimulate the mice to produce an immune response, and the use of an ISA 201 adjuvant with SUDV VLPs might cause low-dose immunogens to initiate a response, and using with adjuvant can save antigen amount (Figure 4C). The same SUDV VLPs antigen dose (20 µg) mix different adjuvants could stimulate mice to produce different degrees humoral immune responses. The ISA 201 adjuvant showed the most effective adjuvant than other adjuvants in mice. The detection of neutralizing antibody showed that sera from mice immunized with 20 µg SUDV VLPs mixed with ISA 201 adjuvant could neutralize approximately 50% of the pseudotyped viruses at an average 1:320 dilution, while individual samples were effective with as high as a 1:640 dilution (Figure 4D).

Both humoral and cellular responses are indispensable for providing protection. IFN-γ is a Th1-type cytokine involved in the antiviral action of cellular immune responses and IL-4 is mainly produced by Th2 cells and is associated with humoral immune responses. The SUDV VLPs could effectively stimulate Th1 and Th2 cytokine production in vaccinated mice (Figure 5A,B). The levels of cytokines secreted from splenocytes, such as IL-2, IFN-γ, and TNF-α secreted from Th1 cells (Figure 6A–C) and IL-4 and IL-10 were secreted from Th2 cells (Figure 6D,E), were significantly increased after vaccination. Moreover, our study revealed that the B cells of mice of the group vaccinated with SUDV VLPs mixed with an ISA 201 adjuvant were activated at one week after the primary immunization (Figure 7).

In addition, we have conducted research on therapeutic antibodies with the goal of preparing antibodies for the post-exposure treatment of SUDV. Horses were selected as immunized animals in our study because horse anti-immunoglobulins have been used in the treatment of various viral infections [37–39]. SUDV VLPs can effectively induce humoral immune responses in horses after immunization (Figure 8A). The horse anti-SUDV immunoglobulin was obtained from #392 horse serum through crude extraction and purification. The neutralizing antibody titer of purified #392 IgG was 1:5120, which was one time lower than that before purification (Figure 8B).

We tested the SUDV VLPs immunogenicity and neutralizing activity of hours purified IgG by using pseudo typed viruses, and efficacy studies still need to be performed. These SUDV VLPs vaccine and horse purified IgG provide ideas for the development of vaccines and therapeutic antibodies that could prevent and treat SUDV infections. The Ebola epidemic is still spreading, vaccination is an effective means of preventing and controlling the outbreaks, and effective antibodies represent key drugs for the treatment of Ebola patients. Therefore, further Ebola vaccine and therapeutic antibody research are needed.

Supplementary Materials: The following are available online at http://www.mdpi.com/1999-4915/12/1/64/s1, Figure S1: Diagram of the plasmids encoding the pFastBacDual-GP-GP (A) and pFastBacDual-VP40-VP40 (B).

Author Contributions: Y.Z. and S.Y. designed the experiments. F.W., Y.Z., R.M., F.Y., H.C. performed the experiment. H.W., N.F., T.W. analyzed the data. S.Z. and F.W. wrote the manuscript. X.X., Y.Z., S.Y., G.W. and Y.G. reviewed the manuscript. All authors have read and agreed to the published version of the manuscript.

Funding: This research was funded by the National Project for Prevention and Control of Transboundary Animal Diseases (Grant No. 2017YFD0501804).

Conflicts of Interest: The authors declare no conflict of interest.

References

1. Baseler, L.; Chertow, D.S.; Johnson, K.M.; Feldmann, H.; Morens, D.M. The Pathogenesis of Ebola Virus Disease. *Annu. Rev. Pathol.* **2017**, *12*, 387–418. [CrossRef] [PubMed]
2. Brès, P. Ebola haemorrhagic fever in Zaire, 1976. *Bull. World Health Organ.* **1978**, *56*, 271–293.
3. Brès, P. Ebola haemorrhagic fever in Sudan, 1976. Report of a WHO/International Study Team. *Bull. World Health Organ.* **1978**, *56*, 247–270.
4. Adams, M.J.; Lefkowitz, E.J.; King, A.M.Q.; Harrach, B.; Harrison, R.L.; Knowles, N.J.; Kropinski, A.M.; Krupovic, M.; Kuhn, J.H.; Mushegian, A.R.; et al. Changes to taxonomy and the International Code of Virus Classification and Nomenclature ratified by the International Committee on Taxonomy of Viruses (2017). *Arch. Virol.* **2017**, *162*, 2505–2538. [CrossRef] [PubMed]
5. Pourrut, X.; Kumulungui, B.; Wittmann, T.; Moussavou, G.; Delicat, A.; Yaba, P.; Nkoghe, D.; Gonzalez, J.P.; Leroy, E.M. The natural history of Ebola virus in Africa. *Microbes Infect.* **2005**, *7*, 1005–1014. [CrossRef]

6. Bukreyev, A.A.; Chandran, K.; Dolnik, O.; Dye, J.M.; Ebihara, H.; Leroy, E.M.; Muhlberger, E.; Netesov, S.V.; Patterson, J.L.; Paweska, J.T.; et al. Discussions and decisions of the 2012–2014 International Committee on Taxonomy of Viruses (ICTV) Filoviridae Study Group, January 2012–June 2013. *Arch. Virol.* **2014**, *159*, 821–830. [CrossRef]

7. Coltart, C.E.; Lindsey, B.; Ghinai, I.; Johnson, A.M.; Heymann, D.L. The Ebola outbreak, 2013–2016: Old lessons for new epidemics. *Philos. Trans. R Soc. Lond. B Biol. Sci.* **2017**, *372*, 20160297. [CrossRef]

8. Leroy, E.; Baize, S.; Gonzalez, J.P. Ebola and Marburg hemorrhagic fever viruses: Update on filoviruses. *Med. Trop.* **2011**, *71*, 111–121.

9. Ascenzi, P.; Bocedi, A.; Heptonstall, J.; Capobianchi, M.R.; Di Caro, A.; Mastrangelo, E.; Bolognesi, M.; Ippolito, G. Ebolavirus and Marburgvirus: Insight the Filoviridae family. *Mol. Asp. Med.* **2008**, *29*, 151–185. [CrossRef]

10. Carette, J.E.; Raaben, M.; Wong, A.C.; Herbert, A.S.; Obernosterer, G.; Mulherkar, N.; Kuehne, A.I.; Kranzusch, P.J.; Griffin, A.M.; Ruthel, G.; et al. Ebola virus entry requires the cholesterol transporter Niemann-Pick C1. *Nature* **2011**, *477*, 340–343. [CrossRef]

11. Jones, S.M.; Feldmann, H.; Stroher, U.; Geisbert, J.B.; Fernando, L.; Grolla, A.; Klenk, H.D.; Sullivan, N.J.; Volchkov, V.E.; Fritz, E.A.; et al. Live attenuated recombinant vaccine protects nonhuman primates against Ebola and Marburg viruses. *Nat. Med.* **2005**, *11*, 786–790. [CrossRef] [PubMed]

12. Wong, G.; Audet, J.; Fernando, L.; Fausther-Bovendo, H.; Alimonti, J.B.; Kobinger, G.P.; Qiu, X. Immunization with vesicular stomatitis virus vaccine expressing the Ebola glycoprotein provides sustained long-term protection in rodents. *Vaccine* **2014**, *32*, 5722–5729. [CrossRef] [PubMed]

13. Sullivan, N.J.; Geisbert, T.W.; Geisbert, J.B.; Shedlock, D.J.; Xu, L.; Lamoreaux, L.; Custers, J.H.; Popernack, P.M.; Yang, Z.Y.; Pau, M.G.; et al. Immune protection of nonhuman primates against Ebola virus with single low-dose adenovirus vectors encoding modified GPs. *PLoS Med.* **2006**, *3*, e177. [CrossRef]

14. Sullivan, N.J.; Sanchez, A.; Rollin, P.E.; Yang, Z.Y.; Nabel, G.J. Development of a preventive vaccine for Ebola virus infection in primates. *Nature* **2000**, *408*, 605–609. [CrossRef] [PubMed]

15. Bukreyev, A.; Rollin, P.E.; Tate, M.K.; Yang, L.; Zaki, S.R.; Shieh, W.J.; Murphy, B.R.; Collins, P.L.; Sanchez, A. Successful topical respiratory tract immunization of primates against Ebola virus. *J. Virol.* **2007**, *81*, 6379–6388. [CrossRef] [PubMed]

16. Blaney, J.E.; Marzi, A.; Willet, M.; Papaneri, A.B.; Wirblich, C.; Feldmann, F.; Holbrook, M.; Jahrling, P.; Feldmann, H.; Schnell, M.J. Antibody quality and protection from lethal Ebola virus challenge in nonhuman primates immunized with rabies virus based bivalent vaccine. *PLoS Pathog.* **2013**, *9*, e1003389. [CrossRef]

17. Shuai, L.; Wang, X.; Wen, Z.; Ge, J.; Wang, J.; Zhao, D.; Bu, Z. Genetically modified rabies virus-vectored Ebola virus disease vaccines are safe and induce efficacious immune responses in mice and dogs. *Antivir. Res.* **2017**, *146*, 36–44. [CrossRef]

18. Quinn, M.; Erkes, D.A.; Snyder, C.M. Cytomegalovirus and immunotherapy: Opportunistic pathogen, novel target for cancer and a promising vaccine vector. *Immunotherapy* **2016**, *8*, 211–221. [CrossRef]

19. Pushko, P.; Bray, M.; Ludwig, G.V.; Parker, M.; Schmaljohn, A.; Sanchez, A.; Jahrling, P.B.; Smith, J.F. Recombinant RNA replicons derived from attenuated Venezuelan equine encephalitis virus protect guinea pigs and mice from Ebola hemorrhagic fever virus. *Vaccine* **2000**, *19*, 142–153. [CrossRef]

20. Sunay, M.M.E.; Martins, K.A.O.; Steffens, J.T.; Gregory, M.; Vantongeren, S.A.; Van Hoeven, N.; Garnes, P.G.; Bavari, S. Glucopyranosyl lipid adjuvant enhances immune response to Ebola virus-like particle vaccine in mice. *Vaccine* **2019**, *37*, 3902–3910. [CrossRef]

21. Suder, E.; Furuyama, W.; Feldmann, H.; Marzi, A.; de Wit, E. The vesicular stomatitis virus-based Ebola virus vaccine: From concept to clinical trials. *Hum. Vaccines Immunother.* **2018**, *14*, 2107–2113. [CrossRef] [PubMed]

22. Furuyama, W.; Marzi, A. Ebola Virus: Pathogenesis and Countermeasure Development. *Annu. Rev. Virol.* **2019**, *6*, 435–458. [CrossRef] [PubMed]

23. Qiu, X.; Wong, G.; Audet, J.; Bello, A.; Fernando, L.; Alimonti, J.B.; Fausther-Bovendo, H.; Wei, H.; Aviles, J.; Hiatt, E.; et al. Reversion of advanced Ebola virus disease in nonhuman primates with ZMapp. *Nature* **2014**, *514*, 47–53. [CrossRef] [PubMed]

24. Zheng, X.; Wong, G.; Zhao, Y.; Wang, H.; He, S.; Bi, Y.; Chen, W.; Jin, H.; Gai, W.; Chu, D.; et al. Treatment with hyperimmune equine immunoglobulin or immunoglobulin fragments completely protects rodents from Ebola virus infection. *Sci. Rep.* **2016**, *6*, 24179. [CrossRef] [PubMed]

25. Boisgerault, F.; Moron, G.; Leclerc, C. Virus-like particles: A new family of delivery systems. *Expert Rev. Vaccines* **2002**, *1*, 101–109. [CrossRef] [PubMed]

26. Naslund, J.; Lagerqvist, N.; Habjan, M.; Lundkvist, A.; Evander, M.; Ahlm, C.; Weber, F.; Bucht, G. Vaccination with virus-like particles protects mice from lethal infection of Rift Valley Fever Virus. *Virology* **2009**, *385*, 409–415. [CrossRef]

27. Warfield, K.L.; Aman, M.J. Advances in virus-like particle vaccines for filoviruses. *J. Infect. Dis.* **2011**, *204* (Suppl. 3), S1053–S1059. [CrossRef]

28. Zeltins, A. Construction and characterization of virus-like particles: A review. *Mol. Biotechnol.* **2013**, *53*, 92–107. [CrossRef]

29. Ohimain, E.I. Recent advances in the development of vaccines for Ebola virus disease. *Virus Res.* **2016**, *211*, 174–185. [CrossRef]

30. Pastor, A.R.; Gonzalez-Dominguez, G.; Diaz-Salinas, M.A.; Ramirez, O.T.; Palomares, L.A. Defining the multiplicity and time of infection for the production of Zaire Ebola virus-like particles in the insect cell-baculovirus expression system. *Vaccine* **2019**, *37*, 6962–6969. [CrossRef]

31. Medina, M.F.; Kobinger, G.P.; Rux, J.; Gasmi, M.; Looney, D.J.; Bates, P.; Wilson, J.M. Lentiviral vectors pseudotyped with minimal filovirus envelopes increased gene transfer in murine lung. *Mol. Ther.* **2003**, *8*, 777–789. [CrossRef] [PubMed]

32. Perera, R.A.; Wang, P.; Gomaa, M.R.; El-Shesheny, R.; Kandeil, A.; Bagato, O.; Siu, L.Y.; Shehata, M.M.; Kayed, A.S.; Moatasim, Y.; et al. Seroepidemiology for MERS coronavirus using microneutralisation and pseudoparticle virus neutralisation assays reveal a high prevalence of antibody in dromedary camels in Egypt, June 2013. *Eurosurveillance* **2013**, *18*, 7. [CrossRef] [PubMed]

33. Ewer, K.; Rampling, T.; Venkatraman, N.; Bowyer, G.; Wright, D.; Lambe, T.; Imoukhuede, E.B.; Payne, R.; Fehling, S.K.; Strecker, T.; et al. A Monovalent Chimpanzee Adenovirus Ebola Vaccine Boosted with MVA. *N. Engl. J. Med.* **2016**, *374*, 1635–1646. [CrossRef] [PubMed]

34. Marzi, A.; Engelmann, F.; Feldmann, F.; Haberthur, K.; Shupert, W.L.; Brining, D.; Scott, D.P.; Geisbert, T.W.; Kawaoka, Y.; Katze, M.G.; et al. Antibodies are necessary for rVSV/ZEBOV-GP-mediated protection against lethal Ebola virus challenge in nonhuman primates. *Proc. Natl. Acad. Sci. USA* **2013**, *110*, 1893–1898. [CrossRef] [PubMed]

35. Warfield, K.L.; Swenson, D.L.; Olinger, G.G.; Kalina, W.V.; Aman, M.J.; Bavari, S. Ebola virus-like particle-based vaccine protects nonhuman primates against lethal Ebola virus challenge. *J. Infect. Dis.* **2007**, *196* (Suppl. 2), S430–S437. [CrossRef] [PubMed]

36. Martins, K.A.; Steffens, J.T.; van Tongeren, S.A.; Wells, J.B.; Bergeron, A.A.; Dickson, S.P.; Dye, J.M.; Salazar, A.M.; Bavari, S. Toll-like receptor agonist augments virus-like particle-mediated protection from Ebola virus with transient immune activation. *PLoS ONE* **2014**, *9*, e89735. [CrossRef]

37. Squaiella-Baptistao, C.C.; Magnoli, F.C.; Marcelino, J.R.; Sant'Anna, O.A.; Tambourgi, D.V. Quality of horse F (ab') 2 antitoxins and anti-rabies immunoglobulins: Protein content and anticomplementary activity. *J. Venom. Anim. Toxins Incl. Trop. Dis.* **2018**, *24*, 16. [CrossRef]

38. Zhang, X.J.; Li, H.L.; Deng, D.Y.; Ji, C.; Yao, X.D.; Liu, J.X. Functional and proteomic comparison of different techniques to produce equine anti-tetanus immunoglobulin F (ab') 2 fragments. *J. Chromatogr. B Anal. Technol. Biomed. Life Sci.* **2018**, *1092*, 29–39. [CrossRef]

39. Perry, A.L.; Hayes, A.J.; Cox, H.A.; Alcock, F.; Parker, A.R. Comparison of five commercial anti-tetanus toxoid immunoglobulin G enzyme-linked immunosorbent assays. *Clin. Vaccine Immunol.* **2009**, *16*, 1837–1839. [CrossRef]

Article

Simultaneous Immunization with Multivalent Norovirus VLPs Induces Better Protective Immune Responses to Norovirus than Sequential Immunization

Maria Malm, Timo Vesikari and Vesna Blazevic *

Vaccine Research Center, Faculty of Medicine and Health Technology, Tampere University, Biokatu 10, FI-33520 Tampere, Finland; maria.malm@tuni.fi (M.M.); timo.vesikari@tuni.fi (T.V.)
* Correspondence: vesna.blazevic@tuni.fi; Tel.: +358-50-4211-054

Received: 23 September 2019; Accepted: 31 October 2019; Published: 2 November 2019

Abstract: Human noroviruses (NoVs) are a genetically diverse, constantly evolving group of viruses. Here, we studied the effect of NoV pre-existing immunity on the success of NoV vaccinations with genetically close and distant genotypes. A sequential immunization as an alternative approach to multivalent NoV virus-like particles (VLPs) vaccine was investigated. Mice were immunized with NoV GI.3, GII.4-1999, GII.17, and GII.4 Sydney as monovalent VLPs or as a single tetravalent mixture combined with rotavirus VP6-protein. Sequentially immunized mice were primed with a trivalent vaccine candidate (GI.3 + GII.4-1999 + VP6) and boosted, first with GII.17 and then with GII.4 Sydney VLPs. NoV serum antibodies were analyzed. Similar NoV genotype-specific immune responses were induced with the monovalent and multivalent mixture immunizations, and no immunological interference was observed. Multivalent immunization with simultaneous mix was found to be superior to sequential immunization, as sequential boost induced strong blocking antibody response against the distant genotype (GII.17), but not against GII.4 Sydney, closely related to GII.4-1999, contained in the priming vaccine. Genetically close antigens may interfere with the immune response generation and thereby immune responses may be differently formed depending on the degree of NoV VLP genotype identity.

Keywords: norovirus; VLP; vaccine; genotype; pre-existing immunity; cross-reactivity; blocking antibodies; original antigenic sin (OAS)

1. Introduction

Noroviruses (NoVs) are the most common cause of epidemic acute gastroenteritis (AGE) in all age groups globally. NoV AGE leads to an estimated 212,000 deaths per year, mainly in young children in developing countries [1]. In developed countries, NoV AEG may cause deaths in the elderly and is associated with economic and societal costs [2]. The NoV has a single-stranded, positive-sense, RNA genome divided into three open reading frames (ORFs) that encode non-structural proteins (ORF1), a major structural capsid protein, VP1 (ORF2), and a minor capsid protein, VP2 (ORF3) [3]. NoV virus-like particles (VLPs) are spontaneously self-assembled by the main capsid protein, VP1, and can be produced in different expression systems for use as candidate vaccines [4–6]. Human NoVs belong mainly to genogroups (GI) I and II, which are further classified to genotypes GI.1–GI.9 and GII.1–GII.27 [7]. Approximately 90% of NoV outbreaks are caused by GII viruses, most belonging to the GII.4 genotype [8].

NoV particles bind in a genotype-specific manner to a versatile group of histo-blood group antigens (HBGAs) [9,10] that have been shown to be important for NoV entry and infection of the cells, functioning as attachment factors [11]. NoV blocking antibody assay, which measures the ability of antibodies to block the binding of VLPs to cell surface carbohydrates, HBGAs, is a surrogate for the standard neutralization assay [10,12–14]. Induction of blocking antibodies is one of the most important correlates of protection identified so far [15,16]. Blocking antibody responses to NoVs are largely genogroup-specific [12,17,18]. Variable levels of cross-blocking is observed between viruses inside the genogroup, depending on genetic and antigenic distance [16,19,20]. Thereby, even though >90% of children above five years of age have generated NoV-specific antibodies to several genotypes [14], repeated infections commonly occur [21–23]. The immune escape is most evident with predominant GII.4 genotype viruses that share >95% identity in their VP1 aa sequence [24]. Despite the close genetic relationship and pre-existing immunity to previously encountered strains, new GII.4 genotypes have emerged periodically every few years by epochal evolution of VP1 [20,25]. GII.4 genotype NoVs have caused seven pandemics since the mid-1990s, including Grimsby (1995/96 US), Farmington Hills (2002), Hunter (2004), Yerseke (2006a), Den Haag (2006b), New Orleans (NO) (2009), and Sydney (SYD) (2012) [25]. Since the 2012 GII.4 SYD pandemic, novel predominant GII.4 viruses have acquired different non-structural regions through recombination, but have retained the pandemic GII.4 SYD capsid [26]. Exceptionally, in the 2014–2015-winter season, concern over a global pandemic was raised when major non-GII.4 genotype, GII.17 Kawasaki outbreaks were reported on several continents, and GII.17 became the predominant genotype in several Asian countries [27]. Spread of the novel GII.17 strain was enhanced by lack of pre-existing GII.17-specific immune responses in the population and low cross-reactivity of GII.17 with other circulating NoVs [28,29].

As no cross-protective immunity exists between NoV genogroups, a multivalent vaccine or a bivalent VLP vaccine composition containing one GI (e.g., GI.1 or GI.3) and one GII genotype VLP (e.g., GII.4 or GII.12) is considered to be a minimum requirement [5,15,30]. The most advanced NoV VLP-based vaccine in clinical development is a bivalent vaccine containing GI.1 and GII.4 VLPs as a mixture [31]. Other vaccine candidates, combining two or more NoV VLP genotypes (a multivalent vaccine), have also been proposed [30,32], including a trivalent combination vaccine developed by our laboratory, containing a bivalent NoV VLP, GI.3, and GII.4-1999, and a rotavirus (RV) VP6 protein [5,32], targeted at two important causative agents of childhood acute gastroenteritis. Our NoV RV combination vaccine candidate is based on non-live subunit antigens and could improve the low efficacy observed with currently used live RV vaccines in developing countries [33]. RV VP6-specific IgA antibodies and CD4$^+$ T cells have been associated with protection against RV infection [34,35].

The work described here investigates possible immunological interference among different VLPs combined as a mixture formulation with RV VP6. Also, sequential immunization as an alternative approach for NoV VLP vaccine immunization strategy has been explored.

2. Materials and Methods

2.1. Recombinant Proteins Production and Purification

Production and purification of NoV genotypes GI.3 (reference strain Genebank accession no: AF414403) (Figure 1a), GII.4-1999 (AF080551) (Figure 1b), GII.17 (BAR42289.1) (Figure 1c, and GII.4 SYD-2012 (AFV08795.1) (Figure 1d), as well as RV VP6 (GQ477131), used for immunizations, were produced in Sf9 insect cells by a recombinant baculovirus technology and purified in our laboratory as described in details previously [5,36]. VLPs used for analytical methods were produced either in a baculovirus expression system (GI.1-2001 and GII.4 NO-2010) [5,36] or in Nicotiana benthamiana plants (GI.4 and GII.4-2006 VLPs), as previously described [4]. The purity, integrity, and morphology of the VLPs were determined by SDS-polyacrylamide gel electrophoresis, immunoblotting, densitometric analysis, and electron microscopy (Figure 1a–d), as described elsewhere [5]. Protein concentration was determined by using a Pierce™ BCA Protein Assay Kit (Thermo Scientific, Rockford, IL, USA).

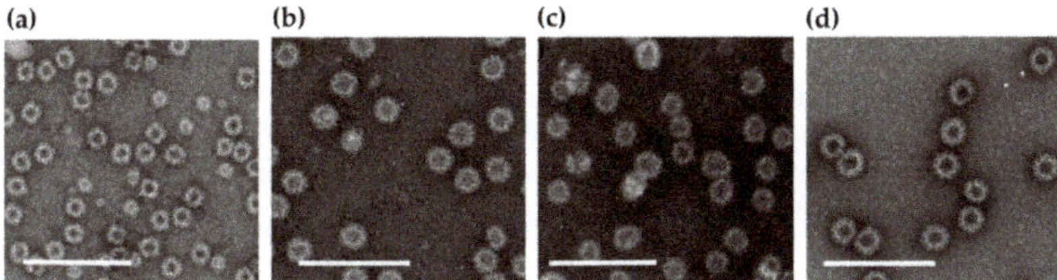

Figure 1. Electron microscopy images of the purified norovirus (NoV) genotypes: (**a**) GI.3, (**b**) GII.4, (**c**) GII.17, and (**d**) GII.4 SYD virus-like particles (VLPs). VLPs were examined by FEI Tecnai F12 electron microscope (Philips 487 Electron Optics, Holland) after negative staining with 3% uranyl acetate, pH 4.6. Bar, 200 nm.

2.2. Mice Immunization

Female BALB/c (H-2d) mice were obtained from Envigo RMS BV and randomly divided into seven groups (Gr) of five mice (Figure 2). At seven weeks of age, the mice were immunized intramuscularly (IM) at the right caudal thigh muscle with a single, monovalent NoV VLP vaccine (Gr I–IV, Figure 2a), or as a multivalent formulation mixture of four NoV genotype VLPs and RV VP6 (Gr V, Figure 2b) two times, at week 0 and week 3 according to the established optimal immunization schedule used in our laboratory [5]. A 10 µg dose per each VLP was administrated, and control (Ctrl, Figure 2a) mice received phosphate-buffered saline (PBS) carrier only. The mice in the sequential immunization group (VI, Figure 2c) were primed with GII.4-1999 + GI.3 VLPs + RV VP6 (the trivalent NoV-RV combination vaccine [32]) at week 0 and week 3 and boosted with GII.17 VLPs at week 5, and with GII.4 SYD at week 7, using a 10 µg dose of each antigen per injection. Tail blood samples were collected at weeks 0 (pre-dose) and 3 from all mice and additionally at weeks 5 and 7 from the mice in the Gr V and Gr VI. Mice were sacrificed, and blood samples (serum) were collected at week 5 (Gr I-IV, Ctrl) or at week 9 (Gr V and VI). Immunizations were conducted under general anesthesia by inhalation of isoflurane (Attane vet, Vet Medic Animal Health Oy), and a formulation of medetomidine (Dorbene® vet, Laboratorios Syva, Leon, Spain) and ketamine (Ketaminol® vet, Intervet International B.V., Boxmeer, The Netherlands) was used for euthanasia. All procedures were carried out in accordance with the regulations and guidelines of the Finnish National Experiment Board (Permission number ESAVI/10800/04.10.07/2016) and mouse welfare was monitored throughout the experiment on a daily basis.

2.3. NoV-Specific ELISA

Homologous and cross-reactive serum IgG binding antibodies were detected by enzyme-linked immunosorbent assay (ELISA) as described elsewhere in detail [5,32]. In brief, individual mouse serum samples of the experimental groups and the control mice were analyzed two-fold diluted, starting at a dilution of 1:200. NoV- and RV VP6-specific IgG antibodies were detected with horse-radish peroxidase (HRP)-conjugated anti-mouse IgG (Sigma–Aldrich, Saint Louis, MO, USA) followed by o-Phenylenediamine dihydrochloride (OPD)-substrate (Sigma–Aldrich) and the optical density (OD) values were measured at 490 nm. The end-point titers of serum IgG were determined as the reciprocal of the highest dilution of serum, giving an OD above the set cut-off value (mean OD of negative control mice serum wells + 3 × SD) and at least 0.100 OD.

2.4. Blocking Assay

NoV VLP blocking assay, a surrogate neutralization assay, was used to determine the presence of serum IgG antibodies that block the binding of NoV VLPs to the HBGA carbohydrates according to the previously described method [37,38]. Pig gastric mucin (PGM, type III, Sigma–Aldrich, Cat. M1778) was coated on microwell plates (Corning Inc, Corning, NY, USA) and blocked with 5% milk in PBS. Starting serum dilution was 1:100 for homologous blocking and 1:20 for heterologous cross-blocking assay. Two-fold diluted serum samples were mixed and pre-incubated with NoV VLPs for 1 h at +37 °C prior to plating on PGM coated microwell plates. Bound VLPs were detected with anti-NoV polyclonal antisera (human [39] or rabbit [4]) followed by secondary IgG-HRP antibody (goat anti-human IgG, Novex, Invitrogen or goat anti-rabbit, Abcam, Cambridge, UK) and OPD-substrate (Sigma–Aldrich). Maximum binding was determined by VLP sample lacking mouse sera, and maximum binding OD 490 nm had to be ≥0.7 for each VLP to be acceptable [13]. The blocking index (%) was calculated as follows: 100% − [(OD490 of wells with VLP and serum/OD490 of maximum binding wells) × 100%]. Blocking titer 50 (BT50) was determined as the reciprocal of the highest serum dilution blocking at least 50% of the maximum binding.

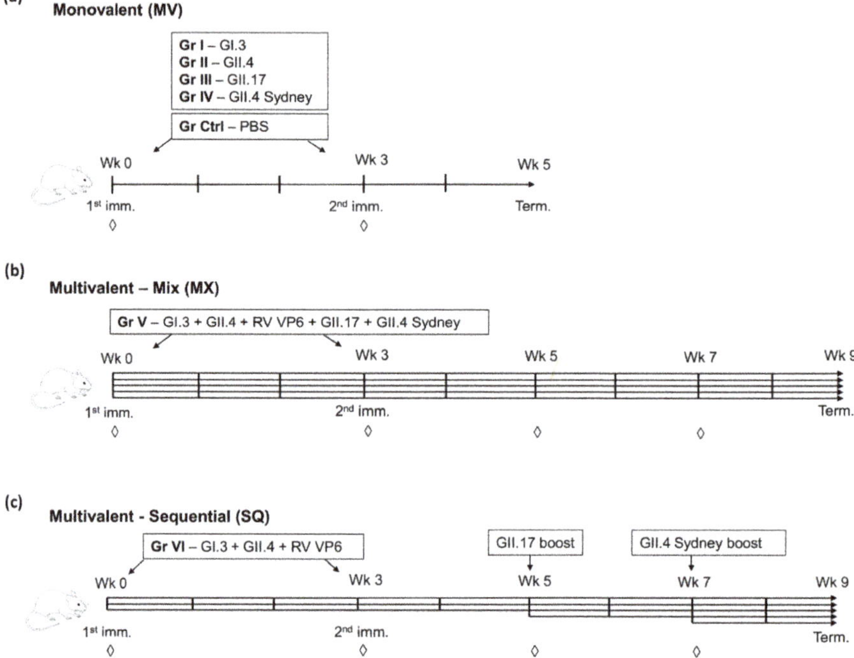

Figure 2. Immunization schemes of the study groups. Experimental groups of mice receiving monovalent (MV) (**a**) NoV VLPs (Gr I–IV) or the control group (Ctrl), receiving carrier only, were intramuscularly immunized twice at study weeks 0 and 3 and terminated at week 5. (**b**) Mice receiving a multivalent mix (MX, Gr V) of the four NoV VLPs and the RV VP6 protein was immunized using the same schedule but terminated at week 9. (**c**) Sequentially immunized mice (SQ, Gr VI) were primed twice at weeks 0 and 3 with the trivalent mix of NoV VLPs and RV VP6, boost immunized at weeks 5 and 7 with heterologous NoV VLPs, and terminated at week 9. Tail blood samples (◊) were collected at the indicated time points. Each horizontal arrow represents one injected antigen.

2.5. Statistics

A non-parametric Mann–Whitney U test was employed to assess the statistical differences between observations of two independent groups. Multiple datasets were compared using a non-parametric Kruskal–Wallis test. Fisher's exact test was used to assess the intergroup differences in the titers. Analyses were conducted by IBM SPSS Statistics (SPSS Inc., Chicago, IL, USA), Version 25.0 and with GraphPad Prism version 8 (GraphPad Software, La Jolla, CA, USA). The statistically significant difference was defined as $p < 0.05$.

3. Results

3.1. Simultaneous Immunization with Multivalent VLP Mixture Formulation

Similar NoV genotype-specific IgG binding antibodies were detected when comparing termination sera of mice immunized twice, either with 10 µg of each monovalent NoV VLP alone (Gr I-IV, Figure 2a) or as a component of a multivalent mixture (Gr V, Figure 2b) (Figure 3). No significant differences ($p < 0.05$) were observed when comparing genotype-specific IgG responses of monovalent or multivalent mix immunized mice at serum dilution 1:200 (Figure 3a). The results showed induction of equal levels of IgG antibodies against GI.3 (Figure 3b), GII.4-1999 (Figure 3c), GII.17 (Figure 3d), and GII.4 SYD (Figure 3e), irrespective of the presence or absence of other co-administrated antigens. NoV-specific IgG was not detected in any of the control animal sera (Gr VII) that received carrier (PBS) only (Figure 3b–e).

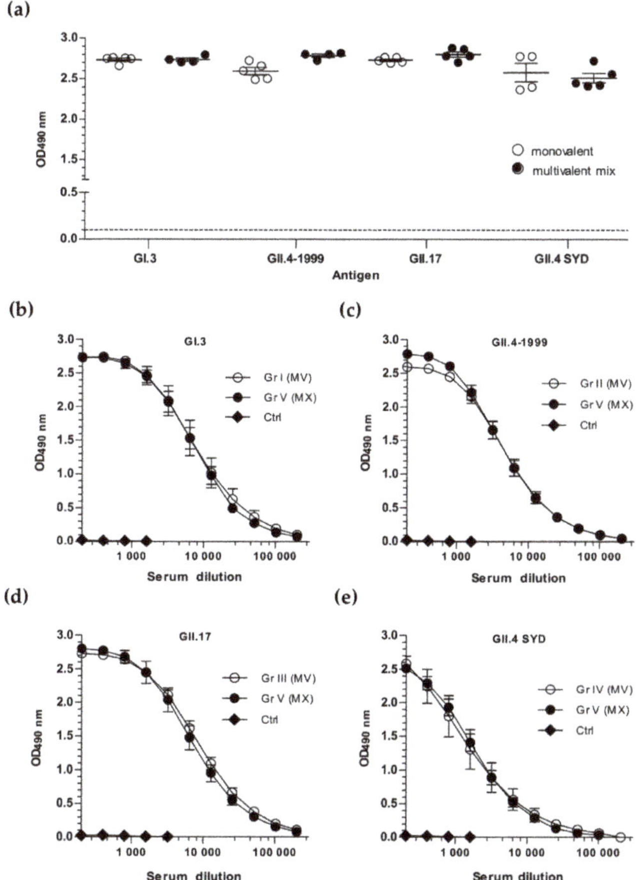

Figure 3. NoV genotype-specific IgG antibody responses induced by monovalent NoV VLPs or multivalent VLP mix immunization. Termination sera of mice immunized with monovalent (MV) VLPs, GI.3 (Gr I), GII.4-1999 (Gr II), GII.17 (Gr III), or GII.4 SYD (Gr IV), or with the multivalent NoV VLP mix (MX, Gr V), or the carrier only (Control group, Ctrl) were analyzed with enzyme-linked immunosorbent assay (ELISA). (**a**) The mean optical density (OD490) at the serum dilution 1:200 of individual mice is illustrated with group mean and the standard error of the mean (SEM). The horizontal dashed line indicates maximum background level (cut-off limit). Mean IgG end-point titration curves specific for NoV (**b**) GI.3, (**c**) GII.4, (**d**) GII.17, and (**e**) GII.4 SYD VLPs with the SEM are shown.

3.2. Sequential Immunization with Genetically Distant and Closely Related NoV VLPs

As an alternative immunization strategy and to study the effect of pre-existing immunity, the sequential immunization schedule was employed (Gr VI, Figure 2). The mice primed twice (week 0 and week 3) with the trivalent combination vaccine formulation (GI.3 + GII.4 + RV VP6) [32] were further immunized with GII.17 VLPs at week 5, followed by a GII.4 SYD VLP boost at week 7, and termination sera IgG was analyzed for all four NoV genotype-specific IgG levels (Figure 4). When compared to genotype-specific immune responses of mice immunized twice with monovalent VLPs, no significant ($p > 0.05$) differences in IgG responses were observed (Figure 4a). Similarly, strong serum IgG titers to GI.3 (Figure 4b), GII.4-1999 (Figure 4c), GII.17 (Figure 4d), and GII.4 SYD (Figure 4e) were measured following one boost immunization, with no significant difference ($p > 0.05$) to corresponding monovalent immunization groups. The response to GII.17 (Figure 4d) was very strong considering that the mice received only one GII.17 VLP dose at week 5 (Gr VI). On the contrary, GII.4 SYD-specific IgG response in Gr VI mice sera (Figure 4e) consists of the genotype-specific antibodies as well as cross-reactive antibodies to GII.4 SYD induced by closely related GII.4-1999 VLPs.

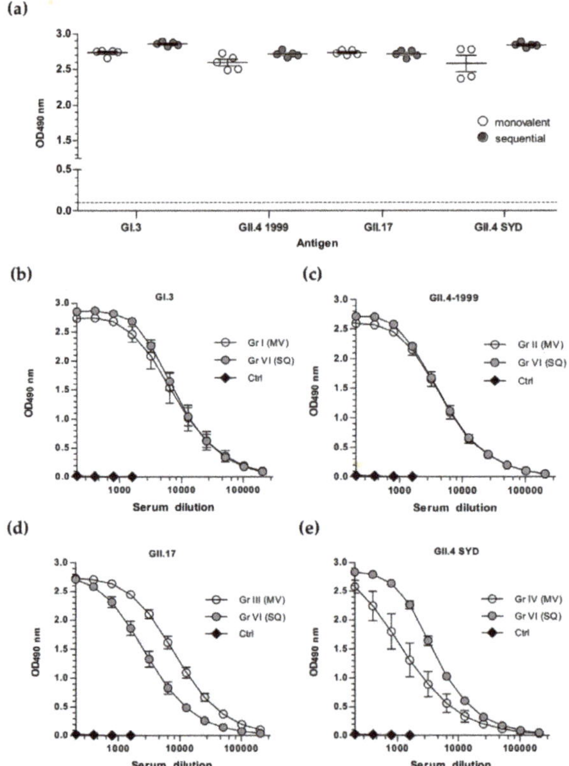

Figure 4. NoV genotype-specific IgG antibody responses induced by monovalent NoV VLPs or sequential VLP immunization. Termination sera of mice immunized with monovalent (MV) VLPs, GI.3 (Gr I), GII.4-1999 (Gr II), GII.17 (Gr III), or GII.4 SYD (Gr IV), or sequentially (SQ) immunized with heterologous VLP boosts (Gr VI) or the carrier only (Control group, Ctrl) were analyzed with enzyme-linked immunosorbent assay (ELISA). (**a**) The mean optical density (OD490) at the serum dilution 1:200 of individual mice is illustrated with the group mean and the standard error of the mean (SEM). The horizontal dashed line indicats maximum background level (cut-off limit). Mean IgG end-point titration curves specific for NoV (**b**) GI.3, (**c**) GII.4, (**d**) GII.17, and (**e**) GII.4 SYD VLPs with the SEM are shown.

3.3. Simultaneous and Sequential Immunizations Induce Different Level of Blocking Antibodies

To investigate if the neutralizing capability of IgG antibodies induced by simultaneous multivalent mixture immunizations (Gr V, Figure 2) is different from response to sequential immunization (Gr VI, Figure 2), genotype-specific blocking activity of the serum was analyzed. Equal levels of GI.3-specific blocking antibodies (Figure 5a) were generated by multivalent mixture (mean BT50 = 720) and sequential immunization (BT50 = 640). Similar levels of GII.4-1999-specific blocking antibodies were measured for both groups (mean BT50 = 640 Gr V; mean BT50 = 480 Gr VI) (Figure 5b). A single boost injection of GII.17 VLPs (Figure 5c) was sufficient to induce a comparable level of GII.17-specific blocking antibodies (mean BT50 = 560) to immunization with multivalent mix (mean BT50 = 720). In contrast, a single GII.4 SYD VLP boost injection (Figure 5d) generated significantly lower blocking antibodies (mean BT50 = 70) than immunization with multivalent mixture (mean BT50 = 400) (p = 0.024). Furthermore, the sera of mice immunized with the monovalent VLP vaccine (Gr I-IV, Figure 2) were tested for homologous blocking titers against all four genotypes (GI.3, GII.4-1999, GII.17, and GII.4 SYD) and blocking antibody levels similar to the levels induced by the multivalent VLP mix (Gr V) were observed (BT50 ≥ 400).

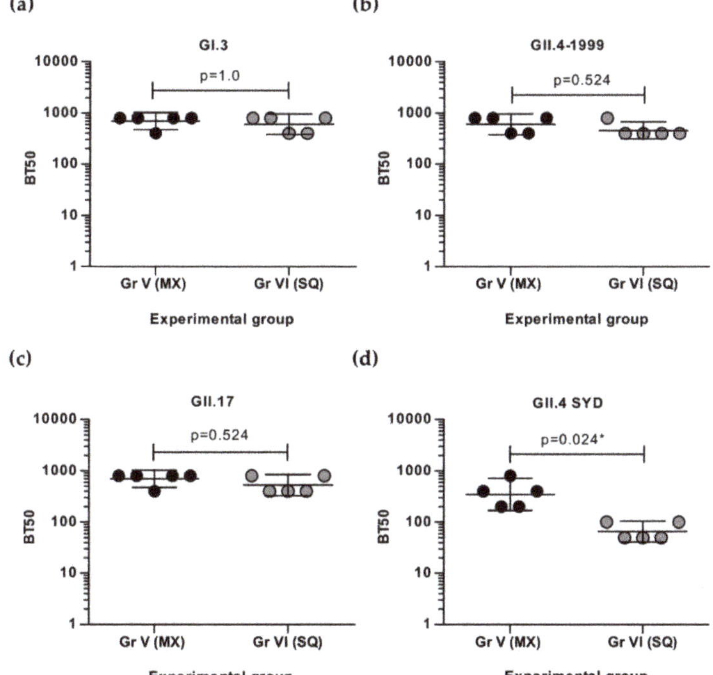

Figure 5. NoV genotype-specific blocking antibodies after immunization with multivalent NoV VLP as a mix or sequentially. Individual sera of mice immunized with multivalent mix (MX, Gr V) or sequentially (SQ, Gr VI) were 2-fold diluted starting at 1:100 dilution and assayed for the blocking of homologous NoV (**a**) GI.3, (**b**) GII.4-1999, (**c**) GII.17, or (**d**) GII.4 SYD VLP binding to histo-blood group antigens present in pig gastric mucin. The blocking index (%) was calculated as [100% − [OD (wells with serum)/OD(wells without serum, maximum VLP binding)] × 100%] and shown are individual mouse 50% blocking titer (BT50) and group geometric mean titer with 95% confidence interval. An arbitrary titer, BT50 of 5, was assigned to samples with <50% blocking index at the lowest serum dilution 1:100. Statistical differences were determined using Fisher's exact test, and a p value of ≤0.05 was considered statistically significant (*).

3.4. Cross-Protective Blocking Antibody Responses

The protective potential of serum antibodies induced by multivalent mixture (Gr V), or sequential immunization (Gr VI) against heterologous NoV VLPs belonging to GI (GI.1 and GI.4) or GII (GII.4-2006 and GII.4 NO) was tested by cross-blocking analysis (Figure 6). Blocking of GII.4-2006 VLP binding (Figure 6a) was equal in both groups (BT50 = 320). A considerable four-fold higher cross-blocking titer of GII.4 NO VLP binding was observed with multivalent mixture immunization sera (Gr V, BT50 = 640) than with sera of the sequential immunization group (Gr VI, BT50 = 160) (Figure 6b). Very low levels of GI.1 VLP cross-blocking antibodies were detected in both the Gr V and Gr VI immune sera (BT50 = 40) (Figure 6c), whereas no blocking antibodies to GI.4 VLP binding were observed (Figure 6d).

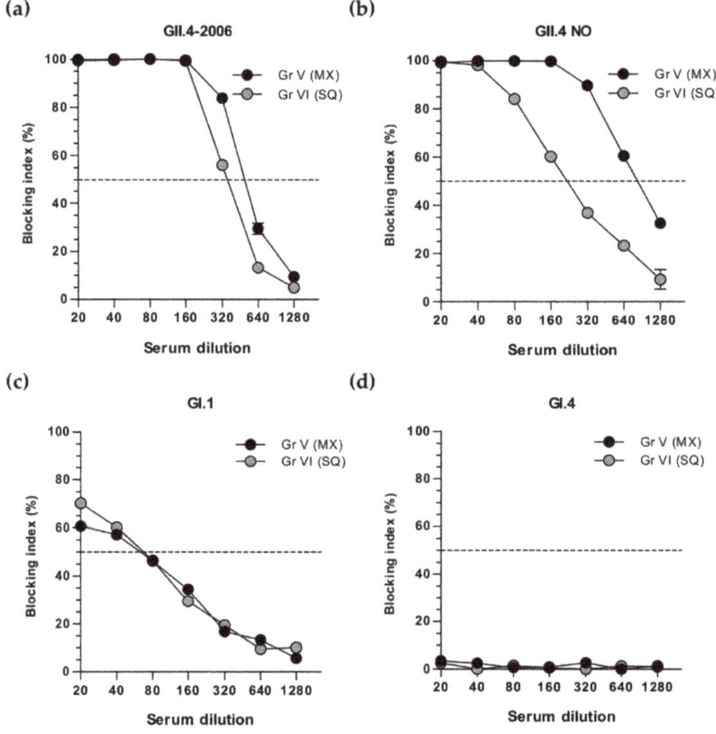

Figure 6. NoV cross-blocking antibodies after multivalent NoV VLP mix or sequential immunization. Group-wise pooled sera of mice immunized with multivalent mix (MX, Gr V) or sequentially (SQ, Gr VI) were 2-fold diluted starting at 1:20 dilution and assayed for the cross-blocking of heterologous NoV (**a**) GII.4-2006, (**b**) GII.4 NO, (**c**) GI.1, or (**d**) GI.4 VLP binding to histo-blood group antigens present in pig gastric mucin. The blocking index (%) was calculated as [100% − [OD (wells with serum)/OD(wells without serum, maximum VLP binding)] × 100%] and shown are group mean blocking indeces (%) with the standard errors of the mean of repeated assays. The horizontal dashed line represents a blocking titer of 50% (BT50).

Cross-blocking of heterologous GII.4-2006 (BT50 = 320) and GII.4 NO (BT50 = 640) VLPs binding by termination sera of mice simultaneously immunized with the VLP mix (Gr V, Figure 6a,b) was similar to genotype-specific blocking of GII.4-1999 (BT50 = 640) and GII.4 SYD (BT50 = 400) VLP binding (Figure 5b,d), respectively. To confirm this finding we measured the kinetics of blocking antibodies of Gr V to homologous GII.4 SYD (Figure 7a) and heterologous GII.4 NO (Figure 7b) in tail blood collected prior to immunization (week 0), after the first immunization (week 3) and two weeks after the second immunization (week 5). Low homologous (Figure 7a) and heterologous (Figure 7b) blocking activities were observed after the first immunization at week 3, followed by equally robust generation of blocking antibodies to GII.4 SYD (BT50 = 400) and GII.4 NO VLPs (BT50 = 400) following the second dose at week 5. Similar levels of blocking antibodies to GII.4 SYD (BT50 = 200) and GII.4 NO VLPs (BT50 = 400) were measured at week 7 (Figure 7a,b).

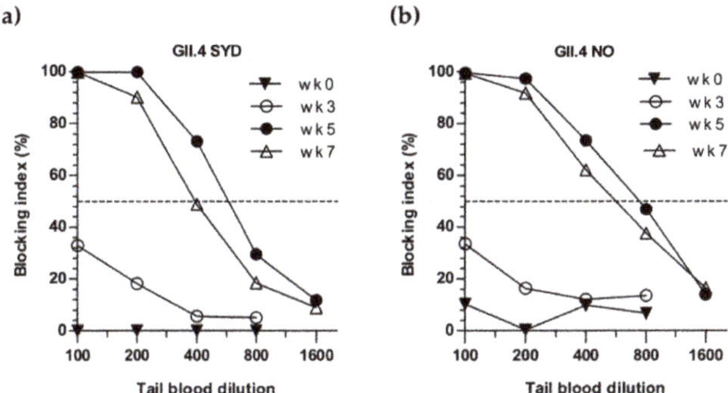

Figure 7. Kinetics of NoV type-specific and cross-reactive blocking antibodies after multivalent NoV VLP mix immunization. Group-wise pooled tail blood samples of mice were collected at the time of the first immunization (week 0), the second immunization (week 3), and two and four weeks after the second immunization (weeks 5 and 7) and 2-fold diluted starting at 1:100 dilution, for assaying the blocking antibodies of (**a**) homologous GII.4 SYD VLPs and (**b**) heterologous GII.4 NO VLPs binding to histo-blood group antigens present in pig gastric mucin. The blocking index (%) was calculated as [100% − [OD (wells with serum)/OD(wells without serum, maximum VLP binding)] × 100%] and shown are titration curves of each time-point blocking index (%). The horizontal dashed line represents a blocking titer of 50% (BT50).

We also wanted to test the cross-blocking potential of GII.17 VLP immunized mice sera (Gr III, Figure 2), because it has been recently shown that GII.17-specific antibodies suppressed the replication of GII.4 SYD in intestinal epithelial cell cultures [40]. As shown above (Figure 3d, Figure 4d, and Figure 5c) monovalent GII.17 VLP immunization induced high genotype-specific IgG antibody levels and blocking antibodies; however, no cross-blocking antibodies to any of the seven VLP genotypes tested (GII.4-1999, GII.4-2006, GII.4 NO, GII.4 SYD, GI.1, GI.3, GI.4) were observed (Figure 8).

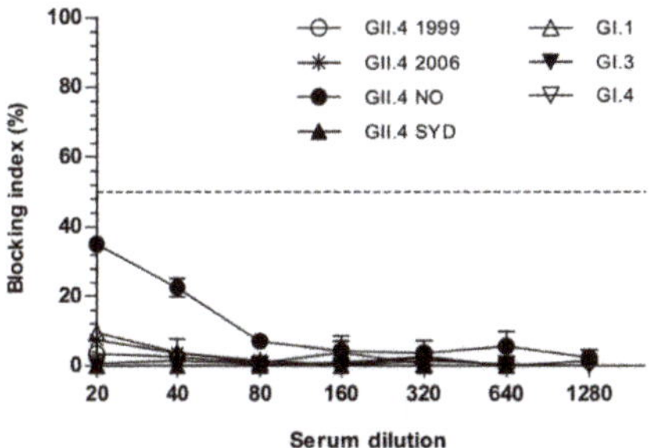

Figure 8. NoV cross-blocking antibodies after monovalent NoV GII.17 VLP immunization. Termination sera was pooled and serially 2-fold diluted starting at a 1:20 dilution and analyzed in a blocking assay against seven heterologous GI and GII VLPs. The blocking index (%) was calculated as [100% − [OD (wells with serum)/OD(wells without serum, maximum VLP binding)] × 100%] and shown are the blocking index (%) of the two analysis. The horizontal dashed line represents a blocking titer of 50% (BT50).

4. Discussion

The design of this study aimed to define the optimal immunization strategy with multivalent NoV VLPs and to predict the effects of pre-existing NoV immunity on the development of immune responses to novel NoV infections that may be caused by genetically very close or distant NoV genotypes. The basis for antigen selection and inclusion of RV VP6 into the multivalent mixture used in this study, and the sequential prime immunization, lies in the trivalent NoV-RV combination vaccine designed in our laboratory, consisting of NoV GI.3 and GII.4-1999 VLPs and RV VP6 protein [5,32,37,41]. Representative of GII, GII.4-1999 was selected as an ancestral genotype with a broad cross-blocking profile [20] and GI representative GI.3 VLP due to relatively high incidence in children [42]. By including RV VP6, the most abundant and highly immunogenic RV antigen, into the combination vaccine, this vaccine candidate is targeted to prevent two main causes of AGE in young children [5,32,37]. Even though the NoV VLP vaccine candidates tested in phase I/II clinical trials are either monovalent or bivalent VLP formulations [43,44], it is not known if it might be necessary to upgrade the vaccines according to emerging new variants, such as the sudden emergence of a novel non-GII.4 variant GII.17 in 2014 might suggest. Currently developed vaccines might not provide protection to GII.17 strains due to low cross-reactivity observed by ourselves and others [28,29].

We have comprehensively studied the immunogenicity of the trivalent NoV VLP and RV VP6 vaccine candidate in preclinical studies, demonstrating induction of strong NoV- and RV-specific immune responses [5,32,37,41]. No interference of the immune responses to any antigens of the combination vaccine have been observed when NoV- and RV-specific humoral and cellular responses have been assessed in mice immunized with bivalent [5] or trivalent [32] NoV-RV combination vaccine. Although VP6 induced immune response generation was not in the scope of the present study, both simultaneous and sequential immunizations induced similar VP6-specific antibody levels (data not shown). We have previously shown that NoV VLPs are highly immunogenic delivered either IM, intradermally (ID), intranasally (IN), or primed IM and boosted IN [45]. Here, we have extended our investigation of different NoV VLP vaccine delivery strategies [17,39] by combining four different NoV genotypes, two GII.4 VLPs, an ancient and a recent pandemic strain (GII.4-1999 and GII.4 SYD), one recently emerging non-GII.4 VLP (GII.17), and one GI genotype (GI.3). To determine the requirements for delivery of the multivalent vaccine antigens, we used NoV VLPs as a mixture delivered at the same time or at separated times, using sequential immunization. The results show that multiple NoV genotype VLPs can be delivered simultaneously as a multivalent VLP vaccine mixture (a cocktail) and immune responses to all vaccine antigens are induced. Congruent to our earlier findings [39], there was no inhibition observed by multivalent mixture immunization, as IgG binding and blocking antibodies were at an equal level to the responses induced by monovalent VLPs. This is in contrast to a previous report by Leroux–Roels et al. [44] showing that GI.1 VLP equivalent dose in a bivalent GI.1 + GII.4 VLP vaccine formulation interfered with GII.4-specific immune responses in humans. It could be speculated that the discrepancy here might be related to differences between NoV genotypes regarding the immunological interference they might possess in relation to other genotypes.

We also investigated if NoV-specific immune responses induced by vaccination may be differently formed depending on the degree of VLP genotype identity with NoV VLPs previously encountered. In the sequential immunization schedule, it could be assumed that the priming of mice with a combination of NoV GI.3 and GII.4 VLPs and RV VP6 mimics the baseline pre-existing immunity to these viruses, which exist in the pediatric population [22,46]. In addition, the boosting of mice with the GII.17 and GII.4 SYD VLPs at later time points reflects the children's exposure to novel NoVs. The sequential immunization with NoV VLPs derived from genetically distant GII.17 genotype was successful in inducing high NoV GII.17-specific immune response. These results are in line with our earlier studies showing that sequential immunization with NoV VLPs derived from genetically diverse GII.4 and GII.12 genotypes has been successful in inducing high NoV-specific immune response to both genotypes [47]. In contrast, sequential boost immunization with GII.4 SYD VLPs, genetically closely related to GII.4-1999 (41% capsid protein identity, by the Protein Basic Local Alignment Search Tool) failed to induce strong blocking GII.4 SYD-specific responses when compared to mice immunized with multivalent VLPs mixed simultaneously. Noteworthy, even though similar level of GII.4 SYD-specific IgG titers were measured in all immunization groups, the GII.4 SYD blocking activity of sequentially immunized group sera was significantly lower, addressing the importance of measuring blocking NoV-specific IgG antibodies. The presence of blocking antibodies reveals the protective potential and specificity of these IgG antibodies, as previously reported in mice [17] and in humans [16], where the lack of cross-blocking antibodies was observed despite the presence of cross-reactive IgG antibodies. The lack of GII.4 SYD blocking antibodies in sequentially immunized mice was surprising, as it could be assumed that priming with closely related GII.4-1999 VLPs would increase GII.4 SYD-specific blocking antibodies. These results indicate that there might be an original antigenic sin (OAS) for closely related antigens (GII.4-1999 and GII.4 SYD, respectively), which could be an obstacle when considering frequently updated NoV VLPs vaccinations. Similarly, results in mice immunized sequentially with variant influenza viruses have suggested that OAS could be a potential strategy by which variant influenza viruses subvert the immune system [48]. Similarly, in humans, higher antibody titers to influenza strains encountered earlier in life following repeated exposures with new variants has been reported, indicating antigenic seniority [49]. Moreover, the antigenic distance hypothesis by Smith et al. in 1999 [50] suggesting that antigenic distance between circulating influenza strains and vaccine strains partly account for the variable vaccine efficacy, has also been supported by recent report [51]. If the same phenomenon would apply to NoVs, it should be taken into consideration when designing vaccines for NoVs. The lack of cross-blocking between GI.1 and GI.4 and vaccine genotypes used in the present study was congruent with our previous results showing low or absent cross-blocking between, e.g., GI.1 and GI.3 NoV VLPs [17].

It has been recently published [40] that GII.17 VLPs induce antibodies able to inhibit GII.4 SYD genotype replication in in vitro cell culture models. For the above reason and as we observed that GII.17 VLPs were highly immunogenic even when delivered only once in a sequential regime, we assumed that GII.17-specific sera might efficiently block the binding of heterologous NoV VLPs not included in the vaccine formulation. Unfortunately, we did not observe such cross-blocking antibodies in mice immunized with GII.17 VLPs. The discrepancy in the results may come from the differences in the assays used to measure the cross-blocking activity. Sato et al. used a novel in vitro cell culture infection model while we employed the widely used ELISA-based blocking assay as a surrogate neutralization assay [10,12,15]. Sato et al. have also shown that GII.4 SYD–specific antibodies did not block GII.17 virus replication. Congruently, we have published that the ancestor GII.4-1999 VLPs are superior to more recent GII.4 variants in inducing cross-reactive responses, including the cross-blocking response to GII.17 VLPs [17,20,28,52]. On the other hand, we observed that when mice are immunized with the quadrivalent NoV VLP mixture containing GI.3, GII.4-1999 (an ancestor GII.4 variant), GII.4 SYD (the most recent pandemic variant), and GII.17, cross-blocking of GII.4 variants 2006 and 2009 (NO) was comparable to homologous GII.4-1999 and GII.4 SYD VLP blocking. These results indicate that a combination of GII.4 VLPs from the ancestral and the most novel variant could suffice to protect against most GII.4 NoV infections. This is a relevant observation as GII.4 NoVs cause over 70% of all NoV AGE worldwide [53].

The protective immunity may be evaded by small changes in the immunodominant epitopes and, therefore, NoV vaccine antigens might need periodic updates, similar to seasonal influenza vaccines. Instead of using multivalent VLP mixtures, one option is to use monovalent or bivalent NoV VLPs to prime the immune system and to then boost with the diverse monovalent VLP. However, this study indicates that OAS might interfere with the immune response generation for genetically closely related NoV genotype VLPs used as a boost. Furthermore, in this study, by combining an ancestor GII.4 (1999) and a novel GII.4 (SYD), it was possible to induce great cross-reactive responses to other GII.4 variants, indicating that vaccine-induced protection could extend to genotypes not included in the vaccine.

Author Contributions: Conceptualization, V.B.; data curation, M.M. and V.B.; formal analysis, M.M. and V.B.; funding acquisition, T.V. and V.B.; investigation, M.M. and V.B.; methodology, M.M. and V.B.; project administration,

V.B.; resources, T.V. and V.B.; supervision, V.B.; visualization, M.M. and V.B.; writing—original draft, M.M. and V.B.; writing—review and editing, T.V. and V.B.

Funding: This research received no external funding.

Acknowledgments: We are grateful to Icon Genetics GmbH for providing magnICON®-produced GI.4 and GII.4-2006 NoV VLPs. The technical assistance given by the personnel of the Vaccine Research Center of Tampere is gratefully acknowledged, and Kirsi Tamminen and Suvi Heinimäki are especially thanked for their expertise. This research did not receive any specific grant from funding agencies in the public, commercial, or not-for-profit sectors.

Conflicts of Interest: The authors declare no conflict of interest.

References

1. Lopman, B.A.; Steele, D.; Kirkwood, C.D.; Parashar, U.D. The Vast and Varied Global Burden of Norovirus: Prospects for Prevention and Control. *PLoS Med.* **2016**, *13*, e1001999. [CrossRef] [PubMed]

2. Cortes-Penfield, N.W.; Ramani, S.; Estes, M.K.; Atmar, R.L. Prospects and Challenges in the Development of a Norovirus Vaccine. *Clin. Ther.* **2017**, *39*, 1537–1549. [CrossRef] [PubMed]

3. Jiang, X.; Wang, M.; Graham, D.Y.; Estes, M.K. Expression, self-assembly, and antigenicity of the Norwalk virus capsid protein. *J. Virol.* **1992**, *66*, 6527–6532. [PubMed]

4. Malm, M.; Diessner, A.; Tamminen, K.; Liebscher, M.; Vesikari, T.; Blazevic, V. Rotavirus VP6 as an Adjuvant for Bivalent Norovirus Vaccine Produced in Nicotiana benthamiana. *Pharmaceutics* **2019**, *11*, 229. [CrossRef]

5. Blazevic, V.; Lappalainen, S.; Nurminen, K.; Huhti, L.; Vesikari, T. Norovirus VLPs and rotavirus VP6 protein as combined vaccine for childhood gastroenteritis. *Vaccine* **2011**, *29*, 8126–8133. [CrossRef]

6. Lucero, Y.; Vidal, R.; O'Ryan, G.M. Norovirus vaccines under development. *Vaccine* **2017**, *36*, 5435–5441. [CrossRef]

7. Chhabra, P.; de Graaf, M.; Parra, G.I.; Chan, M.C.; Green, K.; Martella, V.; Wang, Q.; White, P.A.; Katayama, K.; Vennema, H.; et al. Updated classification of norovirus genogroups and genotypes. *J. Gen. Virol.* **2019**, *100*, 1393–1406. [CrossRef]

8. Vega, E.; Barclay, L.; Gregoricus, N.; Shirley, S.H.; Lee, D.; Vinje, J. Genotypic and epidemiologic trends of norovirus outbreaks in the United States, 2009 to 2013. *J. Clin. Microbiol.* **2014**, *52*, 147–155. [CrossRef]

9. Uusi-Kerttula, H.; Tamminen, K.; Malm, M.; Vesikari, T.; Blazevic, V. Comparison of human saliva and synthetic histo-blood group antigens usage as ligands in norovirus-like particle binding and blocking assays. *Microbes. Infect.* **2014**, *16*, 472–480. [CrossRef]

10. Harrington, P.R.; Lindesmith, L.; Yount, B.; Moe, C.L.; Baric, R.S. Binding of Norwalk virus-like particles to ABH histo-blood group antigens is blocked by antisera from infected human volunteers or experimentally vaccinated mice. *J. Virol.* **2002**, *76*, 12335–12343. [CrossRef]

11. Tan, M.; Jiang, X. Norovirus and its histo-blood group antigen receptors: An answer to a historical puzzle. *Trends Microbiol.* **2005**, *13*, 285–293. [CrossRef] [PubMed]

12. Reeck, A.; Kavanagh, O.; Estes, M.K.; Opekun, A.R.; Gilger, M.A.; Graham, D.Y.; Atmar, R.L. Serological correlate of protection against norovirus-induced gastroenteritis. *J. Infect. Dis.* **2010**, *202*, 1212–1218. [CrossRef] [PubMed]

13. Malm, M.; Tamminen, K.; Blazevic, V. Assessment of Functional Norovirus Antibody Responses by Blocking Assay in Mice. *Methods Mol. Biol.* **2016**, *1403*, 259–268. [CrossRef] [PubMed]

14. Nurminen, K.; Blazevic, V.; Huhti, L.; Rasanen, S.; Koho, T.; Hytonen, V.P.; Vesikari, T. Prevalence of norovirus GII-4 antibodies in Finnish children. *J. Med. Virol.* **2011**, *83*, 525–531. [CrossRef] [PubMed]

15. Ramani, S.; Estes, M.K.; Atmar, R.L. Correlates of Protection against Norovirus Infection and Disease-Where Are We Now, Where Do We Go? *PLoS Pathog.* **2016**, *12*, e1005334. [CrossRef]

16. Malm, M.; Uusi-Kerttula, H.; Vesikari, T.; Blazevic, V. High serum levels of norovirus genotype-specific blocking antibodies correlate with protection from infection in children. *J. Infect. Dis.* **2014**, *210*, 1755–1762. [CrossRef]

17. Malm, M.; Tamminen, K.; Lappalainen, S.; Uusi-Kerttula, H.; Vesikari, T.; Blazevic, V. Genotype considerations for virus-like particle-based bivalent norovirus vaccine composition. *Clin. Vaccine Immunol.* **2015**, *22*, 656–663. [CrossRef]

18. Hansman, G.S.; Natori, K.; Shirato-Horikoshi, H.; Ogawa, S.; Oka, T.; Katayama, K.; Tanaka, T.; Miyoshi, T.; Sakae, K.; Kobayashi, S.; et al. Genetic and antigenic diversity among noroviruses. *J. Gen. Virol.* **2006**, *87*, 909–919. [CrossRef]

19. Blazevic, V.; Malm, M.; Vesikari, T. Induction of homologous and cross-reactive GII.4-specific blocking antibodies in children after GII.4 New Orleans norovirus infection. *J. Med. Virol.* **2015**, *87*, 1656–1661. [CrossRef]

20. Tamminen, K.; Malm, M.; Vesikari, T.; Blazevic, V. Immunological Cross-Reactivity of an Ancestral and the Most Recent Pandemic Norovirus GII.4 Variant. *Viruses* **2019**, *11*, 91. [CrossRef]

21. Blazevic, V.; Malm, M.; Honkanen, H.; Knip, M.; Hyoty, H.; Vesikari, T. Development and maturation of norovirus antibodies in childhood. *Microbes. Infect.* **2015**, *18*, 263–269. [CrossRef] [PubMed]

22. Blazevic, V.; Malm, M.; Salminen, M.; Oikarinen, S.; Hyoty, H.; Veijola, R.; Vesikari, T. Multiple consecutive norovirus infections in the first 2 years of life. *Eur. J. Pediatr.* **2015**, *174*, 1679–1683. [CrossRef] [PubMed]

23. Saito, M.; Goel-Apaza, S.; Espetia, S.; Velasquez, D.; Cabrera, L.; Loli, S.; Crabtree, J.E.; Black, R.E.; Kosek, M.; Checkley, W.; et al. Multiple norovirus infections in a birth cohort in a Peruvian Periurban community. *Clin. Infect. Dis.* **2014**, *58*, 483–491. [CrossRef] [PubMed]

24. Zheng, D.P.; Widdowson, M.A.; Glass, R.I.; Vinje, J. Molecular epidemiology of genogroup II-genotype 4 noroviruses in the United States between 1994 and 2006. *J. Clin. Microbiol.* **2010**, *48*, 168–177. [CrossRef] [PubMed]

25. Eden, J.S.; Tanaka, M.M.; Boni, M.F.; Rawlinson, W.D.; White, P.A. Recombination within the pandemic norovirus GII.4 lineage. *J. Virol.* **2013**, *87*, 6270–6282. [CrossRef] [PubMed]

26. Hasing, M.E.; Lee, B.E.; Qiu, Y.Y.; Xia, M.; Pabbaraju, K.; Wong, A.; Tipples, G.; Jiang, X.; Pang, X.L.L. Changes in norovirus genotype diversity in gastroenteritis outbreaks in Alberta, Canada: 2012-2018. *BMC Infect. Dis.* **2019**, *19*, 177. [CrossRef] [PubMed]

27. De Graaf, M.; Van Beek, J.; Vennema, H.; Podkolzin, A.T.; Hewitt, J.; Bucardo, F.; Templeton, K.; Mans, J.; Nordgren, J.; Reuter, G.; et al. Emergence of a novel GII.17 norovirus - End of the GII.4 era? *Euro. Surveill.* **2015**, *20*, 21178. [CrossRef]

28. Malm, M.; Tamminen, K.; Vesikari, T.; Blazevic, V. Norovirus GII.17 Virus-Like Particles Bind to Different Histo-Blood Group Antigens and Cross-React with Genogroup II-Specific Mouse Sera. *Viral Immunol.* **2018**, *31*, 649–657. [CrossRef]

29. Dai, Y.C.; Xia, M.; Huang, Q.; Tan, M.; Qin, L.; Zhuang, Y.L.; Long, Y.; Li, J.D.; Jiang, X.; Zhang, X.F. Characterization of Antigenic Relatedness between GII.4 and GII.17 Noroviruses by Use of Serum Samples from Norovirus-Infected Patients. *J. Clin. Microbiol.* **2017**, *55*, 3366–3373. [CrossRef]

30. LoBue, A.D.; Lindesmith, L.; Yount, B.; Harrington, P.R.; Thompson, J.M.; Johnston, R.E.; Moe, C.L.; Baric, R.S. Multivalent norovirus vaccines induce strong mucosal and systemic blocking antibodies against multiple strains. *Vaccine* **2006**, *24*, 5220–5234. [CrossRef]

31. Treanor, J.J.; Atmar, R.L.; Frey, S.E.; Gormley, R.; Chen, W.H.; Ferreira, J.; Goodwin, R.; Borkowski, A.; Clemens, R.; Mendelman, P.M. A Novel Intramuscular Bivalent Norovirus Virus-Like Particle Vaccine Candidate-Reactogenicity, Safety, and Immunogenicity in a Phase 1 Trial in Healthy Adults. *J. Infect. Dis.* **2014**, *210*, 1763–1771. [CrossRef] [PubMed]

32. Tamminen, K.; Lappalainen, S.; Huhti, L.; Vesikari, T.; Blazevic, V. Trivalent combination vaccine induces broad heterologous immune responses to norovirus and rotavirus in mice. *PLoS ONE* **2013**, *8*, e70409. [CrossRef] [PubMed]

33. Clark, A.; van Zandvoort, K.; Flasche, S.; Sanderson, C.; Bines, J.; Tate, J.; Parashar, U.; Jit, M. Efficacy of live oral rotavirus vaccines by duration of follow-up: A meta-regression of randomised controlled trials. *Lancet Infect. Dis.* **2019**, *19*, 717–727. [CrossRef]

34. Ward, R.L.; McNeal, M.M. VP6: A candidate rotavirus vaccine. *J. Infect. Dis.* **2010**, *202* (Suppl. 1), S101–S107. [CrossRef]

35. Lappalainen, S.; Pastor, A.R.; Tamminen, K.; Lopez-Guerrero, V.; Esquivel-Guadarrama, F.; Palomares, L.A.; Vesikari, T.; Blazevic, V. Immune responses elicited against rotavirus middle layer protein VP6 inhibit viral replication in vitro and in vivo. *Hum. Vaccin. Immunother.* **2014**, *10*, 2039–2047. [CrossRef]

36. Huhti, L.; Blazevic, V.; Nurminen, K.; Koho, T.; Hytonen, V.P.; Vesikari, T. A comparison of methods for purification and concentration of norovirus GII-4 capsid virus-like particles. *Arch. Virol.* **2010**, *155*, 1855–1858. [CrossRef]

37. Malm, M.; Heinimäki, S.; Vesikari, T.; Blazevic, V. Rotavirus capsid VP6 tubular and spherical nanostructures act as local adjuvants when co-delivered with norovirus VLPs. *Clin. Exp. Immunol.* **2017**, *189*, 331–341. [CrossRef]
38. Lindesmith, L.C.; Debbink, K.; Swanstrom, J.; Vinje, J.; Costantini, V.; Baric, R.S.; Donaldson, E.F. Monoclonal antibody-based antigenic mapping of norovirus GII.4-2002. *J. Virol.* **2012**, *86*, 873–883. [CrossRef]
39. Malm, M.; Tamminen, K.; Heinimaki, S.; Vesikari, T.; Blazevic, V. Functionality and avidity of norovirus-specific antibodies and T cells induced by GII.4 virus-like particles alone or co-administered with different genotypes. *Vaccine* **2018**, *36*, 484–490. [CrossRef]
40. Sato, S.; Hisaie, K.; Kurokawa, S.; Suzuki, A.; Sakon, N.; Uchida, Y.; Yuki, Y.; Kiyono, H. Human Norovirus Propagation in Human Induced Pluripotent Stem Cell-Derived Intestinal Epithelial Cells. *Cell. Mol. Gastroenterol. Hepatol.* **2019**, *7*, 686–688.e5. [CrossRef]
41. Malm, M.; Tamminen, K.; Lappalainen, S.; Vesikari, T.; Blazevic, V. Rotavirus Recombinant VP6 Nanotubes Act as an Immunomodulator and Delivery Vehicle for Norovirus Virus-Like Particles. *J. Immunol. Res.* **2016**, *2016*, 9171632. [CrossRef] [PubMed]
42. Puustinen, L.; Blazevic, V.; Huhti, L.; Szakal, E.D.; Halkosalo, A.; Salminen, M.; Vesikari, T. Norovirus genotypes in endemic acute gastroenteritis of infants and children in Finland between 1994 and 2007. *Epidemiol. Infect.* **2012**, *140*, 268–275. [CrossRef] [PubMed]
43. Kim, L.; Liebowitz, D.; Lin, K.; Kasparek, K.; Pasetti, M.F.; Garg, S.J.; Gottlieb, K.; Trager, G.; Tucker, S.N. Safety and immunogenicity of an oral tablet norovirus vaccine, a phase I randomized, placebo-controlled trial. *JCI Insight* **2018**, *3*, 1–12. [CrossRef] [PubMed]
44. Leroux-Roels, G.; Cramer, J.P.; Mendelman, P.M.; Sherwood, J.; Clemens, R.; Aerssens, A.; De Coster, I.; Borkowski, A.; Baehner, F.; Van Damme, P. Safety and Immunogenicity of Different Formulations of Norovirus Vaccine Candidate in Healthy Adults: A Randomized, Controlled, Double-Blind Clinical Trial. *J. Infect. Dis.* **2018**, *217*, 597–607. [CrossRef]
45. Malm, M.; Tamminen, K.; Vesikari, T.; Blazevic, V. Comparison of Intramuscular, Intranasal and Combined Administration of Norovirus Virus-Like Particle Subunit Vaccine Candidate for Induction of Protective Immune Responses in Mice. *J. Clin. Cell. Immunol.* **2015**, *6*, 1–7. [CrossRef]
46. Malm, M.; Hyoty, H.; Knip, M.; Vesikari, T.; Blazevic, V. Development of T cell immunity to norovirus and rotavirus in children under five years of age. *Sci. Rep.* **2019**, *9*, 3199. [CrossRef]
47. Tamminen, K.; Huhti, L.; Vesikari, T.; Blazevic, V. Pre-existing immunity to norovirus GII-4 virus-like particles does not impair de novo immune responses to norovirus GII-12 genotype. *Viral Immunol.* **2013**, *26*, 167–170. [CrossRef]
48. Kim, J.H.; Skountzou, I.; Compans, R.; Jacob, J. Original antigenic sin responses to influenza viruses. *J. Immunol.* **2009**, *183*, 3294–3301. [CrossRef]
49. Lessler, J.; Riley, S.; Read, J.M.; Wang, S.; Zhu, H.; Smith, G.J.; Guan, Y.; Jiang, C.Q.; Cummings, D.A. Evidence for antigenic seniority in influenza A (H3N2) antibody responses in southern China. *PLoS Pathog.* **2012**, *8*, e1002802. [CrossRef]
50. Smith, D.J.; Forrest, S.; Ackley, D.H.; Perelson, A.S. Variable efficacy of repeated annual influenza vaccination. *Proc. Natl. Acad. Sci. USA* **1999**, *96*, 14001–14006. [CrossRef]
51. Skowronski, D.M.; Chambers, C.; De Serres, G.; Sabaiduc, S.; Winter, A.L.; Dickinson, J.A.; Gubbay, J.B.; Fonseca, K.; Drews, S.J.; Charest, H.; et al. Serial Vaccination and the Antigenic Distance Hypothesis: Effects on Influenza Vaccine Effectiveness During A(H3N2) Epidemics in Canada, 2010–2011 to 2014–2015. *J. Infect. Dis.* **2017**, *215*, 1059–1099. [CrossRef] [PubMed]
52. Malm, M.; Tamminen, K.; Vesikari, T.; Blazevic, V. Type-specific and cross-reactive antibodies and T cell responses in norovirus VLP immunized mice are targeted both to conserved and variable domains of capsid VP1 protein. *Mol. Immunol.* **2016**, *78*, 27–37. [CrossRef] [PubMed]
53. Siebenga, J.J.; Vennema, H.; Zheng, D.P.; Vinje, J.; Lee, B.E.; Pang, X.L.; Ho, E.C.; Lim, W.; Choudekar, A.; Broor, S.; et al. Norovirus illness is a global problem: Emergence and spread of norovirus GII.4 variants, 2001-2007. *J. Infect. Dis.* **2009**, *200*, 802–812. [CrossRef] [PubMed]

Article

Medical Outcomes in Women Who Became Pregnant after Vaccination with a Virus-Like Particle Experimental Vaccine against Influenza A (H1N1) 2009 Virus Tested during 2009 Pandemic Outbreak

Arturo Cérbulo-Vázquez [1], Lourdes Arriaga-Pizano [2], Gabriela Cruz-Cureño [3], Ilka Boscó-Gárate [2], Eduardo Ferat-Osorio [4], Rodolfo Pastelin-Palacios [5], Ricardo Figueroa-Damian [6], Denisse Castro-Eguiluz [7], Javier Mancilla-Ramirez [8], Armando Isibasi [2] and Constantino López-Macías [2,9,10,*]

1   Facultad de Medicina, Plan de Estudios Combinados en Medicina (MD, PhD Program), Universidad Nacional Autónoma de México, Mexico City CP 04510, Mexico; cerbulo@unam.mx
2   Unidad de Investigación Médica en Inmunoquímica, Hospital de Especialidades del Centro Médico Nacional Siglo XXI, Instituto Mexicano del Seguro Social (IMSS), Mexico City CP 06720, Mexico; landapi@hotmail.com (L.A.-P.); ibosco45@hotmail.com (I.B.-G.); isibasi@prodigy.net.mx (A.I.)
3   Escuela Nacional de Ciencias Biológicas, Programa de Inmunología, Instituto Politécnico Nacional, Mexico City CP 11340, Mexico; gabrielacruz30@gmail.com
4   Servicio de Cirugía Gastrointestinal, Unidad Médica de Alta Especialidad, Hospital de Especialidades Dr Bernardo Sepúlveda Gutiérrez, Centro Médico Nacional Siglo XXI, Instituto Mexicano del Seguro Social (IMSS), Mexico City CP 06720, Mexico; eduardoferat@prodigy.net.mx
5   Departamento de Biología, Facultad de Química, Universidad Nacional Autónoma de México, Mexico City CP 04510, Mexico; rodolfop@unam.mx
6   Departamento de Infectología, Instituto Nacional de Perinatología, Mexico City CP 11000, Mexico; rfd6102@yahoo.com.mx
7   Consejo Nacional de Ciencia y Tecnología (CONACYT)- Departamento de Investigación Clínica, Instituto Nacional de Cancerología, Mexico City CP 14080, Mexico; angeldenisse@gmail.com
8   Escuela Superior de Medicina, Instituto Politécnico Nacional; Hospital de la Mujer, Secretaria de Sauld, Mexico City CP 11340, Mexico; javiermancilla@hotmail.com
9   Visiting Professor of Immunology, Nuffield Department of Medicine, University of Oxford, Oxford OX3 7LF, UK
10 Mexican Translational Immunology Research Group, Federation of Clinical Immunology Societies Centers of Excellence, National Autonomous University of Mexico, Mexico City 04510, Mexico
*   Correspondence: constantino@sminmunologia.mx or constantino.lopez@imss.gob.mx

Received: 20 July 2019; Accepted: 15 September 2019; Published: 17 September 2019

Abstract: The clinical effects and immunological response to the influenza vaccine in women who later become pregnant remain to be thoroughly studied. Here, we report the medical outcomes of 40 women volunteers who became pregnant after vaccination with an experimental virus-like particle (VLP) vaccine against pandemic influenza A(H1N1)2009 (influenza A(H1N1)pdm09) and their infants. When included in the VLP vaccine trial, none of the women were pregnant and were randomly assigned to one of the following groups: (1) placebo, (2) 15 µg dose of VLP vaccine, or (3) 45 µg dose of VLP vaccine. These 40 women reported becoming pregnant during the follow-up phase after receiving the placebo or VLP vaccine. Women were monitored throughout pregnancy and their infants were monitored until one year after birth. Antibody titers against VLP were measured in the mothers and infants at delivery and at six months and one year after birth. The incidence of preeclampsia, fetal death, preterm delivery, and premature rupture of membranes was similar among groups. All vaccinated women and their infants elicited antibody titers ($\geq$1:40). Women vaccinated prior to pregnancy had no adverse events that were different from the nonvaccinated population. Even though this study is limited by the sample size, the results suggest that the anti-influenza

A(H1N1)pdm09 VLP experimental vaccine applied before pregnancy is safe for both mothers and their infants.

Keywords: virus-like particle; influenza A(H1N1)pdm09; vaccination; pregnant women; antibody titers

1. Introduction

Vaccination against pandemic influenza A is crucial for limiting the incidence and severity of infection in the general population [1,2]. Influenza A infection in pregnant women has been associated with higher admission rates to intensive care units and severe disease compared with nonpregnant women [3]. Studies in animal models show that the seasonal influenza virus infection in pregnant mice initiates a powerful pro-abortive mechanism and adverse outcomes in fetal health [4], highlighting the importance of vaccination in protecting both pregnant women and their infants to whom immunity may also be passively transferred [5,6].

In 2009, our group conducted a phase 2 clinical trial following Good Clinical Practices according to the International Conference of Harmonization (ICH-GCP) to evaluate the safety and immunogenicity of a virus-like particle (VLP) vaccine against influenza A(H1N1)pdm09. That study was a randomized, placebo-controlled, double-blind clinical trial that included 4555 male and female volunteers. Healthy nonpregnant women (confirmed by a negative pregnancy test) were included in the study [7,8], and by written informed consent, women were requested to avoid pregnancy for the duration of the trial. The use of a contraceptive method was instructed and provided to female participants upon request; however, some of the volunteers became pregnant in the postvaccination follow-up period. As indicated by our institutional Ethics Committee recommendations and local regulations, the pregnant women and their infants remained under the intention to treat protocol, so they were closely monitored. The medical outcomes and antibody response to the vaccine in both the women and their infants were registered. Here, we report, in a nested cohort study, the results of the surveillance of the 40 pregnant women who agreed to take part and provided their informed consent.

2. Materials and Methods

While conducting an ICH-GCP phase 2 clinical trial using a VLP vaccine against pandemic influenza A at the Mexican Institute of Social Security (IMSS, Mexico City, Mexico), 4555 male and nonpregnant female (confirmed by a negative pregnancy test) volunteers were included. Despite the use of contraceptive methods—as stated by the participants—40 women reported pregnancy in the postvaccination period (Table 1). We reported these events to the IMSS Ethics Committee and Comisión Federal para la Prevención de Riesgos Sanitarios (COFEPRIS) (the Mexican equivalent of the US Food and Drug Administration), and they both recommended that the pregnant women and their infants should remain under medical surveillance and be closely monitored according to local regulations.

The design of this study was a nested cohort study that included the 40 women who became pregnant after the influenza A(H1N1)pdm09 virus vaccination and their infants (Figure 1).

Table 1. Characteristics of the study population (women).

	Placebo (*n* = 16)	VLP 15 µg (*n* = 23)	*p*	VLP 45 µg (*n* = 1)
Age (mean ± SD)	24.25 ± 2.43	27.44 ± 5.79	0.14 *	19
Personal history of diseases				
Denied, *n* (%)	16 (100%)	23 (100%)	1 **	1 (100%)

Table 1. *Cont.*

	Placebo (*n* = 16)	VLP 15 µg (*n* = 23)	*p*	VLP 45 µg (*n* = 1)
Family history				
Type 2 diabetes mellitus, *n* (%)	1 (6.25%)	4 (17.4%)		1 (100%)
Hypertension, *n* (%)				1 (100%)
Cancer, *n* (%)		1 (4.34%)		
Cardiovascular disease, *n* (%)				1 (100%)
Renal insufficiency, *n* (%)				1 (100%)
Hypothyroidism, *n* (%)				1 (100%)
Epilepsy, *n* (%)		1 (4.34%)		
Time elapsed from vaccination to pregnancy				
1 month, *n* (%)		6 (26%)		1 (100%)
2 months, *n* (%)		3 (13%)		
3 months, *n* (%)		3 (13%)		
4 months, *n* (%)		2 (8.6%)		
5 months, *n* (%)		2 (8.6%)		
6 months, *n* (%)		3 (13%)		
7–9 months, *n* (%)		2 (8.6%)		
>9 months, *n* (%)		2 (8.6%)		
Breastfed the child				
Yes, *n* (%)	4 (25%)	14 (60.8%)	0.049 **	1 (100%)
No, *n* (%)	12 (75%)	9 (39.2%)		

* Student's *t* test comparing placebo vs. VLP 15 µg; ** Fisher's exact test comparing placebo vs. VLP 15 µg. VLP—virus-like particle.

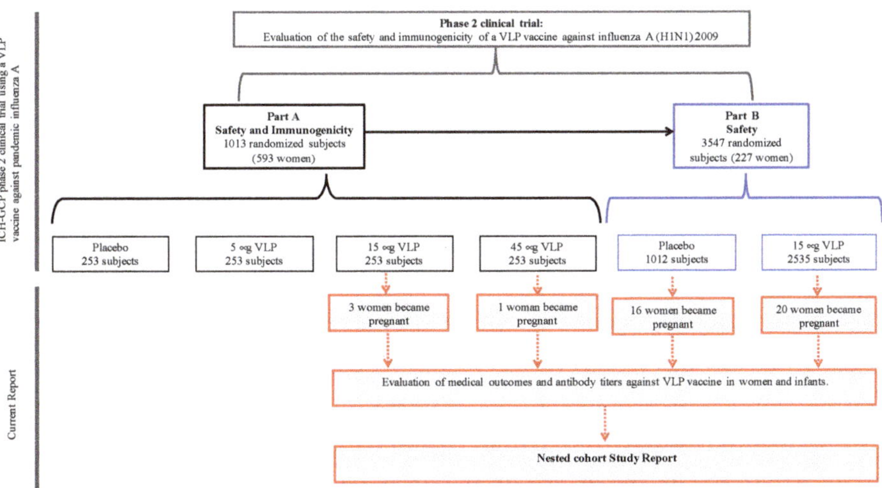

Figure 1. Flow and details of the subjects in the trial. A total of 820 women volunteers participated in the phase 2 clinical trial to evaluate the safety and immunogenicity (part A) and safety (part B) of the VLP vaccine against influenza A(H1N1)pdm09 [6]. After vaccination, 40 women became pregnant—16 from the placebo group, 23 from the 15 µg VLP vaccine dose, and 1 from the 45 µg VLP vaccine dose. All these volunteers were provided with medical surveillance and close monitoring; clinical outcomes and VLP vaccine specific antibody titers in both mothers and their infants were evaluated. Both the Mexican Institute of Social Security and the National Institute of Perinatology Ethic Committees approved the study (IMSS:R-2011-785-040, INPer:212250-06181).

2.1. Participant Characteristics

The 40 women who became pregnant after vaccination were recruited from the VLP vaccine clinical trial groups—16 (40%) pregnant women from the placebo group, 23 (57.5%) from the 15 µg dose of VLP vaccine group, and 1 (2.5%) woman from the 45 µg dose of VLP vaccine group (Figure 1). None of the women had documented an infection with pandemic influenza A (H1N1) 2009 or a vaccination against seasonal or pandemic influenza A (other than the experimental vaccine), and none of them reported a medical history of chronic diseases.

The pregnant women were monitored medically until delivery, following the standard protocol for medical care in Mexico [9]. Both mothers and infants remained under medical surveillance and safety follow-up at 3, 6, and 12 months after delivery. Any adverse medical or perinatal event experienced by the mothers or infants was recorded in detail. The newborn surveillance included anthropometry, gestational age at birth, nutritional status, and congenital disease.

The IMSS and the National Institute of Perinatology Ethic Committees approved the study (IMSS:R-2011-785-040, INPer:212250-06181). All participants signed a written informed consent for the study.

2.2. Sample Collection

Whole blood samples (5 mL) from the pregnant women or umbilical cord blood (10 mL) were collected at birth. At 3, 6, and 12 months of age, 5 mL of peripheral blood was collected from the mothers and 1 mL from the infants. Serum was obtained from the blood samples by centrifugation, and the aliquots were stored at −20 °C until analysis.

2.3. Hemagglutination Inhibition (HAI) Test

The serum samples were tested in triplicate. Assays were conducted as previously described [10,11]. The aliquots of each serum sample were treated using the receptor-destroying enzyme. The samples were diluted (1:2) into V-bottom 96-well microtiter plates. Eight units of 50 µL of hemagglutinin (HA) were added to each well, plated, and incubated for 30 min at room temperature, followed by the addition of in-house, freshly prepared turkey erythrocytes (1% in phosphate-buffered saline). The plates were mixed by agitation, covered, and allowed to set for 30 min at 25 °C. The HAI titer was determined by the reciprocal of the last dilution of nonagglutinated red blood cells. Samples with HAI titers of ≥1:40 were considered to have a positive antibody response to the influenza A(H1N1)pdm09 virus.

2.4. Statistical Analysis

Descriptive statistics were used for the characterization of the patients studied. We compared the placebo and VLP 15 µg dose. The statistical analysis was performed using StataCorp software (2011 Stata Statistical Software—Release 12, College Station, TX: StataCorp LP). Descriptive statistics were used to characterize the categorical variables, expressed as a percentage. A bivariate analysis was performed for comparison among groups, using Fisher's exact test or chi-square test; $p < 0.05$ was considered statistically significant.

3. Results

3.1. Medical Outcomes in Women Who Became Pregnant after Vaccination with VLP Vaccine against Influenza A(H1N1)pdm09 Virus and Their Infants

Table 1 describes the characteristics of the women studied. All volunteers included in the study were residents of the metropolitan area of Mexico City and were rightful claimants of full cover medical attention provided by the public health system; all were homemakers, literate, and none reported a medical history of chronic diseases (Table 1). The mean age was 26.1 ± 5.4 years (see Table 1 for the differences in characteristics among the placebo and 15 µg VLP groups). The mean time to become pregnant after vaccination was 130 ± 115 days; the details on the elapsed time from vaccination to pregnancy are described in Table 1.

The gestational age of the newborns and the frequency of the delivery type were similar between the placebo and 15 µg VLP vaccine groups (Table 2). A similar average birth weight of newborns was observed between the placebo group (2878 ± 554 g) and the 15 µg VLP vaccine group (3081 ± 398 g; $p > 0.05$). The woman from the 45 µg VLP vaccine group delivered via cesarean section at 38 weeks of gestational age, the newborn had a birth weight of 3000 g and an Apgar score typical for a healthy newborn (8/9, 1 min and 5 min).

Table 2. Gestational age and delivery type in the placebo and 15 mg VLP vaccine groups.

Condition	Placebo, *n* (%)	VLP 15 µg, *n* (%)	*p* *
	16 (40)	23 (57.5)	
Gestational age			
Preterm	1 (5.5)	2 (8.7)	0.2
Term	15 (83.3)	20 (86.9)	0.8
Post-term	0	1 (4.3)	0.4
Delivery type			
Vaginal	8 (50)	11 (47.8)	0.9
Cesarean	8 (50)	12 (52.2)	0.9

* Fisher's test.

None of the pregnant women from the placebo group were diagnosed with preeclampsia, fetal death, premature rupture of membranes, oligohydramnios, or gestational hypertension (Table 3). Two of the women from the VLP 15 µg group received a second immunization (28 days after the first dose); no obstetric complications or adverse events developed in these two women. Six (26%) of the women in the 15 µg VLP group experienced adverse events. These events were preeclampsia ($n = 2$), fetal death ($n = 1$), premature rupture of membranes ($n = 1$), oligohydramnios ($n = 1$), and gestational hypertension ($n = 1$). Table 3 describes the obstetric complications for both the placebo and 15 µg VLP groups; the epidemiological reference of obstetric adverse events reported for the Mexican population is also provided [12]. In all cases, the frequency of the above-mentioned events was not significantly different among the placebo and 15 µg VLP groups (Table 3). The patient who received the 45 µg VLP vaccine did not have any pathology during her pregnancy. In the placebo group, one patient had a preterm birth delivered by cesarean section at 36 weeks, whereas 2 out of 23 (8.7%) women in the 15 µg VLP vaccine group had a cesarean preterm birth at 35 and 36 weeks of gestational age. The woman in the 45 µg VLP vaccine group delivered at term gestational age.

Table 3. Obstetric complications in placebo and vaccinated women compared with the Mexican population.

Obstetric Complication	Placebo, *n* (%)	15 µg VLP, *n* (%)	*p* **	Reference Value *
	16 (40)	23 (57.5)		%
Preeclampsia	0	2 (8.6)	0.6	8
Fetal death	0	1 (4.3)	0.7	1.5
Premature rupture of membranes	0	1 (4.3)	0.7	5
Oligohydramnios	0	1 (4.3)	0.7	1.5
Gestational hypertension	0	1 (4.3)	0.7	10

* COMEGO: Clinic Gynecology and Obstetrics Guides in Mexico 2015, Mexico, Nieto Eds.; ** Fisher's test comparing the placebo and the 15 µg VLP groups.

We evaluated 32 infants during the first year of life. Seven abandoned the study during follow-up, and in one case, consent was withdrawn. One infant (6.2%) from the placebo group was diagnosed with pneumonia during the surveillance period. In the 15 µg VLP group, one infant (4.3%) was diagnosed with a hydrocele, one infant (4.3%) was diagnosed with an umbilical hernia, and one infant (4.3%) was diagnosed with gastroesophageal reflux and bacterial pneumonia during medical surveillance (Table 4). Low birth weight and fetal distress were not statistically different between the placebo and 15 µg VLP vaccine groups (Table 4). The infant exposed to the 45 µg VLP dose did not present any pathology.

Table 4. Neonatal and one-year surveillance adverse events in infants of women in the vaccinated and placebo groups.

Condition	Placebo, n (%)	15 µg VLP, n (%)	p *
	16 (40)	23 (57.5)	
Neonatal diagnosis			
Low birth weight (<2500 g)	0	3 (13.0)	0.51
Fetal distress	0	2 (8.7)	0.6
One-year surveillance period			
Pneumonia	1 (6.25)	1 (4.3)	0.7
Hydrocele	0	1 (4.3)	0.1
Umbilical hernia	0	1 (4.3)	0.1
Gastroesophageal reflux	0	1 (4.3)	0.1

* Fisher's test.

3.2. Vaccination Elicited Humoral Response in Pregnant Women and Infants

Figure 2 shows the titer of antibodies against pandemic influenza in the peripheral blood of women who were vaccinated before becoming pregnant. Women in the placebo group showed low levels of antibodies against pandemic influenza at the beginning of the trial; however, these levels increased from delivery up to one year after delivery, and the titer reached high levels (>1:40) at 12 months after delivery. The titer was less than 1:40 in the placebo infants, reaching the highest levels at six months after birth. In the 15 µg VLP vaccine group, the antibody titer in women increased as early as 20 days after vaccination and reached high levels at 36 days after the vaccination, whereas at delivery, the titer was higher than 1:80 and then decreased to 1:40 at 12 months after delivery. The titer in the infants from the 15 µg VLP group was 1:40 at the moment of birth and remained elevated at six months of age; however, the titer fell below 1:40 at 12 months after birth (Figure 2). In the woman from the 45 µg VLP vaccine group, the initial titer was 1:160 and remained protective from the moment of vaccination until one year after delivery; the infant from this group showed a similar titer as the mother at birth, but it diminished at six months of age and remained below 1:40 at 12 months after birth (Figure 2).

To understand the elevated antibody titers observed in women and infants from the placebo group, we determined the antibody titers against pandemic influenza in 249 pregnant women who did not participate in the vaccination trial (Figure 3). About 30% of Mexican women who became pregnant in the period from July 2009 to August 2010 expressed natural immunity and antibody titers of 1:40 against pandemic influenza, which is similar to the antibody titers described for the placebo group.

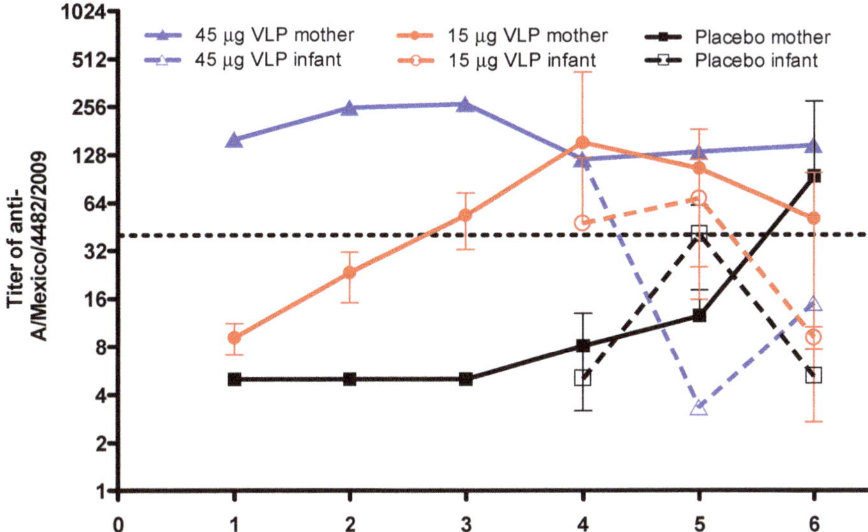

Figure 2. VLP vaccination induces humoral response against the anti-A/Mexico/4482/2009 virus in mothers and infants. Antibody titers in women volunteers vaccinated with 15 µg VLP (*n* = 23), 45 µg VLP (*n* = 1), or placebo (*n* = 16). D0—day of vaccination (first dose); D22—day 22 after vaccination (also, two women in the 15 µg group received a second dose); D36—day 36 after vaccination; birth—day of delivery of the newborns; 6 months—6 months after birth; and 12 months—12 months after birth. Closed symbols and continuous lines show antibody titers in women. Open symbols and dashed lines show antibody titers in infants. Mean ± SD.

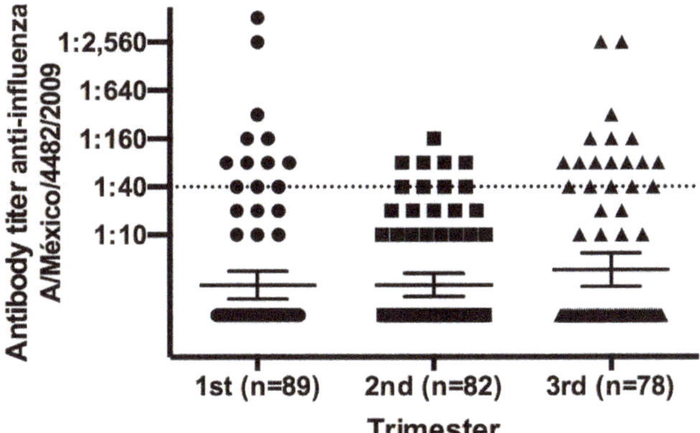

Figure 3. Protective antibody titer against pandemic influenza in healthy Mexican pregnant women. We collected 249 peripheral blood samples from healthy pregnant women who did not participate in the vaccination clinical trial—12 out of 89 women (13%) in the first trimester, 9 out of 82 (12%) in the second trimester, and 18 out of 78 (30%) in the third trimester had an antibody titer ≥1:40. Mean ± SD.

4. Discussion

Here we have reported the results of a nested cohort study that evaluated the potential effect of the anti-influenza A(H1N1)pdm09 VLP experimental vaccine in women who later became pregnant on medical outcomes as well as on the antibody response of the women and their infants [7]. Tavares et al. [13] reported similar results in their study of the effect of a pandemic vaccine against influenza A (H1N1) in pregnant women using an AS03-adjuvanted split virion H1N1 (2009) pandemic influenza vaccine. They did not observe an increase in the frequency of obstetric complications, infection, or teratogenicity. Tamma and Muñoz [14,15] reported no adverse effects in the infants of mothers vaccinated using an inactivated virus.

We monitored 40 women from early gestation to delivery and 32 infants during the first year of life. Although this study may lack statistical power due to the small sample size, our results showed that most women in this cohort reached term gestational age (37–42 weeks); this could suggest that the VLP vaccination does not result in a higher risk of preterm birth than occurs in the nonvaccinated population, as the average gestational age was similar to the Mexican national epidemiological data of term newborns [16].

We did not observe an increase in the rates of maternal obstetric complications or infection. The incidences of preeclampsia, fetal death, and premature rupture of membranes were not significantly higher in the VLP-vaccinated women compared to nonvaccinated women, suggesting that the VLP vaccine is safe in the pregnant population even under conditions of influenza outbreak. A prospective and larger study should be conducted to ascertain the safety of the VLP vaccine in the pregnant population.

The rates of congenital disease, fetal death, preterm delivery, and low birth weight in the neonates of VLP-vaccinated women were similar to that of the nonvaccinated population, suggesting that the VLP vaccine does not result in a higher risk to the fetus. In the VLP vaccine group, there were three cases of low birth weight and two cases of fetal distress, even though no statistical significance was reached, this finding may be clinically relevant and considered for further studies. Pediatric monitoring showed no negative effects on the exposed infants; cases of hydroceles, umbilical hernias, and gastroesophageal reflux are common in Mexican newborn infants, and it was not possible to establish a causal factor between the application of the influenza vaccine prior to pregnancy and the occurrence of these complications. However, the effect of the VLP vaccine applied in pregnant women must be studied to provide confidence in the safety of this type of vaccination.

Vaccination with an inactivated or attenuated virus for influenza A has some disadvantages, such as the risk of anaphylactic reactions and requirements for annual boosters or high doses to reach protection [17–20]. We observed specific antibody titers to A/Mexico/4482/2009 in pregnant women in the placebo group; these antibodies may be passively transferred to the fetus. As we observed, these antibody titers are maintained up to 12 months after birth. About 30% of Mexican women who became pregnant during the period of the pandemic showed elevated antibody titers; accordingly, our data suggest that some women in the placebo group may have been naturally infected by the influenza virus and developed an antibody response that could be transferred to the fetus. One dose of 15 µg VLP vaccine was shown to elicit a transitory but protective titer in the perinatal period, which was maintained 12 months after delivery. A similar titer was detected in infants during the newborn period up to six months after birth, suggesting that one dose of 15 µg VLP is sufficient to elicit an antibody response. Considering 60% of these infants were breastfed throughout the first year of life, it is possible that antibodies may have been passively transferred during this period; thus, the 15 µg VLP dose may be considered sufficient to support passive immunity in the fetus and infant. Finally, the high titer that we found in the woman who was exposed to one dose of 45 µg VLP vaccine suggests an asymptomatic infection with the pandemic influenza virus, and this condition may have interfered with the response to the 45 µg VLP vaccine. However, the titer remained elevated and the antibodies were passively transferred to the neonate at the moment of delivery but were not sustained at six months. Taken together, our data suggest that 15 µg may be an adequate dose of VLP vaccine to elicit an adequate antibody titer in both mother and infant. Another clinical trial is necessary to demonstrate if the 15 µg dose of the VLP vaccine could be considered protective against pandemic influenza in women who become pregnant.

Although vaccination was performed prior to pregnancy, our data indicate that this VLP vaccine does not induce negative side effects in either women or their infants during pregnancy. The vaccine induced a specific antibody response that was observed 21 months after immunization (the last time point observed) and was transferred to the newborns, showing that the VLP vaccine induced a long-lasting antibody response. Although more studies must be conducted to determine if VLP vaccination is safe during pregnancy, these data could be useful as a precedent to support safety and immunogenicity studies of this vaccine platform during pregnancy.

There were limitations in this study. Firstly, the sample size was limited in the placebo and intervention arms, so the statistical power was insufficient to measure the differences in pregnancy complications and birth outcomes. A second limitation was the collection of pregnancy-specific information, since this was not part of the initial clinical trial. Therefore, the information was heterogeneous and the results inconclusive. The most important confounding factor was the exposure to the natural influenza A(H1N1)pdm09 virus infection. Volunteers did not present clinical symptoms suggesting an influenza infection; however, since they were residents of the Mexico City metropolitan area during the pandemic outbreak, they were likely to have been exposed to the virus. This study was importantly limited by the lack of design—the clinical trial was not designed to include pregnant women. Avoiding pregnancy during the study was requested in the written informed consent. Here, we have reported the observations that we could perform on the volunteers who gained the added risk of becoming pregnant as well as observations on their infants. Nonetheless, this study represents a unique opportunity to understand the effect of an experimental anti-influenza VLP vaccine administered prior to pregnancy. Notably, the trial was performed during the particular conditions of a pandemic outbreak at the epicenter of the emergency. Although this study was limited by its sample size, these results could suggest that it is safe to be vaccinated with the anti-influenza A(H1N1)pdm09 VLP experimental vaccine before pregnancy.

Author Contributions: Conceptualization, C.L.-M.; data curation, G.C.-C., I.B.-G., and R.F.-D.; formal analysis, A.C.-V., L.A.A.-P., E.F.-O., R.P.-P., R.F.-D., and D.C.-E.; funding acquisition, C.L.-M. and A.C.-V.; methodology, L.A.A.-P., G.C.-C., and I.B.-G.; project administration, A.I. and C.L.-M.; resources, R.P.-P., J.M.-R., C.L.-M., and A.I.; supervision, A.C.-V., E.F.-O., and J.M.-R.; validation, L.A.A.-P. and A.I.; writing—original draft, A.C.-V. and D.C.-E.; writing—review and editing, C.L.-M.

Funding: This research was funded by the Instituto Mexicano del Seguro Social (grant numbers FIS/IMSS/PROT/PRIO/11/013 and FIS/IMSS/PROT/G12/1152 awarded to Constantino López-Macias) and CONACYT (grant number SALUD-2009-C02-127102 awarded to Arturo Cérbulo-Vázquez).

Acknowledgments: The authors thank the neonatologists and pediatricians in the labor room of the INPer who helped us to obtain cord blood samples. We also appreciate the technical assistance provided by R. Montecillo-Sandoval.

References

1. Milne, G.J.; Halder, N.; Kelso, J.K. The cost effectiveness of pandemic influenza interventions: A pandemic severity based analysis. *PLoS ONE* **2013**, *8*, e61504. [CrossRef] [PubMed]

2. Kelso, J.K.; Halder, N.; Milne, G.J. Vaccination strategies for future influenza pandemics: A severity-based cost effectiveness analysis. *BMC Infect Dis.* **2013**, *13*, 81. [CrossRef] [PubMed]

3. Creanga, A.A.; Johnson, T.F.; Graitcer, S.B.; Hartman, L.K.; Al-Samarrai, T.; Schwarz, A.G.; Chu, S.Y.; Sackoff, J.E.; Jamieson, D.J.; Fine, A.D.; et al. Severity of 2009 pandemic influenza A (H1N1) virus infection in pregnant women. *Obstet. Gynecol.* **2010**, *115*, 717–726. [CrossRef] [PubMed]

4. Littauer, E.Q.; Esser, E.S.; Antao, O.Q.; Vassilieva, E.V.; Compans, R.W.; Skountzou, I. H1N1 influenza virus infection results in adverse pregnancy outcomes by disrupting tissue-specific hormonal regulation. *PLoS Pathog.* **2017**, *13*, e1006757. [CrossRef] [PubMed]

5. Linder, N.; Ohel, G. In utero vaccination. *Clin. Perinatol.* **1994**, *21*, 663–674. [CrossRef]

6. Zuccotti, G.; Pogliani, L.; Pariani, E.; Amendola, A.; Zanetti, A. Transplacental antibody transfer following maternal immunization with a pandemic 2009 influenza A(H1N1) MF59-adjuvanted vaccine. *JAMA* **2010**, *304*, 2360–2361. [CrossRef] [PubMed]

7. Lopez-Macias, C.; Ferat-Osorio, E.; Tenorio-Calvo, A.; Isibasi, A.; Talavera, J.; Arteaga-Ruiz, O.; Arriaga-Pizano, L.; Hickman, S.P.; Allende, M.; Lenhard, K.; et al. Safety and immunogenicity of a virus-like particle pandemic influenza A (H1N1) 2009 vaccine in a blinded, randomized, placebo-controlled trial of adults in Mexico. *Vaccine* **2011**, *29*, 7826–7834. [CrossRef] [PubMed]
8. Lopez-Macias, C. Virus-like particle (VLP)-based vaccines for pandemic influenza: performance of a VLP vaccine during the 2009 influenza pandemic. *Hum. Vaccin Immunother.* **2012**, *8*, 411–414. [CrossRef] [PubMed]
9. Mexicanos, S.d.S.E.U. Atención de la mujer durante el embarazo, parto y puerperio y del recién nacido. Criterios y procedimientos para la prestación del servicio. In *NOM-007-SSA2-1993*; Salud, S.d., Ed.; Comité Consultivo Nacional de Normalización de Servicios de Salud. Available online: http://www.salud.gob.mx/unidades/cdi/nom/007ssa23.html (accessed on 6 January 1995).
10. Pushko, P.; Tumpey, T.M.; Bu, F.; Knell, J.; Robinson, R.; Smith, G. Influenza virus-like particles comprised of the HA, NA, and M1 proteins of H9N2 influenza virus induce protective immune responses in BALB/c mice. *Vaccine* **2005**, *23*, 5751–5759. [CrossRef] [PubMed]
11. Yan, D.; Wei, Y.Q.; Guo, H.C.; Sun, S.Q. The application of virus-like particles as vaccines and biological vehicles. *Appl. Microbiol. Biotechnol.* **2015**, *99*, 10415–10432. [CrossRef] [PubMed]
12. COMEGO. Clinic Gynecology and Obstetrics Guides in Mexico 2015. Available online: http://tv.comego.org.mx/session.php?pag=session05.php&BUSCAR001=3 (accessed on 15 September 2019).
13. Tavares, F.; Nazareth, I.; Monegal, J.S.; Kolte, I.; Verstraeten, T.; Bauchau, V. Pregnancy and safety outcomes in women vaccinated with an AS03-adjuvanted split virion H1N1 (2009) pandemic influenza vaccine during pregnancy: a prospective cohort study. *Vaccine* **2011**, *29*, 6358–6365. [CrossRef] [PubMed]
14. Tamma, P.D.; Steinhoff, M.C.; Omer, S.B. Influenza infection and vaccination in pregnant women. *Expert Rev. Respir. Med.* **2010**, *4*, 321–328. [CrossRef] [PubMed]
15. Munoz, F.M.; Greisinger, A.J.; Wehmanen, O.A.; Mouzoon, M.E.; Hoyle, J.C.; Smith, F.A.; Glezen, W.P. Safety of influenza vaccination during pregnancy. *Am. J. Obstet. Gynecol.* **2005**, *192*, 1098–1106. [CrossRef] [PubMed]
16. Flores-Huerta S, M.-S.H. Birth weight of male and female infants born in hospitals affiliated with the Instituto Mexicano del Seguro Social. *Bol. Med. Hosp. Infect. Mex.* **2012**, *69*, 30–39.
17. McNeil, M.M.; Weintraub, E.S.; Duffy, J.; Sukumaran, L.; Jacobsen, S.J.; Klein, N.P.; Hambidge, S.J.; Lee, G.M.; Jackson, L.A.; Irving, S.A.; et al. Risk of anaphylaxis after vaccination in children and adults. *J. Allergy. Clin. Immunol.* **2016**, *137*, 868–878. [CrossRef] [PubMed]
18. McNeil, M.M.; DeStefano, F. Vaccine-associated hypersensitivity. *J. Allergy Clin. Immunol.* **2018**, *141*, 463–472. [CrossRef]
19. Houser, K.; Subbarao, K. Influenza vaccines: challenges and solutions. *Cell Host Microbe* **2015**, *17*, 295–300. [CrossRef]
20. Shinjoh, M.; Sugaya, N.; Yamaguchi, Y.; Iibuchi, N.; Kamimaki, I.; Goto, A.; Kobayashi, H.; Kobayashi, Y.; Shibata, M.; Tamaoka, S.; et al. Inactivated influenza vaccine effectiveness and an analysis of repeated vaccination for children during the 2016/17 season. *Vaccine* **2018**, *36*, 5510–5518. [CrossRef]

Review

Factors That Govern the Induction of Long-Lived Antibody Responses

Bryce Chackerian * and David S. Peabody

Department of Molecular Genetics and Microbiology, University of New Mexico School of Medicine, Albuquerque, NM 87106, USA; dpeabody@salud.unm.edu
* Correspondence: bchackerian@salud.unm.edu

Received: 9 December 2019; Accepted: 4 January 2020; Published: 7 January 2020

Abstract: The induction of long-lasting, high-titer antibody responses is critical to the efficacy of many vaccines. The ability to produce durable antibody responses is governed by the generation of the terminally differentiated antibody-secreting B cells known as long-lived plasma cells (LLPCs). Once induced, LLPCs likely persist for decades, providing long-term protection against infection. The factors that control the generation of this important class of B cells are beginning to emerge. In particular, antigens with highly dense, multivalent structures are especially effective. Here we describe some pathogens for which the induction of long-lived antibodies is particularly important, and discuss the basis for the extraordinary ability of multivalent antigens to drive differentiation of naïve B cells to LLPCs.

Keywords: long-lived plasma cells; antibodies; multivalency; virus-like particles

---

1. Introduction—B-Cell Responses to Vaccination

Most of the effective vaccines in use today work by inducing humoral immune responses. Activation of naïve B cells by vaccine antigens leads to a series of events which ultimately result in the production of memory B cells and differentiation into two classes of antibody secreting cells, short-lived plasmablasts and long-lived plasma cells (LLPCs). The induction of immunological memory is classically invoked to explain the effectiveness of many vaccines. Memory B cells express the membrane-bound B-cell receptor (BCR) and are primed to respond to antigen exposure. They emerge early from germinal centers and often their immunoglobulin molecules have already undergone isotype switching and some degree of affinity maturation, meaning that they can potentially produce more effective antibodies upon re-stimulation. Memory of vaccination can increase the magnitude of the antibody response and drive the more rapid production of class-switched, high affinity antibodies. Thus, for many pathogens, the existence of memory cells provides a kinetic upperhand that tips the balance from the pathogen to the immune system.

Nevertheless, there are a number of important pathogens for which vaccine-based induction of memory cells alone is insufficient to provide protective immunity. In these cases, protection depends on a pre-existing level of antibodies high enough to entirely prevent the initial infection event, because once one of these agents initiates an infection, it establishes a beachhead which is then partially or completely resistant to antibody effector activity. Consider three common human pathogens, human immunodeficiency virus (HIV), human papillomavirus (HPV), and *Plasmodium*, as examples of diseases in which an effective vaccine is dependent on the ability to induce high-titer and long-lasting antibodies: (i) Once transmitted, HIV quickly traffics to regional lymph nodes and other immune tissues where it can infect T cells and macrophages [1]. The HIV provirus integrates into the genomes of these infected cells, where it establishes a persistent virus reservoir. While anti-retroviral drugs can successfully control active HIV infection, this latent reservoir is refractory to both antiretroviral drugs and anti-HIV antibodies, allowing the virus to reemerge if drugs are discontinued. Thus, antibodies can potentially control infection, but cannot eliminate HIV, so prevention of initial infection is critical. (ii) Similarly, after HPV infects its target cells, basal epithelial cells, the virus remains intracellular and not accessible to antibodies. The HPV lifecycle is intimately tied to the differentiation of basal cells into terminally differentiated epithelial cells [2]. Only when these cells differentiate are HPV virions produced and released. However, HPV-associated cancers do not result from productive infection. HPV can also establish latency in infected cells, and, in some cases, integration or episomal maintenance of the HPV genome can lead to dysregulated expression of the viral oncogenes, E6 and E7 [3]. These HPV-infected basal epithelial cells are the origin of neoplastic lesions that can ultimately progress to invasive tumors. Thus, once HPV infection occurs, antibodies have no value in protecting against cancer [4]. Accordingly, vaccines that elicit HPV-neutralizing antibodies can protect against initial infection, but fail to induce the regression of previously established lesions [5]. (iii) Liver infection by the *Plasmodium* parasite, which causes malaria, occurs within a few hours of feeding by an infected *Anopheles* mosquito. Injected sporozoites rapidly migrate through the bloodstream to the liver where they infect hepatocytes [6]. Once inside hepatocytes, *Plasmodium* is resistant to the effects of anti-sporozoite antibodies, meaning that antibodies only have a brief window to exert their effects. Thus, for these (and many other) pathogens, vaccines can be effective only if they elicit neutralizing antibodies maintained at high levels over time. For these pathogens, vaccine-mediated induction of LLPCs is required.

2. The Molecular Events That Mediate B-Cell Activation and the Role of Multivalency

The activation of B cells and the subsequent downstream events that result in antibody production are consequences of the initial interaction between an antigen and the BCR. The signaling events initiated by this interaction, which are the subject of many excellent reviews [7–9], stimulate B cell proliferation and upregulate MHC Class II and the costimulatory molecules that permit subsequent interactions with T helper cells. B-cell activation is a quantitative phenomenon, in which the degree of activation is dependent on both the affinity of the BCR for its cognate antigen [10,11] and the valency of the antigen. The critical role of antigen valency in B-cell responses was first recognized by Renee and Howard Dintzis at the Johns Hopkins School of Medicine [12–14], who assessed the immunogenicity of a T-cell independent antigen consisting of a polymer (polyacrylamide) displaying a model hapten (dinitrophenol; DNP). The modular nature of this system allowed the dissection of antibody responses as a function of the valency and density of DNP display. They concluded that " ... the fundamental molecular event in the induction of the primary immune response is the linking together by a single antigen molecule of a critical number of separate hapten receptors into a molecularly connected entity", which they termed an immunon [12]. Numerous subsequent studies established this relationship between antigen valency and B-cell responsiveness, particularly in the context of T-cell-independent antibody responses [15,16].

The series of events that begins with the activation of naïve B cells by a T-cell-dependent antigen and ultimately result in differentiation to LLPCs are more complicated than what occurs with a T-cell independent antigen, yet antigen valency and density also play an important role in this process. Multivalent interactions promote BCR clustering and the formation of lipid rafts [17–20]. These, in turn, promote signaling to the B cell and receptor-mediated internalization of the antigen complex [21], steps critical for B cells to present antigen on MHC Class II and receive help from CD4 T cells. Accordingly, multivalency enhances BCR clustering [22], BCR/antigen internalization and antigen presentation [23], as well as the upregulation of costimulatory molecules that are important for subsequent interactions with T helper cells [24]. Thus, these multivalent interactions have a profound effect on early steps in B-cell activation and ultimately influence antibody production and other downstream events.

While the influence of multivalency on the early steps in B-cell activation have been extensively studied, less is known about how these events influence the establishment of germinal centers (GCs) and production of LLPCs. GCs are discrete anatomical sites within B-cell follicles in which B cells proliferate and undergo somatic hypermutation and affinity maturation. It is here that they differentiate to memory cells and LLPCs. In GCs, B cells compete for binding with antigens displayed on follicular dendritic cells, and then present antigens to follicular T helper cells, which in turn provide survival signals to the B cell. Although this process does not require a multivalent antigen, enhanced B-cell crosslinking in GCs leads to increased GC B-cell proliferation and promotes differentiation to plasma cells, particularly in scenarios in which T help is limiting [25]. Thus, it is likely that multivalency can exert its stimulatory effects at multiple steps during the B-cell activation and differentiation process.

3. Immune Responses Elicited by Multivalent Vaccines

The basic immunological studies described above were initiated, in part, to explain the potent immunogenicity of multivalent antigens in a vaccine setting. Many different multivalent display strategies have been employed. They include synthetic nanoparticles [26,27], liposomes [28], micelles [29], and polymers [30]. But one of the most common is to display antigens on platforms based on virus-like particles (VLPs). Many viral proteins have an intrinsic ability, when overexpressed, to self-assemble into VLPs. VLPs lack viral genomes, meaning that they are absolutely non-infectious. VLPs can be derived from diverse viruses, including viruses that infect humans, animals, insects, plants, and bacteria (for recent reviews on this subject, see [31,32] and elsewhere in this Special Issue of *Viruses*). VLPs intrinsically display their own antigens in highly dense, repetitive arrays, and they can also be used as platforms for the high valency display of heterologous antigens. Their multivalency is not the only feature contributing to VLP immunogenicity. Most VLPs have a diameter (between 10 and 100–200 nm) that is optimal for uptake into the lymphatic system. Entry into the lymphatics promotes trafficking to regional lymph nodes, deposition in the subcapsular spaces of the lymph nodes that are key sites for B-cell activation, and interactions with antigen-presenting cells, such as dendritic cells (DCs) [33]. Some VLPs (and some engineered nanoparticles as well) also interact with soluble components of the innate immune system to facilitate this trafficking [34,35]. Moreover, VLPs fall into the size range (<500 nm in diameter) of particulate antigens most readily taken up by DCs [36]. Many VLPs, natural [37] or engineered [38], package adjuvants that can also enhance immunogenicity. Lastly, because they are protein-based, VLPs are naturally a rich source of T helper epitopes (or can be engineered to be so [39]), which are critical for the induction of T-cell-dependent antibody responses. Taken together, these immunological features underlie the use of VLPs as effective stand-alone vaccines (against the virus that they were derived from) and as the basis of vaccine platform technologies to display heterologous antigen targets.

The ability of multivalent display to enhance target antigen immunogenicity is well established. Figures 1 and 2 illustrate the potent immunogenicity conferred by VLP display. In Figure 1, we assessed the immunogenicity of a peptide representing a neo-epitope specific to a mutated version of the epidermal growth factor receptor (EGFRvIII) that is particularly prevalent in glioblastoma multiforme brain cancers and is associated with poor outcomes [40]. The EGFRvIII peptide was chemically conjugated either to Qß bacteriophage VLPs or to a commonly used carrier protein, keyhole limpet hemocyanin (KLH), and then used to immunize mice. After a single dose administered without exogenous adjuvant, we showed that the VLP-based immunogen elicited higher titer anti-peptide IgG antibody responses than the KLH-conjugated antigen, and that these antibody responses were elicited more rapidly (Figure 1), despite using a 5-fold lower dose of the VLP-conjugated material. Using VLPs, anti-peptide IgG was detected as early as one week after immunization. Moreover, remarkably stable levels of antibody are induced using VLP-based immunogens. In a separate study, shown in Figure 2, we longitudinally measured antibody responses in groups of mice immunized with recombinant MS2 bacteriophage VLPs, displaying a broadly neutralizing epitope from the HPV type 16 minor capsid protein (L2; the vaccine is described in [41]). Mice were given one or three doses of vaccine and then were followed for nearly two years after vaccination. Three doses elicited higher antibody titers than a single dose, but this difference became less pronounced over time. Notably, even a single dose gave remarkably stable antibody titers; levels were virtually unchanged for nearly two years after vaccination (essentially the life span of the mouse). This is likely due to the potent induction of LLPCs.

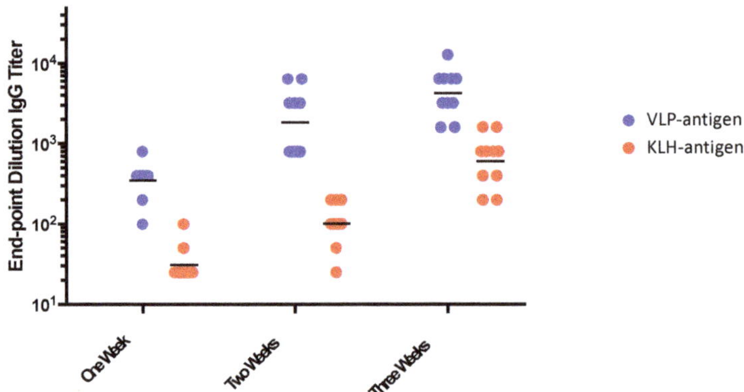

Figure 1. Mice were immunized with a single dose of an EGFRvIII peptide conjugated to Qß VLPs or the same peptide conjugated to keyhole limpet hemocyanin (KLH). Mice were immunized with 5 µg of VLP-EGFRvIII (displaying ~270 EGFvIII peptides per 2520kD VLP) or 25 µg of KLH-EGFRvIII (displaying ~50 EGFvIII peptides per 390 kD molecule of KLH) without exogenous adjuvant. Anti-peptide IgG antibody responses were measured by end-point dilution ELISA, one, two, or three weeks after the immunization. Each dot represents an individual mouse and lines represent the geometric mean of each group.

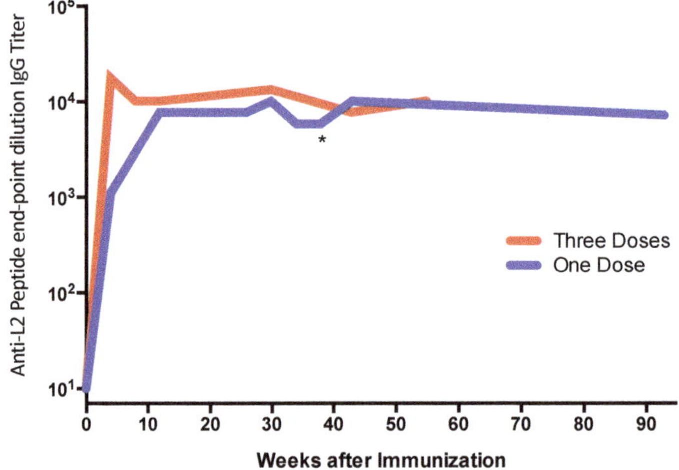

Figure 2. Mice were immunized with one (blue line; immunized at week zero) or three (red line; immunized at weeks zero, four, and eight) doses of 10 µg of recombinant MS2 bacteriophage VLPs displaying a peptide derived from the L2 minor capsid protein of HPV16 without exogenous adjuvant. Anti-peptide IgG antibody responses were measured by end-point dilution ELISA. The line represents the geometric mean of each group (*n* = 5/group). One of the mice in the "One Dose" group died at week 41 (indicated with an asterisk) for a reason that was unrelated to the vaccine. Antibody levels from the first six months after vaccination in the "One Dose" group were originally reported in Reference [42].

Human clinical data from trials of the VLP-based HPV vaccines Cervarix (GlaxoSmithKline) and Gardasil (Merck) also strongly support the concept that VLPs can efficiently induce LLPCs. A longitudinal study of antibody responses induced by these vaccines [43] revealed that both vaccines induce anti-HPV antibody responses that peak approximately one month after the third dose. After an initial period in which antibody levels decay rapidly (with a half-life of ~3.6 months, probably reflecting the loss of short-lived plasmablasts), these levels stabilize—there was no evidence of a further decline in antibody titers over a 4-year follow-up period. Even a single dose of HPV VLPs elicits exceedingly long-lived antibody responses. A subset of women enrolled in the phase III Costa Rica Vaccine Trial of Cervarix received just a single dose of the vaccine (this was due to various reasons, but most commonly these women discovered that they were pregnant after they received their first dose). Compared to the usual three-dose regimen, a single dose gave lower anti-HPV antibody levels, but the profile of antibody decay in the two groups was quite similar [44]. Antibody levels over the six-year period spanning one to seven years after vaccination were extraordinarily stable; these antibodies were essentially maintained at a constant level during this period. Moreover, no HPV16 or HPV18 infections were detected in this group of women during the observation period [45]. Based on these results, the US National Cancer Institute, in collaboration with a Costa Rican partner, is currently conducting a large randomized, controlled trial (the ESCUDDO study) to assess the immunogenicity and efficacy of a single-dose vaccine regimen.

The longevity of antibody responses elicited by the HPV vaccine parallels the humoral immune responses generated by other multivalent viral vaccines. In a seminal study, Amanna and colleagues longitudinally measured human antibody responses to a diverse panel of vaccine antigens, including viral (i.e., multivalent) vaccines and non-viral (toxoid; monovalent) vaccine antigens [46]. They found that the antibodies elicited by multivalent antigens were extremely durable—for example, antibodies against measles had a calculated half-life of ~3000 years—whereas the half-life of antibodies against the monomeric tetanus and diphtheria toxoid vaccines was much shorter (~11–19 years). Interestingly, there was no relationship between the frequency of vaccine-elicited memory B cells and antibody levels, suggesting that LLPCs were elicited upon the initial exposure to the antigen, and were not due to re-activation of memory cells. Taken together, these data strongly support the idea that multivalent vaccines can elicit long-lasting responses in humans (for a review of these and other human studies, see Reference [47]).

4. The Special Case of Using Multivalent Vaccines to Break B-Cell Tolerance

One of the most striking features of VLP display is its ability to elicit antibody responses against self-antigens. The immune system has erected a set of barriers that normally prevent the induction of autoantibody responses. During the early stages of B-cell development in the bone marrow, central B-cell tolerance mechanisms act to change the specificity (through a process referred to as receptor editing) or eliminate (through apoptosis) a percentage of potentially autoreactive B cells (shown schematically in Figure 3). However, these mechanisms are inefficient, and as a consequence a substantial percentage of naïve B cells that move into the periphery are potentially self-reactive [48,49]. Fortunately, when these self-reactive B cells are exposed to a soluble self-antigen in the periphery in the absence of T help, they undergo a number of changes that establish a state of unresponsiveness to subsequent antigen stimulation, known as anergy. Relative to non-anergic cells, anergic B cells are defined by decreased BCR surface expression, competitive exclusion from lymphoid follicles, and a short half-life [50,51]. Nevertheless, even these B cells are susceptible to activation by multivalent antigens. In a seminal study, Bachmann and colleagues showed that B cells from transgenic mice expressing a soluble form of the vesicular stomatitis virus glycoprotein (VSV-G) responded to vaccination with particulate, ordered forms of VSV-G (such as inactivated virions), but not to immunization with soluble monomeric VSV-G. This study suggested that multivalent antigens could effectively activate anergic B cells. Subsequent studies demonstrated that vaccines consisting of self-antigens arrayed on the surface of the VLPs could effectively induce anti-self antibody responses [52,53], that this ability is critically dependent on the density of the self-antigens displayed on VLPs [54,55], and that these interactions may be mediated by binding to surface-expressed IgD on naïve B cells [56]. These findings have led to the development of vaccine candidates that induce antibodies against self-antigens involved in chronic diseases, including angiotensin II (hypertension) [57], PCSK9 (cardiovascular disease) [58], amyloid-beta and hyper-phosphorylated tau (Alzheimer's Disease) [59,60], and others [61]. Several of these vaccines have been tested in human clinical trials [62].

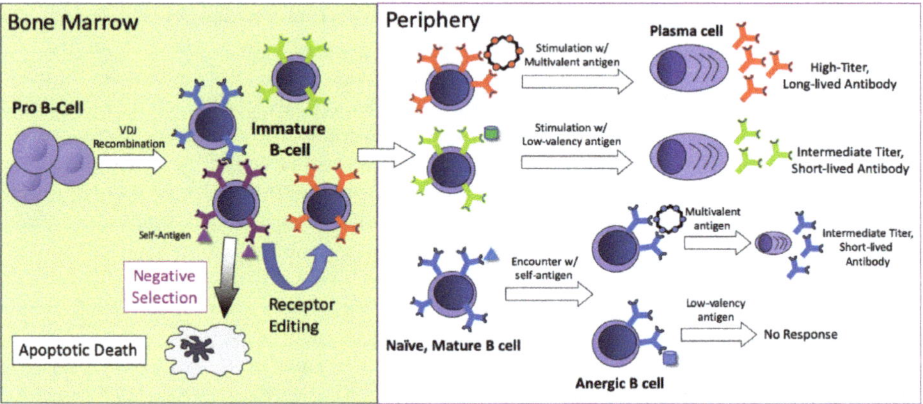

Figure 3. An overview of B-cell development, B-cell tolerance mechanisms, and the outcomes of stimulation with different forms of antigen (self-, low-valency, or multivalent antigen). B-cell responses are dependent on the nature of the antigen and the stage of B-cell development. In the bone marrow, immature B cells that encounter self-antigen either undergo apoptosis or receptor editing, which potentially alters receptor specificity. In the periphery, B cells that encounter multivalent antigens respond strongly, leading to the production of LLPCs that produce large amounts of antibody. Stimulation with low-valency antigens leads to lower titer and less durable antibody responses. If naïve B cells encounter self-antigen they become anergized. Anergic B cells do not respond to stimulation with low-valency antigen, but can be activated by multivalent antigens.

One interesting observation from human clinical trials of VLP-based vaccines targeting amyloid-beta [63] and angiotensin II [64] was that the half-life of the antibodies induced against self-antigens was fairly short, 15–20 weeks. This is in contrast to the durable antibody responses that are observed in vaccination studies using VLPs that target pathogen-derived (foreign) epitopes. Why do VLP-based vaccines that target foreign antigens strongly elicit LLPCs, but generate more transient antibody responses when targeting self-antigens? Studies comparing the reactivity of anergic and non-anergic transgenic B cells may provide an explanation. B cells respond to antigenic stimulation by upregulating a suite of molecules important in activating (i.e., CD86) and receiving help from (CD40) T helper cells. We showed that upregulation of these molecules upon stimulation by VLPs is attenuated in anergic transgenic B cells, relative to non-anergic cells [24]. It is possible that the downregulated expression of the BCR on anergic cells may account for these attenuated responses. Lower levels of BCR are likely to reduce the potential for extensive BCR crosslinking and, consequently, lead to weaker B-cell activation. Alternatively, anergic B cells may have other defects in the signaling pathways involved in B-cell activation. Regardless of the mechanism, these data indicate that reduced B-cell activation has profound downstream consequences. Weaker stimulation results in reduced numbers of LLPCs and less durable antibody responses.

5. Conclusions

VLPs have been exploited as a platform technology to develop vaccines for diverse targets, including specific antigens or epitopes derived from infectious disease targets, self-antigens involved in chronic diseases, and even small molecules, such as drugs of abuse [65]. In general, VLP display is regarded as a method for enhancing the magnitude of an antibody response, similar to how adjuvants can increase the immunogenicity of vaccines. However, as described in this article, and as summarized in Figure 3, the multivalent nature of VLPs also confers special immunostimulatory properties not typically conferred by adjuvants, particularly the ability to elicit durable antibody responses through the induction of LLPCs. For many targets, the induction of long-lived antibody responses is likely to be a key factor in vaccine efficacy. A deeper understanding of the mechanisms whereby multivalent antigens can lead to the induction and maintenance of LLPCs will influence vaccine design and likely lead to improved vaccines in the future.

Author Contributions: B.C., writing—original draft preparation; D.S.P., writing—review and editing; all authors have read and agree to the published version of the manuscript. All authors have read and agreed to the published version of the manuscript.

Funding: This research was funded by the US National Institutes of Health (NIH), grants number R01 HL131696 and U19 AI113187.

Acknowledgments: Thanks to Teal Clocksin, Ebenezer Tumban, and Julianne Peabody for generating the data shown in Figures 1 and 2.

Conflicts of Interest: BC and DSP have equity in Agilvax, Inc.

References

1. Cicala, C.; Nawaz, F.; Jelicic, K.; Arthos, J.; Fauci, A.S. HIV-1 gp120: A Target for Therapeutics and Vaccine Design. *Curr. Drug Targets* **2016**, *17*, 122–135. [CrossRef] [PubMed]

2. Graham, S.V. Keratinocyte Differentiation-Dependent Human Papillomavirus Gene Regulation. *Viruses* **2017**, *9*, 245. [CrossRef] [PubMed]

3. Mittal, S.; Banks, L. Molecular mechanisms underlying human papillomavirus E6 and E7 oncoprotein-induced cell transformation. *Mutat. Res. Rev. Mutat. Res.* **2017**, *772*, 23–35. [CrossRef]

4. Schiller, J.T.; Lowy, D.R. Understanding and learning from the success of prophylactic human papillomavirus vaccines. *Nat. Rev. Microbiol.* **2012**, *10*, 681–692. [CrossRef]

5. Hildesheim, A.; Herrero, R.; Wacholder, S.; Rodriguez, A.C.; Solomon, D.; Bratti, M.C.; Schiller, J.T.; Gonzalez, P.; Dubin, G.; Porras, C.; et al. Effect of human papillomavirus 16/18 L1 viruslike particle vaccine among young women with preexisting infection: A randomized trial. *JAMA* **2007**, *298*, 743–753. [CrossRef] [PubMed]

6. Sinnis, P.; Zavala, F. The skin: Where malaria infection and the host immune response begin. *Semin. Immunopathol.* **2012**, *34*, 787–792. [CrossRef] [PubMed]

7. Dal Porto, J.M.; Gauld, S.B.; Merrell, K.T.; Mills, D.; Pugh-Bernard, A.E.; Cambier, J. B cell antigen receptor signaling 101. *Mol. Immunol.* **2004**, *41*, 599–613. [CrossRef] [PubMed]

8. Kwak, K.; Akkaya, M.; Pierce, S.K. B cell signaling in context. *Nat. Immunol.* **2019**, *20*, 963–969. [CrossRef]

9. Tolar, P.; Sohn, H.W.; Liu, W.; Pierce, S.K. The molecular assembly and organization of signaling active B-cell receptor oligomers. *Immunol. Rev.* **2009**, *232*, 34–41. [CrossRef]

10. Kouskoff, V.; Famiglietti, S.; Lacaud, G.; Lang, P.; Rider, J.E.; Kay, B.K.; Cambier, J.C.; Nemazee, D. Antigens varying in affinity for the B cell receptor induce differential B lymphocyte responses. *J. Exp. Med.* **1998**, *188*, 1453–1464. [CrossRef]

11. Schwickert, T.A.; Victora, G.D.; Fooksman, D.R.; Kamphorst, A.O.; Mugnier, M.R.; Gitlin, A.D.; Dustin, M.L.; Nussenzweig, M.C. A dynamic T cell-limited checkpoint regulates affinity-dependent B cell entry into the germinal center. *J. Exp. Med.* **2011**, *208*, 1243–1252. [CrossRef] [PubMed]

12. Dintzis, H.M.; Dintzis, R.Z.; Vogelstein, B. Molecular determinants of immunogenicity: The immunon model of immune response. *Proc. Natl. Acad. Sci. USA* **1976**, *73*, 3671–3675. [CrossRef] [PubMed]

13. Dintzis, R.Z.; Middleton, M.H.; Dintzis, H.M. Studies on the immunogenicity and tolerogenicity of T-independent antigens. *J. Immunol.* **1983**, *131*, 2196–2203. [PubMed]

14. Dintzis, R.Z.; Vogelstein, B.; Dintzis, H.M. Specific cellular stimulation in the primary immune response: Experimental test of a quantized model. *Proc. Natl. Acad. Sci. USA* **1982**, *79*, 884–888. [CrossRef] [PubMed]

15. Mond, J.J.; Lees, A.; Snapper, C.M. T cell-independent antigens type 2. *Annu. Rev. Immunol.* **1995**, *13*, 655–692. [CrossRef]

16. Vos, Q.; Lees, A.; Wu, Z.Q.; Snapper, C.M.; Mond, J.J. B-cell activation by T-cell-independent type 2 antigens as an integral part of the humoral immune response to pathogenic microorganisms. *Immunol. Rev.* **2000**, *176*, 154–170.

17. Batista, F.D.; Neuberger, M.S. Affinity dependence of the B cell response to antigen: A threshold, a ceiling, and the importance of off-rate. *Immunity* **1998**, *8*, 751–759. [CrossRef]

18. Cheng, P.C.; Brown, B.K.; Song, W.; Pierce, S.K. Translocation of the B cell antigen receptor into lipid rafts reveals a novel step in signaling. *J. Immunol.* **2001**, *166*, 3693–3701. [CrossRef]

19. Mattila, P.K.; Feest, C.; Depoil, D.; Treanor, B.; Montaner, B.; Otipoby, K.L.; Carter, R.; Justement, L.B.; Bruckbauer, A.; Batista, F.D. The actin and tetraspanin networks organize receptor nanoclusters to regulate B cell receptor-mediated signaling. *Immunity* **2013**, *38*, 461–474. [CrossRef]

20. Pierce, S.K. Lipid rafts and B-cell activation. *Nat. Rev. Immunol.* **2002**, *2*, 96–105. [CrossRef]

21. Gazumyan, A.; Reichlin, A.; Nussenzweig, M.C. Ig beta tyrosine residues contribute to the control of B cell receptor signaling by regulating receptor internalization. *J. Exp. Med.* **2006**, *203*, 1785–1794. [CrossRef]

22. Puffer, E.B.; Pontrello, J.K.; Hollenbeck, J.J.; Kink, J.A.; Kiessling, L.L. Activating B cell signaling with defined multivalent ligands. *ACS Chem. Biol.* **2007**, *2*, 252–262. [CrossRef] [PubMed]

23. Bennett, N.R.; Zwick, D.B.; Courtney, A.H.; Kiessling, L.L. Multivalent Antigens for Promoting B and T Cell Activation. *ACS Chem. Biol.* **2015**, *10*, 1817–1824. [CrossRef] [PubMed]

24. Chackerian, B.; Durfee, M.R.; Schiller, J.T. Virus-like display of a neo-self antigen reverses B cell anergy in a B cell receptor transgenic mouse model. *J. Immunol.* **2008**, *180*, 5816–5825. [CrossRef] [PubMed]

25. Turner, J.S.; Ke, F.; Grigorova, I.L. B Cell Receptor Crosslinking Augments Germinal Center B Cell Selection when T Cell Help Is Limiting. *Cell Rep.* **2018**, *25*, 1395–1403. [CrossRef]

26. Brouwer, P.J.M.; Antanasijevic, A.; Berndsen, Z.; Yasmeen, A.; Fiala, B.; Bijl, T.P.L.; Bontjer, I.; Bale, J.B.; Sheffler, W.; Allen, J.D.; et al. Enhancing and shaping the immunogenicity of native-like HIV-1 envelope trimers with a two-component protein nanoparticle. *Nat. Commun.* **2019**, *10*, 4272. [CrossRef]

27. Havenar-Daughton, C.; Carnathan, D.G.; Boopathy, A.V.; Upadhyay, A.A.; Murrell, B.; Reiss, S.M.; Enemuo, C.A.; Gebru, E.H.; Choe, Y.; Dhadvai, P.; et al. Rapid Germinal Center and Antibody Responses in Non-human Primates after a Single Nanoparticle Vaccine Immunization. *Cell Rep.* **2019**, *29*, 1756–1766. [CrossRef]

28. Dubrovskaya, V.; Tran, K.; Ozorowski, G.; Guenaga, J.; Wilson, R.; Bale, S.; Cottrell, C.A.; Turner, H.L.; Seabright, G.; O'Dell, S.; et al. Vaccination with Glycan-Modified HIV NFL Envelope Trimer-Liposomes Elicits Broadly Neutralizing Antibodies to Multiple Sites of Vulnerability. *Immunity* **2019**, *51*, 915–929. [CrossRef]

29. Moon, J.J.; Suh, H.; Bershteyn, A.; Stephan, M.T.; Liu, H.; Huang, B.; Sohail, M.; Luo, S.; Um, S.H.; Khant, H.; et al. Interbilayer-crosslinked multilamellar vesicles as synthetic vaccines for potent humoral and cellular immune responses. *Nat. Mater.* **2011**, *10*, 243–251. [CrossRef]

30. Zhang, F.; Lu, Y.J.; Malley, R. Multiple antigen-presenting system (MAPS) to induce comprehensive B- and T-cell immunity. *Proc. Natl. Acad. Sci. USA* **2013**, *110*, 13564–13569. [CrossRef]

31. Balke, I.; Zeltins, A. Use of plant viruses and virus-like particles for the creation of novel vaccines. *Adv. Drug Deliv. Rev.* **2019**, *145*, 119–129. [CrossRef] [PubMed]

32. Donaldson, B.; Lateef, Z.; Walker, G.F.; Young, S.L.; Ward, V.K. Virus-like particle vaccines: Immunology and formulation for clinical translation. *Expert Rev. Vaccines* **2018**, *17*, 833–849. [CrossRef] [PubMed]

33. Manolova, V.; Flace, A.; Bauer, M.; Schwarz, K.; Saudan, P.; Bachmann, M.F. Nanoparticles target distinct dendritic cell populations according to their size. *Eur. J. Immunol.* **2008**, *38*, 1404–1413. [CrossRef] [PubMed]

34. Link, A.; Zabel, F.; Schnetzler, Y.; Titz, A.; Brombacher, F.; Bachmann, M.F. Innate immunity mediates follicular transport of particulate but not soluble protein antigen. *J. Immunol.* **2012**, *188*, 3724–3733. [CrossRef] [PubMed]

35. Tokatlian, T.; Read, B.J.; Jones, C.A.; Kulp, D.W.; Menis, S.; Chang, J.Y.H.; Steichen, J.M.; Kumari, S.; Allen, J.D.; Dane, E.L.; et al. Innate immune recognition of glycans targets HIV nanoparticle immunogens to germinal centers. *Science* **2019**, *363*, 649–654. [CrossRef]

36. Foged, C.; Brodin, B.; Frokjaer, S.; Sundblad, A. Particle size and surface charge affect particle uptake by human dendritic cells in an in vitro model. *Int. J. Pharm.* **2005**, *298*, 315–322. [CrossRef]

37. Tumban, E.; Peabody, J.; Peabody, D.S.; Chackerian, B. A universal virus-like particle-based vaccine for human papillomavirus: Longevity of protection and role of endogenous and exogenous adjuvants. *Vaccine* **2013**, *31*, 4647–4654. [CrossRef]

38. Senti, G.; Johansen, P.; Haug, S.; Bull, C.; Gottschaller, C.; Muller, P.; Pfister, T.; Maurer, P.; Bachmann, M.F.; Graf, N.; et al. Use of A-type CpG oligodeoxynucleotides as an adjuvant in allergen-specific immunotherapy in humans: A phase I/IIa clinical trial. *Clin. Exp. Allergy* **2009**, *39*, 562–570. [CrossRef]

39. Zeltins, A.; West, J.; Zabel, F.; El Turabi, A.; Balke, I.; Haas, S.; Maudrich, M.; Storni, F.; Engeroff, P.; Jennings, G.T.; et al. Incorporation of tetanus-epitope into virus-like particles achieves vaccine responses even in older recipients in models of psoriasis, Alzheimer's and cat allergy. *NPJ Vaccines* **2017**, *2*, 30. [CrossRef]

40. Wong, A.J.; Ruppert, J.M.; Bigner, S.H.; Grzeschik, C.H.; Humphrey, P.A.; Bigner, D.S.; Vogelstein, B. Structural alterations of the epidermal growth factor receptor gene in human gliomas. *Proc. Natl. Acad. Sci. USA* **1992**, *89*, 2965–2969. [CrossRef]

41. Tumban, E.; Peabody, J.; Tyler, M.; Peabody, D.S.; Chackerian, B. VLPs displaying a single L2 epitope induce broadly cross-neutralizing antibodies against human papillomavirus. *PLoS ONE* **2012**, *7*, e49751. [CrossRef] [PubMed]

42. Tumban, E.; Muttil, P.; Escobar, C.A.A.; Peabody, J.; Wafula, D.; Peabody, D.S.; Chackerian, B. Preclinical refinements of a broadly protective VLP-based HPV vaccine targeting the minor capsid protein, L2. *Vaccine* **2015**, *33*, 3346–3353. [CrossRef] [PubMed]

43. Einstein, M.H.; Takacs, P.; Chatterjee, A.; Sperling, R.S.; Chakhtoura, N.; Blatter, M.M.; Lalezari, J.; David, M.P.; Lin, L.; Struyf, F.; et al. Comparison of long-term immunogenicity and safety of human papillomavirus (HPV)-16/18 AS04-adjuvanted vaccine and HPV-6/11/16/18 vaccine in healthy women aged 18-45 years: End-of-study analysis of a Phase III randomized trial. *Hum. Vaccin Immunother.* **2014**, *10*, 3435–3445. [CrossRef] [PubMed]

44. Safaeian, M.; Porras, C.; Pan, Y.; Kreimer, A.; Schiller, J.T.; Gonzalez, P.; Lowy, D.R.; Wacholder, S.; Schiffman, M.; Rodriguez, A.C.; et al. Durable antibody responses following one dose of the bivalent human papillomavirus L1 virus-like particle vaccine in the Costa Rica Vaccine Trial. *Cancer Prev. Res.* **2013**, *6*, 1242–1250. [CrossRef] [PubMed]

45. Kreimer, A.R.; Herrero, R.; Sampson, J.N.; Porras, C.; Lowy, D.R.; Schiller, J.T.; Schiffman, M.; Rodriguez, A.C.; Chanock, S.; Jimenez, S.; et al. Evidence for single-dose protection by the bivalent HPV vaccine-Review of the Costa Rica HPV vaccine trial and future research studies. *Vaccine* **2018**, *36*, 4774–4782. [CrossRef] [PubMed]

46. Amanna, I.J.; Carlson, N.E.; Slifka, M.K. Duration of humoral immunity to common viral and vaccine antigens. *N. Engl. J. Med.* **2007**, *357*, 1903–1915. [CrossRef]

47. Slifka, M.K.; Amanna, I.J. Role of Multivalency and Antigenic Threshold in Generating Protective Antibody Responses. *Front. Immunol.* **2019**, *10*, 956. [CrossRef]

48. Merrell, K.T.; Benschop, R.J.; Gauld, S.B.; Aviszus, K.; Decote-Ricardo, D.; Wysocki, L.J.; Cambier, J.C. Identification of anergic B cells within a wild-type repertoire. *Immunity* **2006**, *25*, 953–962. [CrossRef]

49. Wardemann, H.; Yurasov, S.; Schaefer, A.; Young, J.W.; Meffre, E.; Nussenzweig, M.C. Predominant autoantibody production by early human B cell precursors. *Science* **2003**, *301*, 1374–1377. [CrossRef]

50. Cyster, J.G.; Hartley, S.B.; Goodnow, C.C. Competition for follicular niches excludes self-reactive cells from the recirculating B-cell repertoire. *Nature* **1994**, *371*, 389–395. [CrossRef]

51. Fulcher, D.A.; Basten, A. Reduced life span of anergic self-reactive B cells in a double-transgenic model. *J. Exp. Med.* **1994**, *179*, 125–134. [CrossRef] [PubMed]

52. Chackerian, B.; Lowy, D.R.; Schiller, J.T. Induction of autoantibodies to mouse CCR5 with recombinant papillomavirus particles. *Proc. Natl. Acad. Sci. USA* **1999**, *96*, 2373–2378. [CrossRef] [PubMed]

53. Chackerian, B.; Lowy, D.R.; Schiller, J.T. Conjugation of a self-antigen to papillomavirus-like particles allows for efficient induction of protective autoantibodies. *J. Clin. Investig.* **2001**, *108*, 415–423. [CrossRef] [PubMed]

54. Chackerian, B.; Lenz, P.; Lowy, D.R.; Schiller, J.T. Determinants of autoantibody induction by conjugated papillomavirus virus-like particles. *J. Immunol.* **2002**, *169*, 6120–6126. [CrossRef]

55. Jegerlehner, A.; Storni, T.; Lipowsky, G.; Schmid, M.; Pumpens, P.; Bachmann, M.F. Regulation of IgG antibody responses by epitope density and CD21-mediated costimulation. *Eur. J. Immunol.* **2002**, *32*, 3305–3314. [CrossRef]

56. Übelhart, R.; Hug, E.; Bach, M.P.; Wossning, T.; Dühren-von Minden, M.; Horn, A.H.; Tsiantoulas, D.; Kometani, K.; Kurosaki, T.; Binder, C.J.; et al. Responsiveness of B cells is regulated by the hinge region of IgD. *Nat. Immunol.* **2015**, *16*, 534–543. [CrossRef]

57. Maurer, P.; Bachmann, M.F. Immunization against angiotensins for the treatment of hypertension. *Clin. Immunol.* **2010**, *134*, 89–95. [CrossRef]

58. Crossey, E.; Amar, M.J.; Sampson, M.; Peabody, J.; Schiller, J.T.; Chackerian, B.; Remaley, A.T. A cholesterol-lowering VLP vaccine that targets PCSK9. *Vaccine* **2015**, *33*, 5747–5755. [CrossRef]

59. Maphis, N.M.; Peabody, J.; Crossey, E.; Jiang, S.; Ahmad, F.A.J.; Alvarez, M.; Mansoor, S.K.; Yaney, A.; Yang, Y.; Sillerud, L.O.; et al. Qß Virus-like particle-based vaccine induces robust immunity and protects against tauopathy. *NPJ Vaccines* **2019**, *4*, 26. [CrossRef]

60. Vandenberghe, R.; Riviere, M.E.; Caputo, A.; Sovago, J.; Maguire, R.P.; Farlow, M.; Marotta, G.; Sanchez-Valle, R.; Scheltens, P.; Ryan, J.M.; et al. Active Abeta immunotherapy CAD106 in Alzheimer's disease: A phase 2b study. *Alzheimers Dement.* **2017**, *3*, 10–22.

61. Bachmann, M.F.; Whitehead, P. Active immunotherapy for chronic diseases. *Vaccine* **2013**, *31*, 1777–1784. [CrossRef]

62. Mohsen, M.O.; Zha, L.; Cabral-Miranda, G.; Bachmann, M.F. Major findings and recent advances in virus-like particle (VLP)-based vaccines. *Semin. Immunol.* **2017**, *34*, 123–132. [CrossRef] [PubMed]

63. Farlow, M.R.; Andreasen, N.; Riviere, M.E.; Vostiar, I.; Vitaliti, A.; Sovago, J.; Caputo, A.; Winblad, B.; Graf, A. Long-term treatment with active Abeta immunotherapy with CAD106 in mild Alzheimer's disease. *Alzheimers Res. Ther.* **2015**, *7*, 23. [CrossRef] [PubMed]

64. Tissot, A.C.; Maurer, P.; Nussberger, J.; Sabat, R.; Pfister, T.; Ignatenko, S.; Volk, H.D.; Stocker, H.; Müller, P.; Jennings, G.T.; et al. Effect of immunisation against angiotensin II with CYT006-AngQb on ambulatory blood pressure: A double-blind, randomised, placebo-controlled phase IIa study. *Lancet* **2008**, *371*, 821–827. [CrossRef]

65. Maurer, P.; Jennings, G.T.; Willers, J.; Rohner, F.; Lindman, Y.; Roubicek, K.; Renner, W.A.; Müller, P.; Bachmann, M.F. A therapeutic vaccine for nicotine dependence: Preclinical efficacy, and Phase I safety and immunogenicity. *Eur. J. Immunol.* **2005**, *35*, 2031–2040. [CrossRef] [PubMed]

 *viruses*

Review

Influenza Virus Like Particles (VLPs): Opportunities for H7N9 Vaccine Development

Peter Pushko * and Irina Tretyakova

Medigen, Inc., 8420 Gas House Pike, Suite S, Frederick, MD 21701, USA; itretyakova@medigen-usa.com
* Correspondence: ppushko@medigen-usa.com

Received: 2 April 2020; Accepted: 27 April 2020; Published: 8 May 2020

Abstract: In the midst of the ongoing COVID-19 coronavirus pandemic, influenza virus remains a major threat to public health due to its potential to cause epidemics and pandemics with significant human mortality. Cases of H7N9 human infections emerged in eastern China in 2013 and immediately raised pandemic concerns as historically, pandemics were caused by the introduction of new subtypes into immunologically naïve human populations. Highly pathogenic H7N9 cases with severe disease were reported recently, indicating the continuing public health threat and the need for a prophylactic vaccine. Here we review the development of recombinant influenza virus-like particles (VLPs) as vaccines against H7N9 virus. Several approaches to vaccine development are reviewed including the expression of VLPs in mammalian, plant and insect cell expression systems. Although considerable progress has been achieved, including demonstration of safety and immunogenicity of H7N9 VLPs in the human clinical trials, the remaining challenges need to be addressed. These challenges include improvements to the manufacturing processes, as well as enhancements to immunogenicity in order to elicit protective immunity to multiple variants and subtypes of influenza virus.

Keywords: H7N9, pandemic influenza A; avian flu; IAV; VLP vaccine

1. Introduction

In addition to pandemic COVID-19 virus infections, influenza virus remains a major threat to public health and causes significant morbidity and mortality worldwide from annual seasonal epidemics and periodic pandemics [1,2]. According to WHO, seasonal influenza affects 5–10% of the global population annually and results in 3-5 million hospitalizations and about half a million deaths per year. Pandemic influenza A virus (IAV) represents an even greater concern. Due to the rapid mutation rate of human IAVs, antigenically novel viruses are emerging periodically, which are difficult to predict, prevent, or treat [3]. Zoonotic influenza viruses of H5N1, H9N2 and recently, H7N9 and H10N8 subtypes have been identified as potentially pandemic [4,5].

Influenza viruses belong to the *Orthomyxoviridae* family and comprise negative-sense, single stranded, segmented RNA genome. The RNA genome segments are loosely encapsidated by the nucleoprotein into virus particle. There are four types of influenza virus—types A, B, C, and D. Influenza A viruses (IAV) and type B viruses are clinically relevant for humans.

IAV are further sub-divided based on the antigenic properties of surface glycoproteins into 18 hemagglutinin (HA) and 11 neuraminidase (NA) subtypes. Only a few IAV subtypes have been known to infect humans, while the majority of them are harbored in their natural hosts such as waterfowl, shorebirds, and other species [6]. Cases of H7N9 human infections caused by an avian-origin H7N9 virus emerged in eastern China in March 2013 [7,8]. This novel virus has immediately raised pandemic concerns as historically, pandemics were caused by the introduction of new subtypes into immunologically naïve human populations [9]. Phylogenetic results indicate that novel H7N9 virus was a triple reassortant derived from avian influenza viruses [7]. Since 2013, surveillance of live poultry markets routinely detected H7N9 virus [10]. Human infections with H7N9 virus were associated mainly with the exposure to infected poultry [11] and were identified in many cities in China [12]. Both low pathogenicity avian influenza (LPAI) and high pathogenicity avian influenza (HPAI) H7N9 viruses have been recorded. The first wave of H7N9 was associated with LPAI virus and lasted from March until September 2013. The following four waves occurred annually until 2017 (Figure 1). During the fifth wave in the 2016/17 season, the emergence of HPAI H7N9 viruses was detected. After no reported human cases of HPAI H7N9 for over a year, another HPAI H7N9 case with severe disease was reported in mainland China in late March 2019, indicating the continuing public health threat from the H7N9 subtype [13]. HPAI subtype H5 and H7 proteins contain multiple basic amino acid cleavage sites between HA1 and HA2 domains within HA proteins, which can be cleaved by furin-like proteases [14] in many host cells and organs that can lead to the efficient spread of the virus and severe disease in humans. In contrast, HA of LPAI viruses does not have the furin cleavage site.

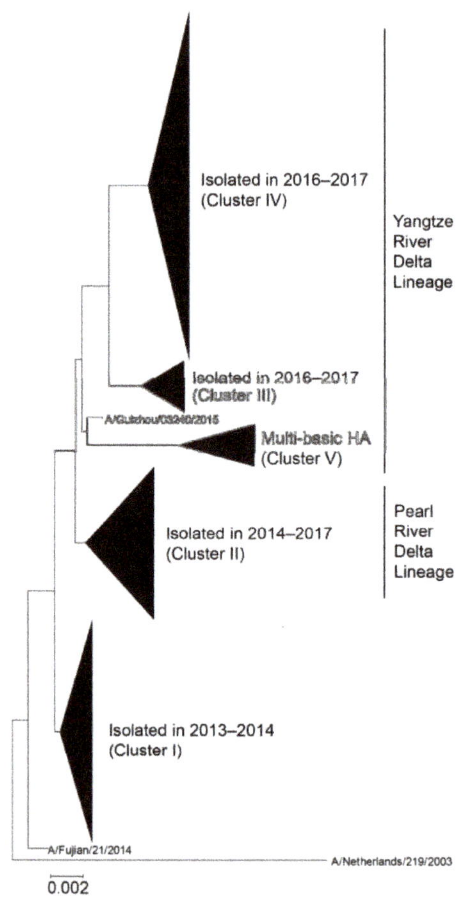

Figure 1. Phylogenetic tree of hemagglutinin (HA) sequences derived from human H7N9 viruses [15]. The evolutionary history was inferred using the neighbor-joining method with Kimura distances. Five major clusters are shown as a collapsed branch. A/Netherlands/219/2003 is defined as an outgroup. The Yangtze River Delta and Pearl River Delta lineages are circulating in China. HPAI H7N9 viruses, which harbor multiple basic amino acids in the HA cleave site, are included in the Yangtze River Delta lineage. Permission: Viruses https://doi.org/10.3390/v11020149.

A fatality rate of up to 38% was reported for H7N9 viruses [16], which highlights the need for a safe and effective vaccine [17]. Several candidate H7N9 vaccine viruses have been prepared and listed by WHO (Table 1). These candidate vaccine viruses are available to vaccine developers for the preparation of H7N9 vaccine in the case of a pandemic. The majority of current vaccine manufacturers prepare vaccines either as split subvirions or live-attenuated viruses, and they are mostly dependent on fertilized chicken eggs as production "bioreactors". This technology is unlikely to meet the vaccine production demand during the rapid pandemic spread [18]. Scalability issues (one vaccine dose/egg), the relatively long 6-month time period from strain isolation to final dose formulation and the requirement of biosecurity facilities for HPAI are the major obstacles for egg-based production [19]. In addition, IAV can acquire adaptive mutations when grown in eggs, which can interfere with the vaccine performance and efficacy. According to the action plan published by WHO in 2006, more than 2.34 billion monovalent vaccine doses will be needed in the case of a global pandemic, which justifies the development of novel technologies capable of supporting surge demand for pandemic influenza vaccine within a short period of time [20].

Table 1. WHO-recommended vaccine strains for H7N9 virus (adapted from [21]).

Vaccine Prototype Strain	Vaccine Candidate	Examples
A/Guangdong/17SF003/2016		*Wild Type Virus*
	Reverse genetics	CBER-RG7C, IDCDC-RG56N, CBER-RG7D, NIBRG-375
A/Hong Kong/125/2017		*Wild type virus*
	Reverse genetics	IDCDC-RG56B
A/Shanghai/2/2013		*Wild type virus*
	Reverse genetics	IDCDC-RG32A, IDCDC-RG32A.3, NIBRG-267, CBER-RG4A
A/Anhui/1/2013		*Wild type virus*
	Reverse genetics	NIBRG-268, NIIDRG-10.1, IDCDC-RG33A, SJ005

2. Recombinant Influenza Virus-Like Particles (VLPs) as Vaccines

Recently, several new platform technologies including recombinant VLPs have been developed in order to facilitate the scale-up of vaccine production including overcoming the drawbacks of the egg-based vaccine production method. The first recombinant influenza vaccine FluBlok based on HA antigen has received FDA approval [22]. The surface envelope glycoproteins such as influenza HA and NA generally are viewed as the primary targets for vaccine development. HA currently represents the major target for vaccine development including approved and experimental vaccines [22]. Recombinant HA-based vaccines have been shown to be efficacious against influenza including H7N9 [17,23–31]. Experimental NA-based vaccines have been reported [32,33]. The induction of immune responses against the surface envelope proteins has advantages, because virus-neutralizing antibody response to the viral envelope proteins can prevent early steps of viral infection. Human monoclonal antibodies targeting the HA glycoprotein can neutralize H7N9 influenza virus [34]. Importantly, the neutralizing antibodies protected against A/Shanghai/2/2013 (H7N9) virus challenge [35].

The novel recombinant VLP vaccine platforms included VLPs prepared in mammalian, insect, and plant expression systems. Recombinant VLPs are morphologically and biochemically similar to the wild type influenza virus (Figure 2); however, they lack viral genetic material and are unable to replicate and cause infection. Table 2 shows examples of various VLPs prepared from H7N9 virus antigens. Cell culture-based production of VLPs can potentially overcome limitations of classic influenza vaccines [36].

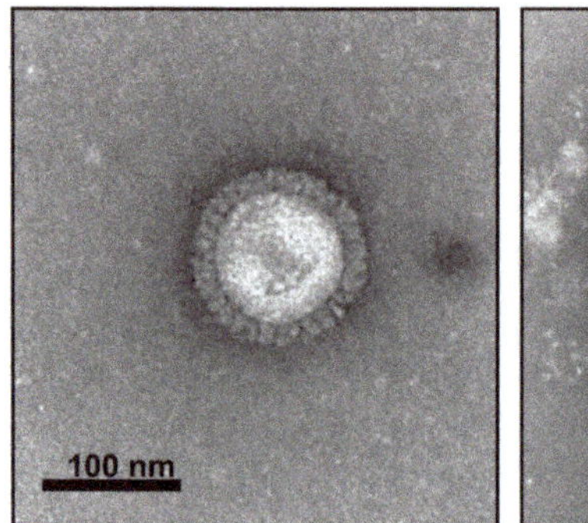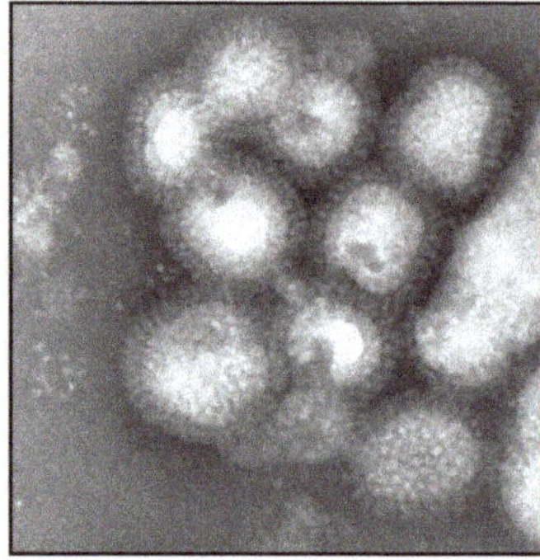

Figure 2. Electron microscopy images of influenza VLPs prepared in Sf9 cells using baculovirus expression system. Recombinant VLPs contain three subtypes (H5, H7, H9), morphologically and biochemically are similar to the wild type influenza virus, and can be found as individual particles (left) or groups of particles (right). Characterization of triple-subtype VLPs was done by negative staining transmission electron microscopy. (Adapted from [50], Copyright 2013, with permission from Elsevier, license 4798470004544).

Table 2. Examples of virus-like particle (VLP) vaccines against H7N9 influenza.

VLP Antigen	Expression System	Vaccination			Protection from Intranasal Challenge	References
		Animals	Dose	Route		
HA	High Five insect cells	mice	10 µg protein plus 10 µg poly(I·C) in PBS	IM	100 mLD$_{50}$	[37]
HA	Insect Sf9	mice ferrets	5 µg 15 µg	IN IM	10 mLD$_{50}$ 10^6 PFU	[38]
HA	Insect Sf9	ferrets	15 µg	SC	10^7 PFU	[31]
HA	Plants	mice ferrets	3 µg 15 µg	IM IM	100 mLD$_{50}$ 10^6 TCID$_{50}$	[39]
HA, M1	Insect *Trichoplusia ni* cell line High Five (BTI-TN-5B1-4)	mice	0.03, 0.3, 3 µg	IM	fully protected against 100 mLD$_{50}$	[40]
HA, M1	Insect Sf9	mice	1.5 µg	IN	10 mLD$_{50}$	[41]
HA, NA, M1	Insect Sf9	mice	6 µg	IM	4.4X10^3 TCID$_{50}$ PFU	[42]
HA, NA, M1	HEK293T	mice	40 µg total protein	IM	NA	[43]
HA, NA, M1	Insect Sf9	mice	3 µg	SC	10^6 TCID$_{50}$	[44]
HA, NA, M1	Insect Sf9	mice chickens	10 µg	IM	NA	[45]
HA, NA, M1	Insect Sf9	mice	40 µg	IM, IN		[46]
HA, NA, BIV Gag	Insect Sf9	chickens	1536 HA units per dose of VLPs	SC	10^6 EID$_{50}$	[47]
HA/M1, M2/NA, BAFF-HA$_{tm}$/M1, APRILHA$_{tm}$/M1	Insect Sf9	mice	0.5 µg HA content	IM	10 mLD$_{50}$ of NIBRG-14 vaccine strain	[48]
Combined HA and M2e5x (Universal)	Insect Sf9	mice	5 µg HA + 10 µg M2e5x VLPs	IM	7.5 x 10^2 PFU = 5 mLD$_{50}$	[49]

mLD$_{50}$, mouse median lethal dose.

As VLPs closely resemble viruses, they contain immunological epitopes in the natural conformation, are highly immunogenic, but they are non-infectious because they are prepared in the absence of viral genomic RNA segments. Each VLP represents a repetitive and highly organized molecular array of antigens displayed on the VLP surface. VLPs have been shown to be better immunogens than subunit vaccines because of their self-adjuvant properties [51]. The antigens present in VLP bind pattern recognition receptors on innate immune effector cells and trigger the innate immunity of host cells [52,53]. The molecular arrays displayed on VLP's surface are also potent inducers of Type 2 T-independent B-cell response [54]. Cryo-electron microscopy of VLPs from 1918 pandemic influenza virus demonstrated a uniform distribution of HA molecules on the surface of VLPs along with prefusion state confirmation [55]. Two types of spike on the surface of virus are formed by trimers of HA and tetramers of NA. It has been estimated that a spherical influenza virion of average diameter 120 nm has ~375 spikes [56].

3. Recombinant Platforms for Expression of H7N9 VLPs

The recombinant platforms using either mammalian, insect, or plant cells have many potential advantages including scalability and intrinsic safety, especially for HPAI viruses such as H7N9 (Table 2). Mammalian cells, such as MDCK, Vero and PERC.C6 cell lines have been extensively studied and applied to influenza vaccine production. The use of VLPs expressed from mammalian cell platform has advantages for VLP production such as flexibility and a similar glycosylation pattern to the human virus. In some reports, VLPs expressed from HEK293T cells containing HA and NA proteins on their surface with a HA:NA ratio of 1:1, which is higher than natural influenza viruses, where the HA:NA ratio is 4:1 [57]. VLPs isolated from HEK293SF cells were also characterized by nano-LC-MS/MS analysis to identify the protein cargo. Nucleolin was the most abundant cellular protein present in VLPs, in addition to the recombinant influenza VLP proteins [58]. However, production of VLPs in mammalian cell expression system is difficult to scale up to the commercial scale, especially when using the transient transfection method. This limitation can be addressed by preparing stable cell line of corresponding genes. For example, a stable cell line expressing H1 and N1 genes under the cumate inducible promoter was generated to create 293HA-NA cells. Plasmid with either Gag or M1 gene was transiently transfected in 293HA-NA cells and VLP yield was compared with Gag or M1 protein. 293HA-NA cells produced seven times more HA and NA with Gag plasmid transfected cells than its counterpart matrix M1-encoding plasmid [59]. Gag VLPs produced by this approach elicited strong antibody response and provided complete protection against the homologous virus strain. Optimization of cell density at the time of induction in inducible stable HEK293 cell lines also increased VLP yield by five-fold [60].

Insect cells such as Sf9 can be maintained in serum-free suspension cultures and have been often used to prepare recombinant VLPs including H7N9. When influenza VLPs were compared between insect Sf9 cells and mammalian HEK293 suspension cells, Thompson et al. found that Sf9 cells produce approximately 35 times more VLP than HEK293 cells [61]. VLPs produced from Sf9 cells also showed higher HA titer and more homogenous VLPs than HEK293 cells derived VLPs [61]. As expected, the high titer of baculovirus vector was also present in VLPs prepared in Sf9 cells. However, VLPs generated in HEK293 cells contained extracellular vesicles, such as exosomes and microvesicles. VLPs produced from Sf9 insect cells may have a safety advantage over other expression system because insect cells are unlikely to contain human pathogens, and baculovirus vector cannot replicate in human cells. VLPs produced from insect cells show higher yields as compared to mammalian cell-produced VLPs, which is beneficial for commercial production. In the above study, Sf9 cells produced up to 35 times higher yield than the HEK293 cells' expression system [61]. In addition, VLPs had higher HA activity and they were more homogenous in morphology and size than mammalian cells produced VLPs [61]. Sf9 cells are usually maintained in a serum free, animal product-free medium. Some of the chemical compounds known as M2 inhibitors, such as amantadine-enhanced expression of HA protein in VLPs by seven-fold in Sf9 expression system [62]. However, the challenge for VLPs expressed using baculovirus expression system is that it is difficult to separate VLPs from the baculovirus expression vector during vaccine manufacturing.

H7N9 VLPs have been prepared from the full-length HA, NA, and matrix M1 genes in Sf9 cells. The H7 HA was derived from A/Shanghai/2/2013 (H7N9) virus, while both NA and M1 proteins were derived from A/Puerto Rico/8/1934 (H1N1) virus, a standard donor virus that is often used for preparation of influenza reassortant viruses used in human influenza vaccine production [63]. VLPs have been isolated from Sf9 cells and their morphology resembling native H7N9 virus envelope was confirmed.

In another study, recombinant H7N9 VLP vaccine was prepared from HA and NA derived from the A/Anhui/1/2013 (H7N9) virus and the matrix M1 protein derived from the H5N1 virus (A/Indonesia/05/2005) [42]. As control vaccines, VLPs were prepared from A/chicken/Jalisco/CPA1/2012 (H7N3) and A/Indonesia/05/2005 (H5N1) viruses. The H7N9 VLPs elicited hemagglutination-inhibition (HAI) antibody response against the homologous H7N9 virus, cross-reactive HAI against the heterologous H7N3 virus, and three- to four-fold higher HAI response if VLPs were administered with ISCOMATRIX™ saponin-based adjuvant. Similarly, all doses of H7N9 VLP elicited anti-NA antibody, with three- to four-fold higher responses measured in the corresponding ISCOMATRIX subgroups. A lethal murine wild-type A/Anhui/1/2013 (H7N9) challenge demonstrated 100% survival of all animals receiving H7N9 or H7N3 VLP vaccines, versus 0% survival in A(H5N1) vaccine and placebo groups [42].

Another study reported a VLP from Sf9 cells that consisted of HA, NA and M1 proteins derived from the human isolate A/Taiwan/S02076/2013(H7N9) as a potential vaccine. In animal experiments, BALB/c mice and specific-pathogen-free chickens receiving the VLP immunization developed HAI serum titer and antibodies against NA and M1 proteins [45].

The immunogenicity and protective efficacy of a H7N9 VLP vaccine were evaluated in the ferret challenge model considered the most appropriate animal model for influenza [31]. Purified recombinant H7N9 VLPs morphologically resembled influenza virions and elicited high-titer serum HAI and neutralizing antibodies to A/Anhui/1/2013 (H7N9) virus. H7N9 VLP-immunized ferrets were challenged with homologous virus. After challenge, VLP-vaccinated animals displayed reductions in fever, weight loss, and virus shedding as compared to the same parameters in the unimmunized control ferrets. H7N9 VLP was also effective in protecting against lung and tracheal infection. The addition of either ISCOMATRIX or Matrix-M1 adjuvant further improved immunogenicity and protection of the VLP vaccine against H7N9 virus [31].

Another recombinant influenza vaccine was developed by expressing H7 from H7N9 (A/Shanghai/2/2013) on the surface of recombinant baculovirus. Although this approach would not result in a typical influenza VLP morphology, the spatial conformation of H7 in the resulting vaccine is expected to be similar to that in an influenza virion or in a cognate VLP. Mice were immunized twice, either intranasally or subcutaneously, with the vaccine. The immunogenicity and cross-protective efficacy of the vaccine were evaluated against H7N9 or H7N7 subtype challenges. The authors concluded that intranasal administration of H7 protein expressed on the baculovirus envelope can be an alternative way to prime the immune system against influenza infection during a pandemic situation [64]. Recently, a baculovirus vaccine expressing the HA of H7N9 strain A/Chicken/Jiaxing/148/2014 was prepared [65]. The recombinant baculovirus-HA generated in this study showed favorable growth characteristics in insect cells, good safety profile, and induced high-level hemagglutination inhibition antibody titer. Moreover, this vaccine demonstrated better efficacy than inactivated whole-virus vaccine JX148, provided complete protection of chickens against challenge with HPAI H7N9 virus, and effectively inhibited viral shedding.

Plant-based VLP vaccines can be produced from *Agrobacterium*-mediated transient expression of influenza HA proteins in *Nicotiana benthamiana*. Plant-made H1 or H5 VLPs mimicked the structure of influenza virions to some extent, they are immunogenic and elicit both humoral and cellular responses [66,67]. Each HA-only particle contains 30–50 homo-trimer HA incorporated into lipid bilayer envelope of plant cell origin [68]. Plant-based recombinant VLP vaccines elicited humoral and cellular responses and protected against challenges. For example, a plant-derived VLP vaccine based on the HA of influenza H7N9 A/Hangzhou/1/2013 was prepared, with no other influenza proteins included in the vaccine. The immunogenicity of such H7-only VLP vaccine was assessed in mice and ferrets after one or two intramuscular dose(s) with, and without, adjuvant (alum or GLA-SE™). In ferrets, H7-specific cell-mediated immunity was also evaluated. The mice and ferrets were challenged with H7N9 A/Anhui/1/2013 influenza virus. A single immunization with the adjuvanted vaccine elicited a strong humoral response and protected mice against an otherwise lethal challenge. Two doses of unadjuvanted vaccine significantly increased humoral response and resulted in 100% protection with significant reduction in clinical signs, leading to nearly asymptomatic infections. In ferrets, a single immunization with the alum-adjuvanted H7 VLP vaccine induced strong humoral and CMI responses with antigen-specific activation of CD3(+) T cells. This plant-made H7 vaccine therefore induced protective responses after either one adjuvanted or two unadjuvanted doses [39]. Potentially, the plant-based transient expression system can allow the production of VLP structures, which contain HA only, without other influenza proteins.

The recombinant HA-only H7 vaccine against H7N9 virus has been also prepared in insect cells [30]. The study has shown that purified HA formed rosette-shaped nanoparticles of approximately 30–50 nm in diameter [69]. Similarly, subviral particles containing the full-length H7 rHA (A/Anhui/1/2013 (H7N9) were isolated from Sf9 cells treated with mild detergent [38]. Similarly to a previous study [30], purified rHA formed particulate structures; the particles were approximately 20 nm in diameter, exhibited hemagglutination activity, and consisted of approximately 3–4 trimers of the full-length HA molecules.

4. Considerations for Immune Responses and Protection

As noted above, immunization with VLPs have shown promising results in protecting from influenza infections with H7N9 [27,70,71] (Table 2). Potentially, antibody induced by VLPs can be more effective than those of subunit vaccines containing recombinant protein antigens [72]. In addition, experimental VLP vaccines generally show higher protective rates to high-risk groups such as children and the elderly [26]. However, the full spectrum of factors affecting immunogenicity of VLPs requires additional studies. For example, it has been shown that amino acid residues of HA that are related to receptor specificity can affect the protective efficacy of H5N1 and H7N9 vaccines in mice [44]. H7N9 VLP vaccine that contained L226 (mammalian specificity) and G228 (avian specificity) in HA showed better immunogenicity and protection efficacy than VLP containing HA with either L226 + S228 or Q226 + S228. This observation indicated that specific HA residues could enhance a vaccine's protection efficacy and HA glycoproteins with both avian-type and human-type receptor specificities may produce better pandemic influenza vaccines for humans [44].

While inactivated vaccines have been shown to induce predominantly systemic humoral response, VLP vaccines stimulate both humoral and cellular immune responses. H7N9 VLPs secreted from 293T cells triggered both humoral and cellular immune responses in mice. This vaccine produced higher levels and antibody and isotypes of IgG, as well as cross-reactive HAI titer against heterologous H1N1 and H1N3 subtypes [43].

The presence of NA and potentially other IAV proteins in VLP could be beneficial. The NA was demonstrated to protect host from influenza infection [73]. Mice immunized with NA VLPs (without HA) were protected against lethal challenge of homologous A/PR/8/34 virus without any weight loss [74].

Furthermore, both the exterior and interior of VLPs can be altered to enhance their stability and immunogenicity. Various adjuvants such as alum, CpG DNA, monophosphoryl lipid A (MPL), poly IC, gardiquimod, cholera toxin (CT) can be encapsulated into VLPs by exogenous and endogenous methods [75,76]. These adjuvanted VLPs show higher levels of antibodies in both sera and mucosa. In another study, adjuvants that stimulate TLR3 or NLPR3 pathways showed higher efficiency of influenza VLP vaccine in aged group of mice [77]. The CpG-adjuvanted intranasal immunization with an egg-derived split H7N9 vaccine offered a high level of protection against H7N9 infection in mice [78].

5. Human Clinical Trials with H7N9 VLP Vaccines

Two phase I clinical trials of experimental H7N9 VLP vaccines with, and without, adjuvant have been completed, NCT01897701 and NCT02078674. The first study enrolled 284 adults (≥18 years of age) in a randomized, observer-blinded, placebo-controlled clinical trial [79]. Interestingly enough, significant increases in N9 neuraminidase-inhibiting antibodies occurred in up to 71.9% of recipients of the vaccine without adjuvant, 92.0% of recipients of vaccine with 30 units of adjuvant, and 97.2% of recipients of vaccine with 60 units of adjuvant. Remarkably, the vaccines that were studied were released for human use within 3 months after the availability of the HA and NA sequences. This illustrates the possibility of rapid response to the public health emergency situations, such as in the case of pandemic influenza outbreaks, as well as other health emergencies such as the ongoing COVID-19 coronavirus pandemic.

In another previously reported phase I clinical trial, subjects vaccinated with two doses of an unadjuvanted H7N9 VLP vaccine responded poorly (15.6% seroconversion rates with 45 µg HA dose). In contrast, 80.6% of subjects receiving H7N9 VLP vaccine (5µg HA) with ISCOMATRIX adjuvant developed HAI responses [80]. The results suggest that adjuvants can be important component for the development of safe and effective H7N9 human vaccines [81].

Plant-derived quadrivalent seasonal VLP vaccine has been evaluated and elicited cross-reactive antibody and T cell response in healthy adults [82]. Although H7N9 antigens have not been included in this study, the confirmed safety profile of VLP in humans suggests an attractive alternative manufacturing method for producing effective and HA-strain matching influenza vaccines.

6. Broadly Protective Influenza VLP Vaccines

Current annual vaccines (inactivated, live attenuated, and recombinant subunit) protect from circulating, antigenically matching IAV and influenza B virus strains. Broad protection is difficult to achieve because of the frequent emergence of new strains [83]. Therefore, current approved and experimental vaccines have limited protection against antigenically mismatched variants and newly emerging viral strains. For new strains, antigenically matching new vaccines have to be developed within short periods of time. Therefore, vaccines are needed that can protect from an antigenically divergent strain, especially from potentially pandemic viruses with a new HA subtype. In addition to H7N9 virus described above, multiple other viruses with pandemic potential, including H5N1, H9N2, and H10N8 subtypes, that have caused human infections in the past, continue to circulate in birds and other animals. Therefore, vaccines capable of protecting against multiple potentially pandemic influenza strains would be advantageous for pandemic preparedness and public health.

One way to achieve protection against multiple strains and increase the breadth of immune response afforded by VLPs can be a blended formulation of monovalent vaccines. A vaccine was prepared by mixing VLPs that display H1, H3, H5 or H7 HA molecules. Mice vaccinated with these VLPs intranasally showed significant protection (94% aggregate survival following vaccination) against 1918 H1, 1957 H2, and avian H5, H6, H7, H10 and H11 HA subtypes [41]. These experiments suggest a promising and practical strategy for developing a broadly protective influenza vaccine.

Potentially, HA molecules from different subtypes can be co-localized within recombinant VLP envelope for broader immune coverage. A multi-subtype, mosaic VLP design containing three or four subtypes of the full-length rHA within the envelope has also been described [50,71]. A triple-subtype VLP contained HA proteins from the potentially pandemic H5N1, H7N2, and H9N2 subtypes [50]. A recombinant baculovirus vector was prepared to co-express the H5, H7, and H9 genes from A/Viet Nam/1203/2004 (H5N1), A/New York/107/2003 (H7N2) and A/Hong Kong/33982/2009 (H9N2) viruses, respectively, as well as NA and M1 genes from A/Puerto Rico/8/1934 (H1N1) virus. VLPs were prepared in Sf9 cells, and the immunogenicity and efficacy of the resulting H5/H7/H9 VLPs were evaluated in a ferret animal model following intranasal (i.n.) vaccination. We showed that i.n. vaccination with the H5/H7/H9 triple-subtype VLP induced immune responses and protected ferrets from experimental challenges with three subtypes of avian influenza viruses [50]. We also prepared triple-HA mosaic VLPs that co-localized A/Vietnam/1203/2004 (H5N1), A/Hong Kong/33982/2009 (H9N2) and A/Shanghai/2/2013 (H7N9) rHA proteins, as well as quadri-HA VLPs (Figure 3).

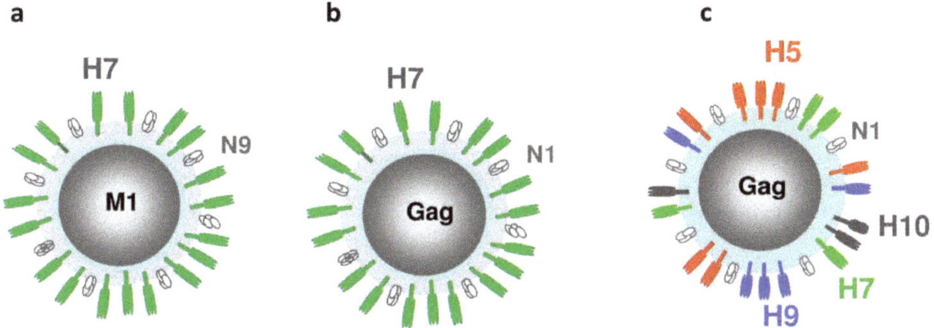

Figure 3. Structural proteins in mono- and quadri-subtype VLPs. (**a**) Mono-subtype H7N9 VLP with H7, N9 and M1 genes; (**b**) Mono-subtype chimeric H7N1 VLP with H7, N1, and Gag genes; (**c**) Quadri-subtype VLPs co-localizing H5, H7, H9 and H10 subtypes along with N1 and Gag within a VLP.

By using Group antigen (Gag) derived from bovine retrovirus as a VLP core in place of M1 protein, quadri-subtype VLPs were prepared, which co-expressed within the VLP the four HA subtypes derived from avian-origin H5N1, H7N9, H9N2 and H10N8 viruses. VLPs showed hemagglutination and neuraminidase activities and reacted with specific antisera [84]. Quadri-subtype vaccine elicited serum antibody in ferrets to the homologous H5, H7, H9, and H10 antigens. Antiserum was also evaluated for cross-reaction with multiple clades of H5N1 virus, and cross-reactivity has been confirmed. The level of immune response suggests protection against multiple influenza subtypes, which was experimentally confirmed by challenge with H10 IAV. Ferrets were protected from challenge with H10 virus [85]. Overall, such multi-subtype, mosaic VLPs that co-localize distinct HA subtypes in the envelope showed broader protection range against different influenza viruses [47,85,86]. Multi-subtype mosaic VLPs combine advantages of conserved HA epitope and blended VLPs, as VLPs contain both subtype-specific head epitopes and the conserved stem epitopes.

Recently, a VLP preparation consisting of retroviral Gag-VLPs pseudo-typed with the HA was expressed using the novel *Trichoplusia ni* (*T.ni*)-derived insect cell line *Tnms*42 and tested successfully to assess the sole contribution of anti-HA immunity in limiting post-influenza secondary *Staphylococcus aureus* bacterial infection, morbidity and mortality in a situation of a vaccine match and mismatch [87]. The results demonstrate that matched anti-HA immunity elicited by a VLP preparation may suffice to prevent morbidity and mortality caused by lethal secondary bacterial infection.

It should also be noted that cross-reactive antibodies to the H7N9 virus were also induced by recombinant viral vectors, such as Newcastle disease virus (NDV) [88–90] and parainfluenza virus PIV5 [91].

7. Expression of H7N9 Influenza Epitopes Using VLP Carriers

The emerging strains potentially can be also targeted by "universal" vaccines consisting of conserved viral proteins or epitopes, such as stem region of HA, or extracellular domain M2e of the ion channel protein M2, which are both considered capable of inducing cross-reactive immune responses. Targeting the stem region of HA for antibody production could be a promising approach to generate a broadly protective influenza vaccine. The immune response can be elicited against both the head and stem region of HA protein. However, because of the constantly changing nature of the head region due to the antigenic drift, new vaccine candidates need to be updated frequently. Since the stem region is evolutionarily more stable and more conserved among different influenza strains, vaccine candidates targeting the stem region might not need to be updated every year and could induce a broad range of protective immunity against different influenza strains [92,93]. However, the stem region is less immunogenic in the virus. Therefore, chimeric VLP approaches have been used to elicit antibodies to the stem region of HA. The long alpha-helix (LAH) region located in influenza virus hemagglutinin (HA) shows conservation among different influenza A strains, and could be used as a candidate target for influenza vaccines. The hepatitis B virus core protein (HBc) was used as a carrier for heterologous LAH epitopes to elicit effective immune responses [94]. The spatial conformation of LAH epitope cloned in the major immunodominant region (MIR) of the HBc molecule and expressed in yeast is shown on Figure 4. The LAH region of the H7N9 influenza virus was inserted into the HBc to prepare chimeric LAH-HBc protein, which was capable of self-assembly into VLPs in *E. coli* expression system [95]. Intranasal immunization of the LAH-HBc VLP in combination with chitosan adjuvant or CTB∗ adjuvant in mice induced both humoral and cellular immune responses effectively and conferred complete protection against lethal challenge with homologous H7N9 virus or heterologous H3N2 virus, as well as partial protection against lethal challenge of heterologous H1N1 virus. These results provide a proof of concept for LAH-HBc VLP vaccine that can be rapidly produced and potentially can serve as an antigen against a future influenza pandemic [95].

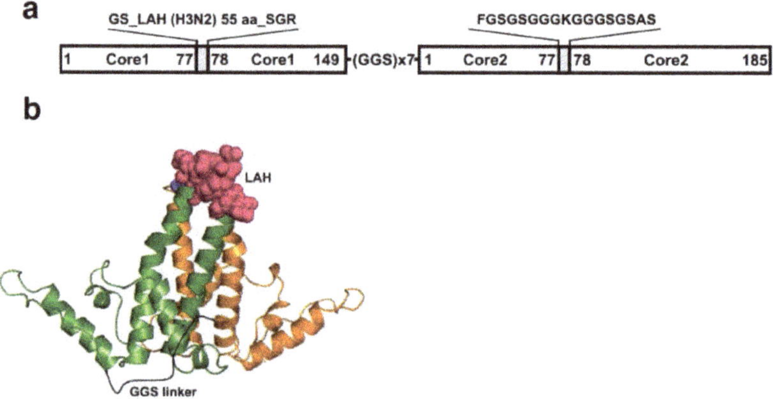

Figure 4. An example of expression of influenza epitope using chimeric VLP based on recombinant LAH3-HBc gene. Design (**a**) and cartoon (**b**) of the LAH3-HBc dimer. Individual hepatitis B virus core protein (HBc) monomers are shown in green and orange. The first major immunodominant region (MIR) contains the LAH domain (the 55 amino acid long influenza H3N2 virus (A/Hong Kong/1/1968, Accession No. AAK51718) HA stalk domain, corresponding to HA amino acids 420–474, shown as pink spheres) while the second MIR contains lysine linker (blue spheres)). The model was created using PyMOL version 1.7rc1. The tandem core dimer is based on the structure from PDB-1QGT, with the linker shown in black.

In another study, the globular head domain (HA1-2, aa 62–284) of the protective H7 HA was fused to the potent TLR5 ligand, *Salmonella typhimurium* flagellin (FliC) [96]. The resulting fusion protein, HA1-2-fliC, which apparently did not form VLPs, was efficiently expressed in *E. coli*, retained the native HA and fliC conformations, and was highly immunogenic in mice by intraperitoneal vaccination. Furthermore, highly immunogenic influenza VLPs have been prepared via the overexpression of four viral proteins, HA, NA, M1, and M2, using M2 fusion with the FliC [97]. The chimeric H5N1 VLPs were further combined with the molecular adjuvant, granulocyte-macrophage colony-stimulating factor (GM-CSF), FliC, or a GM-CSF/FliC [98]. All three forms of the chimeric H5N1 VLPs elicited protective immunity against live virus. Next, the GM-CSF/FliC H5N1 VLPs were obtained to include H7 or H1H7 antigens for developing multi-subtype influenza vaccines [98].

The chimeric norovirus VLPs, P particles, were used to express the trivalent HA2:90-105 epitopes derived from H1, H3 and B subtypes, with 24 copies in total, on the surface loops [99].

Finally, trimeric H7 transiently expressed in *N. benthamiana* was conjugated successfully onto the surface of nanodiamond particles [100]. After two or three immunizations in mice, the mixture of trimeric H7 protein and nanodiamond elicited statistically significant stronger H7-specific-IgG response demonstrated by higher amounts of H7N9-specific IgG.

Other universal influenza vaccine approaches applicable to H7N9 were based on ion channel protein M2. Extracellular part of M2 protein, ~23 aa residues, is highly conserved among human IAV strains, suggesting its potential utility as a broadly protective immunogen for the development of broadly protective influenza vaccines. Tandem repeats of heterologous M2e sequences (M2e5x) derived from human, swine, and avian origin influenza A viruses were expressed on influenza VLPs in a membrane-anchored form. Immunization of mice with M2e5x VLPs induced antibodies that were cross-reactive to antigenically distinct influenza A viruses and conferred cross-protection [101]. Furthermore, the M2e5x VLPs demonstrated a clear advantage in inducing IgG2a isotype antibodies, T cell responses, plasma cells and germinal center B cells as well as in conferring cross protection [102]. These studies paved the way to a novel vaccination strategy by enhancing the cross protective efficacy of live attenuated influenza virus vaccines by supplemented vaccination with M2e5x VLPs [103].

High immunogenicity in mice, even in the absence of adjuvants, was demonstrated by the M2e displayed on VLPs of *Macrobrachium rosenbergii* nodavirus [104,105]. This novel vaccine candidate was tested by the H1N1 and H3N2 challenge in mice, and potentially, it can be universal and applicable to H7N9 virus. In another example, the M2e proteins, as well as NA, were expressed on recombinant *Lactococcus lactis*, which conferred effective mucosal and systemic immune responses [106].

8. Challenges for Recombinant VLPs as Influenza Vaccines

Heterogenous composition is one of the challenges associated with the use of influenza VLPs as a vaccine candidate. VLPs produced by using baculovirus expression system contain live baculovirus vector as impurity. Moreover, considerable fraction of HA protein is bound to viral envelope of baculovirus [107]. Therefore, it is difficult to distinguish whether the HA activity is associated with influenza VLP or with baculovirus-displayed HA in the unpurified samples during vaccine manufacturing, which also causes difficulty for HA quantification. VLPs produced from mammalian expression system do not contain baculovirus contamination but they contain extracellular particles and produce low yields, with VLPs needing to be concentrated up to 200 times to reach measurable titers [72]. The current methods that are used for quantification are mostly applicable to concentrated and purified VLPs; however, better methods need to be developed to characterize the materials at the different stage of bioprocessing during the manufacturing process. Other potential challenges include the presence of endogenous rhabdovirus in standard Sf9 cell lines [108]. Although insect rhabdovirus is not associated with any human disease, the development of rhabdovirus-free cell line can resolve this potential safety concern [108,109].

To overcome manufacturing challenges, technological improvements to production of H7N9 VLPs in baculovirus-Sf9 insect cell cultures have been reported [110–112]. The yield and quality of influenza VLPs produced in insect cells could be improved by inhibiting cytopathic effects of the protein M2 [62]. The ion channel activity of M2 induces significant cytopathic effects in Sf9 insect cells. These effects include altered Sf9 cell morphology and reduced baculovirus replication, resulting in impaired influenza protein expression and thus VLP production. The use of the M2 inhibitor amantadine indeed improved Sf9 cellular expression not only of M2 (~three-fold), but also of HA (~seven-fold) and of matrix protein M1 (~three-fold) when co-expressed to produce influenza VLPs. This increased cellular expression of all three influenza proteins further led to ~two-fold greater VLP yield. The quality of the resulting influenza VLPs was significantly improved, as demonstrated by the ~two-fold, ~50-fold, and ~two-fold increase in the antigen density to approximately 53 HA, 48 M1, and 156 M2 per influenza VLP, respectively [62].

For purification of influenza VLPs, a nitrocellulose membrane-based filtration system [113] and a cascade of ultrafiltration and diafiltration steps, followed by a sterile filtration step [109,114] were used successfully. The bioprocessing of influenza VLPs was recently reviewed [115]. Recently, a fast chromatography step purification method was developed for chimeric HIV-1 Gag influenza-HA VLPs produced in *Tnms*42 insect cells using the baculovirus insect cell expression vector system [116]. The absence of the baculovirus capsid protein p39 in the product fraction was confirmed by HPLC-MS. The process is simple and includes only a few handling steps, which has promising characteristics to become a platform for purification of these types of VLPs [116].

Regarding the detection and quality assessment of influenza vaccines, several methodologies are being used. These include development of standards for potency determination [117]. A high-resolution LC-MS method allows the quantitation of both HA and NA protein concentrations in influenza VLP vaccine candidates [118].

9. Conclusions

Recombinant VLPs were developed with the purpose of rapidly preparing influenza vaccines including against pandemic threat viruses such as H7N9 virus. Multiple designs of VLPs have been evaluated including the use of multivalent VLPs, multi-subtype mosaic VLP, and rationally designed VLPs containing conserved influenza virus epitopes. The use of highly conserved epitopes remains a promising approach for the improved design of highly immunogenic VLPs. Blended VLPs have been used as a classic approach to broaden immune coverage. Multi-subtype mosaic VLPs potentially combine the conserved epitope and blended approaches, as VLPs contain both subtype-specific head epitopes and the conserved stem epitopes. In addition to VLPs, other novel technologies have been used to develop improved influenza vaccines including DNA vaccines, viral vectors, advanced adjuvants, and other approaches. Potentially, the combination of various technologies can prove beneficial for the development of next-generation influenza vaccines including H7N9.

Acknowledgments: We thank Tek Lamichhane and Paul Pumpens for their valuable contributions. This works was supported by the NIH NIAID grant R43AI136220. The views and conclusions in this report are those of the authors and do not necessarily reflect the views of the funding agency.

Conflicts of Interest: Authors declare no conflicts of interest.

References

1. Gostin, L.O.; Phelan, A.; Stoto, M.A.; Kraemer, J.D.; Reddy, K.S. Virus sharing, genetic sequencing, and global health security. *Science* **2014**, *345*, 1295–1296. [CrossRef] [PubMed]
2. Dormitzer, P.R.; Suphaphiphat, P.; Gibson, D.G.; Wentworth, D.E.; Stockwell, T.B.; Algire, M.A.; Alperovich, N.; Barro, M.; Brown, D.M.; Craig, S.; et al. Synthetic generation of influenza vaccine viruses for rapid response to pandemics. *Sci. Transl. Med.* **2013**, *5*, 185ra68. [CrossRef] [PubMed]
3. Mostafa, A.; Abdelwhab, E.M.; Mettenleiter, T.C.; Pleschka, S. Zoonotic Potential of Influenza A Viruses: A Comprehensive Overview. *Viruses* **2018**, *10*, 497. [CrossRef] [PubMed]
4. Philippon, D.A.M.; Wu, P.; Cowling, B.J.; Lau, E.H.Y. Avian influenza human infections at the human-animal interface. *J. Infect. Dis.* **2020**. [CrossRef]
5. Wu, X.; Xiao, L.; Li, L. Research progress on human infection with avian influenza H7N9. *Front. Med.* **2020**, *14*, 8–20. [CrossRef]
6. Yoon, S.W.; Webby, R.J.; Webster, R.G. Evolution and ecology of influenza A viruses. *Curr. Top. Microbiol. Immunol.* **2014**, *385*, 359–375.
7. Gao, R.; Cao, B.; Hu, Y.; Feng, Z.; Wang, D.; Hu, W.; Chen, J.; Jie, Z.; Qiu, H.; Xu, K.; et al. Human infection with a novel avian-origin influenza A (H7N9) virus. *N. Engl. J. Med.* **2013**, *368*, 1888–1897. [CrossRef]

8. Chen, F.; Li, J.; Sun, B.; Zhang, H.; Zhang, R.; Yuan, J.; Ou, X.; Ye, W.; Chen, J.; Liu, Y.; et al. Isolation and characteristic analysis of a novel strain H7N9 of avian influenza virus A from a patient with influenza-like symptoms in China. *Int. J. Infect. Dis.* **2015**, *33*, 130–131. [CrossRef]

9. Pu, J.; Wang, S.; Yin, Y.; Zhang, G.; Carter, R.A.; Wang, J.; Xu, G.; Sun, H.; Wang, M.; Wen, C.; et al. Evolution of the H9N2 influenza genotype that facilitated the genesis of the novel H7N9 virus. *Proc. Natl. Acad. Sci. USA* **2015**, *112*, 548–553. [CrossRef]

10. Jones, J.C.; Sonnberg, S.; Webby, R.J.; Webster, R.G. Influenza A(H7N9) Virus Transmission between Finches and Poultry. *Emerg. Infect. Dis.* **2015**, *21*, 619–628. [CrossRef]

11. Zaraket, H.; Baranovich, T.; Kaplan, B.S.; Carter, R.; Song, M.S.; Paulson, J.C.; Rehg, J.E.; Bahl, J.; Crumpton, J.C.; Seiler, J.; et al. Mammalian adaptation of influenza A(H7N9) virus is limited by a narrow genetic bottleneck. *Nat. Commun.* **2015**, *6*, 6553. [CrossRef] [PubMed]

12. Yang, J.R.; Kuo, C.Y.; Huang, H.Y.; Wu, F.T.; Huang, Y.L.; Cheng, C.Y.; Su, Y.T.; Wu, H.S.; Liu, M.T. Characterization of Influenza A (H7N9) Viruses Isolated from Human Cases Imported into Taiwan. *PLoS ONE* **2015**, *10*, e0119792. [CrossRef] [PubMed]

13. Yu, D.; Xiang, G.; Zhu, W.; Lei, X.; Li, B.; Meng, Y.; Yang, L.; Jiao, H.; Li, X.; Huang, W.; et al. The re-emergence of highly pathogenic avian influenza H7N9 viruses in humans in mainland China, 2019. *Eurosurveillance* **2019**, *24*, 1900273. [CrossRef] [PubMed]

14. Horimoto, T.; Nakayama, K.; Smeekens, S.P.; Kawaoka, Y. Proprotein-processing endoproteases PC6 and furin both activate hemagglutinin of virulent avian influenza viruses. *J. Virol.* **1994**, *68*, 6074–6078. [CrossRef] [PubMed]

15. Ito, M.; Yamayoshi, S.; Murakami, K.; Saito, K.; Motojima, A.; Nakaishi, K.; Kawaoka, Y. Characterization of Mouse Monoclonal Antibodies Against the HA of A(H7N9) Influenza Virus. *Viruses* **2019**, *11*, 149. [CrossRef]

16. Guo, Z.; Xiao, D.; Li, D.; Wang, Y.; Yan, T.; Dai, B. The temporal distribution of new H7N9 avian influenza infections based on laboratory-confirmed cases in Mainland China, 2013–2017. *Sci. Rep.* **2018**, *8*, 4051. [CrossRef]

17. Isakova-Sivak, I.; Rudenko, L. Tackling a novel lethal virus: A focus on H7N9 vaccine development. *Expert Rev. Vaccines* **2017**, *16*, 1–13. [CrossRef]

18. Trombetta, C.M.; Marchi, S.; Manini, I.; Lazzeri, G.; Montomoli, E. Challenges in the development of egg-independent vaccines for influenza. *Expert Rev. Vaccines* **2019**, *18*, 737–750. [CrossRef]

19. Gerdil, C. The annual production cycle for influenza vaccine. *Vaccine* **2003**, *21*, 1776–1779. [CrossRef]

20. WHO. *Department of Immunization, Vaccines and Biologicals. Geneva: Global Pandemic Influenza Action Plan to Increase Vaccine Supply 2006*; WHO: Geneva, Switzerland, 2006.

21. WHO. *Summary of Status of Development and Availability of Avian Influenza A(H7N9) Candidate Vaccine Viruses and Potency Testing Reagents*; WHO: Geneva, Switzerland, 2020.

22. Woo, E.J.; Moro, P.L.; Cano, M.; Jankosky, C. Postmarketing safety surveillance of trivalent recombinant influenza vaccine: Reports to the Vaccine Adverse Event Reporting System. *Vaccine* **2017**, *35*, 5618–5621. [CrossRef]

23. Galarza, J.M.; Latham, T.; Cupo, A. Virus-like particle (VLP) vaccine conferred complete protection against a lethal influenza virus challenge. *Viral Immunol.* **2005**, *18*, 244–251. [CrossRef] [PubMed]

24. Kang, S.M.; Pushko, P.; Bright, R.A.; Smith, G.; Compans, R.W. Influenza virus-like particles as pandemic vaccines. *Curr. Top. Microbiol. Immunol.* **2009**, *333*, 269–289. [PubMed]

25. Perrone, L.A.; Ahmad, A.; Veguilla, V.; Lu, X.; Smith, G.; Katz, J.M.; Pushko, P.; Tumpey, T.M. Intranasal vaccination with 1918 influenza virus-like particles protects mice and ferrets from lethal 1918 and H5N1 influenza virus challenge. *J. Virol.* **2009**, *83*, 5726–5734. [CrossRef] [PubMed]

26. Bright, R.A.; Carter, D.M.; Daniluk, S.; Toapanta, F.R.; Ahmad, A.; Gavrilov, V.; Massare, M.; Pushko, P.; Mytle, N.; Roweet, T.; et al. Influenza virus-like particles elicit broader immune responses than whole virion inactivated influenza virus or recombinant hemagglutinin. *Vaccine* **2007**, *25*, 3871–3878. [CrossRef] [PubMed]

27. Pushko, P.; Tumpey, T.M.; Bu, F.; Knell, J.; Robinson, R.; Smith, G. Influenza virus-like particles comprised of the, H.A.; NA, and M1 proteins of H9N2 influenza virus induce protective immune responses in BALB/c mice. *Vaccine* **2005**, *23*, 5751–5759. [CrossRef] [PubMed]

28. Quan, F.S.; Vunnava, A.; Compans, R.W.; Kang, S.M. Virus-like particle vaccine protects against 2009 H1N1 pandemic influenza virus in mice. *PLoS ONE* **2010**, *5*, e9161. [CrossRef]

29. Ross, T.M.; Mahmood, K.; Crevar, C.J.; Schneider-Ohrum, K.; Heaton, P.M.; Bright, R.A. A trivalent virus-like particle vaccine elicits protective immune responses against seasonal influenza strains in mice and ferrets. *PLoS ONE* **2009**, *4*, e6032. [CrossRef]

30. Buckland, B.; Boulanger, R.; Fino, M.; Srivastava, I.; Holtz, K.; Khramtsov, N.; McPherson, C.; Meghrous, J.; Kubera, P.; Cox, M.M. Technology transfer and scale-up of the Flublok recombinant hemagglutinin (HA) influenza vaccine manufacturing process. *Vaccine* **2014**, *32*, 5496–5502. [CrossRef]

31. Liu, Y.V.; Massare, M.J.; Pearce, M.B.; Sun, X.; Belser, J.A.; Maines, T.R.; Creager, H.M.; Glenn, G.M.; Pushko, P.; Smith, G.E.; et al. Recombinant virus-like particles elicit protective immunity against avian influenza A(H7N9) virus infection in ferrets. *Vaccine* **2015**, *33*, 2152–2158. [CrossRef] [PubMed]

32. Smith, G.E.; Sun, X.; Bai, Y.; Liu, Y.V.; Massare, M.J.; Pearce, M.B.; Belser, J.A.; Maines, T.R.; Creager, H.M.; Glenn, G.M.; et al. Neuraminidase-based recombinant virus-like particles protect against lethal avian influenza A(H5N1) virus infection in ferrets. *Virology* **2017**, *509*, 90–97. [CrossRef]

33. Kingstad-Bakke, B.; Kamlangdee, A.; Osorio, J.E. Mucosal administration of raccoonpox virus expressing highly pathogenic avian H5N1 influenza neuraminidase is highly protective against H5N1 and seasonal influenza virus challenge. *Vaccine* **2015**, *33*, 5155–5162. [CrossRef] [PubMed]

34. Chen, Z.; Wang, J.; Bao, L.; Guo, L.; Zhang, W.; Xue, Y.; Zhou, H.; Xiao, Y.; Wang, J.; Wu, F.; et al. Human monoclonal antibodies targeting the haemagglutinin glycoprotein can neutralize H7N9 influenza virus. *Nat. Commun.* **2015**, *6*, 6714. [CrossRef] [PubMed]

35. Schmeisser, F.; Vasudevan, A.; Verma, S.; Wang, W.; Alvarado, E.; Weiss, C.; Atukorale, V.; Meseda, C.; Weir, J.P. Antibodies to antigenic site A of influenza H7 hemagglutinin provide protection against H7N9 challenge. *PLoS ONE* **2015**, *10*, e0117108. [CrossRef] [PubMed]

36. Perez Rubio, A.; Eiros, J.M. Cell culture-derived flu vaccine: Present and future. *Hum. Vaccines Immunother.* **2018**, *14*, 1874–1882. [CrossRef]

37. Krammer, F.; Albrecht, R.A.; Tan, G.S.; Margine, I.; Hai, R.; Schmolke, M.; Runstadler, J.; Andrews, S.F.; Wilson, P.C.; Cox, R.J.; et al. Divergent H7 immunogens offer protection from H7N9 virus challenge. *J. Virol.* **2014**, *88*, 3976–3985. [CrossRef]

38. Pushko, P.; Pujanauski, L.M.; Sun, X.; Pearce, M.; Hidajat, R.; Kort, T.; Schwartzman, L.M.; Tretyakova, I.; Chunqing, L.; Taubenberger, J.K.; et al. Recombinant H7 Hemagglutinin Forms Subviral Particles that Protect Mice and Ferrets from Challenge with H7N9 Influenza Virus submitted for publication. *Vaccine* **2015**, *33*, 4975–4982. [CrossRef]

39. Pillet, S.; Racine, T.; Nfon, C.; Di Lenardo, T.Z.; Babiuk, S.; Ward, B.J.; Kobinger, G.P.; Landry, N. Plant-derived H7 VLP vaccine elicits protective immune response against H7N9 influenza virus in mice and ferrets. *Vaccine* **2015**, *33*, 6282–6289. [CrossRef]

40. Klausberger, M.; Wilde, M.; Palmberger, D.; Hai, R.; Albrecht, R.A.; Margine, I.; Hirsh, A.; Garcia-Sastre, A.; Grabherr, R.; Krammer, F. One-shot vaccination with an insect cell-derived low-dose influenza A H7 virus-like particle preparation protects mice against H7N9 challenge. *Vaccine* **2014**, *32*, 355–362. [CrossRef]

41. Schwartzman, L.M.; Cathcart, A.L.; Pujanauski, L.M.; Qi, L.; Kash, J.C.; Taubenberger, J.K. An Intranasal Virus-Like Particle Vaccine Broadly Protects Mice from Multiple Subtypes of Influenza A Virus. *MBio* **2015**, *6*, e01044. [CrossRef]

42. Smith, G.E.; Flyer, D.C.; Raghunandan, R.; Liu, Y.; Wei, Z.; Wu, Y.; Kpamegan, E.; Courbron, D.; Fries, L.F., 3rd; Glenn, G.M. Development of influenza H7N9 virus like particle (VLP) vaccine: Homologous A/Anhui/1/2013 (H7N9) protection and heterologous A/chicken/Jalisco/CPA1/2012 (H7N3) cross-protection in vaccinated mice challenged with H7N9 virus. *Vaccine* **2013**, *31*, 4305–4313. [CrossRef]

43. Zhang, L.; Lu, J.; Chen, Y.; Shi, F.; Yu, H.; Huang, C.; Cui, L.; Shi, Z.; Jiao, Y.; Hu, Y. Characterization of Humoral Responses Induced by an H7N9 Influenza Virus-Like Particle Vaccine in BALB/C Mice. *Viruses* **2015**, *7*, 4369–4384. [CrossRef] [PubMed]

44. Xu, L.; Bao, L.; Lau, S.Y.; Wu, W.L.; Yuan, J.; Gu, S.; Li, F.; Lv, Q.; Xu, Y.; Pushko, P.; et al. Hemagglutinin amino acids related to receptor specificity could affect the protection efficacy of H5N1 and H7N9 avian influenza virus vaccines in mice. *Vaccine* **2016**, *34*, 2627–2633. [CrossRef] [PubMed]

45. Hu, C.J.; Chien, C.Y.; Liu, M.T.; Fang, Z.S.; Chang, S.Y.; Juang, R.H.; Chang, S.C.; Chen, H.W. Multi-antigen avian influenza a (H7N9) virus-like particles: Particulate characterizations and immunogenicity evaluation in murine and avian models. *BMC Biotechnol.* **2017**, *17*, 2. [CrossRef] [PubMed]

46. Ren, Z.; Zhao, Y.; Liu, J.; Ji, X.; Meng, L.; Wang, T.; Sun, W.; Zhang, K.; Sang, X.; Yu, Z.; et al. Intramuscular and intranasal immunization with an H7N9 influenza virus-like particle vaccine protects mice against lethal influenza virus challenge. *Int. Immunopharmacol.* **2018**, *58*, 109–116. [CrossRef]

47. Pushko, P.; Tretyakova, I.; Hidajat, R.; Zsak, A.; Chrzastek, K.; Tumpey, T.M.; Kapczynski, D.R. Virus-like particles displaying H5, H7, H9 hemagglutinins and N1 neuraminidase elicit protective immunity to heterologous avian influenza viruses in chickens. *Virology* **2017**, *501*, 176–182. [CrossRef]

48. Hong, J.Y.; Chen, T.H.; Chen, Y.J.; Liu, C.C.; Jan, J.T.; Wu, S.C. Highly immunogenic influenza virus-like particles containing B-cell-activating factor (BAFF) for multi-subtype vaccine development. *Antivir. Res.* **2019**, *164*, 12–22. [CrossRef]

49. Kang, H.J.; Chu, K.B.; Lee, D.H.; Lee, S.H.; Park, B.R.; Kim, M.C.; Kang, S.M.; Quan, F.S. Influenza M2 virus-like particle vaccination enhances protection in combination with avian influenza HA VLPs. *PLoS ONE* **2019**, *14*, e0216871. [CrossRef]

50. Tretyakova, I.; Pearce, M.B.; Florese, R.; Tumpey, T.M.; Pushko, P. Intranasal vaccination with H5, H7 and H9 hemagglutinins co-localized in a virus-like particle protects ferrets from multiple avian influenza viruses. *Virology* **2013**, *442*, 67–73. [CrossRef]

51. Hervas-Stubbs, S.; Rueda, P.; Lopez, L.; Leclerc, C. Insect baculoviruses strongly potentiate adaptive immune responses by inducing type I IFN. *J. Immunol.* **2007**, *178*, 2361–2369. [CrossRef]

52. Buonaguro, L.; Tornesello, M.L.; Tagliamonte, M.; Gallo, R.C.; Wang, L.X.; Kamin-Lewis, R.; Abdelwahab, S.; Lewis, G.K.; Buonaguro, F.M. Baculovirus-derived human immunodeficiency virus type 1 virus-like particles activate dendritic cells and induce ex vivo T-cell responses. *J. Virol.* **2006**, *80*, 9134–9143. [CrossRef]

53. Sailaja, G.; Skountzou, I.; Quan, F.S.; Compans, R.W.; Kang, S.M. Human immunodeficiency virus-like particles activate multiple types of immune cells. *Virology* **2007**, *362*, 331–341. [CrossRef] [PubMed]

54. Bachmann, M.F.; Rohrer, U.H.; Kundig, T.M.; Burki, K.; Hengartner, H.; Zinkernagel, R.M. The influence of antigen organization on B cell responsiveness. *Science* **1993**, *262*, 1448–1451. [CrossRef] [PubMed]

55. McCraw, D.M.; Gallagher, J.R.; Torian, U.; Myers, M.L.; Conlon, M.T.; Gulati, N.M.; Harris, A.K. Structural analysis of influenza vaccine virus-like particles reveals a multicomponent organization. *Sci. Rep.* **2018**, *8*, 10342. [CrossRef] [PubMed]

56. Harris, A.; Cardone, G.; Winkler, D.C.; Heymann, J.B.; Brecher, M.; White, J.M. Influenza virus pleiomorphy characterized by cryoelectron tomography. *Proc. Natl. Acad. Sci. USA* **2006**, *103*, 19123–19127. [CrossRef]

57. Buffin, S.; Peubez, I.; Barriere, F.; Nicolai, M.C.; Tapia, T.; Dhir, V.; Forma, E.; Seve, N.; Legastelois, I. Influenza A and B virus-like particles produced in mammalian cells are highly immunogenic and induce functional antibodies. *Vaccine* 2019. [CrossRef]

58. Venereo-Sanchez, A.; Fulton, K.; Koczka, K.; Twine, S.; Chahal, P.; Ansorge, S.; Gilbert, R.; Henry, O.; Kamen, A. Characterization of influenza H1N1 Gag virus-like particles and extracellular vesicles co-produced in HEK-293SF. *Vaccine* **2019**, *37*, 7100–7107. [CrossRef] [PubMed]

59. Venereo-Sanchez, A.; Gilbert, R.; Simoneau, M.; Caron, A.; Chahal, P.; Chen, W.; Ansorge, S.; Li, X.; Henry, O.; Kamen, A. Hemagglutinin and neuraminidase containing virus-like particles produced in HEK-293 suspension culture: An effective influenza vaccine candidate. *Vaccine* **2016**, *34*, 3371–3380. [CrossRef]

60. Venereo-Sanchez, A.; Simoneau, M.; Lanthier, S.; Chahal, P.; Bourget, L.; Ansorge, S.; Gilbert, R.; Henry, O.; Kamen, A. Process intensification for high yield production of influenza H1N1 Gag virus-like particles using an inducible HEK-293 stable cell line. *Vaccine* **2017**, *35*, 4220–4228. [CrossRef]

61. Thompson, C.M.; Petiot, E.; Mullick, A.; Aucoin, M.G.; Henry, O.; Kamen, A.A. Critical assessment of influenza VLP production in Sf9 and HEK293 expression systems. *BMC Biotechnol.* **2015**, *15*, 31. [CrossRef]

62. Zak, A.J.; Hill, B.D.; Rizvi, S.M.; Smith, M.R.; Yang, M.; Wen, F. Enhancing the Yield and Quality of Influenza Virus-like Particles (VLPs) Produced in Insect Cells by Inhibiting Cytopathic Effects of Matrix Protein M2. *ACS Synth. Biol.* **2019**, *8*, 2303–2314. [CrossRef]

63. Li, X.; Pushko, P.; Tretyakova, I. Recombinant Hemagglutinin and Virus-Like Particle Vaccines for H7N9 Influenza Virus. *J. Vaccines Vaccin.* **2015**, *6*, 287. [PubMed]

64. Prabakaran, M.; Kumar, S.R.; Raj, K.V.; Wu, X.; He, F.; Zhou, J.; Kwang, J. Cross-protective efficacy of baculovirus displayed hemagglutinin against highly pathogenic influenza H7 subtypes. *Antivir. Res.* **2014**, *109*, 149–159. [CrossRef] [PubMed]

65. Hu, J.; Liang, Y.; Hu, Z.; Wang, X.; Gu, M.; Li, R.; Ma, C.; Liu, X.; Hu, S.; Chen, S.; et al. Recombinant baculovirus vaccine expressing hemagglutinin of H7N9 avian influenza virus confers full protection against lethal highly pathogenic H7N9 virus infection in chickens. *Arch. Virol.* **2019**, *164*, 807–817. [CrossRef] [PubMed]

66. Makarkov, A.I.; Chierzi, S.; Pillet, S.; Murai, K.K.; Landry, N.; Ward, B.J. Plant-made virus-like particles bearing influenza hemagglutinin (HA) recapitulate early interactions of native influenza virions with human monocytes/macrophages. *Vaccine* **2017**, *35*, 4629–4636. [CrossRef] [PubMed]

67. Lindsay, B.J.; Bonar, M.M.; Costas-Cancelas, I.N.; Hunt, K.; Makarkov, A.I.; Chierzi, S.; Krawczyk, C.M.; Landry, N.; Ward, B.J.; Rouiller, I. Morphological characterization of a plant-made virus-like particle vaccine bearing influenza virus hemagglutinins by electron microscopy. *Vaccine* **2018**, *36*, 2147–2154. [CrossRef]

68. Le Mauff, F.; Mercier, G.; Chan, P.; Burel, C.; Vaudry, D.; Bardor, M.; Vezina, L.P.; Couture, M.; Lerouge, P.; Landry, N. Biochemical composition of haemagglutinin-based influenza virus-like particle vaccine produced by transient expression in tobacco plants. *Plant Biotechnol. J.* **2015**, *13*, 717–725. [CrossRef]

69. Holtz, K.M.; Robinson, P.S.; Matthews, E.E.; Hashimoto, Y.; McPherson, C.E.; Khramtsov, N.; Reifler, M.J.; Meghrous, J.; Rhodes, D.G.; Cox, M.M.; et al. Modifications of cysteine residues in the transmembrane and cytoplasmic domains of a recombinant hemagglutinin protein prevents cross-linked multimer formation and potency loss. *BMC Biotechnol.* **2014**, *14*, 1. [CrossRef]

70. Latham, T.; Galarza, J.M. Formation of wild-type and chimeric influenza virus-like particles following simultaneous expression of only four structural proteins. *J. Virol.* **2001**, *75*, 6154–6165. [CrossRef]

71. Pushko, P.; Pearce, M.B.; Ahmad, A.; Tretyakova, I.; Smith, G.; Belser, J.A.; Tumpey, T.M. Influenza virus-like particle can accommodate multiple subtypes of hemagglutinin and protect from multiple influenza types and subtypes. *Vaccine* **2011**, *29*, 5911–5918. [CrossRef]

72. Thompson, C.M.; Petiot, E.; Lennaertz, A.; Henry, O.; Kamen, A.A. Analytical technologies for influenza virus-like particle candidate vaccines: Challenges and emerging approaches. *Virol. J.* **2013**, *10*, 141. [CrossRef]

73. Marcelin, G.; DuBois, R.; Rubrum, A.; Russell, C.J.; McElhaney, J.E.; Webby, R.J. A contributing role for anti-neuraminidase antibodies on immunity to pandemic H1N1 2009 influenza A virus. *PLoS ONE* **2011**, *6*, e26335. [CrossRef] [PubMed]

74. Quan, F.S.; Kim, M.C.; Lee, B.J.; Song, J.M.; Compans, R.W.; Kang, S.M. Influenza M1 VLPs containing neuraminidase induce heterosubtypic cross-protection. *Virology* **2012**, *430*, 127–135. [CrossRef] [PubMed]

75. Gomes, A.C.; Flace, A.; Saudan, P.; Zabel, F.; Cabral-Miranda, G.; Turabi, A.E.; Manolova, V.; Bachmann, M.F. Adjusted Particle Size Eliminates the Need of Linkage of Antigen and Adjuvants for Appropriated T Cell Responses in Virus-Like Particle-Based Vaccines. *Front. Immunol.* **2017**, *8*, 226. [CrossRef] [PubMed]

76. Quan, F.S.; Ko, E.J.; Kwon, Y.M.; Joo, K.H.; Compans, R.W.; Kang, S.M. Mucosal adjuvants for influenza virus-like particle vaccine. *Viral Immunol.* **2013**, *26*, 385–395. [CrossRef] [PubMed]

77. Schneider-Ohrum, K.; Giles, B.M.; Weirback, H.K.; Williams, B.L.; DeAlmeida, D.R.; Ross, T.M. Adjuvants that stimulate TLR3 or NLPR3 pathways enhance the efficiency of influenza virus-like particle vaccines in aged mice. *Vaccine* **2011**, *29*, 9081–9092. [CrossRef] [PubMed]

78. Zhou, Y.; Li, S.; Bi, S.; Li, N.; Bi, Y.; Liu, W.; Wang, B. Long-lasting protective immunity against H7N9 infection is induced by intramuscular or CpG-adjuvanted intranasal immunization with the split H7N9 vaccine. *Int. Immunopharmacol.* **2020**, *78*, 106013. [CrossRef]

79. Fries, L.F.; Smith, G.E.; Glenn, G.M. A recombinant viruslike particle influenza A (H7N9) vaccine. *N. Engl. J. Med.* **2013**, *369*, 2564–2566. [CrossRef]

80. Chung, K.Y.; Coyle, E.M.; Jani, D.; King, L.R.; Bhardwaj, R.; Fries, L.; Smith, G.; Glenn, G.; Golding, H.; Khurana, S. ISCOMATRIX adjuvant promotes epitope spreading and antibody affinity maturation of influenza A H7N9 virus like particle vaccine that correlate with virus neutralization in humans. *Vaccine* **2015**, *33*, 3953–3962. [CrossRef]

81. Zheng, D.; Gao, F.; Zhao, C.; Ding, Y.; Cao, Y.; Yang, T.; Xu, X.; Chen, Z. Comparative effectiveness of H7N9 vaccines in healthy individuals. *Hum. Vaccin. Immunother.* **2019**, *15*, 80–90. [CrossRef]

82. Pillet, S.; Aubin, E.; Trepanier, S.; Bussiere, D.; Dargis, M.; Poulin, J.F.; Yassine-Diab, B.; Ward, B.J.; Landry, N. A plant-derived quadrivalent virus like particle influenza vaccine induces cross-reactive antibody and T cell response in healthy adults. *Clin. Immunol.* **2016**, *168*, 72–87. [CrossRef]

83. Osterholm, M.T.; Kelley, N.S.; Sommer, A.; Belongia, E.A. Efficacy and effectiveness of influenza vaccines: A systematic review and meta-analysis. *Lancet Infect. Dis.* **2012**, *12*, 36–44. [CrossRef]

84. Tretyakova, I.; Hidajat, R.; Hamilton, G.; Horn, N.; Nickols, B.; Prather, R.O.; Tumpey, T.M.; Pushko, P. Preparation of quadri-subtype influenza virus-like particles using bovine immunodeficiency virus gag protein. *Virology* **2016**, *487*, 163–171. [CrossRef] [PubMed]

85. Pushko, P.; Sun, X.; Tretyakova, I.; Hidajat, R.; Pulit-Penaloza, J.A.; Belser, J.A.; Maines, T.R.; Tumpey, T.M. Mono- and quadri-subtype virus-like particles (VLPs) containing H10 subtype elicit protective immunity to H10 influenza in a ferret challenge model. *Vaccine* **2016**, *34*, 5235–5242. [CrossRef] [PubMed]

86. Pushko, P.; Tretyakova, I.; Hidajat, R.; Sun, X.; Belser, J.A.; Tumpey, T.M. Multi-clade H5N1 virus-like particles: Immunogenicity and protection against H5N1 virus and effects of beta-propiolactone. *Vaccine* **2018**, *36*, 4346–4353. [CrossRef] [PubMed]

87. Klausberger, M.; Leneva, I.A.; Egorov, A.; Strobl, F.; Ghorbanpour, S.M.; Falynskova, I.N.; Poddubikov, A.V.; Makhmudova, N.R.; Krokhin, A.; Svitich, O.A.; et al. Off-target effects of an insect cell-expressed influenza HA-pseudotyped Gag-VLP preparation in limiting postinfluenza Staphylococcus aureus infections. *Vaccine* **2020**, *38*, 859–867. [CrossRef] [PubMed]

88. Goff, P.H.; Krammer, F.; Hai, R.; Seibert, C.W.; Margine, I.; Garcia-Sastre, A.; Palese, P. Induction of cross-reactive antibodies to novel H7N9 influenza virus by recombinant Newcastle disease virus expressing a North American lineage H7 subtype hemagglutinin. *J. Virol.* **2013**, *87*, 8235–8240. [CrossRef] [PubMed]

89. Liu, Q.; Mena, I.; Ma, J.; Bawa, B.; Krammer, F.; Lyoo, Y.S.; Lang, Y.; Morozov, I.; Mahardika, G.N.; Ma, W.; et al. Newcastle Disease Virus-Vectored H7 and H5 Live Vaccines Protect Chickens from Challenge with H7N9 or H5N1 Avian Influenza Viruses. *J. Virol.* **2015**, *89*, 7401–7408. [CrossRef]

90. Hu, Z.; Liu, X.; Jiao, X.; Liu, X. Newcastle disease virus (NDV) recombinant expressing the hemagglutinin of H7N9 avian influenza virus protects chickens against NDV and highly pathogenic avian influenza A (H7N9) virus challenges. *Vaccine* **2017**, *35*, 6585–6590. [CrossRef]

91. Li, Z.; Gabbard, J.D.; Johnson, S.; Dlugolenski, D.; Phan, S.; Tompkins, S.M.; He, B. Efficacy of a parainfluenza virus 5 (PIV5)-based H7N9 vaccine in mice and guinea pigs: Antibody titer towards HA was not a good indicator for protection. *PLoS ONE* **2015**, *10*, e0120355. [CrossRef] [PubMed]

92. Mallajosyula, V.V.; Citron, M.; Ferrara, F.; Temperton, N.J.; Liang, X.; Flynn, J.A.; Varadarajan, R. Hemagglutinin Sequence Conservation Guided Stem Immunogen Design from Influenza A H3 Subtype. *Front. Immunol.* **2015**, *6*, 329. [CrossRef]

93. Ellebedy, A.H.; Krammer, F.; Li, G.M.; Miller, M.S.; Chiu, C.; Wrammert, J.; Chang, C.Y.; Davis, C.W.; McCausland, M.; Elbein, R.; et al. Induction of broadly cross-reactive antibody responses to the influenza HA stem region following H5N1 vaccination in humans. *Proc. Natl. Acad. Sci. USA* **2014**, *111*, 13133–13138. [CrossRef] [PubMed]

94. Kazaks, A.; Lu, I.N.; Farinelle, S.; Ramirez, A.; Crescente, V.; Blaha, B.; Ogonah, O.; Mukhopadhyay, T.; de Obanos, M.P.; Krimer, A.; et al. Production and purification of chimeric HBc virus-like particles carrying influenza virus LAH domain as vaccine candidates. *BMC Biotechnol.* **2017**, *17*, 79. [CrossRef] [PubMed]

95. Zheng, D.; Chen, S.; Qu, D.; Chen, J.; Wang, F.; Zhang, R.; Chen, Z. Influenza H7N9 LAH-HBc virus-like particle vaccine with adjuvant protects mice against homologous and heterologous influenza viruses. *Vaccine* **2016**, *34*, 6464–6471. [CrossRef] [PubMed]

96. Song, L.; Xiong, D.; Kang, X.; Yang, Y.; Wang, J.; Guo, Y.; Xu, H.; Chen, S.; Peng, D.; Pan, Z.; et al. An avian influenza A (H7N9) virus vaccine candidate based on the fusion protein of hemagglutinin globular head and Salmonella typhimurium flagellin. *BMC Biotechnol.* **2015**, *15*, 79. [CrossRef] [PubMed]

97. Wei, H.J.; Chang, W.; Lin, S.C.; Liu, W.C.; Chang, D.K.; Chong, P.; Wu, S.C. Fabrication of influenza virus-like particles using M2 fusion proteins for imaging single viruses and designing vaccines. *Vaccine* **2011**, *29*, 7163–7172. [CrossRef]

98. Liu, W.C.; Liu, Y.Y.; Chen, T.H.; Liu, C.C.; Jan, J.T.; Wu, S.C. Multi-subtype influenza virus-like particles incorporated with flagellin and granulocyte-macrophage colony-stimulating factor for vaccine design. *Antivir. Res.* **2016**, *133*, 110–118. [CrossRef]

99. Gong, X.; Yin, H.; Shi, Y.; He, X.; Yu, Y.; Guan, S.; Kuai, Z.; Haji, N.M.; Haji, N.M.; Kong, W.; et al. Evaluation of the immunogenicity and protective effects of a trivalent chimeric norovirus P particle immunogen displaying influenza HA2 from subtypes H1, H3 and, B. *Emerg. Microbes Infect.* **2016**, *5*, e51. [CrossRef]

100. Pham, N.B.; Ho, T.T.; Nguyen, G.T.; Le, T.T.; Le, N.T.; Chang, H.C.; Pham, M.D.; Conrad, U.; Chu, H.H. Nanodiamond enhances immune responses in mice against recombinant HA/H7N9 protein. *J. Nanobiotechnol.* **2017**, *15*, 69. [CrossRef]

101. Kim, M.C.; Lee, J.S.; Kwon, Y.M.; O, E.; Lee, Y.J.; Choi, J.G.; Wang, B.Z.; Compans, R.W.; Kang, S.M. Multiple heterologous M2 extracellular domains presented on virus-like particles confer broader and stronger M2 immunity than live influenza A virus infection. *Antivir. Res.* **2013**, *99*, 328–335. [CrossRef]

102. Kim, K.H.; Kwon, Y.M.; Lee, Y.T.; Kim, M.C.; Hwang, H.S.; Ko, E.J.; Lee, Y.; Choi, H.J.; Kang, S.M. Virus-Like Particles Are a Superior Platform for Presenting M2e Epitopes to Prime Humoral and Cellular Immunity against Influenza Virus. *Vaccines* **2018**, *6*, 66. [CrossRef]

103. Lee, Y.T.; Kim, K.H.; Ko, E.J.; Kim, M.C.; Lee, Y.N.; Hwang, H.S.; Lee, Y.; Jung, Y.J.; Kim, Y.J.; Santos, J.; et al. Enhancing the cross protective efficacy of live attenuated influenza virus vaccine by supplemented vaccination with M2 ectodomain virus-like particles. *Virology* **2019**, *529*, 111–121. [CrossRef] [PubMed]

104. Yong, C.Y.; Yeap, S.K.; Ho, K.L.; Omar, A.R.; Tan, W.S. Potential recombinant vaccine against influenza A virus based on M2e displayed on nodaviral capsid nanoparticles. *Int. J. Nanomed.* **2015**, *10*, 2751–2763.

105. Ong, H.K.; Yong, C.Y.; Tan, W.S.; Yeap, S.K.; Omar, A.R.; Razak, M.A.; Ho, K.L. An Influenza A Vaccine Based on the Extracellular Domain of Matrix 2 Protein Protects BALB/C Mice Against H1N1 and H3N2. *Vaccines* **2019**, *7*, 91. [CrossRef] [PubMed]

106. Lahiri, A.; Sharif, S.; Mallick, A.I. Intragastric delivery of recombinant Lactococcus lactis displaying ectodomain of influenza matrix protein 2 (M2e) and neuraminidase (NA) induced focused mucosal and systemic immune responses in chickens. *Mol. Immunol.* **2019**, *114*, 497–512. [CrossRef] [PubMed]

107. Tang, X.C.; Lu, H.R.; Ross, T.M. Hemagglutinin displayed baculovirus protects against highly pathogenic influenza. *Vaccine* **2010**, *28*, 6821–6831. [CrossRef]

108. Ma, H.; Nandakumar, S.; Bae, E.H.; Chin, P.J.; Khan, A.S. The Spodoptera frugiperda Sf9 cell line is a heterogeneous population of rhabdovirus-infected and virus-negative cells: Isolation and characterization of cell clones containing rhabdovirus X-gene variants and virus-negative cell clones. *Virology* **2019**, *536*, 125–133. [CrossRef]

109. Geisler, C.; Jarvis, D.L. Adventitious viruses in insect cell lines used for recombinant protein expression. *Protein Expr. Purif.* **2018**, *144*, 25–32. [CrossRef]

110. Margine, I.; Palese, P.; Krammer, F. Expression of functional recombinant hemagglutinin and neuraminidase proteins from the novel H7N9 influenza virus using the baculovirus expression system. *J. Vis. Exp.* **2013**, *81*, e51112. [CrossRef]

111. Lai, C.C.; Cheng, Y.C.; Chen, P.W.; Lin, T.H.; Tzeng, T.T.; Lu, C.C.; Lee, M.S.; Hu, A.Y. Process development for pandemic influenza VLP vaccine production using a baculovirus expression system. *J. Biol. Eng.* **2019**, *13*, 78. [CrossRef]

112. Effio, C.L.; Hubbuch, J. Next generation vaccines and vectors: Designing downstream processes for recombinant protein-based virus-like particles. *Biotechnol. J.* **2015**, *10*, 715–727. [CrossRef]

113. Park, Y.C.; Song, J.M. Preparation and immunogenicity of influenza virus-like particles using nitrocellulose membrane filtration. *Clin. Exp. Vaccine Res.* **2017**, *6*, 61–66. [CrossRef] [PubMed]

114. Carvalho, S.B.; Silva, R.J.S.; Moleirinho, M.G.; Cunha, B.; Moreira, A.S.; Xenopoulos, A.; Alves, P.M.; Carrondo, M.J.T.; Peixoto, C. Membrane-Based Approach for the Downstream Processing of Influenza Virus-Like Particles. *Biotechnol. J.* **2019**, *14*, e1800570. [CrossRef] [PubMed]

115. Durous, L.; Rosa-Calatrava, M.; Petiot, E. Advances in influenza virus-like particles bioprocesses. *Expert Rev. Vaccines* **2019**, *18*, 1285–1300. [CrossRef] [PubMed]

116. Reiter, K.; Aguilar, P.P.; Grammelhofer, D.; Joseph, J.; Steppert, P.; Jungbauer, A. Separation of influenza virus-like particles from baculovirus by polymer grafted anion-exchanger. *J. Sep. Sci.* **2020**. [CrossRef]

117. Li, C.; Xu, K.; Hashem, A.; Shao, M.; Liu, S.; Zou, Y.; Gao, Q.; Zhang, Y.; Yuan, L.; Xu, M.; et al. Collaborative studies on the development of national reference standards for potency determination of H7N9 influenza vaccine. *Hum. Vaccin Immunother.* **2015**, *11*, 1351–1356. [CrossRef]

118. Guo, J.; Lu, Y.; Zhang, Y.; Mugabe, S.; Wei, Z.; Borisov, O.V. Development and fit-for-purpose verification of an LC-MS method for quantitation of hemagglutinin and neuraminidase proteins in influenza virus-like particle vaccine candidates. *Anal. Biochem.* **2020**, *592*, 113577. [CrossRef]

 *viruses*

MDPI

Review

Virus-Like Particles as an Immunogenic Platform for Cancer Vaccines

Jerri C. Caldeira [1], Michael Perrine [1], Federica Pericle [2] and Federica Cavallo [3,*]

[1]   AgilVax, Inc., Albuquerque, NM 87110, USA; jcaldeira@agilvax.com (J.C.C.); mperrine@agilvax.com (M.P.)
[2]   Border Biomedical Research Center, University of Texas at El Paso, El Paso, TX 79968, USA;
      federica_pericle@outlook.com
[3]   Department of Molecular Biotechnology and Health Sciences, Molecular Biotechnology Center,
      University of Torino, 10124 Torino, Italy
*   Correspondence: federica.cavallo@unito.it; Tel.: +39-011-670-6457

Received: 6 March 2020; Accepted: 24 April 2020; Published: 27 April 2020

Abstract: Virus-like particles (VLP) spontaneously assemble from viral structural proteins. They are naturally biocompatible and non-infectious. VLP can serve as a platform for many potential vaccine epitopes, display them in a dense repeating array, and elicit antibodies against non-immunogenic substances, including tumor-associated self-antigens. Genetic or chemical conjugation facilitates the multivalent display of a homologous or heterologous epitope. Most VLP range in diameter from 25 to 100 nm and, in most cases, drain freely into the lymphatic vessels and induce antibodies with high titers and affinity without the need for additional adjuvants. VLP administration can be performed using different strategies, regimens, and doses to improve the immunogenicity of the antigen they expose on their surface. This article summarizes the features of VLP and presents them as a relevant platform technology to address not only infectious diseases but also chronic diseases and cancer.

Keywords: virus-like particles; vaccine; cancer; immunotherapy

---

1. Introduction

Vaccines are among humanity's most significant achievements. By the end of the 20th century, our efforts to prevent infectious diseases through vaccination and public sanitation were so successful that life expectancies had increased by approximately 30 years. Historically, vaccines have been used primarily as measures to prevent infectious diseases, but that focus has begun to shift as new strategies emerged on their possible use as treatments for certain chronic infectious diseases and cancer. The challenge is to find technical approaches that maximize immunogenicity without compromising safety, tolerability, and efficacy. Virus-like particles (VLP) are an attractive option.

Over the last three decades, genetically or chemically modified VLP have been used to present different classes of epitopes on their surface for diverse applications. Their ability to induce humoral and cellular response has been confirmed by many preclinical and clinical studies [1] and has offered a substantial advancement in the vaccine field. VLP vaccines are used to induce immune response against several diseases in animals [2] and humans. Some examples of treatment are osteoporosis [3], chronic pain [4], type 2 diabetes [5], and tauopathy [6]. Further, other studies have addressed the possibility of using VLP to stimulate the immune response against cancer antigens.

Here, we describe the main features of VLP as a vaccination strategy and their interaction with the immune system before focusing on VLP as cancer vaccines.

2. VLP as a Vaccination Strategy

VLP are nanoparticles that are spontaneously assembled from viral structural proteins. Structurally, they are practically indistinguishable from their corresponding viruses and, a VLP derived from the structural proteins of a viral pathogen can often serve as a highly effective vaccine for that pathogen. The structural protein of Human Papillomavirus (HPV), for example, assembles into a VLP that elicits antibodies that protect against HPV infection and thereby prevent cervical cancer. Similarly, the Hepatitis B Virus (HBV) surface antigen self-assembles into a VLP vaccine that effectively prevents HBV infection [7,8]. Further, in addition to serving as vaccines against the viruses from which they are derived, VLP can function as scaffolds for the presentation of epitopes from any source [9–13]. Most VLP vaccines are formulated with adjuvants and the immunogenicity observed in the preclinical studies need to be confirmed in human trials. Hence, thanks to the high immunogenicity typical of virus particles, VLP serve as platforms for the presentation of a wide variety of potential vaccine epitopes.

2.1. VLP-based Technologies

VLP are naturally biocompatible. They are originated from two groups of viruses, non-enveloped and enveloped, and both groups have been used to display foreign antigens [14]. They have no viral genome and are, therefore, not contagious. VLP are effectively eliminated or degraded, which limits the occurrence of side effects [15]. They present epitopes in dense repetitive arrays, making them effective scaffolds that can elicit antibodies to multiple substances (Table 1).

Table 1. Summary of the virus-like particle (VLP) vaccines described.

Platforms	Targets	Antigens	Types of Vaccines	References
Bacteriophage	Influenza	M2	Preventive	[9]
AP205	Breast cancer	HER2 protein	Preventive	[16]
	Cervical cancer	L2 epitope	Preventive	[17]
PP7	Cervical cancer	L2 (epitope 17–31)	Preventive	[18]
	HCG	C-terminus	Preventive	[11]
	Nicotine abuse	Nicotine	Therapeutic	[19]
	Cholesterol	huPCSK9	Therapeutic	[10]
	Alzheimer	pT181	Therapeutic	[6]
Qβ	Osteoporosis	TRANCE/RANKL	Preventive	[3]
	Chronic pain	aa 19–241NGF	Therapeutic	[4]
	Diabetes type 2	IL-1β	Therapeutic	[5]
	Diabetes type 2	h IL-1β	Therapeutic	[20]
	Melanoma	GL/mutated-MTV Mix-MTV	Therapeutic	[21]
MS2	Breast cancer	xCT	Therapeutic	[22]
	Cervical cancer	L2 (epitope17-31)	Preventive	[23]
CuMV	Melanoma	TT830–843 epitope	Therapeutic	[24]
RHDV	HPV16 tumor	MHC I-restricted (aa 48–57) HPV16 E6	Therapeutic	[25]
	Melanoma	H-2Db	Therapeutic	[26]
	Colorectal cancer	Topoisomerase IIα, survivin	Preventive	[27]
HBV	Hepatocellular Cancer	HBV X protein-derived epitopes	Preventive	[28]
eCPMV	Melanoma	empty	Therapeutic	[29]
SHIV	Pancreatic cancer	hMSLN	Therapeutic	[30]
SIV	Pancreatic cancer	mTrop2	Therapeutic	[31]

Peptides, small-molecule haptens, and self-antigens can elicit high-titer antibody responses when presented on the surface of a VLP [32–35]. VLP immunogenicity, however, is not due to multivalence alone. Most VLP are small particles with a diameter in the range of 25 to 100 nanometer (nm), a size that allows optimal entry into lymphatic vessels, passive drainage to the subcapsular region of lymph nodes (LN), and uptake by professional antigen-presenting cells. Studies in mice conditionally depleted of dendritic cells (DCs) confirm that small particles drain freely to the LN [36,37], while DC transport large particles from the injection site to the LN [38].

There are two primary approaches for the presentation of epitopes on VLP. First, synthetic peptides containing the desired target epitope can be cross-linked chemically to the VLP surface.

VLP-based vaccines produced in this way are very effective in generating an immune response against display molecules and inducing high titers of neutralizing antibodies [19,20,39,40]. In this approach, the VLP, the epitope, and the linker are synthesized separately and combined later in the cross-linking reaction to produce the vaccine. The chemical cross-linking process increases the cost of production [41] but avoids potential protein folding/self-assembly issues that sometimes occur due to genetic fusion. The chemical approach also sometimes allows for higher display valences. Variations on this cross-linking theme include the linkage of foreign peptides and proteins to VLP using enzymatic methods such as Sortase [42] or the so-called Spycatcher [43] technologies.

In the second approach, foreign peptides can be inserted genetically into the coding sequence of a viral structural protein. When expressed in an appropriate host, the protein self-assembles into a VLP with the peptide exposed on its surface. The most widely used VLP platforms for genetic peptide display are based on the woodchuck hepatitis virus (WHBV) [44] and several RNA bacteriophages [17]. Genetic insertion of foreign sequences sometimes prevents the recombinant protein from folding properly [2]. In the case of HBV core proteins, this problem has been largely overcome through the use of a series of genetic variants that can be employed in a combinatorial fashion to find a VLP that tolerates almost any desired insertion. In the case of RNA phage MS2 VLP, peptides are inserted into a surface loop of coat protein (the so-called AB loop). The wild-type coat protein is nearly always destabilized by such insertions, but a genetically engineered single-chain dimer version tolerates the vast majority of small peptide insertions, making it a nearly universal peptide display platform [22,32,45]. It is a consequence of the increased thermodynamic stability conferred by the covalent joining of two subunits associated non-covalently [46].

Whether displayed by genetic insertion or by chemical conjugation, a peptide linked to a VLP has the ability to elicit a high-titer, long-lived, epitope-specific antibody response. Naturally, each method has its advantages and liabilities. As mentioned above, genetic insertion sometimes fails to yield a VLP as a result of improper recombinant protein folding [2]. Chemical cross-linking of a synthetic peptide to a preformed VLP, on the other hand, obviously avoids this problem. Genetic fusions, however, can yield a more uniform product and can be produced biosynthetically in a single step and often in very high yields. Further, at least in the case of MS2 VLP, the genetic insertion has the added advantage of enabling epitope identification in a process akin to phage display by affinity-selection on antibody targets from complex random-sequence or antigen fragment libraries [47].

VLP can be recombinantly expressed in a variety of hosts, including bacteria, yeast, plant, insect, and mammalian cells [48], and is manufactured economically in large amounts with high purity under the current Good Manufacturing Practices (GMP) (Figure 1). Due to their biocompatibility, solubility, efficient uptake, and nanoscale dimensions, VLP also have potential as drug delivery vehicles [49]. The choice of expression host and fermentation process can affect the VLP yield, the integrity, the scale, the cost of production, and the purity of the final product, all factors that can have a significant impact on vaccine stability, efficacy, and safety [22,50,51].

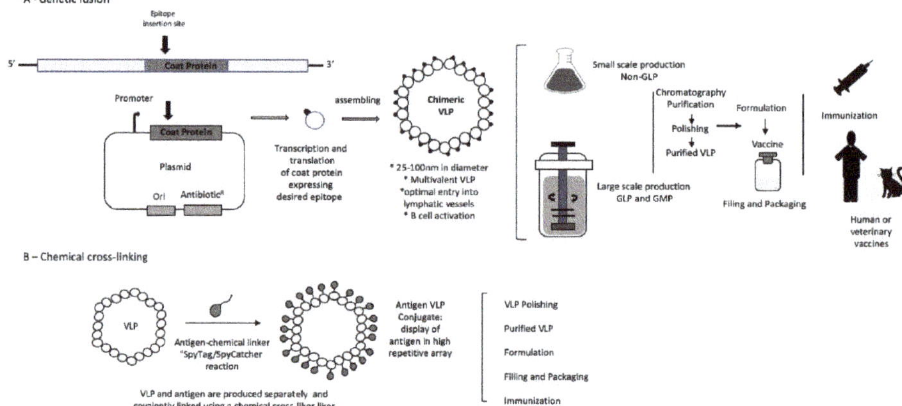

Figure 1. Illustration of virus-like particles (VLPs) production using different approaches. (**A**) Production of chimeric VLP using genetic insertion. The foreign antigen is fused to the coat protein by genetic engineering, and then chimeric VLP is expressed in a suitable host system. (**B**) Chimeric VLP is generated by chemical conjugation of foreign peptides to the surface of the VLP. The VLP production can be carried out on a small scale for scientific research, while under the current GMP, it can be produced on a large scale for human or veterinary.

Even though some VLP vaccines have been approved by the Food and Drug Administration (FDA) for human use, every decoration of a VLP with a new epitope creates an unknown particle that must be well characterized. Different versions of a VLP vaccine can be generated based on its size, and the nature or length of the epitope displayed, so the analysis of their molecular nature becomes essential and requires an appropriate method to confirm its structural properties. The most utilized methods for VLP characterization are analytic ultracentrifugation, high-performance chromatography, cryo-electron microscopy, atomic force microscopy, transmission electron microscopy and dynamic light scattering. These methods confirm the integrity of the VLP for epitope display, which is a key factor for inducing a functional immune response against the target antigen [52].

2.2. VLP Toxicology

The source of the potential toxicity of the VLP vaccine comes from the immunization or the immune response it induces. Since VLP-based vaccines are potent inducers of the immune response, safety studies are necessary to examine the potential toxicities of its components and pharmacodynamic. Any of a VLP's main components—the protein capsid (coat protein), any packaged nucleic acids, the chemical cross-linker, and the displayed antigen—could present some potential toxicity. If an adjuvant is used for vaccination, its potential toxicity must also be considered.

As mentioned above, sometimes, VLP from different viruses such as Q-β, Cowpea chlorotic Mottle Virus (CCMV), and MS2, can incorporate nucleic acids during their expression in the host [53,54]. In the case of the MS2 VLP, this event represents as much as 25% of the particle's mass [45]. Rather than a liability, this RNA component enhances the immune response because of its adjuvant properties as an activator of TLR7 and TLR8 [55,56]. The two prophylactic HPV vaccines approved by the FDA are highly immunogenic and safe [52]. Imiquimod and Resiquimod are ssRNA analogs used as potent anti-viral and anti-tumor therapy with a high degree of safety and no organ toxicity [35].

The succinimidyl (SMPH), an amine-to-sulfhydryl cross-linker used to conjugate the antigen to the VLP chemically, forms an irreversible thioether between the cross-linker and cysteine residue of the antigen. The lability of SMPH in aqueous solution and relatively low concentration (1–2 mM) used in the peptide conjugation process minimizes the possibility of a toxic effect [35]. As an alternative, the use of the bio-orthogonal copper (Cu)-free click chemistry represents an excellent choice for binding conjugated epitopes to VLP. Compared to SMPH, this method has proven to be more effective, safer, and more immunogenic [21,24].

The display of a chemically conjugated antigen can present a potential for toxicity, especially when it is a self-antigen. Unconjugated antigen may bind to healthy cells and produce some adverse effects. The manufacturing process should remove or reduce the concentration of unconjugated antigen to the levels where no biological effect occurs. However, when the removal of self-antigen excess is not possible, its toxicity must be addressed. This problem does not exist with autoantigen genetic conjugates [35].

Adjuvants, such as inorganic salts (alum), oil emulsions (MF5), and lipid A component (MPL), can be used to improve vaccine immunity and may cause adverse reactions. Luckily, VLP vaccines by themselves induce such a robust immune response that the adjuvant may be unnecessary, but if the use of an adjuvant is required for optimal immune response, its potential toxicity must be addressed [57].

The immunization process or the resultant immune response might be a source for local or general systemic toxicities and, therefore, preclinical studies must be representative of the vaccine route and formulation intended for clinical use. The constituents required to modulate the immune response should also be present. When possible, batches of a vaccine developed for clinical vaccination should be used in toxicology study, so the product is close as much as possible to the one that enters practice. Production scale up, for example, can introduce subtle changes that can have significant effects on product quality.

Appropriate animal toxicology assessments can determine the viability of candidate VLP vaccines. Repeated dose toxicity studies in individual animal species are sufficient to validate new prophylactic vaccine products [58,59]. However, for VLP that present autoantigens, non-clinical studies in two species may be required. To characterize the cytotoxic T lymphocyte (CTL) response, the rodent is an appropriate model. Some mouse strains show a differential tendency for a Th1 or Th2 response, a critical consideration in choosing an animal recipient [60], especially for VLP vaccines designed for tumor therapy where the precise nature of the immune response can have vital importance. Rabbits represent a good choice for toxicology studies because they show the required humoral response and are large enough to be used for full-dose administration to human recipients [61].

Toxicology studies are usually performed in several doses. However, single doses may be necessary if the immune response induced by the first administration alters the reaction of the subsequent ones. Such repeated dose studies are critical to support the safety profile of vaccines under development [61]. These studies provide useful information for VLP platforms inducing B or T cell responses, as this application involves more than one dose or immunization. The dose, the route of administration, the system distribution, and the degradation of the VLP must also be considered for toxicology studies. Physiological effects outside the immune system are of concern, especially for vaccines with autoantigens [24,62].

2.3. VLP and the Lymphatic System

The draining of nanoparticles to LN is an essential property of VLP vaccines. As mentioned above, most VLP platforms have diameters ranging from 25 to 100 nm, making them small enough to be drained into the lymphatic vessels, whose diameters vary from 10–60 µm up to as much as 2 mm. Ultra-small nanoparticles such as Pluronic-stabilized polypropylene sulfide (PPS) (~25–100 nm) injected intradermally are transported highly efficiently into the lymphatic capillaries and their draining LN [63]. EαRFP, a recombinant protein injected subcutaneously (s.c.), is transported to the LN in 18 h [64]. In one lymphatic trafficking study, VLP of bacteriophage Qβ (30 nm of diameter) labeled with Alexa-488 were injected in C57BL/6 mice. Fluorescent polystyrene nanoparticles with diameters of 20, 500, and 1000 nm were also injected. Within two hours, VLP and 20 nm beads were detected in the subscapular sinus of the popliteal LN, where they were associated with LN-resident DCs, B cells and macrophages. This pattern is compatible with drainage of afferent lymphatic capillaries. By 48 h, they were localized at the subcapsular, paracortex, and cortex areas of the LN in the vicinity with B cell follicles [36]. Larger 500 and 1000 nm beads needed 24 to 48 h to reach the popliteal LN and colocalized to areas where DCs reside. They did not efficiently enter lymphatic vessels and were possibly carried into the LN by specialized cells immigrating from the skin. One week after injection, 20, 500 and 1000 nm particles were still present at the injection sites and popliteal LN. At this time, the presence of VLP was reduced, which may be due to its degradation. The transport differences of these nanoparticles indicate size-dependent entry and distribution into the lymphatic system [36].

By freely entering the afferent lymphatic vessels and the LN subcapsular sinus where they encounter B cells and resident DCs, the VLP can efficiently initiate a humoral immune response. The VLP then spreads through the small gaps (0.1–1 μm) of the sinus floor to the follicular area and interacts with naive B cells [65]. The VLP can be kept for an extended period in the germinal center by follicular DCs, leading to the clonal expansion and the development of a long-lived antibody response. Alternatively, the VLP can be processed by DCs in the paracortex region, where a substantial fraction of the DC population is immature and able to process new antigens [66,67], initiating an immune response and the development of effector mechanisms [37]. DCs are also involved in the transport of large particles from the interstitial space to the lymphatic vessels through an active mechanism involving cell adhesion molecules [38]. Larger particles confined to the interstitial space before entering the LN are susceptible to the action of phagocytic cells [36], reducing their ability to be drained or transported.

VLP represent a reliable antigen delivery system whose route of administration determines the strength of the immune response [37]. Studies of different immune routes using simian human immunodeficiency SHIV VLP (approximately 90 nm in diameter) and near-infrared (NIR) fluorescent dyes in SKH-1 hairless immunocompetent mice show significant differences in particles trafficked into LN. Five minutes after intradermal injection, the VLP can be detected in the inguinal, popliteal, lumber, and sciatic LN, remaining detectable for up to six days [37]. When the VLP reach the subcapsular sinus, a larger population of B cells is activated and migrate from the border between the T cell zone and follicle to receive proliferation signals from antigen-specific T helper cells [68]. The increase in B cell proliferation can lead to increased levels of antibody-secreting plasma cells. Hence, an increase in VLP uptake in the LN can result in a more robust immune response. Further studies using SHIV intradermal (i.d.) immunization in C57Bl/6 mice showed a healthy antibody level production, possibly due to the increase in LN involvement leading to a better overall immune response [37].

Through effective drainage, the VLP can generate antibodies with higher affinity for specific epitopes through somatic hypermutation (SHM) at the germinal center. VLP-stimulation of gene regulation, cytokine production, and antigen-specific antibody production provide important information for the development of VLP-based vaccine [69]. Intradermal vaccination seems to produce a higher activated germinal center B cells [37].

Spleen cells from C57BL/6 mice i.d. immunized with SHIV VLP were analyzed by flow cytometry for the expression of various activation markers. The results demonstrated that most B cells (B220$^+$) were also positive for Fas and GL7 proteins, typically expressed on activated germinal center B cells. The population of B cells was also highly positive for the expression of CD80, a co-stimulatory molecule required for T cell activation and survival. When up-regulated on activated B cell, CD80 provides T cell co-stimulation through CD28 signaling. In the same experiment, the B220$^+$PD1$^+$ population increased after VLP immunization. PD-1 is a member of the CD28/CTLA4 family that is up-regulated on activated macrophages, T cells, and B cells. This increase in the population of activated and germinal center B cells, as evidenced by the expression of specific surface markers, might explain why i.d. vaccination leads to significantly higher affinity levels of antibody production [37]. VLP immunization was also effective in inducing an expansion of CTLs, as demonstrated by a cytotoxicity assay using a simian immunodeficiency virus (SIV) gag peptide pool. CTLs are an essential component of the cellular immune response, where they induce the death of virally infected cells and tumor cells. The i.d. SHIV VLP immunization produced higher numbers of CTLs than other vaccination pathways [37].

Humoral and cell-mediated responses require the activation of multiple effector mechanisms to identify virus infected cells and tumor cells, which may not be achieved by all vaccines. Cancer vaccines rely mostly on the development of antigen-specific antibodies and on the generation of antigen-specific CTLs to recognize distinct antigens present on the cancer cell surface to eliminate them. It is well established that through cross-linking B cell receptors (BCRs), VLP activate the B cells, originating a strong activation response [70]. VLP can be modified to present a variety of epitopes, and to display B cell epitopes in a rigid, organized, and repetitive manner to effectively induce the production of neutralizing antibodies [71].

2.4. VLP Vaccination Can Directly Activate B Cells

B cells are an integral part of the adaptive immune response. The identification of the subtype of B cells that binds to VLP can help to understand the mechanisms of B cell activation, differentiation, gene regulation, cytokine, and antigen-specific antibody production. In an explicative study, the SIV, SHIV, and chimeric influenza HA/SIV (SIV gag plus influenza hemagglutinin) VLP, known to induce high titers of antibodies, were incubated with mouse splenocytes [72]. All three types of VLP bound directly to $CD19^+$ naive B cells but not to $CD3^+$ T cells, as expected by the fact that the mouse CD4 cannot be properly recognized by HIV envelope protein. Dose-dependent binding and unlabeled VLP competition assays showed that the VLP binding to naïve B cells is specific [72]. Indeed, B cells may be specific for VLP antigens, and binding occurs via their antigen receptor (BCR) rather than toll like receptors (TLR), and induces up-regulation of the activation markers CD69 and CD86 [72].

Among the population of naïve B cells ($B220^+IgM^+$), conventional B2 cells ($CD43^-CD5^-$) increased after treatment with VLP, compared with PBS control, while the B1 cell subpopulations B1a ($B220^+IgM^+CD43^+CD5^+$) and B1b ($B220^+IgM^+CD43^+CD5^-$) did not change after VLP exposure [72]. This data indicates that incubation of naïve B cells with VLP can stimulate B2 cells to expand in vitro.

Therefore, some VLP are capable of directly inducing the activation of the humoral response through specific binding to naïve B2 cells in vitro [72]. However, immunization with other VLPs may produce different results. A study with HPV16 vaccination, for instance, showed the ability of HPV16 VLP to activate naïve B cells, although, in this case, an increase in B1 cell subpopulation was reported [73]. Another study using Qβ and AP205 VLP induced the expansion of B1 cells from the marginal zone of the spleen [74]. The nature of the VLP structure and the type of infection are speculated to be involved in these differences.

VLP purification after expression in *Escherichia Coli* does not remove all the bacterial endotoxin, leaving traces of lipopolysaccharide (LPS) in the VLP formulation. Are the VLP alone responsible for inducing the expansion of naïve B cells, or is the residual endotoxin involved? Spleen cells incubated with VLP, LPS, or anti-CD40 antibody in the presence or absence of polymyxin B (PMB—an antibiotic that blocks LPS activity) helped to answer this question. The naïve B cell proliferation was reduced in the presence of LPS and PMB but was not affected when treated with VLP and anti-CD40 in the presence or absence of PMB, showing that activation of naïve B2 cells by VLP is not dependent upon the presence of endotoxin [72]. The same study showed that in the supernatant of naïve mouse splenocytes stimulated by treatment with VLP, the expression of IL-12, MIP-1α, and MIP-1β is elevated, while the expression of IL-4 and MCP-1, which favor IgG1 antibody production, was decreased. Therefore, VLP stimulation is conducive to IgG2a class-switch recombination (Figure 2) [72].

B cells can respond to antigen in a T-dependent or T-independent way. In both cases, besides antigen binding through the BCR, additional signals are required to induce B cells to proliferate and differentiate into plasma cells producing antibodies [75]. VLP bind and activate naive B cells, but can VLP induce B cells to differentiate into plasma cells? Splenocytes incubated for 48 h with VLP were transferred to a SIV VLP-coated polyvinylidene fluoride filter plate for 3 h at 37 °C. The ELISPOT assay showed that VLP treatment induces the differentiation of activated B cells into plasma cells, at least in vitro. These data were confirmed by real-time PCR analysis where the levels of Blimp-1 and XBP-1 increased after splenocytes incubation with VLP; these two proteins are essential for the differentiation of plasma cells. The level of antibodies produced after plasma cell differentiation was evaluated by ELISA, with a remarkable increase in both IgM and IgG2a, confirming that VLP stimulated a humoral response in vitro [72]. VLP immunization can also stimulate B cell differentiation into a plasma cell and class-switch recombination in vivo [72].

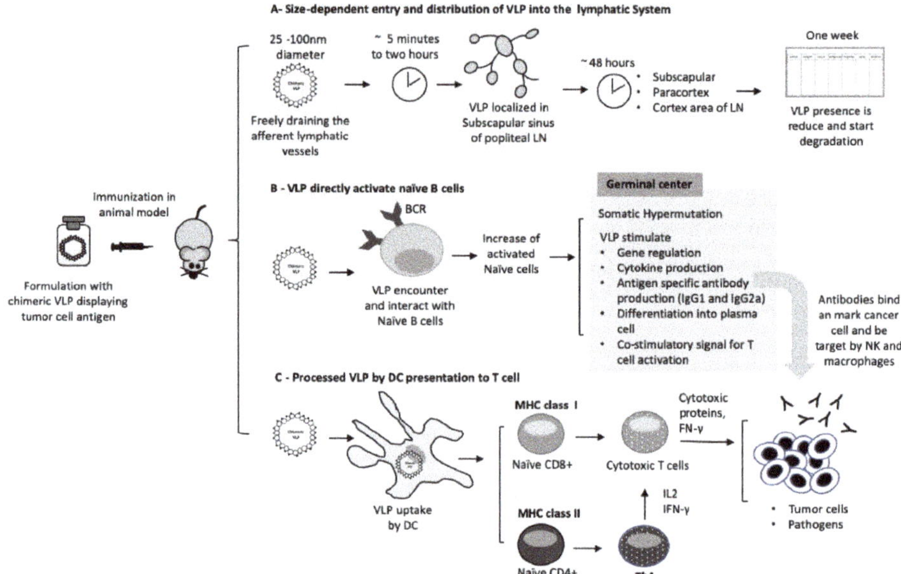

Figure 2. Illustration of virus-like particles (VLP) triggering immune response. (A) The draining of nanoparticles to the lymphatic system is an essential property of nanoparticles. (B) VLP can directly activate naïve B cells and produce a long-lasting immune response. (C) VLPs processed by DC cells trigger immune response and development of effector mechanisms.

2.5. VLP Can Activate the Complement System

Proteins on the surface of VLP, like those of the viruses from which they are derived or other pathogens, are very organized and repetitive. Hence, an active binding to natural IgM antibodies or IgG, can recruit complement component 1q (C1q) and activate the complement cascade. In addition, protein C and other pentraxins can bind to the surface of VLP, also activating the classical complement cascade, and facilitating their uptake by DCs and macrophages. After being taken up by these antigen-presenting cells (APCs), the VLP reaches the endosome-lysosome compartment and is degraded into peptides. These peptides through MHC class II molecules are carried to the cell surface and presented to CD4[+] T helper cells. The vaccine antigen can alternatively be presented by MHC class I molecules to induce CD8[+] T cell responses, an essential requirement for therapeutic vaccine's candidates [76].

2.6. VLP Vaccination Strategy, Regimen, and Dose

Vaccination has the primary purpose of producing long-lasting protection against diseases. The choice of appropriate vaccine strategy, regimen, and dose is crucial for the success of vaccination. It becomes especially concerning when immaturity or senescence of the immune system can affect the efficacy of the immunization [77]. Different strategies of prime-boost vaccination against infectious diseases searching to improve humoral and cellular immunity have been studied [78,79]. These heterologous strategies induce efficient humoral and cellular responses to the same antigen presented by two different delivery systems. Priming with a DNA vaccine or viral vector followed by boosting with a protein-based vaccine usually induces a strong cellular immune response, with higher and more specific antibody production as compared to homologous delivery systems [80]. In the circumstances where homologous protein-based vaccination induces strong humoral response but weak cellular immunity, the heterologous approach can be more productive. Below are examples of the prime-boost regimen:

- DNA prime/viral vector boost, viral vector prime/DNA boost;
- DNA prime/protein boost, DNA prime/peptide boost;
- Protein prime/viral boost, viral prime/protein boost;
- DNA prime/VLP boost, VLP prime/live vector boost.

They combine a better antibody and CD4$^+$ T cell response induced by protein antigen, efficient stimulation of T cell response by DNA vaccines, and improvement of the CD4$^+$/CD8$^+$ T cells and antibody response by recombinant viral vectors. Similarly, different vectors have different immune characteristics and, therefore, induce a unique immune response to the immunodominant epitope, and reduce the immunity against the vector [79].

The choice of the optimal time and the frequency of repeated boosts can impact the quality of the immune response [81]. Understanding the process of establishing immune memory is valuable information for choosing the interval between the primary immunization and booster injections. Memory T cells with high proliferative potential are formed several weeks after the prime immunization, suggesting that the boosting should occur at least two or three months after the first immunization. A similar concept applies to antibody responses, as memory B cells need to undergo a germinal center response to develop [82]. Repeated boosting drives T cell towards terminal differentiation and recruits a subset of a previously generated memory cell; as a result, a heterologous population of memory T cells at several different stages can be found [78]. An appropriate immunization schedule can take advantage of this process.

The establishment of permanent protective T cell memory may depend on the antigen load and its kinetics. The possibility of mimicking pathogen replication enhances the T cell activity. A study using choriomeningitis virus (LCMV) in mice with different doses that prolong the antigen exposure showed a significant effect on the immune response, where high doses of the virus led T cell exhaustion, while low doses induced potent and durable T cell activation [55]. The ideal vaccination schedule starts with low doses, followed by a peak dose and finishes with low doses, instead of repeated doses of equivalent vaccine concentration [83]. This approach appears to be more suitable for therapeutic vaccines. In contrast, for prophylactic vaccines, when an effective B cell response is wanted, a simple vaccination schedule with an injection in a month or more apart may be preferable. The immune system can generate a reservoir of antigens in the germinal center, keeping B cells stimulated, which results in a stronger antibody response. Memory B cells and plasma cells favor this event that may contribute to the success of inactivated virus vaccines [76].

2.7. Usage of Adjuvants

Adjuvants may be needed to increase the antigen reservoir and enhance vaccine efficacy. For almost a century, Aluminum and Freund's adjuvants were used to improve vaccine immunogenicity. Nowadays, many others have been developed [84]. Preclinical studies show that some VLP vaccines do not need additional adjuvants to produce a robust immune response since they have intrinsic adjuvant properties to activate TLR. In mice, the ssRNA into the VLP [85] induces higher antibody titers of IgG2a, IgG2b and IgA isotypes [86], while VLP with the ssRNA removed, elicit IgG1 antibodies, indicative of a Th2 response [18,87]. Overall, VLP are very immunogenic, even in the absence of TLR-mediated stimuli. However, TLR ligands increase the immune response of a vaccine, as observed in Cervarix, an HPV-derived VLP used in combination with MPL, which is a ligand for TLR4 [40]. The TLR9 ligands CpG-containing oligodeoxynucleotides (CpG-ODNs), are also used in vaccines to enhance B and T cell responses to a given recombinant epitope. CpG package into VLP is resistant to DNase I digestion and does not induce the undesirable side effect (splenomegaly and lethal toxic shock) caused by the free form of CpG. Vaccination with CpG-loaded VLP protected from infection with vaccinia virus and eradicated established solid fibrosarcoma tumors [84]. Therefore, VLP presenting the CpG might represent a valuable approach to induce CTL responses [76]. Another way to enhance the VLP cell-mediated immune response is to present the universal T-assisted pan-HLA-DR-binding epitope (PADRE) on the nanoparticles. The RHDV VLP carrying PADRE and displaying HPV16 E6 peptides successfully prolonged survival of mice with pre-existing HPV tumors [25].

3. VLP as Cancer Vaccines

In the past decade, cancer immunotherapy with checkpoint inhibitors has achieved a remarkable clinical outcome in patients with advanced stages of cancer [88]. However, the high cost for such therapies and the low percentage of patient responders highlight the need to find new immune-based strategies to fight cancer, which are both cost-effective and efficacious for a broad patient base [89,90]. Human cancer emerges through a combination of genetic and epigenetic changes that favor its survival. The tumor-associated antigen and tumor-specific antigens (neo-antigens) also create an opportunity for tumor detection and destruction by the immune system. However, cancer cells have adopted several strategies to evade immune recognition, leading to a state of tumor tolerance [91]. Biological response modifiers (such as vaccines) have the potential to fight cancer and can be used to break tumor tolerance. The vaccine must deliver a large number of selected cancer antigens to promote B cell and DC activation and antigen presentation to T cells. As mentioned above, with the VLP platform, even short peptides identified in tumor cells (that otherwise would be quickly cleared) can be presented to DCs to induce cellular or humoral response against cancer. Favorable results obtained by preventive VLP-based vaccines against cancer caused by infectious diseases—for instance, HBV and HPV—reinforce this concept, and many attempts are now ongoing for the development of VLP-based vaccines against many non-infection-related cancers. In the following sections, we review some relevant examples.

3.1. Cervical Cancer

The current FDA-approved HPV vaccines are derived from HPV capsid protein L1 formulated with aluminum-containing adjuvants—aluminum hydroxyphosphate sulfate for Gardasil and the AS04 adjuvant system, composed of aluminum hydroxide and the TLR ligand 3-O-desacyl-4.-monophosphoryl lipid A (MPL) for Cervarix [92]. These vaccines are highly immunogenic and induce long-acting neutralizing antibodies. Their formulation is complex and expensive, which prohibits affordability in developing countries, where 85% of cervical cancer cases occur.

A preclinical study showed that another VLP-HPV vaccine using a neutralizing epitope from the HPV minor protein L2 could offer a robust immune response using a lower-cost production methodology and improved clinical applicability [23]. For this new approach, HPV16L2 was displayed at the N-terminus of MS2 VLP and at the AB loop of PP7 VLP. The L2-MS2 and L2-PP7 VLP were produced in a dry powder version using Spray Dryer (SD) and had their thermostability storage testing at 4 °C, 37 °C, and room temperature for one month. Spray Transmission Electron Microscopy (TEM) showed that SD and storage at the tested temperatures do not affect the nanoparticles' integrity or stability, preserving their ability to generate HPV-neutralizing antibodies. Immunization with reconstituted L2-MS2 VLP elicited high titers of anti-L2 IgG. The immunized mice were challenged with HPV Pseudo Virus (PsV), and their protection was compared with that induced by Gardasil immunization. When tested with two homologous PsV types (PsV31 and PsV45), L2-MS2-immunized mice were efficiently protected from both infections, while Gardasil failed to protect against the PsVs45 challenge. This work showed that L2-MS2 vaccines could be produced in a dry powder version with stability up to seven months at room temperature [23]. For other VLP-based vaccines against HPV cervical cancer, see [62].

3.2. Hepatocellular Carcinoma

The acute and chronic hepatitis caused by HBV can progress to cirrhosis and eventually to hepatocellular carcinoma (HCC). The current HBV vaccines are based on VLP of the S antigen; the most recently approved, HEPLISAV-B, has CpG as adjuvant [92]. These vaccines reduce the occurrence of

HBV infection but are ineffective in treating existing HCC. The use of multiple epitopes' VLP was tested as a vaccine for HCC treatment in a preclinical study. The HBV X protein is highly expressed in HCC. High frequency of epitopes of HBV X displayed at HBV core protein elicited epitope-specific CD8+ T cells and a more forceful response than a single peptide showing that multiepitope-loaded VLP augment immunogenicity for each epitope by increasing the number of CTLs resulting in a significant anti-tumor response [28]. It is a proof of concept of the potential efficacy of VLP-based vaccines for HCC therapy.

3.3. Skin Cancer

Melanoma is an antigenic and immunogenic cancer with well-characterized tumor-associated antigens. Ideally, a melanoma-preventing vaccine should expand the number of spontaneous CTLs produced against the tumor at the tumor site and throughout the body as long-term immune surveillance [93].

A notable example of the potentiality of VLP against melanoma comes from a preclinical study using VLP not decorated with tumor antigens but with inherent immunogenic properties.

In situ vaccination, using the VLP generated by the plant virus Cowpea Mosaic Virus (CPMV), was shown to reduce B16F10 melanoma and generate systemic anti-tumor immunity [29]. The approach of in situ vaccination using VLP is limited, but local immunization can strongly modulate the local microenvironment and the anti-tumor immunity. Empty CPMV (eCPMV) VLP lacking any nucleic acid were tested in in vitro cultures of bone marrow-derived DCs (BMDCs) and primary macrophages from C57BL/6 mice. Twenty-four hours later, an increase in pro-inflammatory cytokines (IL-6, IL-1β, IL-12p40, IFN-γ and TNK-α) was observed, providing evidence that the empty VLP is intrinsically immunostimulatory. Notably, the VLP were produced in plants and, therefore, were endotoxin free. When tested in non-tumor-bearing mice, inhalation of eCPMV activated Ly6G+ neutrophils that up-regulated the CD11b activation marker and the CD86 co-stimulatory molecule. Inhalation of eCMVP by mice bearing lung melanoma metastases, induced by an intravenous (i.v.) injection of B16F10 cells, alters the lung microenvironment, with a significant increase in tumor-infiltrating neutrophils (TINs) and $CD11b^+Ly6G^+$-activated neutrophils as well as a reduction in $CD11b^-Ly6G^+$ quiescent neutrophils. In addition, weekly intratracheal administration of 100 μg of eCPMV reduced the number of metastatic foci both in mice i.v. injected with B16F10 cells and in mice developing spontaneous lung metastases after a subcutaneous challenge with 4T1 mammary cancer cells [29]. Mice depleted of neutrophils and those lacking IL-12, IFN-γ or T, B, and NK cells failed to respond to the eCPMV inhalation therapy, confirming that neutrophils can disturb the tolerogenic nature of the tumor microenvironment and orchestrate an immune response that leads to tumor elimination. It probably occurs due to activation of pre-existing or de novo-induced anti-tumor responses, or both [29]. Administration of eCPMV in situ was also effective for treating dermal tumors. When the VLP were injected intra-tumorally in mice bearing cutaneous tumors induced by B16F10 melanoma cells or CT26 colon cancer cells, tumor growth was delayed. In the case of B16F10 melanomas, 50% of the treated mice rejected the tumor. These surviving mice were also protected from a re-injection of B16F10 cells, indicating the induction of a protective systemic immune response.

Another important preclinical study was based on the use of VLP decorated with the $H-2D^b$ immunodominant epitope of the human glycoprotein 100 (gp10025-33) [26], a melanoma-associated antigen [94]. The epitope was incorporated into rabbit hemorrhagic disease virus (RHDV) VLP in one, two, or three copies, resulting in gp100.1L, gp100.2L, and gp100.3L VLP, respectively. All three VLP induced a robust $CD8^+$ T cell proliferation, and gp100.2L and gp100.3L also induced a significant enhancement of IFN-γ production [26]. The activation of T cell proliferation was confirmed in in vivo experiments for gp100.2L and gp100.3L. These two versions of VLP also showed a therapeutic effect against murine melanomas. Mice vaccinated with the VLP remained tumor free over 60 days, which is an indication of the induction of a specific anti-tumor immunity. The mono or dimannosylation of gp100-VLP enhanced VLP uptake into APCs, but the anti-tumor T cell response was only enhanced by dimannosylation [26].

Another proof of the ability of VLP-based vaccines to prevent B16F10 melanoma comes from a study in which a recombinant cucumber-mosaic VLP (CuMV) (~ 30 nm) was used. [24]. The universal tetanus toxoid (TT) epitope TT830-843 has been genetically fused to CuMV (CuMV$_{TT}$) and is displayed in the interior surface to avoid interference with TT-specific antibodies. The TT epitope serves as an additional T helper cell epitope, especially in elderly patients. CuMV$_{TT}$ was labeled with AF488 and was used in the study of drainage dynamics. The study showed that fast and effective drainage of VLP, that reached the regional LN in just one minute. A version of these VLP expressing the H-2D^b restricted p33 peptide from LCMV antigen (CuMV$_{TT}$p33-VLP) was formulated with the micron-sized microcrystalline tyrosine (MCT) adjuvant. MCT facilitated depot formation and prolonged the release of the VLP up to 9 days compared to 4 days of the free VLP form. With only one vaccination, CuMV$_{TT}$p33 VLP formulated with MCT can induce high levels of p33-specific T cells and release of the cytokines IFN-γ and TNF-α. When this formulation was tested in mice bearing tumors from B16F10 melanoma cells expressing the p33 antigen, tumor progression was inhibited with a significant increase in total CD8$^+$ T cells and p33-specific T cell infiltration, showing proper anti-tumor protection [24].

A recent study proposed a therapeutic vaccine for melanoma, in which the loaded CpG Qβ-VLP and the desired epitope are combined with bio-orthogonal copper-free click chemistry; Dibenzocyclooctyne NHS ester (DBCO) is used as a cross-linking agent. This new method for peptide conjugation was shown to be superior to SMPH not only for coupling efficiency but also for inciting CTL and IFN-γ response in vivo. The developed vaccine was created by the conjugation of peptides from tumor-specific germline (immunopeptidomics) and mutated (whole-exome sequencing) CTL epitopes from B16F10 melanoma cells. Three different multitarget (MTV) vaccines were originated: 1) germline-multitarget vaccine (GL-MTV), 2) mutated epitope vaccine (mutated-MTV), and 3) a combination of both (Mix-MTV). Subsequently, they were tested in mice with transplanted B16F10 melanoma tumors in combination with anti-CD25. It was observed that the three vaccines inhibited the tumor progression, the Mix-MTV vaccination being more effective for anti-tumor protection. It was also the only one that showed significant tumor growth reduction. Compared with the control group and mutant MTV, the Mix-MTV vaccine was more effective in increasing CD8$^+$ T cell density. For Mix-MTV vaccination, higher IFN-γ production was also observed. Data show that the combination of germline and mutant epitopes in a vaccine is more effective against growing B16F10 melanoma tumors [21].

3.4. Pancreatic Cancer

Pancreatic cancer is resistant to many forms of treatment, highlighting the urgent need to develop novel therapies to fight this neoplastic disease.

Mesothelin (MSLN), a glycoprotein membrane whose expression is limited to mesothelial cells, is overexpressed in most pancreatic cancers and is involved in tumor adhesion and dissemination, making it a robust therapeutic target [95,96]. By presenting the human (h) MSLN on simian-human immunodeficiency VLP (SHIV), a pancreatic VLP-hMSLN vaccine was created [30]. The efficacy of VLP-hMSLN was tested in Panc02 cells implanted orthotopically in mice, showing significantly lower tumor progression and reduction in tumor mass and the survival of 60% of vaccinated mice. Protection was due to the induction of high concentrations of specific anti-hMSLN antibodies and the release of IFN-γ, accompanied by a strong and specific CTL response and a reduction in regulatory T cells (Treg; CD4$^+$FOXP3$^+$ICOS$^-$) [30]. The hMSLN was a xenogenic antigen in mice, even if human and mouse MSLN has 60% homology. To verify whether the mouse (m)MSLN can break self-tolerance, mMSLN was displayed in the SHIV VLP (VLP-mMSLN). Immunization with this murine VLP version efficiently induced CD8$^+$ T cell activation, reduced tumor volume, and survival of tumor-bearing mice [30].

Another pancreatic vaccine candidate was previously developed using SIV expressing mTrop2 (mTrop2 VLP) [31]. Trop2 is a surface glycoprotein overexpressed in pancreatic cancer and is associated with poor patient survival [97]. The efficacy of mTrop2 VLP vaccination was tested in the syngeneic pancreatic cancer murine model. The immunization was able to break tolerance and generate a cellular and humoral response overcoming some of the suppressive tumor microenvironment agents. The results showed a reduction in tumor growth, activation and infiltration of CD4$^+$, CD8$^+$ T cells, and NK cells. VLP vaccination, in combination with gemcitabine, increased the survival of tumor-bearing mice, suggesting that mTrop2 VLP might be the right approach for pancreatic cancer treatment [31].

3.5. Colorectal Cancer

Colorectal cancer (CRC) is distinguished by a high inter-patient and intra-tumor heterogeneity that makes the identification of possible target antigens challenging. Hence, targeting a combination of CRC antigens seems to be a more promising approach. A preclinical study developed a VLP from the VP60 capsid proteins of RHDV displaying at the N-terminus peptide of the murine topoisomerase IIα (T.VP60), survivin (S.VP60) or both peptides (TS.VP60) [27]. These two antigens are involved in relevant mechanisms for tumor growth in both murine and human CRC. CRC cell line MC38-OVA was injected subcutaneously on day 0, and vaccination was performed over three consecutive days (days 7, 8, and 9) and monitored for 100 days. Synthetic CpGs (25 μg) were used as an adjuvant. T.VP60, S.VP60, and TS.VP60 vaccination delayed tumor growth; the first two treatments (T.VP60, S.VP60) improved survival by 60%, while TS.VP60 enhanced it by 73%. All mice that survived were tumor free at day 100; these mice received a second challenge of MC38-OVA, and none of the mice developed the tumor with 100% of survival up to the completion of the 200 day trial. The results suggest the development of a systemic memory response and demonstrate that a VLP vaccine with multiple epitopes can be an effective treatment strategy for CRC [27].

Other CRC-associated antigens are under investigation as possible targets of VLP-based vaccination. The Cluster of Differentiation 44 (CD44) is an adhesion molecule in the extracellular matrix that induces tumor cell invasion and metastasis. The variant form CD44v6 is a CRC cell marker and its positiveness is correlated with poor survival in CRC patients. CD44 variants stabilize cystine/glutamate antiporter protein xCT (see below) on the cell surface of several human carcinomas. The presence of xCT itself is an independent predictor of recurrence in CRC patients and is correlated with lymphatic and venous invasion. xCT is associated with CRC cancer cell growth, invasion and metastasis, making xCT a reliable target for the development of a CRC therapeutic vaccine [98]. Recently developed VLP-based vaccines targeting xCT [22,99] would be an ideal candidate for preclinical testing in CRC models.

3.6. Breast Cancer

Breast cancer is a heterogeneous disease. Triple-negative breast cancer (TNBC) is an invasive form of breast cancer, with high propensity to relapse, and characterized by the absence of epidermal growth factor receptor 2 (HER2), progesterone, estrogen. The presence of cancer stem cells (CSCs) is related to the aggressiveness of TNBC. These CSCs are resistant to conventional therapies and are conducive to tumor development and progression [100,101]. Currently, there is no successful clinical therapy for this subtype of breast cancer. TNBC resists current cytotoxic therapies due to its unique detoxification mechanism, which is greatly promoted by the overexpression of the cystine-glutamic acid transporter xCT (SLC7A11). The xCT protein is overexpressed in several human tumors but not in healthy mammary gland tissue, suggesting that the protein up-regulation occurs only upon oncogenic transformation, contributing to a decrease in patient survival [102]. xCT exports glutamate at a ratio of 1:1 in the exchange of extracellular cystine [103]. Intracellularly, cystine is reduced to cysteine, a limiting precursor in the biosynthesis of glutathione (GSH), which plays an essential role in cellular defense against oxidative stress, reducing diverse reactive oxygen species (ROS) [104]. xCT increases the intracellular concentration of GSH, reduces p38/mitogen-activated Protein Kinase activation, inhibiting ROS and preventing cell apoptosis. As a result, xCT protects CSCs from ROS, making them more resistant to conventional therapies, and promotes tumor growth [105].

Sulfasalazine (SASP) is an FDA-approved drug used to treat chronic inflammatory diseases that inhibits xCT protein function [106]; however, SASP has numerous side effects and, as a result, is not a viable therapeutic option. A vaccine using a genetic fusion of the extracellular domain 6 (ECD6) from xCT protein on the AB loop of bacteriophage MS2-VLP, named AX09-0M6, was used as a prophylactic and therapeutic vaccine in a preclinical TNBC model. The adopted vaccination regimen was able to induce high levels of IgG2a antibodies that bound to mouse and human xCT breast CSCs. Anti-AX09-0M6 antibodies reduced the number and dimension of spheres and increased ROS concentration in all breast cancer cell lines tested, demonstrating that the vaccination with AX09-0M6 prevents BCSC self-renewal. The in vivo experiments showed a significant reduction in 4T1 tumor growth and the number of spontaneous pulmonary metastases, increased natural killer (NK) cell infiltration, and CD8+ T cells, which is indicative of positive tumor microenvironment changes and antibody-dependent cellular toxicity (ADCC). No adjuvants were used in this vaccination approach, and no toxicity was detected [22,99]. Similar results were obtained with an MS2-VLP expressing the ECD3 of xCT, named AX09-0M3 (Rolih et al., data in publication).

The epidermal growth factor receptor-2 (HER2) protein is overexpressed in another aggressive form of breast cancer in humans. Nearly 30% of all invasive breast cancers overexpress HER2. Like many other tumors, HER2+ breast cancer also expresses xCT, and the AX09 VLP against xCT also demonstrated a significant activity against this type of cancer [22,99]. A VLP-based vaccine approach for HER2+ breast cancer vaccination uses Acinetobacter phage AP205 to covalently display the HER2 protein (HER2-VLP) [16]. This vaccine is able to break B cell tolerance and induces a high level of anti-HER2-neutralizing antibodies. Tumor growth was inhibited after HER2-VLP immunization in FVB mice injected with HER2+ transplantable breast cancer cells or HER2+ tumor fragments. In addition, HER2 transgenic mice that spontaneously develop aggressive HER2 mammary carcinomas were tumor free until one year of age after HER2-VLP vaccination, with high levels of anti-HER2 antibodies lasting for up to six months after vaccination [16]. The generated anti-HER2 antibodies bind to mouse and human HER2 in the same way as commercial anti-HER2 monoclonal antibodies.

Although the level of endotoxin was not shown for the AX09 VLP vaccine and not removed from the HER2-VLP vaccine, these vaccines represent an excellent example of how versatile and efficient VLP can be in overcoming B cell tolerance to tumor-associated self-antigens.

4. Conclusions

Recent advances in cancer treatment have confirmed that immunotherapy is useful in the ever-evolving fight against cancer. Preclinical studies such as the ones listed above clearly illustrate that VLP can efficiently stimulate the immune system not only against the antigens of the microbial world but also against tumor-associated self-antigens. VLP can target solid tumors or CSCs with the potential to be used as prophylactic or therapeutic cancer vaccines, alone or in combination with chemotherapy, checkpoint inhibitors, or future therapies.

Author Contributions: J.C.C.: conceptualization and writing (original draft, figures and tables and editing); M.P., F.P. and F.C.: conceptualization and writing (review). All authors have read and agreed to the published version of the manuscript.

Acknowledgments: This work was supported by grants from the Fondazione Ricerca Molinette Onlus.

Conflicts of Interest: The authors declare no potential conflicts of interest.

References

1. Raghunandan, R. Virus-like particles: Innate immune stimulators. *Expert Rev. Vaccines* **2011**, *10*, 409–411. [CrossRef]
2. Wang, G.; Liu, Y.; Feng, H.; Chen, Y.; Yang, S.; Wei, Q.; Wang, J.; Liu, D.; Zhang, G. Immunogenicity evaluation of ms2 phage-mediated chimeric nanoparticle displaying an immunodominant b cell epitope of foot-and-mouth disease virus. *PeerJ* **2018**, *6*, e4823. [CrossRef]
3. Spohn, G.; Schwarz, K.; Maurer, P.; Illges, H.; Rajasekaran, N.; Choi, Y.; Jennings, G.T.; Bachmann, M.F. Protection against osteoporosis by active immunization with trance/rankl displayed on virus-like particles. *J. Immunol.* **2005**, *175*, 6211–6218. [CrossRef] [PubMed]

4. Rohn, T.A.; Ralvenius, W.T.; Paul, J.; Borter, P.; Hernandez, M.; Witschi, R.; Grest, P.; Zeilhofer, H.U.; Bachmann, M.F.; Jennings, G.T. A virus-like particle-based anti-nerve growth factor vaccine reduces inflammatory hyperalgesia: Potential long-term therapy for chronic pain. *J. Immunol.* **2011**, *186*, 1769–1780. [CrossRef] [PubMed]

5. Spohn, G.; Schori, C.; Keller, I.; Sladko, K.; Sina, C.; Guler, R.; Schwarz, K.; Johansen, P.; Jennings, G.T.; Bachmann, M.F. Preclinical efficacy and safety of an anti-il-1beta vaccine for the treatment of type 2 diabetes. *Mol. Methods Clin. Dev.* **2014**, *1*, 14048. [CrossRef] [PubMed]

6. Maphis, N.M.; Peabody, J.; Crossey, E.; Jiang, S.; Jamaleddin Ahmad, F.A.; Alvarez, M.; Mansoor, S.K.; Yaney, A.; Yang, Y.; Sillerud, L.O.; et al. Qss virus-like particle-based vaccine induces robust immunity and protects against tauopathy. *NPJ Vaccines* **2019**, *4*, 26. [CrossRef]

7. Milich, D.R. T- and b-cell recognition of hepatitis b viral antigens. *Immunol. Today* **1988**, *9*, 380–386. [CrossRef]

8. Ho, J.K.; Jeevan-Raj, B.; Netter, H.J. Hepatitis b virus (hbv) subviral particles as protective vaccines and vaccine platforms. *Viruses* **2020**, *12*, 126. [CrossRef]

9. Tissot, A.C.; Renhofa, R.; Schmitz, N.; Cielens, I.; Meijerink, E.; Ose, V.; Jennings, G.T.; Saudan, P.; Pumpens, P.; Bachmann, M.F. Versatile virus-like particle carrier for epitope based vaccines. *PLoS ONE* **2010**, *5*, e9809. [CrossRef]

10. Crossey, E.; Amar, M.J.A.; Sampson, M.; Peabody, J.; Schiller, J.T.; Chackerian, B.; Remaley, A.T. A cholesterol-lowering vlp vaccine that targets pcsk9. *Vaccine* **2015**, *33*, 5747–5755. [CrossRef]

11. Caldeira, J.; Bustos, J.; Peabody, J.; Chackerian, B.; Peabody, D.S. Epitope-specific anti-hcg vaccines on a virus like particle platform. *PLoS ONE* **2015**, *10*, e0141407. [CrossRef]

12. Hu, X.; Deng, Y.; Chen, X.; Zhou, Y.; Zhang, H.; Wu, H.; Yang, S.; Chen, F.; Zhou, Z.; Wang, M.; et al. Immune response of a novel atr-ap205-001 conjugate anti-hypertensive vaccine. *Sci. Rep.* **2017**, *7*, 12580. [CrossRef] [PubMed]

13. Lino, C.A.; Caldeira, J.C.; Peabody, D.S. Display of single-chain variable fragments on bacteriophage ms2 virus-like particles. *J. Nanobiotechnology* **2017**, *15*, 13. [CrossRef] [PubMed]

14. Yan, D.; Wei, Y.Q.; Guo, H.C.; Sun, S.Q. The application of virus-like particles as vaccines and biological vehicles. *Appl. Microbiol. Biotechnol.* **2015**, *99*, 10415–10432. [CrossRef] [PubMed]

15. Marintcheva, B. Virtual virus, a semester-long interdisciplinary project on the crossroads of creativity and knowledge integration. *FEMS Microbiol. Lett.* **2017**, *364*, fnx097. [CrossRef] [PubMed]

16. Palladini, A.; Thrane, S.; Janitzek, C.M.; Pihl, J.; Clemmensen, S.B.; de Jongh, W.A.; Clausen, T.M.; Nicoletti, G.; Landuzzi, L.; Penichet, M.L.; et al. Virus-like particle display of her2 induces potent anti-cancer responses. *Oncoimmunology* **2018**, *7*, e1408749. [CrossRef] [PubMed]

17. Caldeira Jdo, C.; Medford, A.; Kines, R.C.; Lino, C.A.; Schiller, J.T.; Chackerian, B.; Peabody, D.S. Immunogenic display of diverse peptides, including a broadly cross-type neutralizing human papillomavirus l2 epitope, on virus-like particles of the rna bacteriophage pp7. *Vaccine* **2010**, *28*, 4384–4393. [CrossRef] [PubMed]

18. Tumban, E.; Peabody, J.; Peabody, D.S.; Chackerian, B. A universal virus-like particle-based vaccine for human papillomavirus: Longevity of protection and role of endogenous and exogenous adjuvants. *Vaccine* **2013**, *31*, 4647–4654. [CrossRef] [PubMed]

19. Maurer, P.; Jennings, G.T.; Willers, J.; Rohner, F.; Lindman, Y.; Roubicek, K.; Renner, W.A.; Muller, P.; Bachmann, M.F. A therapeutic vaccine for nicotine dependence: Preclinical efficacy, and phase i safety and immunogenicity. *Eur. J. Immunol.* **2005**, *35*, 2031–2040. [CrossRef] [PubMed]

20. Cavelti-Weder, C.; Timper, K.; Seelig, E.; Keller, C.; Osranek, M.; Lassing, U.; Spohn, G.; Maurer, P.; Muller, P.; Jennings, G.T.; et al. Development of an interleukin-1beta vaccine in patients with type 2 diabetes. *Mol. Ther.* **2016**, *24*, 1003–1012. [CrossRef]

21. Mohsen, M.O.; Vogel, M.; Riether, C.; Muller, J.; Salatino, S.; Ternette, N.; Gomes, A.C.; Cabral-Miranda, G.; El-Turabi, A.; Ruedl, C.; et al. Targeting mutated plus germline epitopes confers pre-clinical efficacy of an instantly formulated cancer nano-vaccine. *Front. Immunol.* **2019**, *10*, 1015. [CrossRef] [PubMed]

22. Bolli, E.; O'Rourke, J.P.; Conti, L.; Lanzardo, S.; Rolih, V.; Christen, J.M.; Barutello, G.; Forni, M.; Pericle, F.; Cavallo, F. A virus-like-particle immunotherapy targeting epitope-specific anti-xct expressed on cancer stem cell inhibits the progression of metastatic cancer in vivo. *Oncoimmunology* **2018**, *7*, e1408746. [CrossRef] [PubMed]

23. Tumban, E.; Muttil, P.; Escobar, C.A.; Peabody, J.; Wafula, D.; Peabody, D.S.; Chackerian, B. Preclinical refinements of a broadly protective vlp-based hpv vaccine targeting the minor capsid protein, l2. *Vaccine* **2015**, *33*, 3346–3353. [CrossRef] [PubMed]

24. Mohsen, M.O.; Heath, M.D.; Cabral-Miranda, G.; Lipp, C.; Zeltins, A.; Sande, M.; Stein, J.V.; Riether, C.; Roesti, E.; Zha, L.; et al. Vaccination with nanoparticles combined with micro-adjuvants protects against cancer. *J. Immunother. Cancer* **2019**, *7*, 114. [CrossRef]

25. Jemon, K.; Young, V.; Wilson, M.; McKee, S.; Ward, V.; Baird, M.; Young, S.; Hibma, M. An enhanced heterologous virus-like particle for human papillomavirus type 16 tumour immunotherapy. *PLoS ONE* **2013**, *8*, e66866. [CrossRef]

26. Kramer, K.; Al-Barwani, F.; Baird, M.A.; Young, V.L.; Larsen, D.S.; Ward, V.K.; Young, S.L. Functionalisation of virus-like particles enhances antitumour immune responses. *J. Immunol. Res.* **2019**, *2019*, 5364632. [CrossRef]

27. Donaldson, B.; Al-Barwani, F.; Pelham, S.J.; Young, K.; Ward, V.K.; Young, S.L. Multi-target chimaeric vlp as a therapeutic vaccine in a model of colorectal cancer. *J. Immunother Cancer* **2017**, *5*, 69. [CrossRef]

28. Ding, F.X.; Wang, F.; Lu, Y.M.; Li, K.; Wang, K.H.; He, X.W.; Sun, S.H. Multiepitope peptide-loaded virus-like particles as a vaccine against hepatitis b virus-related hepatocellular carcinoma. *Hepatology* **2009**, *49*, 1492–1502. [CrossRef]

29. Lizotte, P.H.; Wen, A.M.; Sheen, M.R.; Fields, J.; Rojanasopondist, P.; Steinmetz, N.F.; Fiering, S. In situ vaccination with cowpea mosaic virus nanoparticles suppresses metastatic cancer. *Nat. Nanotechnol.* **2016**, *11*, 295–303. [CrossRef]

30. Zhang, S.; Yong, L.K.; Li, D.; Cubas, R.; Chen, C.; Yao, Q. Mesothelin virus-like particle immunization controls pancreatic cancer growth through cd8+ t cell induction and reduction in the frequency of cd4+ foxp3+ icos- regulatory t cells. *PLoS ONE* **2013**, *8*, e68303. [CrossRef]

31. Cubas, R.; Zhang, S.; Li, M.; Chen, C.; Yao, Q. Chimeric trop2 virus-like particles: A potential immunotherapeutic approach against pancreatic cancer. *J. Immunother.* **2011**, *34*, 251–263. [CrossRef] [PubMed]

32. Peabody, D.S.; Manifold-Wheeler, B.; Medford, A.; Jordan, S.K.; do Carmo Caldeira, J.; Chackerian, B. Immunogenic display of diverse peptides on virus-like particles of rna phage ms2. *J. Mol. Biol.* **2008**, *380*, 252–263. [CrossRef]

33. Ord, R.L.; Caldeira, J.C.; Rodriguez, M.; Noe, A.; Chackerian, B.; Peabody, D.S.; Gutierrez, G.; Lobo, C.A. A malaria vaccine candidate based on an epitope of the plasmodium falciparum rh5 protein. *Malar. J.* **2014**, *13*, 326. [CrossRef] [PubMed]

34. Frietze, K.M.; Peabody, D.S.; Chackerian, B. Engineering virus-like particles as vaccine platforms. *Curr. Opin. Virol.* **2016**, *18*, 44–49. [CrossRef]

35. Jennings, G.T.; Bachmann, M.F. The coming of age of virus-like particle vaccines. *Biol. Chem.* **2008**, *389*, 521–536. [CrossRef] [PubMed]

36. Manolova, V.; Flace, A.; Bauer, M.; Schwarz, K.; Saudan, P.; Bachmann, M.F. Nanoparticles target distinct dendritic cell populations according to their size. *Eur. J. Immunol.* **2008**, *38*, 1404–1413. [CrossRef]

37. Cubas, R.; Zhang, S.; Kwon, S.; Sevick-Muraca, E.M.; Li, M.; Chen, C.; Yao, Q. Virus-like particle (vlp) lymphatic trafficking and immune response generation after immunization by different routes. *J. Immunother.* **2009**, *32*, 118–128. [CrossRef]

38. Randolph, G.J.; Angeli, V.; Swartz, M.A. Dendritic-cell trafficking to lymph nodes through lymphatic vessels. *Nat. Rev. Immunol.* **2005**, *5*, 617–628. [CrossRef]

39. Roldao, A.; Mellado, M.C.; Castilho, L.R.; Carrondo, M.J.; Alves, P.M. Virus-like particles in vaccine development. *Expert Rev. Vaccines* **2010**, *9*, 1149–1176. [CrossRef]

40. Chroboczek, J.; Szurgot, I.; Szolajska, E. Virus-like particles as vaccine. *Acta Biochim. Pol.* **2014**, *61*, 531–539. [CrossRef]

41. Smith, M.T.; Hawes, A.K.; Bundy, B.C. Reengineering viruses and virus-like particles through chemical functionalization strategies. *Curr. Opin. Biotechnol.* **2013**, *24*, 620–626. [CrossRef]

42. Tang, S.; Xuan, B.; Ye, X.; Huang, Z.; Qian, Z. A modular vaccine development platform based on sortase-mediated site-specific tagging of antigens onto virus-like particles. *Sci. Rep.* **2016**, *6*, 25741. [CrossRef] [PubMed]

43. Janitzek, C.M.; Peabody, J.; Thrane, S.; Carlsen, P.H.R.; Theander, T.G.; Salanti, A.; Chackerian, B.; Nielsen, M.A.; Sander, A.F. A proof-of-concept study for the design of a vlp-based combinatorial hpv and placental malaria vaccine. *Sci. Rep.* **2019**, *9*, 5260. [CrossRef] [PubMed]

44. Menne, S.; Maschke, J.; Tolle, T.K.; Lu, M.; Roggendorf, M. Characterization of t-cell response to woodchuck hepatitis virus core protein and protection of woodchucks from infection by immunization with peptides containing a t-cell epitope. *J. Virol.* **1997**, *71*, 65–74. [CrossRef] [PubMed]

45. Peabody, D.S. Role of the coat protein-rna interaction in the life cycle of bacteriophage ms2. *Mol. Gen. Genet.* **1997**, *254*, 358–364. [CrossRef] [PubMed]

46. Caldeira, J.C.; Peabody, D.S. Thermal stability of rna phage virus-like particles displaying foreign peptides. *J. Nanobiotechnol.* **2011**, *9*, 22. [CrossRef] [PubMed]

47. Chackerian, B.; Caldeira Jdo, C.; Peabody, J.; Peabody, D.S. Peptide epitope identification by affinity selection on bacteriophage ms2 virus-like particles. *J. Mol. Biol.* **2011**, *409*, 225–237. [CrossRef]

48. Fuenmayor, J.; Godia, F.; Cervera, L. Production of virus-like particles for vaccines. *New Biotechnol.* **2017**, *39*, 174–180. [CrossRef]

49. Schott, J.W.; Galla, M.; Godinho, T.; Baum, C.; Schambach, A. Viral and non-viral approaches for transient delivery of mrna and proteins. *Curr. Gene Ther.* **2011**, *11*, 382–398. [CrossRef]

50. Charlton Hume, H.K.; Vidigal, J.; Carrondo, M.J.T.; Middelberg, A.P.J.; Roldao, A.; Lua, L.H.L. Synthetic biology for bioengineering virus-like particle vaccines. *Biotechnol. Bioeng.* **2019**, *116*, 919–935. [CrossRef]

51. Boigard, H.; Cimica, V.; Galarza, J.M. Dengue-2 virus-like particle (vlp) based vaccine elicits the highest titers of neutralizing antibodies when produced at reduced temperature. *Vaccine* **2018**, *36*, 7728–7736. [CrossRef] [PubMed]

52. Zhang, X.; Xin, L.; Li, S.; Fang, M.; Zhang, J.; Xia, N.; Zhao, Q. Lessons learned from successful human vaccines: Delineating key epitopes by dissecting the capsid proteins. *Hum. Vaccines Immunother.* **2015**, *11*, 1277–1292. [CrossRef] [PubMed]

53. Fang, P.Y.; Gomez Ramos, L.M.; Holguin, S.Y.; Hsiao, C.; Bowman, J.C.; Yang, H.W.; Williams, L.D. Functional rnas: Combined assembly and packaging in vlps. *Nucleic Acids Res.* **2017**, *45*, 3519–3527. [CrossRef]

54. Villagrana-Escareno, M.V.; Reynaga-Hernandez, E.; Galicia-Cruz, O.G.; Duran-Meza, A.L.; De la Cruz-Gonzalez, V.; Hernandez-Carballo, C.Y.; Ruiz-Garcia, J. Vlps derived from the ccmv plant virus can directly transfect and deliver heterologous genes for translation into mammalian cells. *BioMed Res. Int.* **2019**, *2019*, 4630891. [CrossRef] [PubMed]

55. Zinkernagel, R.M. On natural and artificial vaccinations. *Annu. Rev. Immunol.* **2003**, *21*, 515–546. [CrossRef]

56. Hou, B.; Saudan, P.; Ott, G.; Wheeler, M.L.; Ji, M.; Kuzmich, L.; Lee, L.M.; Coffman, R.L.; Bachmann, M.F.; DeFranco, A.L. Selective utilization of toll-like receptor and myd88 signaling in b cells for enhancement of the antiviral germinal center response. *Immunity* **2011**, *34*, 375–384. [CrossRef] [PubMed]

57. Petrovsky, N. Comparative safety of vaccine adjuvants: A summary of current evidence and future needs. *Drug Saf.* **2015**, *38*, 1059–1074. [CrossRef] [PubMed]

58. Goetz, K.B.; Pfleiderer, M.; Schneider, C.K. First-in-human clinical trials with vaccines–what regulators want. *Nat. Biotechnol.* **2010**, *28*, 910–916. [CrossRef]

59. Jennings, G.T.; Bachmann, M.F. Immunodrugs: Therapeutic vlp-based vaccines for chronic diseases. *Annu. Rev. Pharm. Toxicol* **2009**, *49*, 303–326. [CrossRef]

60. Darville, T.; Andrews, C.W., Jr.; Sikes, J.D.; Fraley, P.L.; Braswell, L.; Rank, R.G. Mouse strain-dependent chemokine regulation of the genital tract t helper cell type 1 immune response. *Infect. Immun.* **2001**, *69*, 7419–7424. [CrossRef]

61. Forster, R. Study designs for the nonclinical safety testing of new vaccine products. *J. Pharm. Toxicol Methods* **2012**, *66*, 1–7. [CrossRef]

62. Mohsen, M.O.; Speiser, D.E.; Knuth, A.; Bachmann, M.F. Virus-like particles for vaccination against cancer. *Wiley Interdiscip Rev. Nanomed Nanobiotechnol* **2020**, *12*, e1579. [CrossRef] [PubMed]

63. Reddy, S.T.; van der Vlies, A.J.; Simeoni, E.; Angeli, V.; Randolph, G.J.; O'Neil, C.P.; Lee, L.K.; Swartz, M.A.; Hubbell, J.A. Exploiting lymphatic transport and complement activation in nanoparticle vaccines. *Nat. Biotechnol.* **2007**, *25*, 1159–1164. [CrossRef] [PubMed]

64. Itano, A.A.; McSorley, S.J.; Reinhardt, R.L.; Ehst, B.D.; Ingulli, E.; Rudensky, A.Y.; Jenkins, M.K. Distinct dendritic cell populations sequentially present antigen to cd4 t cells and stimulate different aspects of cell-mediated immunity. *Immunity* **2003**, *19*, 47–57. [CrossRef]

65. Pape, K.A.; Catron, D.M.; Itano, A.A.; Jenkins, M.K. The humoral immune response is initiated in lymph nodes by b cells that acquire soluble antigen directly in the follicles. *Immunity* **2007**, *26*, 491–502. [CrossRef] [PubMed]

66. Wilson, N.S.; El-Sukkari, D.; Villadangos, J.A. Dendritic cells constitutively present self antigens in their immature state in vivo and regulate antigen presentation by controlling the rates of mhc class ii synthesis and endocytosis. *Blood* **2004**, *103*, 2187–2195. [CrossRef] [PubMed]

67. Wilson, N.S.; El-Sukkari, D.; Belz, G.T.; Smith, C.M.; Steptoe, R.J.; Heath, W.R.; Shortman, K.; Villadangos, J.A. Most lymphoid organ dendritic cell types are phenotypically and functionally immature. *Blood* **2003**, *102*, 2187–2194. [CrossRef]

68. Garside, P.; Ingulli, E.; Merica, R.R.; Johnson, J.G.; Noelle, R.J.; Jenkins, M.K. Visualization of specific b and t lymphocyte interactions in the lymph node. *Science* **1998**, *281*, 96–99. [CrossRef]

69. Hangartner, L.; Zellweger, R.M.; Giobbi, M.; Weber, J.; Eschli, B.; McCoy, K.D.; Harris, N.; Recher, M.; Zinkernagel, R.M.; Hengartner, H. Nonneutralizing antibodies binding to the surface glycoprotein of lymphocytic choriomeningitis virus reduce early virus spread. *J. Exp. Med.* **2006**, *203*, 2033–2042. [CrossRef]

70. Sallusto, F.; Lanzavecchia, A.; Araki, K.; Ahmed, R. From vaccines to memory and back. *Immunity* **2010**, *33*, 451–463. [CrossRef]

71. Bachmann, M.F.; Zinkernagel, R.M. Neutralizing antiviral b cell responses. *Annu. Rev. Immunol.* **1997**, *15*, 235–270. [CrossRef]

72. Zhang, S.; Cubas, R.; Li, M.; Chen, C.; Yao, Q. Virus-like particle vaccine activates conventional b2 cells and promotes b cell differentiation to igg2a producing plasma cells. *Mol. Immunol.* **2009**, *46*, 1988–2001. [CrossRef] [PubMed]

73. Yang, R.; Murillo, F.M.; Cui, H.; Blosser, R.; Uematsu, S.; Takeda, K.; Akira, S.; Viscidi, R.P.; Roden, R.B. Papillomavirus-like particles stimulate murine bone marrow-derived dendritic cells to produce alpha interferon and th1 immune responses via myd88. *J. Virol.* **2004**, *78*, 11152–11160. [CrossRef] [PubMed]

74. Gatto, D.; Bauer, M.; Martin, S.W.; Bachmann, M.F. Heterogeneous antibody repertoire of marginal zone b cells specific for virus-like particles. *Microbes Infect.* **2007**, *9*, 391–399. [CrossRef]

75. Gatto, D.; Brink, R. The germinal center reaction. *J. Allergy Clin. Immunol.* **2010**, *126*, 898–907; quiz 908–909. [CrossRef]

76. Bachmann, M.F.; Jennings, G.T. Vaccine delivery: A matter of size, geometry, kinetics and molecular patterns. *Nat. Rev. Immunol.* **2010**, *10*, 787–796. [CrossRef]

77. Prelog, M. Differential approaches for vaccination from childhood to old age. *Gerontology* **2013**, *59*, 230–239. [CrossRef]

78. Wirth, T.C.; Xue, H.H.; Rai, D.; Sabel, J.T.; Bair, T.; Harty, J.T.; Badovinac, V.P. Repetitive antigen stimulation induces stepwise transcriptome diversification but preserves a core signature of memory cd8(+) t cell differentiation. *Immunity* **2010**, *33*, 128–140. [CrossRef] [PubMed]

79. Kardani, K.; Bolhassani, A.; Shahbazi, S. Prime-boost vaccine strategy against viral infections: Mechanisms and benefits. *Vaccine* **2016**, *34*, 413–423. [CrossRef]

80. Aurisicchio, L.; Fridman, A.; Mauro, D.; Sheloditna, R.; Chiappori, A.; Bagchi, A.; Ciliberto, G. Safety, tolerability and immunogenicity of v934/v935 htert vaccination in cancer patients with selected solid tumors: A phase i study. *J. Transl. Med.* **2020**, *18*, 39. [CrossRef]

81. Bachmann, M.F.; Beerli, R.R.; Agnellini, P.; Wolint, P.; Schwarz, K.; Oxenius, A. Long-lived memory cd8+ t cells are programmed by prolonged antigen exposure and low levels of cellular activation. *Eur. J. Immunol.* **2006**, *36*, 842–854. [CrossRef] [PubMed]

82. Crotty, S.; Johnston, R.J.; Schoenberger, S.P. Effectors and memories: Bcl-6 and blimp-1 in t and b lymphocyte differentiation. *Nat. Immunol.* **2010**, *11*, 114–120. [CrossRef] [PubMed]

83. Johansen, P.; Storni, T.; Rettig, L.; Qiu, Z.; Der-Sarkissian, A.; Smith, K.A.; Manolova, V.; Lang, K.S.; Senti, G.; Mullhaupt, B.; et al. Antigen kinetics determines immune reactivity. *Proc. Natl. Acad. Sci. USA* **2008**, *105*, 5189–5194. [CrossRef] [PubMed]

84. Storni, T.; Ruedl, C.; Schwarz, K.; Schwendener, R.A.; Renner, W.A.; Bachmann, M.F. Nonmethylated cg motifs packaged into virus-like particles induce protective cytotoxic t cell responses in the absence of systemic side effects. *J. Immunol* **2004**, *172*, 1777–1785. [CrossRef] [PubMed]

85. Pickett, G.G.; Peabody, D.S. Encapsidation of heterologous rnas by bacteriophage ms2 coat protein. *Nucleic Acids Res.* **1993**, *21*, 4621–4626. [CrossRef]

86. Swanson, C.L.; Wilson, T.J.; Strauch, P.; Colonna, M.; Pelanda, R.; Torres, R.M. Type i ifn enhances follicular b cell contribution to the t cell-independent antibody response. *J. Exp. Med.* **2010**, *207*, 1485–1500. [CrossRef]

87. Jegerlehner, A.; Maurer, P.; Bessa, J.; Hinton, H.J.; Kopf, M.; Bachmann, M.F. Tlr9 signaling in b cells determines class switch recombination to igg2a. *J. Immunol* **2007**, *178*, 2415–2420. [CrossRef]

88. Andersson, P.; Ostheimer, C. Editorial: Combinatorial approaches to enhance anti-tumor immunity: Focus on immune checkpoint blockade therapy. *Front. Immunol.* **2019**, *10*, 2083. [CrossRef]

89. Christofi, T.; Baritaki, S.; Falzone, L.; Libra, M.; Zaravinos, A. Current perspectives in cancer immunotherapy. *Cancers* **2019**, *11*, 1472. [CrossRef]

90. Wahid, B.; Ali, A.; Rafique, S.; Waqar, M.; Wasim, M.; Wahid, K.; Idrees, M. An overview of cancer immunotherapeutic strategies. *Immunotherapy* **2018**, *10*, 999–1010. [CrossRef]

91. Muenst, S.; Laubli, H.; Soysal, S.D.; Zippelius, A.; Tzankov, A.; Hoeller, S. The immune system and cancer evasion strategies: Therapeutic concepts. *J. Intern. Med.* **2016**, *279*, 541–562. [CrossRef] [PubMed]

92. Del Giudice, G.; Rappuoli, R.; Didierlaurent, A.M. Correlates of adjuvanticity: A review on adjuvants in licensed vaccines. *Semin. Immunol.* **2018**, *39*, 14–21. [CrossRef] [PubMed]

93. Iancu, E.M.; Baumgaertner, P.; Wieckowski, S.; Speiser, D.E.; Rufer, N. Profile of a serial killer: Cellular and molecular approaches to study individual cytotoxic t-cells following therapeutic vaccination. *J. Biomed. Biotechnol.* **2011**, *2011*, 452606. [CrossRef]

94. Baba, T.; Sato-Matsushita, M.; Kanamoto, A.; Itoh, A.; Oyaizu, N.; Inoue, Y.; Kawakami, Y.; Tahara, H. Phase i clinical trial of the vaccination for the patients with metastatic melanoma using gp100-derived epitope peptide restricted to hla-a*2402. *J. Transl. Med.* **2010**, *8*, 84. [CrossRef]

95. Li, M.; Bharadwaj, U.; Zhang, R.; Zhang, S.; Mu, H.; Fisher, W.E.; Brunicardi, F.C.; Chen, C.; Yao, Q. Mesothelin is a malignant factor and therapeutic vaccine target for pancreatic cancer. *Mol. Cancer Ther.* **2008**, *7*, 286–296. [CrossRef] [PubMed]

96. Bharadwaj, U.; Li, M.; Chen, C.; Yao, Q. Mesothelin-induced pancreatic cancer cell proliferation involves alteration of cyclin e via activation of signal transducer and activator of transcription protein 3. *Mol. Cancer Res. Mcr.* **2008**, *6*, 1755–1765. [CrossRef] [PubMed]

97. Fong, D.; Moser, P.; Krammel, C.; Gostner, J.M.; Margreiter, R.; Mitterer, M.; Gastl, G.; Spizzo, G. High expression of trop2 correlates with poor prognosis in pancreatic cancer. *Br. J. Cancer* **2008**, *99*, 1290–1295. [CrossRef]

98. Sugano, K.; Maeda, K.; Ohtani, H.; Nagahara, H.; Shibutani, M.; Hirakawa, K. Expression of xct as a predictor of disease recurrence in patients with colorectal cancer. *Anticancer. Res.* **2015**, *35*, 677–682.

99. Ruiu, R.; Rolih, V.; Bolli, E.; Barutello, G.; Riccardo, F.; Quaglino, E.; Merighi, I.F.; Pericle, F.; Donofrio, G.; Cavallo, F.; et al. Fighting breast cancer stem cells through the immune-targeting of the xct cystine-glutamate antiporter. *Cancer Immunol. Immunother.* **2019**, *68*, 131–141. [CrossRef]

100. Visvader, J.E.; Lindeman, G.J. Cancer stem cells in solid tumours: Accumulating evidence and unresolved questions. *Nat. Rev. Cancer* **2008**, *8*, 755–768. [CrossRef]

101. Fulawka, L.; Donizy, P.; Halon, A. Cancer stem cells–the current status of an old concept: Literature review and clinical approaches. *Biol. Res.* **2014**, *47*, 66. [CrossRef] [PubMed]

102. Briggs, K.J.; Koivunen, P.; Cao, S.; Backus, K.M.; Olenchock, B.A.; Patel, H.; Zhang, Q.; Signoretti, S.; Gerfen, G.J.; Richardson, A.L.; et al. Paracrine induction of hif by glutamate in breast cancer: Egln1 senses cysteine. *Cell* **2016**, *166*, 126–139. [CrossRef] [PubMed]

103. Lewerenz, J.; Hewett, S.J.; Huang, Y.; Lambros, M.; Gout, P.W.; Kalivas, P.W.; Massie, A.; Smolders, I.; Methner, A.; Pergande, M.; et al. The cystine/glutamate antiporter system x(c)(-) in health and disease: From molecular mechanisms to novel therapeutic opportunities. *Antioxid. Redox Signal.* **2013**, *18*, 522–555. [CrossRef] [PubMed]

104. Galadari, S.; Rahman, A.; Pallichankandy, S.; Thayyullathil, F. Reactive oxygen species and cancer paradox: To promote or to suppress? *Free Radic. Biol. Med.* **2017**, *104*, 144–164. [CrossRef]

105. Nabeyama, A.; Kurita, A.; Asano, K.; Miyake, Y.; Yasuda, T.; Miura, I.; Nishitai, G.; Arakawa, S.; Shimizu, S.; Wakana, S.; et al. Xct deficiency accelerates chemically induced tumorigenesis. *Proc. Natl. Acad. Sci. USA* **2010**, *107*, 6436–6441. [CrossRef]

106. Chen, R.S.; Song, Y.M.; Zhou, Z.Y.; Tong, T.; Li, Y.; Fu, M.; Guo, X.L.; Dong, L.J.; He, X.; Qiao, H.X.; et al. Disruption of xct inhibits cancer cell metastasis via the caveolin-1/beta-catenin pathway. *Oncogene* **2009**, *28*, 599–609. [CrossRef]

 *viruses*

Review

Virus-Like Particle-Mediated Vaccination against Interleukin-13 May Harbour General Anti-Allergic Potential beyond Atopic Dermatitis

John Foerster * and Aleksandra Molęda *

Department of Molecular and Clinical Medicine, University of Dundee, Medical School, Ninewells Hospital, Jacquie Woods Centre, Ninewells Drive, Dundee DD1 9SY, UK
* Correspondence: j.foerster@dundee.ac.uk (J.F.); a.moleda@dundee.ac.uk (A.M.)

Received: 6 March 2020; Accepted: 9 April 2020; Published: 13 April 2020

Abstract: Virus-like particle (VLP)-based anti-infective prophylactic vaccination has been established in clinical use. Although validated in proof-of-concept clinical trials in humans, no VLP-based therapeutic vaccination against self-proteins to modulate chronic disease has yet been licensed. The present review summarises recent scientific advances, identifying interleukin-13 as an excellent candidate to validate the concept of anti-cytokine vaccination. Based on numerous clinical studies, long-term elimination of IL-13 is not expected to trigger target-related serious adverse effects and is likely to be safer than combined targeting of IL-4/IL-13. Furthermore, recently published results from large-scale trials confirm that elimination of IL-13 is highly effective in atopic dermatitis, an exceedingly common condition, as well as eosinophilic esophagitis. The distinctly different mode of action of a polyclonal vaccine response is discussed in detail, suggesting that anti-IL-13 vaccination has the potential of outperforming monoclonal antibody-based approaches. Finally, recent data have identified a subset of follicular T helper cells dependent on IL-13 which selectively trigger massive IgE accumulation in response to anaphylactoid allergens. Thus, prophylactic IL-13 vaccination may have broad application in a number of allergic conditions.

Keywords: VLP; IL-13; interleukin-13; vaccine; Tfh cells

---

1. Introduction: Why Develop an Anti-IL-13 Vaccine?

In an age dominated by an avalanche of monoclonal antibodies (MAb) hitting the market for a dizzying variety of conditions, one may ask why replacing this obvious successful model with a vaccine strategy is sensible at all. The answer to this question is four-fold. Two of these are generic and would apply to many vaccines replacing a MAb treatment. First, it will be much more accessible to patients (economics). Second, it will have much broader applications thanks to the qualitatively different nature of a polyclonal vs. a monoclonal immune response (see below for details). The third reason is more limited to IL-13: it has an excellent safety profile as a target compared to almost all other cytokines, with the exception of IL-17. Fourth, and finally, very recent data suggest that targeting IL-13 with a VLP vaccine may have very broad anti-allergic potential, possibly leading to amelioration of allergies and being able to achieve synergistic effects with other anti-allergic treatments. The present review will discuss these four themes in turn, not aiming to be exhaustive, but rather with the intent to stimulate further efforts to tackle many open questions. By implication, therefore, this review does not intend to deliver a broader review of the signalling, structure, or immunobiology of IL-13 which have been reviewed excellently recently elsewhere [1–3].

2. Virus-Like Particles as a Construct for IL-13 Therapeutic Vaccine

Virus-like particle (VLP) vaccines are nanostructures generated by self-assembly of structural proteins resembling the native version in their morphology and composition but lack the genomic material of infectious capacity [4,5]. VLPs can be acquired from a variety of expression systems and platforms including bacteriophages (MS2, PP7 and AP205 [6–8]), yeast (*Hansenula polymorpha* and *Saccharomyces cerevisiae* [9,10]), bacteria (*Escherichia coli* [8]), mammalian cell lines (Vero, 293T and BHK cell lines [11–13]), plant cell culture (cowpea mosaic virus, cucumber mosaic virus, tobacco mosaic virus, and bean yellow dwarf virus [14–16]) and insect cell lines (Baculovirus and Sf9 cell line [12,17]) [18]. Vaccine development faces a clear challenge: production of sufficient amounts of quality antibodies to target the desired antigen. VLPs provide an excellent vaccine delivery platform due to their composition: their small size (usually 20–200 nm), geometry and flexibility during development [4]. Their size allows easy passage and drainage through the lymph to reach all areas such as secondary lymphoid organs resulting in profound effects in targeting follicular B cells [4,19–21]. Furthermore, CD8+ and plasmacytoid subsets of dendritic cells (DCs) can cross-present small-sized antigens such as VLPs and active B cells and T cells in the lymph nodes to induce cytotoxic effects [20,22,23]. Repetitive multivalent surface arrangement allows cross-linking of B cell receptors, perfect for inducing great amounts and long-lasting antibodies [20,24]. VLPs also act as a template for further engineering, where additional epitopes, proteins and nucleic acids are easily incorporated alongside vaccine targets that can significantly increase immunity such as Toll-like receptor (Tlr) ligands [20,25]. These characteristics can thus provide solutions for vaccine delivery challenges and are readily modified for a vast variety of constructs to boost immune responses in many individuals.

3. The Health Economics of IL-13-Targetable Diseases

In terms of economics, it is obvious that health care systems globally are under huge strain; personal bankruptcies due to health care expenditure in the US alone tell the story: a 2019 study in the American Journal of Public Health found that two-thirds of personal bankruptcies are filed due to medical bills, equating to more than half a million of affected people despite the Affordable Care Act [26]. While health care cost in other economies may not be quite as exorbitant, that fact is offset by the simple unavailability of many high-quality medicines to patients who cannot afford private health care. Given current demographic trends toward increased old age-related morbidity, including the dementia 'epidemic', as well as globally increased longevity, the search for truly affordable health care solutions represents a distinct priority.

The clinical indications amenable to anti-IL-13 vaccination based on documented action of anti-IL-13 MAbs to date include atopic dermatitis, subgroups of asthma, and eosinophilic esophagitis. However, the list of other potential indications is much longer and has been discussed in detail [27]. Crucially, in the context of competitive resource allocation vis-à-vis conditions such as dementia, cancer, and emerging infectious diseases, it is clear that per-case expenditure available by health care providers will not be able to satisfy the profit margins required to offset large-scale manufacture of monoclonal antibodies. Hence, vaccine approaches, which also avoid the need for laboratory monitoring infrastructure, will become eminently competitive in the near future.

4. Monoclonal Antibodies vs. Polyclonal Vaccine Responses

The rate of recent marketing approvals of MAbs and sometimes decoy receptors suggests that they are highly effective in ameliorating disease. However, a closer look prompts the question: do they reach full therapeutic potential? Specifically, it is becoming increasingly evident that the serum concentrations required for monoclonal antibodies to be effective are rather extreme. A striking example for this is the group of monoclonals targeting the p19 subunit of IL-23, currently licensed for psoriasis: guselkumab, risankizumab, and tildrakizumab. While the molecular mode of action is identical between all three antibodies, even the comparatively high affinity of tildrakizumab to its cytokine target (300 pM) is evidently suboptimal based on its inferior clinical activity compared to the competitor MAb, which feature Kd values in the vicinity of a staggering 2 pM (Table 1). Notably, the relatively low efficacy of tildrakizumab exists despite a much higher relative affinity of this MAb for the cytokine compared to the receptor (Table 1).

Table 1. Association of IL-23-targeting monoclonal antibody (MAb) affinity with clinical efficacy.

	Target MAb Affinity	Clinical Efficacy [1]	Fold-Increase Vs. Receptor Binding [2]	Ref
Guselkumab	2 pM	91%	3000-fold	[28,29]
Risankizumab	2 pM	91%	3000-fold	[30,31]
Tildrakizumab	300 pM	61%	20-fold	[32]

[1] Response of psoriasis severity, measured as the so-called PASI75 index, indicating 75% improvement from baseline, measured at twelve weeks after treatment start, in each study cited. [2] Compared to the 6 nM reported affinity of IL-23 to receptor [33].

For MAbs targeting the IL-4/13 system, this overall scenario appears similar. MAbs and their respective molecular targets are graphically summarised in Figure 1. Structural and signalling details have been summarised in detail recently elsewhere [2]. As detailed in Table 2, the picture emerging from an integrated review of clinical success (or failure) of related MAbs is that affinities of Kd values of approximately 60 pM or higher appear to correlate with clinical activity, justifying clinical development all the way through phase III trials. The failure of pascolinumab is mechanistically of interest, as this MAb harbours high affinity. By implication, its lack of satisfactory activity in clinical trials would clearly suggest that isolated IL-4 inhibition, with intact IL-13 signalling, is insufficient to achieve useful suppression of pro-inflammatory signalling.

Table 2. Ligand affinities and affinities relative to ligand–receptor binding of various cytokine-binding clinically used MAbs.

Mab/Blocker	Net Target [1]	Affinity (pM)	Tested in [2]	Effective in	Fold Affinity Vs. Receptor [3]	Ref
			Group 1: IL-13-only blockers			
Lebrikizumab	IL-13Ra1	<10	Asthma, AD	AD, Asthma [4]	3.5	[35,36]
Anrukinzumab	IL-13Ra1	385	UC, Asthma	—	0.1	[37]
RPC4046	IL-13Ra1/Ra2	50	EoE	EoE	0.7	[38,39]
Tralokinumab	IL-13Ra1/Ra2	58	Asthma, AD	AD	0.8	[40]
			Group 2: IL-4-selective and IL4/IL-13 combined blockers			
Pascolizumab	IL-4Ra	60	Asthma	—	1.4	[41]
Dupilumab	IL-4Ra, IL-13Ra1	9	Asthma, AD	Asthma, AD	10	[42]
Pitrankinra	IL-4Ra, IL-13Ra1	100 [5]	Asthma	—	1	[43]
Altrakincept	IL-4Ra, IL-13Ra1	1000	Asthma	—	0.1	[44,45]
AMG317	IL-4Ra, IL-13Ra1	180	Asthma	—	0.8	[46]

[1] Signal-transducing receptor pathway inactivated by monoclonal antibody. The actual targets of each IL-13- and IL-4-inhibiting MAb are shown in Figure 1. [2] Disease abbreviations: UC—ulcerative colitis, EoE—eosinophilic esophagitis, and Pso—psoriasis. [3] Target–receptor affinities are complex, involving sequential ternary complex formation of more than one receptor subunit. The data shown depict the ratio between MAb/cytokine affinity and cytokine/receptor affinity, respectively, for IL-13 -> IL-13Ra1/IL-4Ra (30–40 pM, for group 1), and for IL-4 -> IL-4Ra (100 pM, group two), as detailed in [47]. The highest affinity target/receptor binding reported was chosen as most relevant to gauge the competitive strength of drug circulating in serum. [4] In the high-biomarker patient subgroup. [5] Assuming retained affinity for the dual mutation differentiating pitankinra from native IL-4.

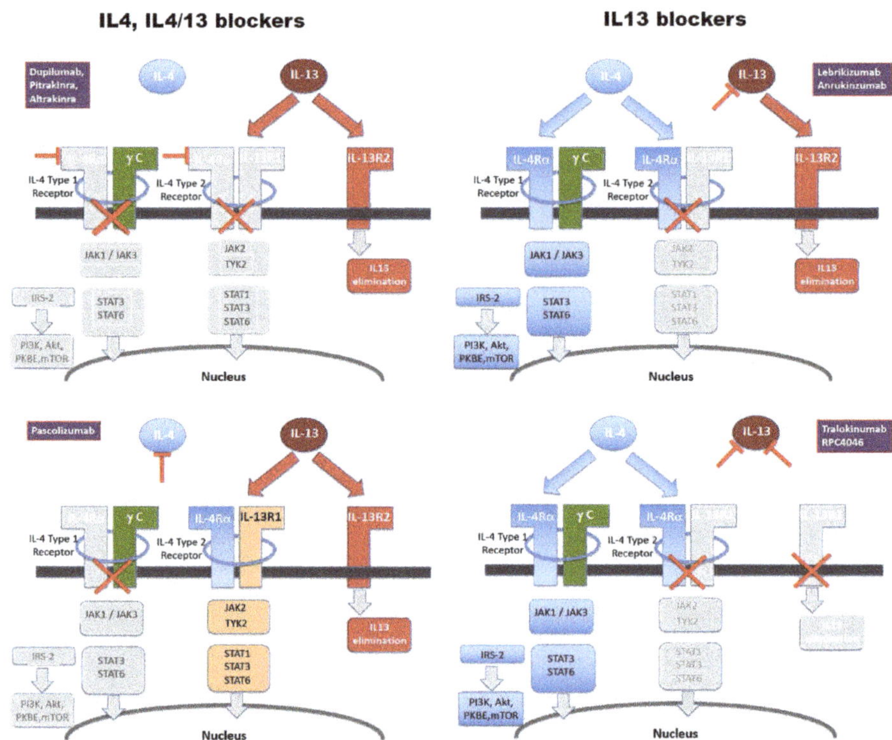

Figure 1. Molecular targets of monoclonal antibodies clinically evaluated to inhibit IL4/IL-13 signalling to date. For a list of antibodies and references, see Table 1. Colour codes used are blue for signalling transduced through IL-4, but not IL-13, beige for either IL-4 or IL-13 (through the IL-4Ra/IL-13Ra1 receptor dimer), and red for IL-13, but not IL-4. IL-13Ra2, also depicted in red, acts primarily as a cellular disposal facility for IL-13 [34]. For details on structure and signalling, see 2.

The apparent requirement for ultra-high cytokine affinity of MAbs is all the more noteworthy, as trough serum concentrations are usually above 5 g/mL. By contrast, serum concentrations reported for IL-13 range from below 1 pg/mL, regardless of whether in healthy probands or symptomatic asthma patients [48], to between 0.16 and 0.24 ng/mL in diverse populations and patient groups [49]. Even if concentrations in target tissues (lung or skin) may be significantly higher than in serum, it is remarkable that molar excess of several orders of magnitude of MAb to cytokine still can prove clinically insufficient.

By contrast, polyclonal responses act by combining various affinities of different antibody species. The quantity of B cell clones, as well as resulting affinities, can vary widely. Thus, a bivalent VLP-based vaccine against norovirus consisting of Gll.1 and Gll.4 sequence strains was found to produce an oligoclonal response, with < 3 antibodies accounting for 58–86% of epitope-specific clones [14]. This study utilised the VEE replicon vector to transfect ORF2 genes of norovirus for production of viral protein 1, a capsid protein that can self-assembly into a VLP when introduced into BHK-21 cells. By contrast, a recent in-depth analysis showed that influenza virus vaccination in humans triggered 36 ± 12 (mean ± SD, range 16–49) B cell lineages expanding >50-fold and, collectively, accounting for 22 ± 12% of each subject's antibody repertoire during peak response [50]. In another influenza vaccine study, vaccination elicited between 40 and 147 clonotypes where the top 6% most abundant clonotypes accounted for > 60% of the entire repertoire [51]. In terms of affinity, Kd values between 2.5 pM and 160 nM were found [51]. A large-scale analysis of influenza vaccination in healthy controls and SLE patients elicited antibody affinities between 1 and 100 nM [52]. A single-protein vaccination of HIV-Env produced final affinity-matured Ig affinities to the antigen between approximately 3 and 10 nM in Rhesus monkeys [53]. Collectively, these data strongly suggest that vaccine-triggered natural effector antibodies show significantly less single-clone affinity compared to the engineered MAbs detailed in Table 1 but bear the potential of significant cooperative action. Although no detailed studies are available on therapeutic VLP vaccines, these likely trigger responses comparable to anti-virus vaccines. Accordingly, the VLP-based vaccines targeting interleukin-1 and angiotensin both met their pharmacodynamic endpoints in clinical trials [54,55]. Likewise, a number of VLP-based vaccines have been shown to be effective in other mammalian species, including treating insect bite hypersensitivity in horses [56]. Here, IL-5 was chemically conjugated to a VLP from cucumber mosaic virus engineered with the tetanus toxoid (TT) epitope, a property to enhance immunological responses of T helper cells and memory B cells [56]. This characteristic was also employed in vaccine constructs against psoriasis, cat allergy and Alzheimer's disease [57]. Designing a VLP vaccine that contains the TT epitope alongside the target antigen can be a useful tool in ensuring enhanced immune protection across majority of immunised individuals, especially those that are aging [57]. Similarly, for influenza vaccines, polyclonal approaches may outperform monoclonal targeting based on broader antigen coverage [58].

Apart from the affinity and polyclonality of a vaccine response, a third fundamental difference between MAbs and polyclonal responses is that MAbs are of a singular—mostly IgG1—isotype, which is efficient in engaging effector functions such as antibody-dependent cytotoxity (ADCC), antibody-dependent cellular phagocytosis (ADCP), or pro-inflammatory cytokine activation, mediated by Fc binding of a variety of Fc gamma receptors (FcγRs). It is noteworthy that dupilumab is of the IgG4 isotype, which is less efficient in engaging FcγRs [59]. By contrast, polyclonal responses produce a mix of isotypes, which will vary between vaccine recipients in addition to cooperative action between various species targeting the cytokine at different epitopes.

5. The Inhibition of Anaphylactic TfH Cells: A Novel Role for IL-13 Neutralisation

Very recently, the paradigm in IL-4/IL-13 immune function in the germinal centre reaction was that IL-4 is central to germinal centre-based affinity maturation, as well as isotype switching toward IgE of B-cells, but not IL-13, through follicular T-helper (TfH) 2 cells characterised by IL-4 secretion (reviewed in [60]). However, a very recent landmark study identified a new Tfh subset characterised by IL-13 secretion which is responsible for the generation of IgE^{high} B cells in response to stimulation with allergens [61] (Figure 2). Crucially, TfH13 cells were not only identified in the newly created hyper-IgE-harbouring mouse model, but also detectable and increased in humans with peanut allergy. Importantly, TfH13 cells did not drive IgE^{high} B cell and peripheral IgE production in response to parasitic (helminth) infection but only in response to anaphylactoid allergen challenge (also reviewed in [62]). Moreover, T cell-specific ablation of IL-13 was found to profoundly suppress IgE synthesis. These data suggest that elimination of IL-13 function may lead to suppression of anaphylactic reaction to a wide range of allergens.

As detailed above, the response to such therapeutic intervention may profoundly differ between anti-IL-13 MAb administration and anti-IL-13 vaccination, in particular since the latter is expected to induce IL-13-targeted memory clones (both B and T cells), which could modulate the germinal centre reaction by cytokine secretion and which would not feature in MAb-mediated IL-13 suppression. In addition, it is intriguing that TfH1 cells, which would mediate class switching in response to a VLP-type vaccine, would be instrumentalised in this way to suppress the activity of TfH13 cells. In fact, this concept appears to already have been validated for TfH2 cells, since blocking of IL-4 caused a profound block in Th2-class switch and improved antibody responses to oxycodone and diptheria-pertussis-tetanus vaccination [64]. Clinically, it is very possible that the routinely observed major drop in serum IgE levels in patients treated with dupilumab is in part mediated by suppression of TfH2 cells. It would therefore be of interest to compare the performance of monoclonal- vs. vaccine-mediated IL-13 inhibition in the novel DOCK8-/- mouse model established in the above cited study. Moreover, given the different mode of action of a polyclonal response, combined with the novel anti-TfH13 mode of action, it is entirely possible that VLP vaccine-based IL-13 targeting will perform different to MAb-based treatment in allergic asthma.

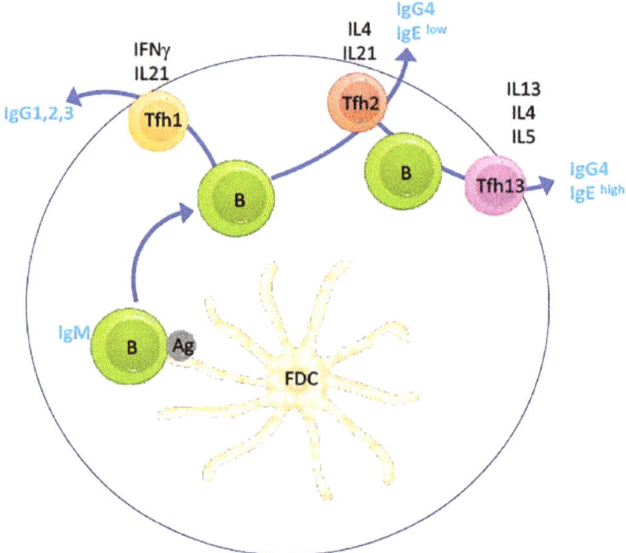

Figure 2. Function of IL-13-dependent TfH cells in IgE production in response to allergens. Simplified schematic showing the light zone of a peripheral lymphoid germinal centre (GC) where a naïve IgM-positive B cell survives apoptosis after antigen (Ag) encounter presented by a follicular dendritic cell (FDC). Prior and additional interaction with Th cells are not shown. Depending on the type of antigen, the B cell may then receive stimulation by TfH1 cells, resulting in class switching to Ig-isoforms, as indicated. Further, sequential stimulation by TfH2 cells may cause secondary class switching to IgG4 or IgE^{low} phenotypes. Finally, the newly identified IL-13-dependent TfH13 cells, may cause tertiary switching to IgE^{high} in the context of allergen stimulation. The schematic is based on figures in [61,63].

6. Safety of Anti-IL-13 Vaccination

In light of the above-detailed roles of IL-13 and IL-4, it is both highly significant and reassuring that a large double-blind controlled study on almost 180 patients did not find any effect of tetanus vaccine-induced specific IgG formation after twelve weeks of dupilumab treatment [65]. Furthermore, emerging post-marketing surveillance data of patients on dupilumab confirms an approximately 10% incidence of conjunctivitis. In one recent real-world review of 54 patients, 2 patients had to discontinue due to, or for reasons related to, conjunctivitis symptoms [66]. Since this drug-specific side effect had already been recognised during pre-marketing phase II–III studies and continues to be non-reported in any anti-IL-13 treated cohort, it is highly likely that conjunctivitis will not be associated with long-term IL-13 elimination. Furthermore, a growing array of meta-analyses, and systematic reviews of anti-IL-13 MAb-treated patients fail to uncover target-related limiting adverse events in large patient groups [27,67–69].

7. Conclusions

Utilising VLPs as vaccine constructs provides flexibility and many prospects in design that can significantly enhance protection and immunogenicity. IL-13 represents an excellent proof-of-concept target to validate anti-cytokine vaccination as a therapeutic strategy in humans based on validated vaccine targets, an excellent target safety profile, and broad vaccine applicability. Treating IL-13-related immunological disorders with a therapeutic vaccine will offer a new avenue in medical intervention.

Author Contributions: J.F. and A.M. wrote and edited the manuscript. All authors have read and agreed to the published version of the manuscript.

Funding: This work was not supported by external funding.

Acknowledgments: The authors would like to acknowledge the support of Jean-Christophe Bourdon in the co-supervision of AM research.

Conflicts of Interest: J.F. is co-owner and CEO of a commercial company (www.healvax.com) that holds licence rights for the clinical development of IL-13 targeting vaccines. Healvax had no role in the design of the study; in the collection, analyses, or interpretation of data; in the writing of the manuscript, or in the decision to publish the results.

References

1. Bao, K.; Reinhardt, R.L. The differential expression of IL-4 and IL-13 and its impact on type-2 immunity. *Cytokine* **2015**, *75*, 25–37. [CrossRef] [PubMed]

2. Junttila, I. Tuning the Cytokine Responses: An Update on Interleukin (IL)-4 and IL-13 Receptor Complexes. *Front. Immunol.* **2018**, *9*, 888. [CrossRef] [PubMed]

3. LaPorte, S.L.; Juo, Z.S.; Vaclavikova, J.; Colf, L.A.; Qi, X.; Heller, N.M.; Keegan, A.D.; Garcia, K.C. Molecular and structural basis of cytokine receptor pleiotropy in the interleukin-4/13 system. *Cell* **2008**, *132*, 259–272. [CrossRef] [PubMed]

4. Cimica, V.; Galarza, J.M. Adjuvant formulations for virus-like particle (VLP) based vaccines. *Clin. Immunol.* **2017**, *183*, 99–108. [CrossRef] [PubMed]

5. Fuenmayor, J.; Gòdia, F.; Cervera, L. Production of virus-like particles for vaccines. *N. Biotechnol.* **2017**, *39*, 174–180. [CrossRef]

6. Lino, C.A.; Caldeira, J.C.; Peabody, D.S. Display of single-chain variable fragments on bacteriophage MS2 virus-like particles. *J. Nanobiotechnol.* **2017**, *15*, 13. [CrossRef]

7. Tumban, E.; Peabody, J.; Peabody, D.S.; Chackerian, B. A Pan-HPV Vaccine Based on Bacteriophage PP7 VLPs Displaying Broadly Cross-Neutralizing Epitopes from the HPV Minor Capsid Protein, L2. *PLoS ONE* **2011**, *6*, e23310. [CrossRef]

8. Janitzek, C.M.; Matondo, S.; Thrane, S.; Nielsen, M.; Kavishe, R.; Mwakalinga, S.B.; Theander, T.; Salanti, A.; Sander, A.F. Bacterial superglue generates a full-length circumsporozoite protein virus-like particle vaccine capable of inducing high and durable antibody responses. *Malar. J.* **2016**, *15*, 545. [CrossRef]

9. Wetzel, D.; Rolf, T.; Suckow, M.; Kranz, A.; Barbian, A.; Chan, J.A.; Leitsch, J.; Weniger, M.; Jenzelewski, V.; Kouskousis, B.; et al. Establishment of a yeast-based VLP platform for antigen presentation. *Microb. Cell Factories* **2018**, *17*, 17. [CrossRef]

10. Kim, H.J.; Kim, H.-J. Yeast as an expression system for producing virus-like particles: What factors do we need to consider? *Lett. Appl. Microbiol.* **2017**, *64*, 111–123. [CrossRef]

11. Huang, X.; Wang, X.; Zhang, J.; Xia, N.; Zhao, Q. Escherichia coli-derived virus-like particles in vaccine development. *NPJ Vaccines* **2017**, *2*, 3. [CrossRef] [PubMed]

12. Wu, C.-Y.; Yeh, Y.-C.; Yang, Y.-C.; Chou, C.; Liu, M.-T.; Wu, H.-S.; Chan, J.-T.; Hsiao, P.-W. Mammalian Expression of Virus-Like Particles for Advanced Mimicry of Authentic Influenza Virus. *PLoS ONE* **2010**, *5*, e9784. [CrossRef]

13. Arevalo, M.T.; Wong, T.M.; Ross, T.M. Expression and Purification of Virus-like Particles for Vaccination. *J. Vis. Exp.* **2016**, *112*, e54041. [CrossRef] [PubMed]

14. Lindesmith, L.C.; McDaniel, J.R.; Changela, A.; Verardi, R.; Kerr, S.A.; Costantini, V.; Brewer-Jensen, P.D.; Mallory, M.L.; Voss, W.N.; Boutz, D.R.; et al. Sera Antibody Repertoire Analyses Reveal Mechanisms of Broad and Pandemic Strain Neutralizing Responses after Human Norovirus Vaccination. *Immunity* **2019**, *50*, 1530–1541. [CrossRef] [PubMed]

15. Sainsbury, F.; Saxena, P.; Aljabali, A.A.; Saunders, K.; Evans, D.; Lomonossoff, G.P. *Genetic Engineering and Characterization of Cowpea Mosaic Virus Empty Virus-Like Particles*; Humana Press: Totowa, NJ, USA, 2014; pp. 139–153.

16. Storni, F.; Zeltins, A.; Balke, I.; Heath, M.D.; Kramer, M.F.; Skinner, M.A.; Zha, L.; Roesti, E.; Engeroff, P.; Muri, L.; et al. Vaccine against peanut allergy based on engineered virus-like particles displaying single major peanut allergens. *J. Allergy Clin. Immunol.* **2019**, *145*, 1240–1253. [CrossRef]

17. Larkin, E.J.; Brown, A.D.; Culver, J.N. Fabrication of Tobacco Mosaic Virus-Like Nanorods for Peptide Display. In *Methods in Molecular Biology (Clifton, N.J.)*; Humana Press: Totowa, NJ, USA, 2018; Volume 1776, pp. 51–60.

18. Abdoli, A.; Soleimanjahi, H.; Fotouhi, F.; Teimoori, A.; Pour Beiranvand, S.; Kianmehr, Z. Human Papillomavirus Type16- L1 VLP Production in Insect Cells. *Iran. J. Basic Med. Sci.* **2013**, *16*, 891–895.

19. Jennings, G.T.; Bachmann, M.F. Immunodrugs: Therapeutic VLP-Based Vaccines for Chronic Diseases. *Annu. Rev. Pharmacol. Toxicol.* **2008**, *49*, 303–326. [CrossRef]

20. Mohsen, M.O.; Zha, L.; Cabral-Miranda, G.; Bachmann, M.F. Major findings and recent advances in virus–like particle (VLP)-based vaccines. *Semin. Immunol.* **2017**, *34*, 123–132. [CrossRef]

21. Bachmann, M.F.; Jennings, G.T. Vaccine delivery: A matter of size, geometry, kinetics and molecular patterns. *Nat. Rev. Immunol.* **2010**, *10*, 787–796. [CrossRef]

22. Bachmann, M.; Rohrer, U.; Kündig, T.; Burki, K.; Hengartner, H.; Zinkernagel, R. The influence of antigen organization on B cell responsiveness. *Science* **1993**, *262*, 1448–1451. [CrossRef]

23. Phan, T.G.; Grigorova, I.; Okada, T.; Cyster, J.G. Subcapsular encounter and complement-dependent transport of immune complexes by lymph node B cells. *Nat. Immunol.* **2007**, *8*, 992–1000. [CrossRef] [PubMed]

24. Keller, S.A.; Bauer, M.; Manolova, V.; Muntwiler, S.; Saudan, P.; Bachmann, M.F. Cutting Edge: Limited Specialization of Dendritic Cell Subsets for MHC Class II-Associated Presentation of Viral Particles. *J. Immunol.* **2010**, *184*, 26–29. [CrossRef] [PubMed]

25. Bachmann, M.F.; El-Turabi, A.; Fettelschoss-Gabriel, A.; Vogel, M. The Prospects of an Active Vaccine Against Asthma Targeting IL-5. *Front. Microbiol.* **2018**, *9*, 2522. [CrossRef] [PubMed]

26. Himmelstein, D.U.; Lawless, R.M.; Thorne, D.; Foohey, P.; Woolhandler, S. Medical Bankruptcy: Still Common Despite the Affordable Care Act. *Am. J. Public Health* **2019**, *109*, 431–433. [CrossRef] [PubMed]

27. Foerster, J.; Molęda, A. Feasibility Analysis of Interleukin-13 as a Target for a Therapeutic Vaccine. *Vaccines (Basel)* **2019**, *7*, 20. [CrossRef] [PubMed]

28. FDA Application Review Guselkumab. Available online: https://www.accessdata.fda.gov/drugsatfda_docs/nda/2017/761061Orig1s000MultidisciplineR.pdf. (accessed on 1 March 2020).

29. Blauvelt, A.; Papp, K.; Griffiths, C.E.M.; Randazzo, B.; Wasfi, Y.; Shen, Y.-K.; Li, S.; Kimball, A. Efficacy and safety of guselkumab, an anti-interleukin-23 monoclonal antibody, compared with adalimumab for the continuous treatment of patients with moderate to severe psoriasis: Results from the phase III, double-blinded, placebo- and active comparator–controlled VOYAGE 1 trial. *J. Am. Acad. Dermatol.* **2017**, *76*, 405–417.

30. Singh, S.; Kroe-Barrett, R.R.; Canada, K.A.; Zhu, X.; Sepulveda, E.; Wu, H.; He, Y.; Raymond, E.L.; Ahlberg, J.; Frego, L.E.; et al. Selective targeting of the IL23 pathway: Generation and characterization of a novel high-affinity humanized anti-IL23A antibody. *MAbs* **2015**, *7*, 778–791. [CrossRef]

31. Gordon, K.B.; Strober, B.; Lebwohl, M.; Augustin, M.; Blauvelt, A.; Poulin, Y.; Papp, K.; Sofen, H.; Puig, L.; Foley, P.; et al. Efficacy and safety of risankizumab in moderate-to-severe plaque psoriasis (UltIMMa-1 and UltIMMa-2): Results from two double-blind, randomised, placebo-controlled and ustekinumab-controlled phase 3 trials. *Lancet* **2018**, *392*, 650–661. [CrossRef]

32. Reich, K.; Papp, K.; Blauvelt, A.; Tyring, S.K.; Sinclair, R.; Thaçi, D.; Nograles, K.; Mehta, A.; Cichanowitz, N.; Li, Q.; et al. Tildrakizumab versus placebo or etanercept for chronic plaque psoriasis (reSURFACE 1 and reSURFACE 2): Results from two randomised controlled, phase 3 trials. *Lancet* **2017**, *390*, 276–288. [CrossRef]

33. Bloch, Y.; Bouchareychas, L.; Merceron, R.; Składanowska, K.; Van den Bossche, L.; Detry, S.; Govindarajan, S.; Elewaut, D.; Haerynck, F.; Dullaers, M.; et al. Structural Activation of Pro-inflammatory Human Cytokine IL-23 by Cognate IL-23 Receptor Enables Recruitment of the Shared Receptor IL-12Rbeta1. *Immunity* **2018**, *48*, 45–58. [CrossRef]

34. Kasaian, M.T.; Raible, D.; Marquette, K.; Cook, T.A.; Zhou, S.; Tan, X.-Y.; Tchistiakova, L. IL-13 Antibodies Influence IL-13 Clearance in Humans by Modulating Scavenger Activity of IL-13Ralpha2. *J. Immunol.* **2011**, *187*, 561–569. [CrossRef] [PubMed]

35. Liu, Y.; Zhang, S.; Chen, R.; Wei, J.; Guan, G.; Zhou, M.; Dong, N.; Cao, Y. Meta-analysis of randomized controlled trials for the efficacy and safety of anti-interleukin-13 therapy with lebrikizumab in patients with uncontrolled asthma. *Allergy Asthma Proc.* **2018**, *39*, 332–337. [CrossRef]

36. Hanania, N.A.; Korenblat, P.; Chapman, K.; Bateman, E.D.; Kopecky, P.; Paggiaro, P.; Yokoyama, A.; Olsson, J.; Gray, S.; Holweg, C.T.J.; et al. Efficacy and safety of lebrikizumab in patients with uncontrolled asthma (LAVOLTA I and LAVOLTA II): Replicate, phase 3, randomised, double-blind, placebo-controlled trials. *Lancet Respir. Med.* **2016**, *4*, 781–796. [CrossRef]

37. Tiwari, A.; Kasaian, M.; Heatherington, A.C.; Jones, H.M.; Hua, F. A mechanistic PK/PD model for two anti-IL13 antibodies explains the difference in total IL-13 accumulation observed in clinical studies. *MAbs* **2016**, *8*, 983–990. [CrossRef] [PubMed]

38. Tripp, C.S.; Cuff, C.; Campbell, A.L.; Hendrickson, B.A.; Voss, J.; Melim, T.; Wu, C.; Cherniack, A.D.; Kim, K. RPC4046, A Novel Anti-interleukin-13 Antibody, Blocks IL-13 Binding to IL-13 alpha1 and alpha2 Receptors: A Randomized, Double-Blind, Placebo-Controlled, Dose-Escalation First-in-Human Study. *Adv. Ther.* **2017**, *34*, 1364–1381. [CrossRef] [PubMed]

39. Hirano, I.; Collins, M.H.; Dayan, Y.A.; Evans, L.; Gupta, S.; Schoepfer, A.M.; Straumann, A.; Safroneeva, E.; Grimm, M.; Smith, H.; et al. RPC4046, a Monoclonal Antibody Against IL13, Reduces Histologic and Endoscopic Activity in Patients With Eosinophilic Esophagitis. *Gastroenterology* **2019**, *156*, 592–603. [CrossRef]

40. Popovic, B.; Breed, J.; Rees, D.; Gardener, M.; Vinall, L.; Kemp, B.; Spooner, J.; Keen, J.; Minter, R.; Uddin, F.; et al. Structural Characterisation Reveals Mechanism of IL-13-Neutralising Monoclonal Antibody Tralokinumab as Inhibition of Binding to IL-13Ralpha1 and IL-13Ralpha2. *J. Mol. Biol.* **2017**, *429*, 208–219. [CrossRef]

41. Hart, T.K.; Blackburn, M.N.; Brigham-Burke, M.; DeDe, K.; Al-Mahdi, N.; Zia-Amirhosseini, P.; Cook, R.M. Preclinical efficacy and safety of pascolizumab (SB 240683): A humanized anti-interleukin-4 antibody with therapeutic potential in asthma. *Clin. Exp. Immunol.* **2002**, *130*, 93–100. [CrossRef]

42. Kim, J.-E.; Jung, K.; Kim, J.-A.; Kim, S.-H.; Park, H.-S.; Kim, Y.S. Engineering of anti-human interleukin-4 receptor alpha antibodies with potent antagonistic activity. *Sci. Rep.* **2019**, *9*, 7772. [CrossRef]

43. Getz, E.B.; Fisher, D.M.; Fuller, R. Human Pharmacokinetics/Pharmacodynamics of an Interleukin-4 and Interleukin-13 Dual Antagonist in Asthma. *J. Clin. Pharmacol.* **2009**, *49*, 1025–1036. [CrossRef]

44. Maliszewski, C.R.; Sato, T.A.; Bos, T.V.; Waugh, S.; Dower, S.K.; Slack, J.; Beckmann, M.P.; Grabstein, K.H. Cytokine receptors and B cell functions. I. Recombinant soluble receptors specifically inhibit IL-1- and IL-4-induced B cell activities in vitro. *J. Immunol.* **1990**, *144*, 3028–3033. [PubMed]

45. Borish, L.C.; Nelson, H.S.; Lanz, M.J.; Claussen, L.; Whitmore, J.B.; Agosti, J.M.; Garrison, L. Interleukin-4 receptor in moderate atopic asthma. A phase I/II randomized, placebo-controlled trial. *Am. J. Respir. Crit. Care Med.* **1999**, *160*, 1816–1823. [CrossRef] [PubMed]

46. Corren, J.; Busse, W.; Meltzer, E.O.; Mansfield, L.; Bensch, G.; Fahrenholz, J.; Wenzel, S.E.; Chon, Y.; Dunn, M.; Weng, H.H.; et al. A randomized, controlled, phase 2 study of AMG 317, an IL-4Ralpha antagonist, in patients with asthma. *Am. J. Respir. Crit. Care Med.* **2010**, *181*, 788–796. [CrossRef] [PubMed]

47. Kraich, M.; Klein, M.; Patiño, E.; Harrer, H.; Nickel, J.; Sebald, W.; Mueller, T.D. A modular interface of IL-4 allows for scalable affinity without affecting specificity for the IL-4 receptor. *BMC Biol.* **2006**, *4*, 13. [CrossRef] [PubMed]

48. Ledger, K.S.; Agee, S.J.; Kasaian, M.T.; Forlow, S.B.; Durn, B.L.; Minyard, J.; Lu, Q.A.; Todd, J.; Vesterqvist, O.; Burczynski, M.E. Analytical validation of a highly sensitive microparticle-based immunoassay for the quantitation of IL-13 in human serum using the Erenna®immunoassay system. *J. Immunol. Methods* **2009**, *350*, 161–170. [CrossRef]

49. Hua, F.; Ribbing, J.; Reinisch, W.; Cataldi, F.; Martin, S. A pharmacokinetic comparison of anrukinzumab, an anti- IL-13 monoclonal antibody, among healthy volunteers, asthma and ulcerative colitis patients. *Br. J. Clin. Pharmacol.* **2015**, *80*, 101–109. [CrossRef]

50. Horns, F.; Vollmers, C.; Dekker, C.L.; Quake, S.R. Signatures of selection in the human antibody repertoire: Selective sweeps, competing subclones, and neutral drift. *Proc. Natl. Acad. Sci. USA* **2019**, *116*, 1261–1266. [CrossRef]

51. Lee, J.; Boutz, D.R.; Chromikova, V.; Joyce, M.G.; Vollmers, C.; Leung, K.; Horton, A.P.; DeKosky, B.J.; Lee, C.-H.; Lavinder, J.J.; et al. Molecular-level analysis of the serum antibody repertoire in young adults before and after seasonal influenza vaccination. *Nat. Med.* **2016**, *22*, 1456–1464. [CrossRef]

52. Kaur, K.; Zheng, N.-Y.; Smith, K.; Huang, M.; Li, L.; Pauli, N.T.; Dunand, C.J.H.; Lee, J.-H.; Morrissey, M.; Wu, Y.; et al. High Affinity Antibodies against Influenza Characterize the Plasmablast Response in SLE Patients After Vaccination. *PLoS ONE* **2015**, *10*, e0125618. [CrossRef]

53. Wang, Y.; Sundling, C.; Wilson, R.; O'Dell, S.; Chen, Y.; Dai, K.; PhaD, G.E.; Zhu, J.; Xiao, Y.; Mascola, J.R.; et al. High-Resolution Longitudinal Study of HIV-1 Env Vaccine-Elicited B Cell Responses to the Virus Primary Receptor Binding Site Reveals Affinity Maturation and Clonal Persistence. *J. Immunol.* **2016**, *196*, 3729–3743. [CrossRef]

54. Cavelti-Weder, C.; Timper, K.; Seelig, E.; Keller, C.; Osranek, M.; Lässing, U.; Spohn, G.; Maurer, P.; Müller, P.; Jennings, G.T.; et al. Development of an Interleukin-1beta Vaccine in Patients with Type 2 Diabetes. *Mol. Ther.* **2016**, *24*, 1003–1012. [CrossRef]

55. Tissot, A.C.; Maurer, P.; Nussberger, J.; Sabat, R.; Pfister, T.; Ignatenko, S.; Volk, H.-D.; Stocker, H.; Müller, P.; Jennings, G.T.; et al. Effect of immunisation against angiotensin II with CYT006-AngQb on ambulatory blood pressure: A double-blind, randomised, placebo-controlled phase IIa study. *Lancet* **2008**, *371*, 821–827. [CrossRef]

56. Fettelschoss-Gabriel, A.; Fettelschoss, V.; Thoms, F.; Giese, C.; Daniel, M.; Olomski, F.; Kamarachev, J.; Birkmann, K.; Bühler, M.; Kummer, M.; et al. Treating insect-bite hypersensitivity in horses with active vaccination against IL-5. *J. Allergy Clin. Immunol.* **2018**, *142*, 1194–1205. [CrossRef]

57. Zeltins, A.; West, J.; Zabel, F.; El-Turabi, A.; Balke, I.; Haas, S.; Maudrich, M.; Storni, F.; Engeroff, P.; Jennings, G.T.; et al. Incorporation of tetanus-epitope into virus-like particles achieves vaccine responses even in older recipients in models of psoriasis, Alzheimer's and cat allergy. *NPJ Vaccines* **2017**, *2*, 30. [CrossRef]

58. Berry, C.M. Antibody immunoprophylaxis and immunotherapy for influenza virus infection: Utilization of monoclonal or polyclonal antibodies? *Hum. Vaccin. Immunother.* **2017**, *14*, 796–799. [CrossRef]

59. Wang, X.; Mathieu, M.; Brezski, R. IgG Fc engineering to modulate antibody effector functions. *Protein Cell* **2017**, *9*, 63–73. [CrossRef]

60. Sahoo, A.; Wali, S.; Nurieva, R. T helper 2 and T follicular helper cells: Regulation and function of interleukin-4. *Cytokine Growth Factor Rev.* **2016**, *30*, 29–37. [CrossRef]

61. Gowthaman, U.; Chen, J.S.; Zhang, B.; Flynn, W.F.; Lu, Y.; Song, W.; Joseph, J.; Gertie, J.A.; Xu, L.; Collet, M.A.; et al. Identification of a T follicular helper cell subset that drives anaphylactic IgE. *Science* **2019**, *365*, 6456. [CrossRef]

62. Gowthaman, U.; Chen, J.S.; Eisenbarth, S.C. Regulation of IgE by T follicular helper cells. *J. Leukoc. Biol.* **2020**, *107*, 409–418. [CrossRef]

63. Heesters, B.A.; Van Der Poel, C.E.; Das, A.; Carroll, M.C. Antigen Presentation to B Cells. *Trends Immunol.* **2016**, *37*, 844–854. [CrossRef]

64. Laudenbach, M.; Baruffaldi, F.; Robinson, C.; Carter, P.; Seelig, D.; Baehr, C.; Pravetoni, M. Blocking interleukin-4 enhances efficacy of vaccines for treatment of opioid abuse and prevention of opioid overdose. *Sci. Rep.* **2018**, *8*, 5508. [CrossRef] [PubMed]

65. Blauvelt, A.; Simpson, E.L.; Tyring, S.K.; Purcell, L.A.; Shumel, B.; Petro, C.D.; Akinlade, B.; Gadkari, A.; Eckert, L.; Graham, N.M.; et al. Dupilumab does not affect correlates of vaccine-induced immunity: A randomized, placebo-controlled trial in adults with moderate-to-severe atopic dermatitis. *J. Am. Acad. Dermatol.* **2019**, *80*, 158–167. [CrossRef] [PubMed]

66. Jo, C.E.; Georgakopoulos, J.R.; Ladda, M.; Ighani, A.; Mufti, A.; Drucker, A.M.; Piguet, V.; Yeung, J. Evaluation of long-term efficacy, safety, and reasons for discontinuation of dupilumab for moderate-to-severe atopic dermatitis in clinical practice: A retrospective cohort study. *J. Am. Acad. Dermatol.* **2020**, *82*. In Press. [CrossRef] [PubMed]

67. Luo, J.; Liu, D.; Liu, C.-T. The Efficacy and Safety of Antiinterleukin 13, a Monoclonal Antibody, in Adult Patients With Asthma: A Systematic Review and Meta-Analysis. *Medicine (Baltimore)* **2016**, *95*, e2556. [CrossRef] [PubMed]

68. Zhang, Y.; Cheng, J.; Li, Y.; He, R.; Pan, P.; Su, X.; Hu, C. The Safety and Efficacy of Anti–IL-13 Treatment with Tralokinumab (CAT-354) in Moderate to Severe Asthma: A Systematic Review and Meta-Analysis. *J. Allergy Clin. Immunol. Pr.* **2019**, *7*, 2661–2671. [CrossRef]

69. Carlsson, M.; Braddock, M.; Li, Y.; Wang, J.; Xu, W.; White, N.; Megally, A.; Hunter, G.; Colice, G. Evaluation of Antibody Properties and Clinically Relevant Immunogenicity, Anaphylaxis, and Hypersensitivity Reactions in Two Phase III Trials of Tralokinumab in Severe, Uncontrolled Asthma. *Drug Saf.* **2019**, *42*, 769–784. [CrossRef]

Review

Recent Advances in the Use of Plant Virus-Like Particles as Vaccines

Ina Balke and Andris Zeltins *

Plant virology group, Latvian Biomedical Research and Study Centre, Ratsupites 1, LV1067 Riga, Latvia; inab@biomed.lu.lv
* Correspondence: anze@biomed.lu.lv

Received: 30 January 2020; Accepted: 26 February 2020; Published: 28 February 2020

Abstract: Vaccination is one of the most effective public health interventions of the 20th century. All vaccines can be classified into different types, such as vaccines against infectious diseases, anticancer vaccines and vaccines against autoimmune diseases. In recent decades, recombinant technologies have enabled the design of experimental vaccines against a wide range of diseases using plant viruses and virus-like particles as central elements to stimulate protective and long-lasting immune responses. The analysis of recent publications shows that at least 97 experimental vaccines have been constructed based on plant viruses, including 71 vaccines against infectious agents, 16 anticancer vaccines and 10 therapeutic vaccines against autoimmune disorders. Several plant viruses have already been used for the development of vaccine platforms and have been tested in human and veterinary studies, suggesting that plant virus-based vaccines will be introduced into clinical and veterinary practice in the near future.

Keywords: plant virus; virus-like; vaccine platform; epitope; antigen; immune response

---

1. Introduction

Vaccination is one of the most powerful public health interventions of the 20th century, preventing an estimated six million deaths a year [1]. From a public point of view, vaccination programs result in cost savings that exceed investments by 16-fold [2].

During the last 200 years, academic scientists and the vaccine industry have developed a large number of vaccines, including approximately 50 vaccines licensed for human use and more than 300 veterinary vaccines [3]. All currently used vaccines can be classified into several types, and the classification also reflects the historical steps in vaccine development [4,5]. These types include vaccines against different infectious agents (bacteria, viruses, parasites); vaccines obtained by different methods of development; vaccines containing single or multiple antigens and recombinant vaccines (Figure 1, Supplementary Table S1).

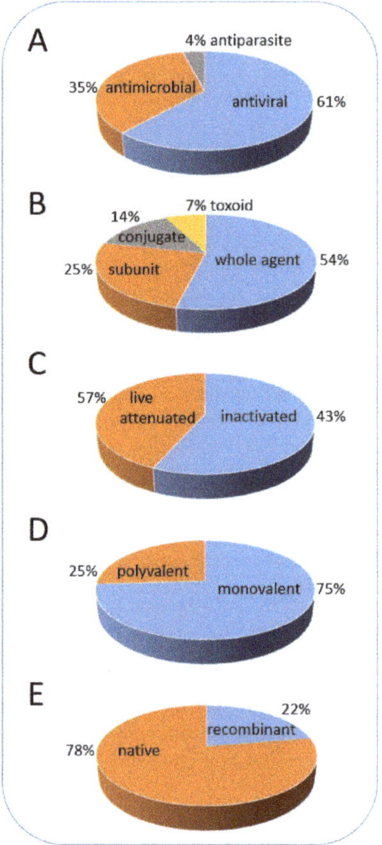

Figure 1. Classification of vaccines. (**A**) vaccines against different infectious agents; (**B,C**) vaccine types according to method of development; (**D**) vaccines containing single or multiple antigen subtypes from the same pathogen (mono-or poly-valent); (**E**) vaccines containing native or artificially generated (recombinant) antigens. The proportion (%) of a corresponding vaccine type is calculated based on the list of diseases and available vaccines (Supplementary Table S1; based on data from World Health Organization [6]).

The very first vaccine was the smallpox vaccine introduced by Jenner more than 200 years ago [7]. He used a heterologous virus containing material originating from infected cows (cowpox) for the protection of humans against smallpox. Further development and international use of the vaccine led to the complete eradication of smallpox, which is one of the greatest achievements in medicine [4].

Targeted vaccine development began with Luis Pasteur's discovery that pathogenic microorganisms are less virulent if cultivated under suboptimal conditions or treated with different chemicals. He introduced the concept of attenuated vaccines, which was later exploited in the generation of many vaccines, including vaccines against anthrax and rabies [4]. Later, progress in cell culture technology enabled the culture of mammalian cells and allowed the propagation of viruses, resulting in effective attenuated vaccines against polio, measles, mumps, rubella, influenza and other infectious diseases [8].

The third group of vaccines was generated by the chemical inactivation of infectious bacteria and viruses as well as bacterial toxins (toxoid vaccines). The application of increased temperatures or inactivating chemicals in the production of different vaccines is still used by industry for vaccines against cholera, polio, rabies, Japanese encephalitis and other infectious diseases [9].

The development of molecular genetics, bacteriology and biochemistry contributed significantly to the understanding of the molecular structure of living cells and allowed the identification of pathogen components potentially suitable for vaccine generation. The exploitation of this idea resulted in a long list of so-called subunit vaccines, including vaccines against *Haemophilus influenza,* meningococcus, pneumococcus (polysaccharides and their conjugates with carrier proteins) and hepatitis B (plasma-derived protein antigen) [9].

The advent of molecular biology resulted in the next breakthrough in the medicinal industry. The achievements in gene engineering have dramatically influenced the construction of vaccines. Since the 1980s, recombinant technologies have been introduced for the development of different vaccine types, such as live and attenuated recombinant bacteria and viruses, as well as the production of toxins and other protein antigens using recombinant hosts [8].

The hepatitis B (HBV) vaccine was the first subunit vaccine, and it was generated by gene engineering techniques more than 30 years ago. The expression of a cloned copy of the HBV surface antigen (HBsAg) in yeast cells resulted in the production of noninfectious virus-like particles (VLPs) [10] and allowed the replacement of a previously generated plasma-derived vaccine [11]. The success of HBV vaccines strongly stimulated the development of recombinant vaccines based on viral structural proteins and marked the beginning of a new era in rationally designed VLP platforms for the generation of prophylactic and therapeutic vaccines [12].

Using the same principle of viral coat protein (CP) expression in heterologous hosts, vaccines against cervical cancer (Gardasil and Cervarix; [13–15]), hepatitis A (Hecolin; [16]) and malaria (RTS,S; [17]) were constructed, clinically tested and licensed for human use in the subsequent decades. It is important to stress that the RTS,S vaccine is the first among licensed vaccines containing a VLP carrier (HBsAg) with an incorporated foreign antigen (CS).

Artificial virus-like structures derived from plant virus proteins are well known due to virus assembly studies performed since the 1950s [18]. Based on the use of carrier proteins with chemically coupled peptide antigens as promising vaccine candidates [19], Haynes et al. [20] generated an experimental vaccine using a gene engineering approach instead of chemical coupling. They combined the tobacco mosaic virus (TMV) *CP* gene with an extension encoding a C-terminally located, 8 AA-long antigenic peptide from poliovirus. The resulting VLPs purified from recombinant bacterial cells were immunogenic and stimulated the formation of antibodies against poliovirus in rats. These results, together with those of HBsAg [21], introduced the use of VLPs as a central carrier element of many experimental and licensed vaccines. Additionally, the study clearly demonstrates that nonpathogenic viruses are also suitable for vaccine generation after the introduction of relevant antigens into their structure.

Viruses and their derivatives possess several characteristics that are highly important for their use as vaccines [22]. Most likely, the most important property of viruses and VLPs is their structural organization. Structurally, viruses are constructed of hundreds or thousands of highly ordered CP molecules, which serve as repeated antigens for the mammalian immune system. These antigens on the virus surface can stimulate B cells by crosslinking B cell receptors and induce long-lasting antibody responses. In addition, most viruses have the optimal size, shape and rigidity to enter the lymphatic system through the pores in lymph vessels. This facilitates the trafficking of viral particles and VLPs to lymph nodes and their uptake by antigen-presenting cells (APCs). Moreover, viruses and VLPs encapsulate specific host-derived nucleic acids (DNA or RNA), which stimulate Toll-like receptors in APCs and serve as natural vaccine adjuvants [23,24].

Plant viruses and VLPs (Figure 2), compared with other VLPs, have additional advantages as vaccine carrier structures. It is well known that plant viruses are not able to infect mammalian organisms. Therefore, the probability of pre-existing immunity against plant viruses is considerably lower compared with that against VLPs derived from mammalian viruses (e.g., HBV and papilloma virus). Most plant viruses are assembled from single or a few CP molecules and demonstrate remarkable structural flexibility, allowing different manipulations, such as disassembly/reassembly, as well as chemical and genetic modifications. This enables the rational design of vaccines and the introduction of different antigens derived from infectious agents, allergens and self-molecules. Additionally, plant VLPs and other artificially generated VLPs do not contain replicating nucleic acids. This prevents the reversion of vaccines into infectious viruses, which is a serious risk factor for attenuated viral vaccines. From a technological viewpoint, plant viruses and VLPs can be produced in different recombinant hosts (bacteria, yeasts, plants, and eukaryotic cells) under cGMP conditions; the mentioned structural flexibility of plant viruses allows the construction of universal vaccine platforms [25,26].

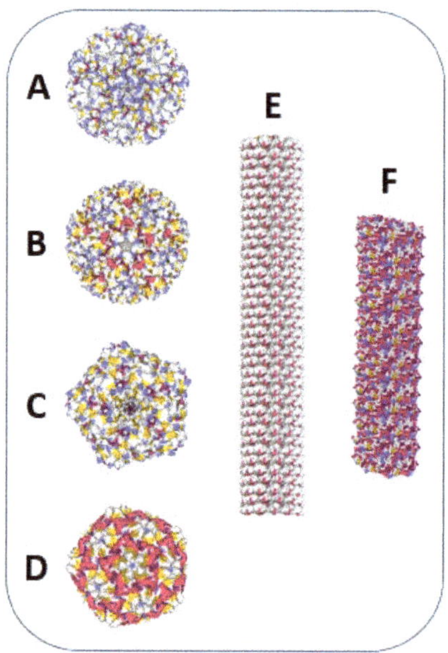

Figure 2. Examples of icosahedral and helical plant viruses, used for vaccine development. Images were created using Protein Data Bank and NGL 3D viewer [27]. α-helices are shown in red, β-sheets–in yellow. (**A**) Cucumber mosaic virus (CMV) structure (T = 3 symmetry, diameter = 28 nm). Image of 5OW6 [28]. (**B**) Cowpea chlorotic mottle virus (CCMV) structure (T = 3 symmetry, diameter = 29 nm). Image of 1CWP [29]. (**C**) Cowpea mosaic virus (CPMV) structure (T = 3 symmetry, diameter = 28 nm). Image of 1NY7 [30]. (**D**) Sesbania mosaic virus (SeMV) structure (T = 3 symmetry, diameter = 28 nm). Image of 1X33 [31]. (**E**) Tobacco mosaic virus (TMV) structure (cryo-EM reconstruction of a TMV fragment; particle length = 300 nm, diameter = 18 nm). Image of 3J06 [32]. (**F**) Bamboo mosaic virus (BaMV) structure (cryo-EM reconstruction of a BaMV fragment; particle length = 490 nm, diameter = 15 nm). Image of 5A2T [33].

In this review, we summarize the publicly available data on vaccines derived from plant viruses, emphasizing the newest developments in the construction of experimental vaccines. We used two databases as a source of information: the PubMed database of the US National Institutes of Health (https://www.ncbi.nlm.nih.gov/pubmed/) and the Web of Science (https://apps.webofknowledge.com). Our literature search demonstrates the rapidly growing interest in plant viruses as vaccine carriers; for example, a cumulative search for the term "plant virus vaccine" in Web of Science yielded 787 publication records through 2009, more than 1400 through 2014, and at least 2204 publications through the end of 2019. As revealed by the search, plant virus derivatives serve as components for at least 71 experimental vaccines against infectious diseases, 16 anti-cancer vaccines and 10 vaccines against allergies and autoimmune diseases (examples and literature citations are summarized in

Supplemental Tables S2, S3 and S4). The increasing interest and recent publications stimulated us to summarize and discuss the latest developments in the design of vaccines based on plant viruses. Descriptions of other plant VLP applications can be found in several recent review articles [34–38].

2. Plant VLP-Derived Vaccines against Infectious Diseases

The efforts of the vaccine industry have resulted in a comparably long list of licensed prophylactic vaccines that effectively prevent infection by different infectious agents (Supplementary Table S1). However, the efficacy of current vaccines is moderate or even poor in some cases and does not ensure long-lasting protection and the eradication of the corresponding diseases [39]. Additionally, there is an urgent need for vaccines against viral and bacterial pathogens, such as human immunodeficiency virus, tuberculosis, malaria, and many other agents, as well as emerging diseases, such as Zika virus, Dengue, Ebola and SARS. Therefore, interest in alternative solutions for the generation of new vaccines and the improvement of existing vaccines is continuously growing.

Experimental vaccines based on plant viruses have been constructed and tested for more than 30 years since the first TMV-based vaccine was developed against poliovirus [20]. In this chapter, we will discuss the recent development of plant virus-derived prophylactic vaccines against important pathogens such as influenza and malaria.

In most cases, plant virus-derived vaccines are peptide vaccines that are genetically fused or chemically coupled to viral CP. Peptide vaccines can stimulate the formation of neutralizing antibodies and ensure protection against the corresponding pathogen in animal models (Suppl. Table S2). The advantages of this approach include the reduced influence of short peptides on the structural integrity of viral carriers and the easy production of experimental vaccines. However, the antigen is frequently not a linear AA sequence but a spatially complex structure involving different parts of the antigenic protein. For such antigens, peptide-based vaccines are not suitable for stimulating protection via the immune response. Therefore, several authors have suggested the usage of whole proteins as antigens, which can result in epitope structures closely related to native ones and elicit neutralizing antibodies after immunizations. Genetic fusions of whole antigenic proteins with viral CPs negatively influence VLP formation in most cases. However, sometimes plant viral coats are able to accommodate even very long antigenic sequences on their surfaces [40,41]. As alternatives to genetic fusions, other approaches can result in antigen presentation, such as chemical coupling or physical binding of the antigen to the VLP surface using binding partner molecules. Successful examples of the generation of vaccines using the chemical or enzymatic coupling of whole antigens include studies of experimental vaccines against *Francisella tularensis* [42], influenza [43], *Yersinia pestis* [44], *Plasmodium vivax* [45], and Zika virus [46].

Methods used for the incorporation of whole antigen sequences into viral CPs, allowing the preservation of viral morphology, are discussed in recent review articles [25,47].

Several plant virus-based vaccines have been tested for tolerability, safety and efficacy in human clinical trials. The first such vaccine is an edible vaccine against rabies containing a recombinant antigen derived from rabies virus proteins (a fusion of peptides derived from the G and N proteins) incorporated in the N-terminus of the Alfalfa mosaic virus (AlMV). Introduction of the chimeric gene into the TMV-based plant vector and transient expression in spinach leaves resulted in AlMV-like particles exposing rabies epitopes. The inclusion of raw spinach leaves containing these VLPs in the diet of human volunteers led to significant antibody responses to rabies virus and plant virus carriers [48,49]. This pioneering study clearly demonstrates the potential of plant virus-based carriers for the generation of human vaccines. Later, several vaccine candidates were constructed based on AlMV carriers (Figure 3A [41,50]).

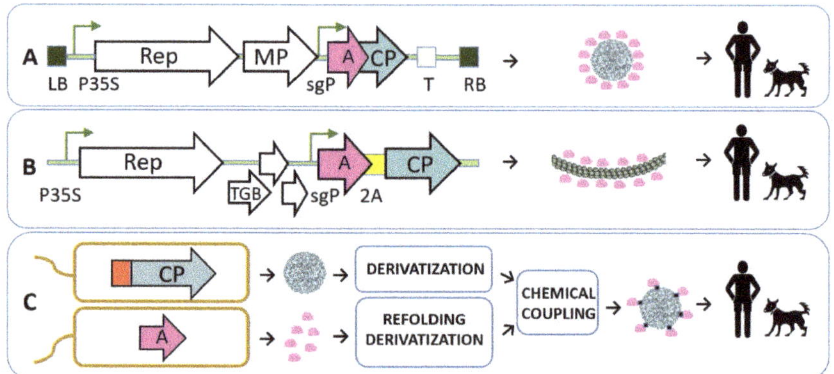

Figure 3. Examples of recombinant vaccine platforms used for introduction of different antigens in plant viral structures. (**A**) «launch» vector [51] based on TMV for production of antigen-containing viruses in plants. LB and RB denotes agrobacterial shuttle vector T-DNA region left and right borders; P35S—CaMV promoter; sgP—subgenomic promoter; T—transcription terminator; Rep, MP—TMV replicase and movement protein; pink and turkish arrows denote the relative position of antigen (A) and AlMV coat protein gene (*CP*), respectively [41]. (**B**) BaMV-based vector for production of antigen (A)-containing mosaic viruses in plants. *sgP*—subgenomic promoter; A—antigen-coding sequence; 2A—self-processing peptide sequence; *Rep, TGB, CP*–replicase, triple gene block and coat protein genes [40]. (**C**) Bacterial system for antigen introduction in plant VLP structure [28]. Antigen (A) and CMV VLPs are obtained from separate bacterial cultures, purified, derivatized and joined in chemical coupling reaction. CMV *CP* gene contains T-cell epitope coding sequence (red box).

Malaria is one of the biggest health threats, causing hundreds of millions of cases and 435,000 deaths in 87 countries, especially in Africa [52]. The existing vaccine, RTS,S, has demonstrated only limited efficacy in clinical trials [17]. A recent publication [53] summarizes the results of a Phase I study assessing a malaria transmission-blocking vaccine based on a recombinant fusion of the *P. falciparum* antigen Pfs25 with the plant virus AlMV. Vaccine candidates produced in whole plants under cGMP conditions demonstrate acceptable safety and tolerability and induce antibodies against Pfs25, depending on the injected vaccine dose. However, the generated antibodies did not produce the expected reduction in *P. falciparum* transmission from mosquitoes to human host cells. The authors concluded that it is necessary to improve the formulation of the vaccine and increase the proportion of the Pfs25 antigen incorporated in plant VLPs.

Recently, other plant virus-derived VLPs have been suggested as active components for vaccines against malaria. The authors recommend the usage of thrombospondin-related adhesive protein (TRAP) as an antigen, which is necessary for *Plasmodium vivax* sporozoite motility and liver cell invasion. The vaccine was designed based on Cucumber mosaic virus (CMV) VLPs with chemically coupled TRAP antigen that were formulated with microcrystalline tyrosine (MCT) as an adjuvant. Vaccine formulations containing VLPs increased antibody production against TRAP compared to that produced by the antigen alone. The VLP-based vaccine formulated with MCT conferred significant protection in the challenge test with recombinant *P. berghei*, suggesting that MCT can be used as an advantageous adjuvant alternative for prophylactic VLP vaccines [45].

The next most important viral infection is influenza, which causes 250,000 to 500,000 deaths globally every year. Today, there are 26 licensed inactivated vaccines, some of which are produced routinely. However, existing vaccines do not ensure complete protection against the circulating influenza strains [54]. The reasons for low vaccine efficacy are the low levels of hemagglutination-inhibiting antibodies and influenza strain variations in the hemagglutinin (HA) sequence. For the improvement of flu vaccines, the use of stronger adjuvants and increased amounts of HA as well as the addition of new antigens (neuraminidase and matrix protein M2e) to influenza vaccines should increase efficacy [39].

Different peptide antigens, including peptides derived from nucleoproteins and M2e proteins, have been tested using plant viruses as vaccine carriers and adjuvants. Some of them demonstrated significant protection against challenges in different animal models (see examples in Supplementary Table S2). One experimental vaccine is currently being tested in a human trial that aims to evaluate the safety and reactogenicity of Papaya mosaic virus (PapMV) VLPs as adjuvants for use in a seasonal flu trivalent vaccine in 48 healthy volunteers [55]. The efficacy of the plant VLP-adjuvanted flu vaccine has not yet been reported. Earlier publications suggest that filamentous PapMV VLPs containing a consensus peptide of the influenza matrix M2e protein increase the survival of immunized mice if used as an adjuvant together with a trivalent inactivated flu vaccine. These results demonstrate the strong adjuvanting properties of filamentous plant VLPs for influenza vaccines [56].

Plant virus-based peptide vaccines are also being evaluated in veterinary trials. A vaccine based on Cowpea mosaic virus (CPMV) protects vaccinated dogs against lethal challenge with canine parvovirus [57]. Another vaccine against foot and mouth disease virus (FMDV) was designed using a bamboo mosaic virus (BaMV) vector and 37 AAs from the FMDV VP1 protein; all immunized swine were protected against FMDV challenge [58]. Another veterinary trial demonstrated vaccine safety and the protection of pigs against porcine circovirus (PCV) after vaccination with an inactivated CMV-based vaccine containing an incorporated PCV CP epitope [59].

3. Plant VLP-Derived Anticancer Vaccines

Cancer is one of the leading causes of death globally and was responsible for an estimated 9.6 million deaths in 2018 [60]. Currently, alongside surgery, chemotherapy and radiation, cancer immunotherapy has become an important component of cancer treatment.

Cancer immunotherapy has a more than 150-year history. It began with the first observations of the significant regression of sarcomas in patients after accidental infection by a *Streptococcus* bacterium. In 1891, William Coley, an American surgeon, used heat-inactivated bacteria to treat a large number of patients suffering from inoperable cancers. His therapy method resulted in the cure of more than 1000 patients. However, the method was later replaced with radiation and chemotherapy due to dangerous infection risks and the absence of reproducible results. The next important milestone in the development of cancer immunotherapy was the finding of Old et al. in 1959, which demonstrated the antitumor activity of the tuberculosis vaccine (BCG) in a mouse model. The vaccine has been introduced for the treatment of bladder cancer and has been used in clinics since the 1970s. These and other key events in cancer immunology are discussed in several recent review articles [61,62].

Several cell- and protein-based anticancer vaccines are approved for use in clinics or are being tested in late-stage clinical trials, such as autologous dendritic cell vaccines, recombinant virus-based vaccines, peptide-based vaccines, DNA vaccines, and human tumor whole-cell vaccines. However, for the most part, clinically tested anticancer vaccines have demonstrated limited or no efficacy in comparison to that of traditional treatments, requiring the development of new strategies and combinations of different approaches [63,64].

One such new approach for vaccination against cancer is based on the usage of nanoparticles, such as liposomes, carbon nanotubes, synthetic biodegradable and biocompatible polymers, inorganic nanoparticles, VLPs and different combinations of such particles [65].

VLPs represent a powerful and flexible tool for generation of the active components of cancer vaccines, as demonstrated in numerous preclinical studies (for review see [66]). We have already discussed the advantages of VLPs as vaccines, such as the multiplicity of antigens, the sizes of VLPs, which allow them to enter the lymphatic system, and their capability to encapsulate nucleic acids stimulating Toll-like receptors. As a result, VLPs are able to induce strong T cell responses, which is the most important requirement for a therapeutic vaccine against cancer. Moreover, VLP technology is already used for the prevention of cancer with Papilloma virus-derived VLPs, which efficiently protect immunized individuals against cervical cancer [66].

Plant viruses, including both native and recombinant viruses as well as their noninfectious derivatives (VLPs), have been considered as nanoparticle structures with antitumor activity since 2006, when McCormick et al. introduced a melanoma-specific peptide into TMV using recombinant fusion and chemical coupling [67,68] and tested both vaccines in tumor challenge models. The authors observed direct TMV uptake by dendritic cells and enhanced production of interferon γ (IFNg). Interestingly, the vaccine prepared by the chemical coupling of peptides to TMV ensured better survival of the animals than the recombinant vaccine when both vaccine variants were formulated with the CpG DNA adjuvant.

The importance of vaccine formulation was also demonstrated in a recent study using icosahedral CMV VLPs with chemically coupled p33 peptide derived from lymphocytic choriomeningitis virus (LCMV). The vaccine was formulated with MCT, CpG and alum adjuvants and tested in an aggressive mouse melanoma model.

VLPs adjuvanted with MCT effectively retarded the development of the tumor. The effect was comparable when CpG DNA was used as the adjuvant, whereas alum was ineffective in slowing down tumor growth [69].

A recent study suggests a highly interesting immunotherapy approach. Three different icosahedral plant virus-derived nanoparticles with chemically coupled breast cancer epitopes efficiently elicited the formation of antibodies against the HER2 receptor, which is overexpressed in breast cancer cells. Additionally, all three vaccines stimulated T-cell-mediated immune responses when tested separately. Sequential use of these vaccines reduced the immune responses against VLP carriers and improved the formation of antibodies against the HER2 peptide. This suggested prime-boost strategy considerably reduced tumor development and enhanced the survival rate in a mouse tumor model. The study clearly demonstrated that the Th1-type immune response, including the formation of IgG2a antibodies, the secretion of IFNg and the activation of CD4+/CD8+ T-cells, is the most important factor for immunotherapy. Interestingly, plant VLPs derived from CPMV demonstrated better immunotherapeutic properties than those derived from Cowpea chlorotic mottle virus (CCMV) and Sesbania mosaic virus (SeMV). Therefore, additional studies are needed to understand the differences between different plant VLPs used as carriers for anticancer vaccines [70]. Other examples of plant virus-based cancer vaccines with the corresponding literature citations are summarized in Supplementary Table S3.

The analysis of several recent publications reveals the potential application of unmodified plant VLPs without any introduced antigens as cancer immunotherapy agents [71,72]. In one study, the authors compared differently prepared nanoparticles derived from CPMV as immunotherapeutic agents in a murine ovarian cancer model. Native CPMV particles containing viral RNA induced a more pronounced therapeutic effect and the survival of experimental animals in comparison to empty CPMV VLPs produced in plant or insect cells. The enhanced immunomodulatory effect apparently is due to the presence of encapsidated ssRNA in native CPMV virions, which activates Toll-like receptors 7/8. Interestingly, the chosen recombinant host can also influence the immune-stimulating properties of produced plant VLPs [72].

In another study, structural variants of plant virus-derived nanoparticles from several species were used as in situ vaccines, such as icosahedral CPMV and three variants of TMV (native, in vitro RNA-templated assembly of short TMV, and spherical TMV produced after thermal treatment of native virions). The results suggest the superior immune-stimulating properties of CPMV compared with those of other structural variants of TMV and confirm that antigen multiplicity is one of the most important factors involved in eliciting a strong immune response. One possible explanation for the enhanced immune-stimulating properties of CPMV in a mouse melanoma model is the ability of VLPs to recruit monocytes into the tumor microenvironment, leading to the infiltration of neutrophils and natural killer cells and resulting in tumor growth inhibition. Authors suggest that the intrinsic properties of some plant viruses allow them to be developed as cancer vaccines for clinical use; however, the detailed mechanisms of immune stimulation remain to be elucidated [71].

4. Plant VLP-Derived Vaccines against Allergies, Autoimmune Diseases and Other Diseases

Vaccinations are also highly promising for the treatment of allergies and autoimmune and neurodegenerative diseases, as demonstrated in the latest studies. Similar to antimicrobial and anticancer vaccines, plant viruses and VLPs can serve as carrier structures for corresponding antigens, which are important in disease development.

One of the most challenging diseases in terms of therapy development is Alzheimer's disease (AD). More than 40 million people worldwide are suffering from dementia, which is caused by the formation of plaques containing proteolytic fragments of amyloid precursor protein Aβ (AA 1-42). Existing therapies and experimental approaches, including the use of monoclonal antibodies and vaccines, have only a small impact on disease progression. According to a recent opinion, vaccination should be prophylactic; the levels of specific antibodies have to be sufficiently high for the targeting of oligomeric species of Aβ peptides, and the sizes of the epitopes included in vaccines have to be less than 8 AA to prevent the stimulation of pathogenic T-cell responses [73].

The concept of AD vaccines has existed for more than 15 years; however, the first clinical trial using a vaccine containing aggregated Aβ peptide (AA 1-42) resulted in cases of T-cell-mediated brain inflammation [74]. Later, an experimental vaccine consisting of Aβ peptide and a bacteriophage Qβ conjugate produced strong antibody responses without significant T-cell responses in mice [75].

Plant viruses and derived VLPs are also considered as epitope carriers for Alzheimer vaccines. Potential vaccine candidates have been generated using infectious CMV fused with Aβ peptides purified from plant biomass [76]. Recently, we constructed a plant VLP platform (Figure 3C) based on the same CMV by the genetic incorporation of a universal T-cell epitope in the interior of particles and the chemical coupling of Aβ1-6 peptide to the VLP surface. Sera obtained from immunized mice were shown to recognize Alzheimer plaques in human brain sections, suggesting that the CMV-Aβ(1-6) vaccine induced the production of specific antibodies [28].

Allergic disorders have been among of the most common chronic diseases in Europe in recent decades [77]. Patients suffering from allergies have to avoid allergens and use anti-histamine medications. Alternatively, allergen-specific immunotherapy (AIT) is the only available treatment for the reduction of allergy symptoms. For therapy, crude allergen extracts containing a mixture of native allergenic proteins are used. The typical disadvantages of AIT are the risk of anaphylactic reactions, the long duration of therapy, the low quality of the natural extracts used in AIT and the unsatisfactory efficacy of the treatment.

Peanut allergy is the most frequent cause of anaphylactic reactions and death among food allergies. There is currently no safe and effective therapy for peanut allergy, especially for patients with a severe allergy. Recently, we constructed several plant virus-based, immunologically optimized vaccines for peanut allergy by chemical coupling of peanut allergens Ara h 1 and Ara h 2 as well as mixture of proteins purified from roasted peanut extract to CMV VLPs. The resulting vaccines did not cause allergic reactions and induced specific IgG antibodies to protect peanut-sensitized mice against anaphylactic shock. Notably, immunizations with single allergen-containing VLPs ensured protection against challenge with the complex allergen mixture, suggesting a new vaccination strategy for the treatment of peanut allergy [78]. Other examples of VLP usage in the treatment of allergic diseases are summarized in a recently published review article [79].

Allergies to cats affect more than 10% of the human population, and the prevalence is increasing [80]. Similar to that of food allergies, current cat allergy treatment includes the avoidance of exposure, the use of anti-histamines and steroids and long-term subcutaneous immunotherapy. All these measures are only partially effective in eliminating allergic reactions; additionally, immunotherapy is bound to result in safety issues. The predominant cat allergen is the secretoglobulin Fel d 1, which is secreted by cat sebaceous and salivary glands. We generated a CMV-based vaccine containing recombinant Fel d 1 allergen using chemical coupling and demonstrated that VLP coupling effectively reduces allergic reactions, stimulates the formation of Fel d 1-specific IgGs and protects sensitized mice against anaphylactic shock [76]. Moreover, Bachmann and collaborators proposed a new strategy for the treatment of Fel d 1 allergy involving immunizing cats against their own Fel d 1 allergen. The induced anti–Fel d 1 antibodies exhibited a strong neutralization ability and might result in reduced symptoms in allergic cat owners [81].

Monoclonal antibodies (mAbs) produced by immortalized hybridoma cells represent a new way to treat different diseases. Since the approval of the first mAb preventing kidney transplant rejection in 1992, numerous mAbs have been used in clinics for treatment of cancers, bacterial and viral infections, and various cardiovascular, respiratory, neurological and autoimmune diseases [82,83]. MAbs have been proven to be efficient agents for treating chronic inflammatory diseases via the selective inhibition of cytokines, which are excessively produced in several disease conditions. Taking into account the high costs and side effects of mAb therapies, there is significant interest in replacing mAb treatments with active immunization against autologous proteins, including interleukins [84,85].

Based on this idea, we generated an IL17-containing plant VLP-based vaccine and tested it in a mice psoriasis model [28]. The vaccinated mice demonstrated a similar reduction in psoriatic symptoms compared to that in animals passively immunized with IL17A antibody. Moreover, the vaccination elicited protective effects in suboptimal conditions, such as those involving older mice or low vaccine doses.

The knowledge obtained from successful studies of human vaccinations and mAb treatments can be transferred to the development of veterinary vaccines. It is well known that, for example, allergic hypersensitivity is linked to the activation of eosinophils and the enhanced production of interleukin 5 (IL5) by Th2 cells. The eosinophil count can be effectively reduced by a specific humanized anti-IL5 mAb, which is used for treatment of human asthma [86]. Active vaccination with a plant VLP-IL5 conjugate instead of a mAb induced a potent IL5 antibody response in horses, reducing the symptoms of insect bite hypersensitivity. Simultaneously, vaccination did not significantly influence the blood eosinophil count and the parasitic load in vaccinated horses [87,88].

Another example of a CMV-based therapeutic vaccine is a vaccine reducing the production of IL-31 in dogs and horses suffering from itching during atopic dermatitis or insect bite hypersensitivity. In both animals, the vaccine is well tolerated and improves the disease symptoms [89,90].

In the last few years, nerve growth factor (NGF), a key molecule involved in the regulation of neuronal regeneration during injury and pain perception, has been suggested as a promising target for osteoarthritis (OA) treatment. In humans, monoclonal antibodies against NGF significantly suppress pain associated with late-stage OA. Based on this, we constructed a CMV-based NGF vaccine and demonstrated its therapeutic efficacy by showing that it alleviates spontaneous pain behavior in surgically induced murine OA [91].

5. Conclusions

Vaccine production is one of the most challenging industrial processes. Several important factors influence vaccine development and marketing. One of them is the time required to bring a new vaccine from the development phase to the market, which can exceed 15 years, including 7 years on average for the design, construction, validation and commencement of industrial manufacture. The next factor is the significant human and financial resources necessary for the development and sustainable on-demand production of vaccines. The vaccine industry has to continuously solve problems, such as long life cycles of vaccines, high facility costs and the complexity of global vaccine demand [92,93]. Long vaccine development periods are an additional challenge negatively influencing the availability of vaccines against emerging infectious diseases, such as Ebola virus, SARS-CoV, Zika virus, Chickungunja virus and others [94]. Moreover, the chosen antigen structure used for vaccine manufacturing can be different from the corresponding native antigen, reducing or even preventing the neutralizing ability of the generated antibodies. This aspect critically influences the development process of vaccines against pathogens, especially against infectious agents with high genetic variation. HIV is an example of such a pathogen, against which there is no effective vaccine despite many years of effort by academic scientists and the vaccine industry [95]. One other challenging aspect that should be mentioned here as a factor that is possibly important for vaccine development in the future is the fact that the latest studies suggest that the microbiome in the digestive track of recipients can influence the efficacy of vaccines. The effect is observed both in humans and in laboratory mice, suggesting the need for further investigation [96].

As shown in this review, plant viruses and their noninfectious derivatives (VLPs) have been intensively studied as immunologically active multivalent structures useful for the generation of new prophylactic and therapeutic vaccines against human or animal infectious agents, cancers and autoimmune diseases. When compared with other VLPs, plant VLP-based vaccines have additional advantages, such as flexibility in vaccine construction, the ease of VLP production and purification, stability and the low risk of preexisting immunity. All these properties make plant viruses an attractive alternative to animal and human VLPs.

Today, no plant virus-based vaccines are commercially available. However, several vaccine platforms have been developed that allow the construction of new vaccines in a comparably short period of time. Some experimental vaccine platforms tolerate the incorporation of large antigens and even full-sized proteins in viral structures without influencing the particle morphology and immune-stimulating properties. As the native spatial structures of recombinant antigens are highly important for eliciting antibodies with neutralizing activity and long-lasting immunity [97,98], future recombinant VLP vaccines must contain multiple copies of correctly folded antigens as well as different immunostimulating components, including T-cell epitopes and nucleic acids.

The achievements discussed here suggest that vaccines based on plant virus-based carriers are very useful for addressing different challenges in vaccine construction and will be developed into new and approved prophylactic and therapeutic vaccines for human and veterinary use in the coming years.

Supplementary Materials: The following are available online at http://www.mdpi.com/1999-4915/12/3/270/s1.

Author Contributions: Data collection and draft writing, I.B.; draft writing and editing of the final manuscript, A.Z. All authors have read and agreed to the published version of the manuscript.

Funding: This work was supported by Latvian Science Council (Grant No. lzp-2019/1-0131).

Conflicts of Interest: The authors declare that they have no competing interests.

References

1. Andre, F.E.; Booy, R.; Bock, H.L.; Clemens, J.; Datta, S.K.; John, T.J.; Lee, B.W.; Lolekha, S.; Peltola, H.; Ruff, T.A.; et al. Vaccination greatly reduces disease, disability, death and inequity worldwide. *Bull. World Health Organ* **2008**, *86*, 140–146. [CrossRef] [PubMed]
2. Ozawa, S.; Clark, S.; Portnoy, A.; Grewal, S.; Brenzel, L.; Walker, D.G. Return on Investment from Childhood Immunization In Low- And Middle-Income Countries, 2011–20. *Health Aff. (Millwood)* **2016**, *35*, 199–207. [CrossRef] [PubMed]
3. Barrett, A.D.T. Vaccinology in the twenty-first century. *NPJ Vaccines* **2016**, *1*, 16009. [CrossRef] [PubMed]
4. Plotkin, S.A.; Plotkin, S.L. The development of vaccines: How the past led to the future. *Nat. Rev. Microbiol.* **2011**, *9*, 889–893. [CrossRef]
5. Karch, C.P.; Burkhard, P. Vaccine technologies: From whole organisms to rationally designed protein assemblies. *Biochem. Pharmacol.* **2016**, *120*, 1–14. [CrossRef]
6. World Health Organization. Vaccines and Diseases. Available online: https://www.who.int/immunization/diseases/en/ (accessed on 21 February 2020).
7. Jenner, E. *An Inquiry into the Causes and Effects of the Variole Vaccinae, a Disease Discovered in Some of the Western Counties of England, Particularly Gloucestershire and Known by the Name of the Cow-Pox*; Sampson Low: London, UK, 1798.
8. Plotkin, S.A. Vaccines: Past, present and future. *Nat. Med.* **2005**, *11*, S5–S11. [CrossRef]
9. Plotkin, S. History of vaccination. *Proc. Natl. Acad. Sci. USA* **2014**, *111*, 12283–12287. [CrossRef]
10. Valenzuela, P.; Medina, A.; Rutter, W.J.; Ammerer, G.; Hall, B.D. Synthesis and assembly of hepatitis B virus surface antigen particles in yeast. *Nature* **1982**, *298*, 347–350. [CrossRef]

11. Hilleman, M.R.; McAleer, W.J.; Buynak, E.B.; McLean, A.A. The preparation and safety of hepatitis B vaccine. *J. Infect.* **1983**, *7* (Suppl. 1), 3–8. [CrossRef]

12. Jennings, G.T.; Bachmann, M.F. The coming of age of virus-like particle vaccines. *Biol. Chem.* **2008**, *389*, 521–536. [CrossRef]

13. Lowy, D.R.; Schiller, J.T. Prophylactic human papillomavirus vaccines. *J. Clin. Investig.* **2006**, *116*, 1167–1173. [CrossRef] [PubMed]

14. Petrosky, E.; Bocchini, J.A., Jr.; Hariri, S.; Chesson, H.; Curtis, C.R.; Saraiya, M.; Unger, E.R.; Markowitz, L.E.; Centers for Disease, C.; Prevention. Use of 9-valent human papillomavirus (HPV) vaccine: Updated HPV vaccination recommendations of the advisory committee on immunization practices. *MMWR Morb. Mortal. Wkly Rep.* **2015**, *64*, 300–304. [PubMed]

15. Paavonen, J.; Naud, P.; Salmeron, J.; Wheeler, C.M.; Chow, S.N.; Apter, D.; Kitchener, H.; Castellsague, X.; Teixeira, J.C.; Skinner, S.R.; et al. Efficacy of human papillomavirus (HPV)-16/18 AS04-adjuvanted vaccine against cervical infection and precancer caused by oncogenic HPV types (PATRICIA): Final analysis of a double-blind, randomised study in young women. *Lancet* **2009**, *374*, 301–314. [CrossRef]

16. Zhang, X.O.; Wei, M.X.; Pan, H.R.; Lin, Z.J.; Wang, K.H.; Weng, Z.S.; Zhu, Y.B.; Xin, L.; Zhang, J.; Li, S.W.; et al. Robust manufacturing and comprehensive characterization of recombinant hepatitis E virus-like particles in Hecolin (R). *Vaccine* **2014**, *32*, 4039–4050. [CrossRef]

17. Tinto, H.; D'Alessandro, U.; Sorgho, H.; Valea, I.; Tahita, M.C.; Kabore, W.; Kiemde, F.; Lompo, P.; Ouedraogo, S.; Derra, K.; et al. Efficacy and safety of RTS,S/AS01 malaria vaccine with or without a booster dose in infants and children in Africa: Final results of a phase 3, individually randomised, controlled trial. *Lancet* **2015**, *386*, 31–45. [CrossRef]

18. Butler, P.J. The current picture of the structure and assembly of tobacco mosaic virus. *J. Gen. Virol.* **1984**, *65*, 253–279. [CrossRef]

19. Emini, E.A.; Jameson, B.A.; Wimmer, E. Priming for and induction of anti-poliovirus neutralizing antibodies by synthetic peptides. *Nature* **1983**, *304*, 699–703. [CrossRef]

20. Haynes, J.R.; Cunningham, J.; von Seefried, A.; Lennick, M.; Garvin, R.T.; Shen, S.-H. Development of a Genetically–Engineered, Candidate Polio Vaccine Employing the Self–Assembling Properties of the Tobacco Mosaic Virus Coat Protein. *Bio/Technology* **1986**, *4*, 637–641. [CrossRef]

21. Valenzuela, P.; Coit, D.; Medina-Selby, M.A.; Kuo, C.H.; Van Nest, G.; Lyn Burke, R.; Bull, P.; Urdea, M.S.; Graves, P.V. Antigen Engineering in Yeast: Synthesis and Assembly of Hybrid Hepatitis B Surface Antigen-Herpes Simplex 1 gD Particles. *Bio/Technology* **1985**, *3*, 323–326. [CrossRef]

22. Bachmann, M.F.; Jennings, G.T. Vaccine delivery: A matter of size, geometry, kinetics and molecular patterns. *Nat. Rev. Immunol.* **2010**, *10*, 787–796. [CrossRef]

23. Benne, N.; van Duijn, J.; Kuiper, J.; Jiskoot, W.; Slutter, B. Orchestrating immune responses: How size, shape and rigidity affect the immunogenicity of particulate vaccines. *J. Control Release* **2016**, *234*, 124–134. [CrossRef] [PubMed]

24. Mohsen, M.O.; Gomes, A.C.; Vogel, M.; Bachmann, M.F. Interaction of Viral Capsid-Derived Virus-Like Particles (VLPs) with the Innate Immune System. *Vaccines (Basel)* **2018**, *6*, 37. [CrossRef] [PubMed]

25. Balke, I.; Zeltins, A. Use of plant viruses and virus-like particles for the creation of novel vaccines. *Adv. Drug Deliv. Rev.* **2019**, *145*, 119–129. [CrossRef] [PubMed]

26. Zeltins, A. Protein Complexes and Virus-Like Particle Technology. *Subcell Biochem.* **2018**, *88*, 379–405. [CrossRef]

27. Rose, A.S.; Hildebrand, P.W. NGL Viewer: A web application for molecular visualization. *Nucleic Acids Res.* **2015**, *43*, W576–W579. [CrossRef]

28. Zeltins, A.; West, J.; Zabel, F.; El Turabi, A.; Balke, I.; Haas, S.; Maudrich, M.; Storni, F.; Engeroff, P.; Jennings, G.T.; et al. Incorporation of tetanus-epitope into virus-like particles achieves vaccine responses even in older recipients in models of psoriasis, Alzheimer's and cat allergy. *NPJ Vaccines* **2017**, *2*, 30. [CrossRef]

29. Speir, J.A.; Munshi, S.; Wang, G.; Baker, T.S.; Johnson, J.E. Structures of the native and swollen forms of cowpea chlorotic mottle virus determined by X-ray crystallography and cryo-electron microscopy. *Structure* **1995**, *3*, 63–77. [CrossRef]

30. Lin, T.; Chen, Z.; Usha, R.; Stauffacher, C.V.; Dai, J.B.; Schmidt, T.; Johnson, J.E. The refined crystal structure of cowpea mosaic virus at 2.8 a resolution. *Virology* **1999**, *265*, 20–34. [CrossRef]

31. Sangita, V.; Lokesh, G.L.; Satheshkumar, P.S.; Saravanan, V.; Vijay, C.S.; Savithri, H.S.; Murthy, M.R. Structural studies on recombinant T = 3 capsids of Sesbania mosaic virus coat protein mutants. *Acta Crystallogr. Sect. D* **2005**, *61*, 1402–1405. [CrossRef]

32. Ge, P.; Zhou, Z.H. Hydrogen-bonding networks and RNA bases revealed by cryo electron microscopy suggest a triggering mechanism for calcium switches. *Proc. Natl. Acad. Sci. USA* **2011**, *108*, 9637–9642. [CrossRef]

33. DiMaio, F.; Chen, C.C.; Yu, X.; Frenz, B.; Hsu, Y.H.; Lin, N.S.; Egelman, E.H. The molecular basis for flexibility in the flexible filamentous plant viruses. *Nat. Struct. Mol. Biol.* **2015**, *22*, 642–644. [CrossRef] [PubMed]

34. Eiben, S.; Koch, C.; Altintoprak, K.; Southan, A.; Tovar, G.; Laschat, S.; Weiss, I.M.; Wege, C. Plant virus-based materials for biomedical applications: Trends and prospects. *Adv. Drug Deliv. Rev.* **2019**, *145*, 96–118. [CrossRef]

35. Steele, J.F.C.; Peyret, H.; Saunders, K.; Castells-Graells, R.; Marsian, J.; Meshcheriakova, Y.; Lomonossoff, G.P. Synthetic plant virology for nanobiotechnology and nanomedicine. *Wiley Interdiscip. Rev. Nanomed. Nanobiotechnol.* **2017**, *9*. [CrossRef] [PubMed]

36. Lam, P.; Steinmetz, N.F. Plant viral and bacteriophage delivery of nucleic acid therapeutics. *Wiley Interdiscip. Rev. Nanomed. Nanobiotechnol.* **2018**, *10*. [CrossRef]

37. Narayanan, K.B.; Han, S.S. Icosahedral plant viral nanoparticles-bioinspired synthesis of nanomaterials/nanostructures. *Adv. Colloid Interface Sci.* **2017**, *248*, 1–19. [CrossRef] [PubMed]

38. Rybicki, E.P. Plant molecular farming of virus-like nanoparticles as vaccines and reagents. *Wiley Interdiscip. Rev. Nanomed. Nanobiotechnol.* **2019**, *12*, e1587. [CrossRef]

39. Plotkin, S.A. Vaccines We Need but Don't Have. *Viral Immunol.* **2018**, *31*, 114–116. [CrossRef]

40. Chen, T.H.; Hu, C.C.; Liao, J.T.; Lee, Y.L.; Huang, Y.W.; Lin, N.S.; Lin, Y.L.; Hsu, Y.H. Production of Japanese Encephalitis Virus Antigens in Plants Using Bamboo Mosaic Virus-Based Vector. *Front. Microbiol.* **2017**, *8*, 788. [CrossRef]

41. Jones, R.M.; Chichester, J.A.; Mett, V.; Jaje, J.; Tottey, S.; Manceva, S.; Casta, L.J.; Gibbs, S.K.; Musiychuk, K.; Shamloul, M.; et al. A plant-produced Pfs25 VLP malaria vaccine candidate induces persistent transmission blocking antibodies against Plasmodium falciparum in immunized mice. *PLoS ONE* **2013**, *8*, e79538. [CrossRef]

42. Banik, S.; Mansour, A.A.; Suresh, R.V.; Wykoff-Clary, S.; Malik, M.; McCormick, A.A.; Bakshi, C.S. Development of a Multivalent Subunit Vaccine against Tularemia Using Tobacco Mosaic Virus (TMV) Based Delivery System. *PLoS ONE* **2015**, *10*, e0130858. [CrossRef]

43. Laliberte-Gagne, M.E.; Bolduc, M.; Therien, A.; Garneau, C.; Casault, P.; Savard, P.; Estaquier, J.; Leclerc, D. Increased Immunogenicity of Full-Length Protein Antigens through Sortase-Mediated Coupling on the PapMV Vaccine Platform. *Vaccines (Basel)* **2019**, *7*, 49. [CrossRef] [PubMed]

44. Arnaboldi, P.M.; Sambir, M.; D'Arco, C.; Peters, L.A.; Seegers, J.F.; Mayer, L.; McCormick, A.A.; Dattwyler, R.J. Intranasal delivery of a protein subunit vaccine using a Tobacco Mosaic Virus platform protects against pneumonic plague. *Vaccine* **2016**, *34*, 5768–5776. [CrossRef] [PubMed]

45. Cabral-Miranda, G.; Heath, M.D.; Mohsen, M.O.; Gomes, A.C.; Engeroff, P.; Flaxman, A.; Leoratti, F.M.S.; El-Turabi, A.; Reyes-Sandoval, A.; Skinner, M.A.; et al. Virus-Like Particle (VLP) Plus Microcrystalline Tyrosine (MCT) Adjuvants Enhance Vaccine Efficacy Improving T and B Cell Immunogenicity and Protection against Plasmodium berghei/vivax. *Vaccines (Basel)* **2017**, *5*, 10. [CrossRef] [PubMed]

46. Cabral-Miranda, G.; Lim, S.M.; Mohsen, M.O.; Pobelov, I.V.; Roesti, E.S.; Heath, M.D.; Skinner, M.A.; Kramer, M.E.; Martina, B.E.E.; Bachmann, M.F. Zika Virus-Derived E-DIII Protein Displayed on Immunologically Optimized VLPs Induces Neutralizing Antibodies without Causing Enhancement of Dengue Virus Infection. *Vaccines (Basel)* **2019**, *7*, 72. [CrossRef]

47. Zeltins, A. Viral nanoparticles: Principles of construction and characterization. In *Viral Nanotechnology*, 1st ed.; Khudyakov, E., Pumpens, P., Eds.; CRC Press: Boca Raton, FL, USA, 2016; pp. 93–119.

48. Yusibov, V.; Hooper, D.C.; Spitsin, S.V.; Fleysh, N.; Kean, R.B.; Mikheeva, T.; Deka, D.; Karasev, A.; Cox, S.; Randall, J.; et al. Expression in plants and immunogenicity of plant virus-based experimental rabies vaccine. *Vaccine* **2002**, *20*, 3155–3164. [CrossRef]

49. Kushnir, N.; Streatfield, S.J.; Yusibov, V. Virus-like particles as a highly efficient vaccine platform: Diversity of targets and production systems and advances in clinical development. *Vaccine* **2012**, *31*, 58–83. [CrossRef]

50. Ortega-Berlanga, B.; Musiychuk, K.; Shoji, Y.; Chichester, J.A.; Yusibov, V.; Patino-Rodriguez, O.; Noyola, D.E.; Alpuche-Solis, A.G. Engineering and expression of a RhoA peptide against respiratory syncytial virus infection in plants. *Planta* **2016**, *243*, 451–458. [CrossRef]

51. Musiychuk, K.; Stephenson, N.; Bi, H.; Farrance, C.E.; Orozovic, G.; Brodelius, M.; Brodelius, P.; Horsey, A.; Ugulava, N.; Shamloul, A.M.; et al. A launch vector for the production of vaccine antigens in plants. *Influenza Other Respir. Viruses* **2007**, *1*, 19–25. [CrossRef]

52. World Health Organization. Malaria. Available online: http://www.who.int/mediacentre/factsheets/fs094/en/ (accessed on 25 November 2019).

53. Chichester, J.A.; Green, B.J.; Jones, R.M.; Shoji, Y.; Miura, K.; Long, C.A.; Lee, C.K.; Ockenhouse, C.F.; Morin, M.J.; Streatfield, S.J.; et al. Safety and immunogenicity of a plant-produced Pfs25 virus-like particle as a transmission blocking vaccine against malaria: A Phase 1 dose-escalation study in healthy adults. *Vaccine* **2018**, *36*, 5865–5871. [CrossRef]

54. Tregoning, J.S.; Russell, R.F.; Kinnear, E. Adjuvanted influenza vaccines. *Hum. Vaccines Immunother.* **2018**, *14*, 550–564. [CrossRef]

55. Folia Biotech Inc. Safety and Reactogenicity of a PAL Combined with Seasonal Flu Vaccine in Healthy Adults. Available online: https://clinicaltrials.gov/ct2/show/record/NCT02188810 (accessed on 25 November 2019).

56. Carignan, D.; Therien, A.; Rioux, G.; Paquet, G.; Gagne, M.L.; Bolduc, M.; Savard, P.; Leclerc, D. Engineering of the PapMV vaccine platform with a shortened M2e peptide leads to an effective one dose influenza vaccine. *Vaccine* **2015**, *33*, 7245–7253. [CrossRef]

57. Langeveld, J.P.; Brennan, F.R.; Martinez-Torrecuadrada, J.L.; Jones, T.D.; Boshuizen, R.S.; Vela, C.; Casal, J.I.; Kamstrup, S.; Dalsgaard, K.; Meloen, R.H.; et al. Inactivated recombinant plant virus protects dogs from a lethal challenge with canine parvovirus. *Vaccine* **2001**, *19*, 3661–3670. [CrossRef]

58. Yang, C.D.; Liao, J.T.; Lai, C.Y.; Jong, M.H.; Liang, C.M.; Lin, Y.L.; Lin, N.S.; Hsu, Y.H.; Liang, S.M. Induction of protective immunity in swine by recombinant bamboo mosaic virus expressing foot-and-mouth disease virus epitopes. *BMC Biotechnol.* **2007**, *7*, 62. [CrossRef]

59. Gellert, A.; Salanki, K.; Tombacz, K.; Tuboly, T.; Balazs, E. A cucumber mosaic virus based expression system for the production of porcine circovirus specific vaccines. *PLoS ONE* **2012**, *7*, e52688. [CrossRef] [PubMed]

60. World Health Organization. Cancer. Available online: https://www.who.int/news-room/fact-sheets/detail/cancer (accessed on 14 December 2019).

61. Oiseth, S.J.; Aziz, M.S. Cancer immunotherapy: A brief review of the history, possibilities, and challenges ahead. *J. Cancer Metastasis Treat.* **2017**, *3*, 250–261. [CrossRef]

62. Sahin, U.; Tureci, O. Personalized vaccines for cancer immunotherapy. *Science* **2018**, *359*, 1355–1360. [CrossRef]

63. Thomas, S.; Prendergast, G.C. Cancer Vaccines: A Brief Overview. *Methods Mol. Biol.* **2016**, *1403*, 755–761. [CrossRef] [PubMed]

64. Mougel, A.; Terme, M.; Tanchot, C. Therapeutic Cancer Vaccine and Combinations with Antiangiogenic Therapies and Immune Checkpoint Blockade. *Front. Immunol.* **2019**, *10*, 467. [CrossRef] [PubMed]

65. Wen, R.; Umeano, A.C.; Kou, Y.; Xu, J.; Farooqi, A.A. Nanoparticle systems for cancer vaccine. *Nanomedicine (Lond.)* **2019**, *14*, 627–648. [CrossRef] [PubMed]

66. Mohsen, M.O.; Speiser, D.E.; Knuth, A.; Bachmann, M.F. Virus-like particles for vaccination against cancer. *Wiley Interdiscip. Rev. Nanomed. Nanobiotechnol.* **2020**, *12*, e1579. [CrossRef]

67. McCormick, A.A.; Corbo, T.A.; Wykoff-Clary, S.; Palmer, K.E.; Pogue, G.P. Chemical conjugate TMV-peptide bivalent fusion vaccines improve cellular immunity and tumor protection. *Bioconjug. Chem.* **2006**, *17*, 1330–1338. [CrossRef] [PubMed]

68. McCormick, A.A.; Corbo, T.A.; Wykoff-Clary, S.; Nguyen, L.V.; Smith, M.L.; Palmer, K.E.; Pogue, G.P. TMV-peptide fusion vaccines induce cell-mediated immune responses and tumor protection in two murine models. *Vaccine* **2006**, *24*, 6414–6423. [CrossRef] [PubMed]

69. Mohsen, M.O.; Heath, M.D.; Cabral-Miranda, G.; Lipp, C.; Zeltins, A.; Sande, M.; Stein, J.V.; Riether, C.; Roesti, E.; Zha, L.S.; et al. Vaccination with nanoparticles combined with micro-adjuvants protects against cancer. *J. Immunother. Cancer* **2019**, *7*. [CrossRef]

70. Cai, H.; Shukla, S.; Wang, C.; Masarapu, H.; Steinmetz, N.F. Heterologous Prime-Boost Enhances the Antitumor Immune Response Elicited by Plant-Virus-Based Cancer Vaccine. *J. Am. Chem. Soc.* **2019**, *141*, 6509–6518. [CrossRef] [PubMed]

71. Murray, A.A.; Wang, C.; Fiering, S.; Steinmetz, N.F. In Situ Vaccination with Cowpea vs Tobacco Mosaic Virus against Melanoma. *Mol. Pharm.* **2018**, *15*, 3700–3716. [CrossRef] [PubMed]

72. Wang, C.; Beiss, V.; Steinmetz, N.F. Cowpea Mosaic Virus Nanoparticles and Empty Virus-Like Particles Show Distinct but Overlapping Immunostimulatory Properties. *J. Virol.* **2019**, *93*. [CrossRef]

73. Bachmann, M.F.; Jennings, G.T.; Vogel, M. A vaccine against Alzheimer's disease: Anything left but faith? *Expert Opin. Biol. Ther.* **2019**, *19*, 73–78. [CrossRef]

74. Bayer, A.J.; Bullock, R.; Jones, R.W.; Wilkinson, D.; Paterson, K.R.; Jenkins, L.; Millais, S.B.; Donoghue, S. Evaluation of the safety and immunogenicity of synthetic Abeta42 (AN1792) in patients with AD. *Neurology* **2005**, *64*, 94–101. [CrossRef] [PubMed]

75. Chackerian, B.; Rangel, M.; Hunter, Z.; Peabody, D.S. Virus and virus-like particle-based immunogens for Alzheimer's disease induce antibody responses against amyloid-beta without concomitant T cell responses. *Vaccine* **2006**, *24*, 6321–6331. [CrossRef] [PubMed]

76. Vitti, A.; Piazzolla, G.; Condelli, V.; Nuzzaci, M.; Lanorte, M.T.; Boscia, D.; De Stradis, A.; Antonaci, S.; Piazzolla, P.; Tortorella, C. Cucumber mosaic virus as the expression system for a potential vaccine against Alzheimer's disease. *J. Virol. Methods* **2010**, *169*, 332–340. [CrossRef] [PubMed]

77. Klimek, L.; Kundig, T.; Kramer, M.F.; Guethoff, S.; Jensen-Jarolim, E.; Schmidt-Weber, C.B.; Palomares, O.; Mohsen, M.O.; Jakob, T.; Bachmann, M. Virus-like particles (VLP) in prophylaxis and immunotherapy of allergic diseases. *Allergo J. Int.* **2018**, *27*, 245–255. [CrossRef] [PubMed]

78. Storni, F.; Zeltins, A.; Balke, I.; Heath, M.D.; Kramer, M.F.; Skinner, M.A.; Zha, L.; Roesti, E.; Engeroff, P.; Muri, L.; et al. Vaccine against peanut allergy based on engineered Virus-Like-Particles displaying single major peanut allergens. *J. Allergy Clin. Immunol.* **2019**. [CrossRef] [PubMed]

79. Anzaghe, M.; Schulke, S.; Scheurer, S. Virus-Like Particles as Carrier Systems to Enhance Immunomodulation in Allergen Immunotherapy. *Curr. Allergy Asthma Rep.* **2018**, *18*, 71. [CrossRef] [PubMed]

80. Chan, S.K.; Leung, D.Y.M. Dog and Cat Allergies: Current State of Diagnostic Approaches and Challenges. *Allergy Asthma Immunol. Res.* **2018**, *10*, 97–105. [CrossRef] [PubMed]

81. Thoms, F.; Jennings, G.T.; Maudrich, M.; Vogel, M.; Haas, S.; Zeltins, A.; Hofmann-Lehmann, R.; Riond, B.; Grossmann, J.; Hunziker, P.; et al. Immunization of cats to induce neutralizing antibodies against Fel d 1, the major feline allergen in human subjects. *J. Allergy Clin. Immunol.* **2019**, *144*, 193–203. [CrossRef]

82. Singh, S.; Kumar, N.K.; Dwiwedi, P.; Charan, J.; Kaur, R.; Sidhu, P.; Chugh, V.K. Monoclonal Antibodies: A Review. *Curr. Clin. Pharmacol.* **2018**, *13*, 85–99. [CrossRef]

83. Hafeez, U.; Gan, H.K.; Scott, A.M. Monoclonal antibodies as immunomodulatory therapy against cancer and autoimmune diseases. *Curr. Opin. Pharmacol.* **2018**, *41*, 114–121. [CrossRef]

84. Foerster, J.; Moleda, A. Feasibility Analysis of Interleukin-13 as a Target for a Therapeutic Vaccine. *Vaccines (Basel)* **2019**, *7*, 20. [CrossRef] [PubMed]

85. Foerster, J.; Bachman, M. Beyond passive immunization: Toward a nanoparticle-based IL-17 vaccine as first in class of future immune treatments. *Nanomedicine (Lond.)* **2015**, *10*, 1361–1369. [CrossRef] [PubMed]

86. Varricchi, G.; Bagnasco, D.; Borriello, F.; Heffler, E.; Canonica, G.W. Interleukin-5 pathway inhibition in the treatment of eosinophilic respiratory disorders: Evidence and unmet needs. *Curr. Opin. Allergy Clin. Immunol.* **2016**, *16*, 186–200. [CrossRef] [PubMed]

87. Fettelschoss-Gabriel, A.; Fettelschoss, V.; Thoms, F.; Giese, C.; Daniel, M.; Olomski, F.; Kamarachev, J.; Birkmann, K.; Buhler, M.; Kummer, M.; et al. Treating insect-bite hypersensitivity in horses with active vaccination against IL-5. *J. Allergy Clin. Immunol.* **2018**, *142*, 1194–1205.e3. [CrossRef] [PubMed]

88. Fettelschoss-Gabriel, A.; Fettelschoss, V.; Olomski, F.; Birkmann, K.; Thoms, F.; Buhler, M.; Kummer, M.; Zeltins, A.; Kundig, T.M.; Bachmann, M.F. Active vaccination against interleukin-5 as long-term treatment for insect-bite hypersensitivity in horses. *Allergy* **2019**, *74*, 572–582. [CrossRef] [PubMed]

89. Bachmann, M.F.; Zeltins, A.; Kalnins, G.; Balke, I.; Fischer, N.; Rostaher, A.; Tars, K.; Favrot, C. Vaccination against IL-31 for the treatment of atopic dermatitis in dogs. *J. Allergy Clin. Immunol.* **2018**, *142*, 279–281.e1. [CrossRef] [PubMed]

90. Olomski, F.; Fettelschoss, V.; Jonsdottir, S.; Birkmann, K.; Thoms, F.; Marti, E.; Bachmann, M.F.; Kundig, T.M.; Fettelschoss-Gabriel, A. Interleukin 31 in insect bite hypersensitivity-Alleviating clinical symptoms by active vaccination against itch. *Allergy* **2019**. [CrossRef]

91. Von Loga, I.S.; El-Turabi, A.; Jostins, L.; Miotla-Zarebska, J.; Mackay-Alderson, J.; Zeltins, A.; Parisi, I.; Bachmann, M.F.; Vincent, T.L. Active immunisation targeting nerve growth factor attenuates chronic pain behaviour in murine osteoarthritis. *Ann. Rheum. Dis.* **2019**, *78*, 672–675. [CrossRef]

92. Plotkin, S.; Robinson, J.M.; Cunningham, G.; Iqbal, R.; Larsen, S. The complexity and cost of vaccine manufacturing-An overview. *Vaccine* **2017**, *35*, 4064–4071. [CrossRef] [PubMed]

93. Robinson, J.M. Vaccine Production: Main Steps and Considerations. In *Vaccine Book*, 2nd ed.; Academic Press: Cambridge, MA, USA, 2016; pp. 77–96.

94. Rottingen, J.A.; Gouglas, D.; Feinberg, M.; Plotkin, S.; Raghavan, K.V.; Witty, A.; Draghia-Akli, R.; Stoffels, P.; Piot, P. New Vaccines against Epidemic Infectious Diseases. *N. Engl. J. Med.* **2017**, *376*, 610–613. [CrossRef] [PubMed]

95. Anderson, R.M. The Impact of Vaccination on the Epidemiology of Infectious Diseases. In *Vaccine Book*, 2nd ed.; Academic Press: Cambridge, MA, USA, 2016; pp. 3–31. [CrossRef]

96. Ciabattini, A.; Olivieri, R.; Lazzeri, E.; Medaglini, D. Role of the Microbiota in the Modulation of Vaccine Immune Responses. *Front. Microbiol.* **2019**, *10*, 1305. [CrossRef] [PubMed]

97. Chackerian, B.; Peabody, D.S. Factors That Govern the Induction of Long-Lived Antibody Responses. *Viruses* **2020**, *12*, 74. [CrossRef] [PubMed]

98. Slifka, M.K.; Amanna, I.J. Role of Multivalency and Antigenic Threshold in Generating Protective Antibody Responses. *Front. Immunol.* **2019**, *10*, 956. [CrossRef] [PubMed]

Review

Advantages and Prospects of Tag/Catcher Mediated Antigen Display on Capsid-Like Particle-Based Vaccines

Kara-Lee Aves [1,†], Louise Goksøyr [1,2,†] and Adam F. Sander [1,2,*]

[1] Faculty of Health Science, Institute for Immunology and Microbiology, University of Copenhagen, 2200 Copenhagen, Denmark; kara-lee@sund.ku.dk (K.-L.A.); louiseg@sund.ku.dk (L.G.)
[2] AdaptVac Aps, Agern Alle 1, 2970 Hørsholm, Denmark
*   Correspondence: asander@sund.ku.dk
†   Both authors contributed equally to this review.

Received: 28 December 2019; Accepted: 4 February 2020; Published: 6 February 2020

Abstract: Capsid-like particles (CLPs) are multimeric, repetitive assemblies of recombinant viral capsid proteins, which are highly immunogenic due to their structural similarity to wild-type viruses. CLPs can be used as molecular scaffolds to enable the presentation of soluble vaccine antigens in a similar structural format, which can significantly increase the immunogenicity of the antigen. CLP-based antigen display can be obtained by various genetic and modular conjugation methods. However, these vary in their versatility as well as efficiency in achieving an immunogenic antigen display. Here, we make a comparative review of the major CLP-based antigen display technologies. The Tag/Catcher-AP205 platform is highlighted as a particularly versatile and efficient technology that offers new qualitative and practical advantages in designing modular CLP vaccines. Finally, we discuss how split-protein Tag/Catcher conjugation systems can help to further propagate and enhance modular CLP vaccine designs.

Keywords: capsid; antigen display; virus-like particle; vaccine; platform

---

1. Rationale for the Development of Capsid-Like Particle-Based Vaccines

The development of vaccines remains the most effective method for preventing and controlling the spread of infectious diseases [1,2]. Traditional live-attenuated vaccines, such as those targeting measles [3], rubella [4] and smallpox [5] are highly immunogenic and can induce potent and long-lived antibody responses even after a single immunization [6]. This was originally ascribed to the ability of the attenuated virus to replicate in the host after vaccination. The high immunogenicity of live-attenuated viruses comes with a cost of increased safety risks and challenging manufacturing processes [7,8]. In contrast, modern recombinant subunit vaccines (i.e., based on a soluble protein antigen) show high safety, but in general fail to induce similar long-lasting antibody responses in humans [6,9–12]. Within recent years, several virus-like particle (VLP) based vaccines have been tested in clinical trials, of which vaccines targeting Hepatitis B (HBV) (Recombivax HB® and Engerix-B®), Human papillomavirus (HPV) (Cervarix®, Gardasil®, and Gardasil 9®) and Hepatitis E (HEV) (Hecolin®) have been licensed [13]. Notably, the HPV vaccine stands as a unique example of a recombinant subunit vaccine with comparable immunogenicity to live-attenuated vaccines, even after a single dose [14–16]. This vaccine is formed by the self-assembly of the HPV major capsid protein into capsid-like particles (CLP), the structure of which is thought to be key to its high potency [17]. CLPs constitute a subclass of VLPs and are rigid, non-lipid, protein-based particles. A large number of studies have jointly established a strong causal link between the high immunogenicity of CLPs and their structural similarities to native viruses. Of these properties, their size (20–200 nm in diameter) and repetitive surface geometry are considered the most important [18–26]. Moreover, it has long been recognized that the immunogenicity of a vaccine antigen can be significantly increased if it is delivered to the immune system in a similar multivalent, repetitive and particulate format [27,28]. Consequently, several strategies have been pursued, exploiting CLPs as scaffolds for the presentation of heterologous antigens, including self-antigens.

Here, we will review key attributes of various CLP-based vaccine technologies in terms of their ability to facilitate a highly immunogenic epitope display. Additionally, practical aspects of the conjugation systems, such as their versatility, manufacturability and scalability will be discussed. In this context, we highlight the Tag/Catcher-AP205 platform as a particularly versatile and effective technology and provide a rationale for further development of this technology for vaccine design.

2. Methods for Antigen Display on CLPs

2.1. Genetic Fusion

The first widely used approach for CLP-based antigen display was by genetic fusion of a heterologous antigen sequence to the capsid subunit protein [27–29]. When successful, each subunit of the CLP will display the inserted antigenic sequence in a pattern matching that of the underlying CLP. Although this approach can deliver foreign epitopes in a viral-like configuration, it is generally limited to small peptide antigens that do not inhibit particle assembly [30–32]. However, in some cases, even small peptide antigens prevent CLP assembly. Thus, the success of genetic fusion is difficult to predict, and structural characteristics of the CLP as well as biochemical properties of the peptide must be taken into account [33,34]. Studies on HBcAg CLPs show that highly charged and hydrophobic peptides tend to prevent CLP assembly [30,35]. Due to the small size of the incorporated antigen, CLP vaccines made by genetic fusion generally induce narrow epitope-specific antibody responses [36]. The challenges associated with genetic fusion have to some extent been relieved by creating mosaic CLPs, consisting of both native and genetically modified CLP subunits. However, this is at the cost of reduced antigen density [37,38].

2.2. Modular Antigen Display

CLP-display of large and complex protein antigens is not easily obtained by genetic fusion. Thus, an alternative strategy has been to attach protein antigens to the surface of preassembled CLPs. Importantly, when using such modular approaches, the resultant epitope display is directly dependent on the applied conjugation method, as well as the specific surface geometry of the employed CLP backbone. In addition, these technologies allow for separate recombinant production of antigens in various expression systems, ensuring high quality and correct protein processing before CLP conjugation [39–42]. The different conjugation systems used for modular CLP vaccine development is described in detail elsewhere [43]. Here, the main techniques are listed in Table 1 to provide a comparative overview of their relative practicality and ability to facilitate a high-quality epitope display. This evaluation is done in recognition that HPV CLPs represent a highly immunogenic viral epitope display, and on that basis, modular platforms should likewise facilitate ordered, high-density and unidirectional presentation of antigens in their native conformation. This cannot, to a similar degree, be achieved by all available antigen conjugation methods.

Table 1. Properties of different conjugation strategies used for capsid-like particles (CLP) antigen display.

	Antigen Display				Versatility			Examples
	High Density	Covalent linkage	Uniform Distribution	Unidirectional	Antigen Size/ Complexity	Expression Systems	Capsid Backbones	
Genetic Fusion	+	+	+	+	−	−	+	[31,44]
Chemical Conjugation	−	+	−	+/−	+/−	+	+	[45,46]
Click Chemistry	+/−	+	+	+	+/−	+	+	[47,48]
Affinity-based	+/−	−	+	+	+	+/−	+	[39,49,50]
Split-protein Systems	+	+	+	+	+	+	+/−	[39,41,51]

+ indicates property can be achieved, − indicates property not readily achieved, +/− indicates property can only partly be achieved or has not been experimentally validated.

2.2.1. Chemical Conjugation and Click Chemistry

One of the most widely used techniques for modular CLP antigen display has been through chemical conjugation, which is compatible with most capsid backbones. A common approach is through cross-linking of lysine residues on the CLP surface to cysteine residues present or incorporated into the antigen [45,46,52]. This vaccine design has been shown to induce high titers of antigen-specific antibodies, and several vaccines have shown promising results in preclinical studies, as well as in clinical testing (e.g., Nicotine-Qβ [53]). However, the CLP surface often contains multiple lysine residues, resulting in uneven antigen distribution, and little control over antigen orientation. In addition, introduction of a reactive cysteine may cause antigen misfolding [54,55]. In some cases, chemical conjugation can lead to destabilization of the CLP, and cannot always facilitate high-density antigen display, due to suboptimal coupling efficiency [46,56]. Some of the limitations associated with standard chemical conjugation have been resolved by the development of click chemistry [35,48,57]. This method uses the incorporation of unnatural amino acids, allowing for increased control over site-specificity. This technique requires minimal change to the VLP and antigen, while providing a highly specific and fast coupling reaction. However, the scalability of this technology remains a concern [33,58].

2.2.2. Affinity-Based Conjugation

Affinity-based conjugation systems, such as streptavidin/biotin [42,49,59], and to a lesser extent His-tag/Ni-NTA [50], have been used to facilitate unidirectional display. This is due to the single attachment site on each antigen and promotes an even display of heterologous epitopes. In addition, these methods have the advantage of being able to display larger and complex protein antigens, enabling the induction of broad, polyclonal humoral responses without compromising CLP formation and stability. For further development of the streptavidin/biotin conjugation system, engineered monomeric streptavidin (mSA) was used for antigen display on HPV CLPs containing a biotin acceptor site (AviTag™) [60]. A potential disadvantage of this strategy, is that the coupling is based on a non-covalent interaction, and thus there is a risk of antigen disengagement. This can affect the coupling efficiency and diminish antigen density. Additionally, it has been hypothesized that dissociation due to an altered chemical environment in vivo, may reduce the biological efficacy of the vaccine.

2.2.3. Split-Protein (Tag/Catcher) Conjugation

Split-protein (Tag/Catcher) conjugation systems have been developed [61–70] and used for covalent anchoring of vaccine antigens onto CLPs [39,51]. The split-protein technology is based on the separation of a bacterial pili protein, into a reactive peptide (Tag) and corresponding protein binding partner (Catcher). Upon mixing in solution, the Tag and Catcher rapidly react to form a spontaneous isopeptide bond [61]. In the following section, we discuss key features of the Tag/Catcher-AP205 technology and emphasize the versatility of this platform, as well as its ability to effectively mediate highly immunogenic antigen display.

The Tag/Catcher-AP205 platform was developed by genetic incorporation of the SpyTag [39] and SpyCatcher [39,51] into the capsid protein of the *Acinetobacter phage* AP205, yielding particles of 180 subunits with a diameter of approximately 36 nm and 43 nm, respectively. Since its development, the Tag/Catcher-AP205 platform has been utilized to display structurally and functionally diverse vaccine antigens, ranging in size from small peptides (e.g., toxins of 19 amino acids [71]) to large (>300 kDa) trimeric proteins [72]. These studies have repeatedly demonstrated a remarkable ability to achieve complete and even decoration of the CLP surface, with coupling efficiencies reaching 100% for smaller vaccine antigens, which are not limited by steric hindrance. Importantly, the resultant CLP-display induces antibody titers of high quality [55], affinity [41] and avidity [39]. The platform is additionally capable of effectively overcome B-cell tolerance and induce strong antibody responses against a variety of self-antigens, including IL-5, CTLA-4, PD-L1 and Her2 [39,41].

Several features of the AP205 CLP make it attractive as a vaccine backbone, including its structure, intrinsic immunogenicity and manufacturability [73,74]. While the overall capsid structure within the RNA bacteriophage family is similar, the surface exposed regions available for genetic fusion differ substantially [74,75]. The AP205 capsid is remarkable in that both the N- and C- termini are surface exposed and evenly distributed on the assembled CLP [74]. Moreover, AP205 CLPs tolerate genetic fusion at both the N- and C-terminus of the subunit protein, while maintaining stable CLP assembly [39,44]. For future large-scale manufacturing and clinical development, the Tag/Catcher-AP205 platform can be cost-effectively produced at very high yield in *E. coli* [39,74]. In fact, although the scalability of the platform has previously been questioned [76,77], our results show that fermentation can enable the production and purification of correctly assembled CLPs in the scale of grams per liter bacterial cell culture (manuscript in preparation).

Combinatorial Antigen Display

The ability to simultaneously display multiple different antigens on the same CLP could have numerous applications, but has so far proven technically challenging, with only a few examples reported [48,78]. However, the unique exposure of both termini of AP205 has made it possible to readily achieve such combinatorial antigen display. A vaccine targeting both HPV and placental malaria was recently described [79]. In this study, concatenated RG1 epitopes (from the HPV L2 protein) were genetically fused to the C-terminus of AP205, while VAR2CSA (a placental malaria antigen) was conjugated via the Tag/Catcher system to the N-terminus of AP205, without hampering vaccine stability. Vaccination induced high titers of functional antibodies targeting both components, thus providing a proof-of-concept for dual antigen display on the Tag/Catcher-AP205 platform.

Control over Antigen Orientation

From the early years of VLP research, the importance of ordered antigen display has been noted [80]. Since then, there has been a further appreciation of the benefits of unidirectional display, which can be achieved with the Tag/Catcher-AP205 technology. This was demonstrated by a study comparing different platforms presenting the malaria Pfs25 antigen with varying degrees of antigen organization [55]. The unidirectional display, obtained by the Tag/Catcher-AP205 technology, induced antibodies of higher biological efficacy, compared to when the antigen was presented in several different orientations, as the result of chemical cross-linking [55]. On that basis, it was hypothesized that unidirectional antigen display can enable induction of a more focused antibody response. Moreover, unidirectional presentation may also be exploited to mask certain regions of an antigen. In a recent study by Escolano et al., a Spy-tagged HIV envelope protein was displayed on SpyCatcher-AP205 CLPs [72]. Here, the dense unidirectional antigen display promoted induction of broadly neutralizing antibodies (bNAb), while masking dominant non-neutralizing epitopes, present near the CLP surface. An additional benefit of the Tag/Catcher conjugation technology is the small size of the SpyTag (13 amino acids), which allows its incorporation into internal antigen positions (e.g., in flexible loops) [81,82]. This provides further opportunities to optimize antigen orientation on the CLP surface.

Multimeric Antigen Display

Many viral antigens, such as the HIV envelope trimer, are multimeric glycoproteins [83]. Broadly neutralizing antibodies often target non-linear, conformational epitopes located at the intersection between protomers [83–85]. For induction of such bnAbs, the antigen thus needs to be delivered in its native quaternary structure [86]. The first study to achieve this through the Tag/Catcher technology, successfully displayed HIV envelope trimers on AP205 [72]. This demonstrates the platform's ability to allow antigen multimerization while providing increased stabilization of the protein complex on the CLP surface. Importantly, such display enables conformational epitopes to be presented in a native-like structural context. This represents an interesting new development for the display of multimeric and highly complex antigens, for a more focused immune activation.

Practicality of the Tag/Catcher-AP205 System

It is possible that affinity-based platforms such as mSA/AviTag^{TM} [60] and His/trisNTA [50] could provide the same opportunities for controlled high density and unidirectional antigen display, as the split protein (Tag/Catcher) technology. However, practical aspects including versatility and ease of use favor the Tag/Catcher-AP205 platform. In our experience, SpyTag and SpyCatcher can more easily be genetically fused to a variety of antigens and allow expression in a greater range of expression systems (e.g., insect [39,41], mammalian [unpublished] and bacterial [79]), compared to mSA [60]. In addition, SpyTag is less likely to negatively affect the fold of the recombinant fusion protein. SpyCatcher has also shown to positively affect the solubility and yield of certain antigens that are otherwise difficult to express [39]. Importantly, the Tag/Catcher reaction is compatible with a broad range of buffer conditions, enabling ample opportunities to optimize antigen/CLP formulation. The recently described Spy&Go method furthermore facilitates protein purification via the SpyTag, thereby enabling a more simple antigen design without the need for a separate purification tag [67]. These practical aspects of the Tag/Catcher-AP205 technology further strengthens the future opportunities of this platform.

3. Prospects for Further Development of Tag/Catcher-Based CLP Display Technologies

The Tag/Catcher conjugation system is able to fulfill many of the criteria important for obtaining a viral-like epitope display, as highlighted in Table 1. This was demonstrated through the development and success of the SpyTag/SpyCatcher-AP205 platform. However, there are more opportunities to explore. Adaptations in the modular vaccine design, such as the choice of backbone, particle size and antigen density, as well as the addition of appropriate intrinsic and extrinsic adjuvants, could aid in broadening the development of CLP-based vaccines. Additionally, practical aspects such as manufacturability, scalability and cost-effectiveness, can determine future downstream success, and therefore need to be taken into account during vaccine development. In the section below, we will discuss several aspects that should be considered when designing future Tag/Catcher CLP-based vaccines and highlight several unanswered questions and hypotheses that are yet to be experimentally validated.

3.1. CLP Backbones

The high diversity of well-characterized viral capsids has provided a range of CLP backbones with varying structural and biological properties. The most common include CLPs derived from animal viruses (e.g., Hepatitis B core antigen (HBcAg) and HPV), plant viruses (e.g., Cowpea mosaic virus (CPMV) and tobacco mosaic virus (TMV)) and bacteriophages (e.g., MS2, Qβ and AP205) [87–89]. Recently, successful expression and assembly of 80 novel RNA bacteriophage CLPs was described, which further broadens the pool of potential vaccine backbones [90]. The simple structure of phage capsids makes them well suited for rapid and cost-effective recombinant bacterial expression [44,90,91].

In contrast, CLPs derived from mammalian viruses often require more complex expression systems, such as yeast [17] (Gardasil®) and insect cells [92] (Cervarix®). However, advancements in transient plant expression may offer a safe and inexpensive alternative to these conventional systems, as it allows for post-translational modifications, such as glycosylation. Thus, several complex CLPs, including HPV, have successfully been produced in plants [93–96].

Different capsid backbones may not only possess different opportunities for antigen-display but may also vary in their intrinsic immune-stimulatory qualities. This can additionally be impacted by the expression system employed [97] or engineered onto the CLP [57]. As an example, T-cell epitopes on the HPV CLP has been shown to contribute to the high immunogenicity of the particle [14,98]. Likewise, the genetic fusion of T-cell epitopes onto CLPs can induce priming of cytotoxic T-cell responses in vivo [59,99,100]. The lumen of CLPs can also be exploited for the effective delivery of intrinsic adjuvants, such as CpG [101]. Additionally, ssRNA bacteriophages such as AP205, are able to encapsulate host RNA during recombinant bacterial expression, which adjuvant the immune response via Toll-like receptor (TLR) 7 and 8 activation [74]. One recent study demonstrated that the origin of the packaged RNA can further modulate the humoral response by directing IgG class switching [97]. While prokaryotic RNA induces a predominantly IgG2 response, eukaryotic RNA elicits an IgG1 dominated response. However, packaged nucleic acids could be considered disadvantageous for further clinical development. Importantly, in the context of developing self-antigen based vaccines for the treatment of non-infectious diseases, the packaged RNA may possess a safety risk by promoting activation of an autoimmune T-cell response [102,103]. It has also been a concern, that pre-existing anti-capsid immunity could have an immunosuppressive effect on the displayed antigen [104,105]. Together, these factors indicate that it would be valuable to exploit Tag/Catcher-based antigen display on a broader range of CLP backbones.

3.2. Particle Size, Valiancy and Spacing

It is well documented that the pharmacokinetics of nanoparticles and their engagement with the innate immune system is affected by the particle size, charge and surface properties [106–108]. While large particles (500 nm–1 μm) need to be processed by antigen-presenting cells prior to transport to the lymph nodes, pathogens and VLPs of 20–200 nm have the ability to rapidly and effectively traffic through the lymphatic system in a cell-free state, and thus interact directly with B-cells in the lymph nodes [19,23,109]. However, it remains to be clarified whether variation in size within this "optimal" range, as well as the resultant valency, will affect the immune response. Thus, it is unknown if the particulate size/valency differences between e.g., HPV CLPs (55 nm particle of 360 subunits), bacteriophage CLPs such as AP205 and Qβ (30 nm particles of 180 subunits), and smaller nanoparticles such as ferritin (12 nm) [110] are significant.

The close spacing of repetitive epitopes is believed to be a key determinant for the high immunogenicity of viruses and VLPs [18,23,111–113]. Accordingly, antigen density has become a common quality measure for modular CLP vaccines. Early studies concluded that narrow epitope spacing of 5–10 nm is a critical determinant in humoral responses (i.e., by facilitating B-cell receptor crosslinking and B-cell activation). However, these first studies were based on hapten-polymer conjugates as repetitive antigens [111,114]. It was additionally shown that the immunogenicity of the native VSV-G protein displayed at high density by the enveloped RNA virus (VSV) was significantly higher compared to less organized forms of recombinant VSV-G protein (i.e., soluble or displayed in micelles) [80]. A study using HBc and Qβ CLPs has since demonstrated a positive correlation between antigen density and the vaccine-induced antigen-specific IgG responses, and observed that high doses of low-density particles could not counteract this effect [115]. However, antigen conjugation in this study was achieved using chemical cross-linking and thus the maximum coupling density tested did not exceed 50%. Given the ability of the Tag/Catcher-based conjugation system to achieve complete decoration of CLP backbones, we propose to use this technology to further investigate whether there exists a threshold, beyond which increasing the density would have no further effect. This could have specific implications in the optimization of modular vaccines, e.g., when greater antigen spacing is required for providing sufficient access to epitopes near the capsid surface [116].

3.3. Effect of Platform Rigidity

The influence of platform rigidity is also a topic of debate, and it is yet to be confirmed whether epitopes that are held as part of a rigid, semi-crystalline structure (e.g., HPV) possess the same immunological properties as epitopes that are part of a more flexible structure (e.g., modular CLPs). If this is the case, the use of long flexible linkers involved in some of the more complex conjugation strategies should be avoided. Likewise, the nature of the underlying VLP should be considered. There are several examples where enveloped VLPs and other lipid-based vaccine platforms have successfully been used for induction of protective immune responses; targeting either the native surface protein (e.g., the hepatitis B virus (HBV) surface antigen (HBsAg) [117]) or a displayed foreign antigen [118–120]. Recently, two studies have performed head-to-head comparisons between lipid- and capsid-based VLP vaccines. Chen et al. showed that vaccination with a self-antigen peptide displayed on Qβ CLPs resulted in a 200-fold increase in antigen-specific IgG titers, compared to antigen presentation on liposome particles [113]. In contrast, Marini et al. compared Pfs25 antigen display via the Tag/Catcher conjugation system on either HBsAg lipid-based particles or AP205 CLPs, and showed no significant differences in antibody titers and functionality between the two vaccines [77]. The duration of the induced antibody responses was however not tested in either study. This could be an important factor to investigate, as the antibody longevity of licensed VLP-based vaccines differ substantially. When targeting many important pathogens, such as HIV, HPV and malaria, neutralizing antibody levels need to be maintained in order for the vaccine to be protective. This is due to the importance of preventing the initial infection of these pathogens. In such cases, memory B-cell dependent humoral responses would not be sufficient [121]. While the HBV vaccine has shown to be effective and has decreased HBV prevalence worldwide [122], the long-term efficacy of the HBV vaccine is memory B-cell dependent. In general, durable antibody responses are not induced, even after three doses [123]. The RTS,S vaccine, which is based on genetic incorporation of a malaria antigen into HBsAg VLPs, only showed short-term efficacy in clinical trials, due to a similar rapid decline in antibody levels [118,124,125]. In contrast, HPV CLPs are strong inducers of long-lived antibody-producing plasma cells [14,126]. These marked differences in the induced immune responses could reflect fundamental differences in the immunological properties of lipid- versus capsid-based backbones [127].

3.4. The Need for Thorough Comparative Studies

As described above, the immune response elicited by CLP-based vaccines is likely influenced by multiple factors. However, direct comparisons of many of these variables have so far been lacking in the field. Firstly, there are no conclusive comparisons between the different conjugation systems. A study by Leneghan et al. indicates that unidirectional antigen display obtained by split-protein conjugation is more favorable than the disordered display elicited by chemical conjugation [55]. However, these results may in part be due to a combination of additional confounding factors such as the particle size, epitope density and chemical modifications of the antigen, all of which varied between the tested vaccines. This study highlights the difficulty with conducting such comparative studies, as it is challenging to separate the many variables intrinsically associated with different CLP platforms. Likewise, the full immunological effect of capsid backbones from divergent origins, containing different intrinsic properties, remains to be defined. When considering the final vaccine formulation, it would moreover be important to compare the effect of extrinsic adjuvants, to ascertain which work best in synergy with the immune-activation obtained by the CLP platform. Improved consensus in these areas would greatly aid in refining and directing the rational design of CLP-based vaccines. This could provide a toolbox of components that can intentionally be combined to modulate or skew the response, to induce the most suitable immune activation against a given disease or antigen.

4. Concluding Remarks

Overall, CLP-based display has proven effective in increasing the immunogenicity of vaccine antigens. This has offered new possibilities for developing vaccines against important infectious diseases, which have proven difficult to target using conventional vaccine strategies. CLP-display has even enabled the induction of strong antibody responses to self-antigens, and thus made it possible to develop therapeutic vaccines against non-infectious diseases, including cancer.

In this review, we have highlighted the unique ability of the split-protein-based Tag/Catcher-AP205 platform to mediate high-density, unidirectional antigen presentation of a broad range of vaccine antigens. We believe this highly versatile technology, combined with the recent discovery of many novel CLP backbones, will create numerous opportunities for strategic design of new CLP vaccines with distinct immune-stimulatory properties.

Funding: This research (incl. APC) was funded by the Innovation Fund Denmark, grant number 8053-00175B and 8088-00032A.

Conflicts of Interest: A.F.S. is listed as co-inventor on a patent application covering the AP205 CLP vaccine platform technology (WO2016112921 A1) licensed to AdaptVac. A.F.S and L.G. are currently partially employed in AdaptVac. K-L.A. declares no conflict of interest.

References

1. Doherty, M.; Buchy, P.; Standaert, B.; Giaquinto, C.; Prado-Cohrs, D. Vaccine impact: Benefits for human health. *Vaccine* **2016**, *34*, 6707–6714. [CrossRef]
2. Lam, E.; McCarthy, A.; Brennan, M. Vaccine-preventable diseases in humanitarian emergencies among refugee and internally-displaced populations. *Hum. Vaccines Immunother.* **2015**, *11*, 2627–2636. [CrossRef]
3. Marin, M.; Nguyen, H.Q.; Langidrik, J.R.; Edwards, R.; Briand, K.; Papania, M.J.; Seward, J.F.; LeBaron, C.W. Measles Transmission and Vaccine Effectiveness during a Large Outbreak on a Densely Populated Island: Implications for Vaccination Policy. *Clin. Infect. Dis.* **2006**, *42*, 315–319. [CrossRef]
4. Greaves, W.L.; Orenstein, W.A.; Hinman, A.R.; Nersesian, W.S. Clinical efficacy of rubella vaccine. *Pediatr. Infect. Dis. J.* **1983**, *2*, 284–286. [CrossRef]
5. Amanna, I.J.; Slifka, M.K.; Crotty, S. Immunity and immunological memory following smallpox vaccination. *Immunol. Rev.* **2006**, *211*, 320–337. [CrossRef]
6. Amanna, I.; Carlson, N.; Slifka, M. Duration of Humoral Immunity to Common Viral and Vaccine Antigens. *N. Engl. J. Med.* **2007**, *357*, 1903–1951. [CrossRef]
7. Plotkin, S.A.; Plotkin, S.L. The development of vaccines: How the past led to the future. *Nat. Rev. Microbiol.* **2011**, *9*, 889–893. [CrossRef]
8. Plotkin, S.; Robinson, J.M.; Cunningham, G.; Iqbal, R.; Larsen, S. The complexity and cost of vaccine manufacturing–An overview. *Vaccine* **2017**, *35*, 4064–4071. [CrossRef]
9. Herrington, D.A.; Nardin, E.H.; Losonsky, G.; Bathurst, I.C.; Barr, P.J.; Hollingdale, M.R.; Edelman, R.; Levine, M.M. Safety and immunogenicity of a recombinant sporozoite malaria vaccine against Plasmodium vivax. *Am. J. Trop. Med. Hyg.* **1991**, *45*, 695–701. [CrossRef]
10. Sirima, S.B.; Mordmüller, B.; Milligan, P.; Ngoa, U.A.; Kironde, F.; Atuguba, F.; Tiono, A.B.; Issifou, S.; Kaddumukasa, M.; Bangre, O.; et al. A phase 2b randomized, controlled trial of the efficacy of the GMZ2 malaria vaccine in African children. *Vaccine* **2016**, *34*, 4536–4542. [CrossRef]
11. Manoff, S.B.; George, S.L.; Bett, A.J.; Yelmene, M.L.; Dhanasekaran, G.; Eggemeyer, L.; Sausser, M.L.; Dubey, S.A.; Casimiro, D.R.; Clements, D.E.; et al. Preclinical and clinical development of a dengue recombinant subunit vaccine. *Vaccine* **2015**, *33*, 7126–7134. [CrossRef] [PubMed]
12. Purcell, A.W.; McCluskey, J.; Rossjohn, J. More than one reason to rethink the use of peptides in vaccine design. *Nat. Rev. Drug Discov.* **2007**, *6*, 404–414. [CrossRef] [PubMed]
13. Kelly, H.G.; Kent, S.J.; Wheatley, A.K. Immunological basis for enhanced immunity of nanoparticle vaccines. *Expert Rev. Vaccines* **2019**, *18*, 269–280. [CrossRef]
14. Schiller, J.; Lowy, D. Explanations for the high potency of HPV prophylactic vaccines. *Vaccine* **2018**, *36*, 4768–4773. [CrossRef]

15. Schiller, J.T.; Castellsagué, X.; Garland, S.M. A review of clinical trials of human papillomavirus prophylactic vaccines. *Vaccine* **2012**, *30*, F123–F138. [CrossRef]

16. De Vincenzo, R.; Conte, C.; Ricci, C.; Scambia, G.; Capelli, G. Long-term efficacy and safety of human papillomavirus vaccination. *Int. J. Womens. Health* **2014**, *6*, 999–1010. [CrossRef]

17. Carter, J.J.; Yaegashi, N.; Jenison, S.A.; Galloway, D.A. Expression of human papillomavirus proteins in yeast Saccharomyces cerevisiae. *Virology* **1991**, *182*, 513–521. [CrossRef]

18. Bachmann, M.F.; Zinkernagel, R.M. The influence of virus structure on antibody responses and virus serotype formation. *Immunol. Today* **1996**, *17*, 553–558. [CrossRef]

19. Manolova, V.; Flace, A.; Bauer, M.; Schwarz, K.; Saudan, P.; Bachmann, M.F. Nanoparticles target distinct dendritic cell populations according to their size. *Eur. J. Immunol.* **2008**, *38*, 1404–1413. [CrossRef]

20. Alexander Titz, B.; Brombacher, F.; Alexander Link, M.F.; Zabel, F.; Schnetzler, Y.; Link, A.; Titz, A.; Bachmann, M.F. Innate Immunity Mediates Follicular Innate Immunity Mediates Follicular Transport of Particulate but Not Soluble Protein Antigen. *J. Immunol.* **2012**, *188*, 3724–3733.

21. Mohsen, M.O.; Gomes, A.C.; Cabral-Miranda, G.; Krueger, C.C.; Leoratti, F.M.; Stein, J.V.; Bachmann, M.F. Delivering adjuvants and antigens in separate nanoparticles eliminates the need of physical linkage for effective vaccination. *J. Control. Release* **2017**, *251*, 92–100. [CrossRef] [PubMed]

22. Keller, S.A.; Bauer, M.; Manolova, V.; Muntwiler, S.; Saudan, P.; Bachmann, M.F. Cutting Edge: Limited Specialization of Dendritic Cell Subsets for MHC Class II-Associated Presentation of Viral Particles. *J. Immunol.* **2010**, *184*, 26–29. [CrossRef] [PubMed]

23. Bachmann, M.F.; Jennings, G.T. Vaccine delivery: A matter of size, geometry, kinetics and molecular patterns. *Nat. Rev. Immunol.* **2010**, *10*, 787–796. [CrossRef] [PubMed]

24. Schadlich, L.; Senger, T.; Gerlach, B.; Mucke, N.; Klein, C.; Bravo, I.G.; Muller, M.; Gissmann, L. Analysis of Modified Human Papillomavirus Type 16 L1 Capsomeres: The Ability To Assemble into Larger Particles Correlates with Higher Immunogenicity. *J. Virol.* **2009**, *83*, 7690–7705. [CrossRef] [PubMed]

25. Thones, N.; Herreiner, A.; Schadlich, L.; Piuko, K.; Muller, M. A Direct Comparison of Human Papillomavirus Type 16 L1 Particles Reveals a Lower Immunogenicity of Capsomeres than Viruslike Particles with Respect to the Induced Antibody Response. *J. Virol.* **2008**, *82*, 5472–5485. [CrossRef] [PubMed]

26. Mohsen, M.; Gomes, A.; Vogel, M.; Bachmann, M. Interaction of Viral Capsid-Derived Virus-Like Particles (VLPs) with the Innate Immune System. *Vaccines* **2018**, *6*, 37. [CrossRef]

27. Pumpens, P.; Borisova, G.P.; Crowther, R.A.; Grens, E. Hepatitis B Virus Core Particles as Epitope Carriers. *Intervirology* **1995**, *38*, 63–74. [CrossRef]

28. Kratz, P.A.; Böttcher, B.; Nassal, M. Native display of complete foreign protein domains on the surface of hepatitis B virus capsids. *Proc. Natl. Acad. Sci. USA.* **1999**, *96*, 1915–1920. [CrossRef]

29. Mastico, R.A.; Talbot, S.J.; Stockley, P.G. Multiple presentation of foreign peptides on the surface of an RNA-free spherical bacteriophage capsid. *J. Gen. Virol.* **1993**, *74*, 541–548. [CrossRef]

30. Karpenko, L.I.; Ivanisenko, V.A.; Pika, I.A.; Chikaev, N.A.; Eroshkin, A.M.; Veremeiko, T.A.; Ilyichev, A.A. Insertion of foreign epitopes in HBcAg: How to make the chimeric particle assemble. *Amino Acid* **2000**, *18*, 329–337. [CrossRef]

31. Peabody, D.S.; Manifold-Wheeler, B.; Medford, A.; Jordan, S.K.; do Carmo Caldeira, J.; Chackerian, B. Immunogenic Display of Diverse Peptides on Virus-like Particles of RNA Phage MS2. *J. Mol. Biol.* **2008**, *380*, 252–263. [CrossRef] [PubMed]

32. Billaud, J.-N.; Peterson, D.; Barr, M.; Chen, A.; Sallberg, M.; Garduno, F.; Goldstein, P.; McDowell, W.; Hughes, J.; Jones, J.; et al. Combinatorial Approach to Hepadnavirus-Like Particle Vaccine Design. *J. Virol.* **2005**, *79*, 13656–13666. [CrossRef] [PubMed]

33. Frietze, K.M.; Peabody, D.S.; Chackerian, B. Engineering virus-like particles as vaccine platforms. *Curr. Opin. Virol.* **2016**, *18*, 44–49. [CrossRef] [PubMed]

34. Tyler, M.; Tumban, E.; Dziduszko, A.; Ozbun, M.A.; Peabody, D.S.; Chackerian, B. Immunization with a consensus epitope from Human Papillomavirus L2 induces antibodies that are broadly neutralizing. *Vaccine* **2014**. [CrossRef] [PubMed]

35. Lu, Y.; Chan, W.; Ko, B.Y.; VanLang, C.C.; Swartz, J.R.; Tirrell, D.A. Assessing sequence plasticity of a virus-like nanoparticle by evolution toward a versatile scaffold for vaccines and drug delivery. *Proc. Natl. Acad. Sci. USA.* **2015**, *112*, 12360–12365. [CrossRef]

36. Chackerian, B. Virus-like particles: Flexible platforms for vaccine development. *Vaccines* **2007**, *6*, 381–390. [CrossRef]

37. Vogel, M.; Diez, M.; Eisfeld, J.; Nassal, M. In vitro assembly of mosaic hepatitis B virus capsid-like particles (CLPs): Rescue into CLPs of assembly-deficient core protein fusions and FRET-suited CLPs. *FEBS Lett.* **2005**, *579*, 5211–5216. [CrossRef]

38. Pokorski, J.K.; Hovlid, M.L.; Finn, M.G. Cell Targeting with Hybrid Qβ Virus-Like Particles Displaying Epidermal Growth Factor. *ChemBioChem* **2011**, *12*, 2441–2447. [CrossRef]

39. Thrane, S.; Janitzek, C.M.; Matondo, S.; Resende, M.; Gustavsson, T.; Jongh, W.A.; De Clemmensen, S.; Roeffen, W.; Bolmer, M.V.D.V.; Gemert, G.J.V.; et al. Bacterial superglue enables easy development of efficient virus - like particle based vaccines. *J. Nanobiotechnol.* **2016**, *14*, 1–16. [CrossRef]

40. Janitzek, C.M.; Matondo, S.; Thrane, S.; Nielsen, M.A.; Kavishe, R.; Mwakalinga, S.B.; Theander, T.G.; Salanti, A.; Sander, A.F.; Gardner, M.; et al. Bacterial superglue generates a full-length circumsporozoite protein virus-like particle vaccine capable of inducing high and durable antibody responses. *Malar. J.* **2016**, *15*, 545. [CrossRef]

41. Palladini, A.; Thrane, S.; Janitzek, C.M.; Pihl, J.; Clemmensen, S.B.; Adriaan De Jongh, W.; Clausen, T.M.; Nicoletti, G.; Landuzzi, L.; Penichet, M.L.; et al. Virus-like particle display of HER2 induces potent anti-cancer responses. *Oncoimmunology* **2017**. [CrossRef] [PubMed]

42. Chackerian, B.; Lowy, D.R.; Schiller, J.T. Conjugation of a self-antigen to papillomavirus-like particles allows for efficient induction of protective autoantibodies. *J. Clin. Investig.* **2001**, *108*, 415–423. [CrossRef] [PubMed]

43. Brune, K.D.; Howarth, M. New Routes and Opportunities for Modular Construction of Particulate Vaccines: Stick, Click, and Glue. *Front. Immunol.* **2018**, *9*, 1–15. [CrossRef] [PubMed]

44. Tissot, A.C.; Renhofa, R.; Schmitz, N.; Cielens, I.; Meijerink, E.; Ose, V.; Jennings, G.T.; Saudan, P.; Pumpens, P.; Bachmann, M.F.; et al. Versatile Virus-Like Particle Carrier for Epitope Based Vaccines. *PLoS ONE* **2010**, *5*, e9809. [CrossRef]

45. Spohn, G.; Schori, C.; Keller, I.; Sladko, K.; Sina, C.; Guler, R.; Schwarz, K.; Johansen, P.; Jennings, G.T.; Bachmann, M.F. Preclinical efficacy and safety of an anti-IL-1β vaccine for the treatment of type 2 diabetes. *Mol. Ther. Methods Clin. Dev.* **2014**, *1*, 1–11. [CrossRef] [PubMed]

46. Jegerlehner, A.; Tissot, A.; Lechner, F.; Sebbel, P.; Erdmann, I.; Kündig, T.; Bächi, T.; Storni, T.; Jennings, G.; Pumpens, P.; et al. A molecular assembly system that renders antigens of choice highly repetitive for induction of protective B cell responses. *Vaccine* **2002**, *20*, 3104–3112. [CrossRef]

47. Yin, Z.; Comellas-Aragones, M.; Chowdhury, S.; Bentley, P.; Kaczanowska, K.; BenMohamed, L.; Gildersleeve, J.C.; Finn, M.G.; Huang, X. Boosting Immunity to Small Tumor-Associated Carbohydrates with Bacteriophage Qβ Capsids. *ACS Chem. Biol.* **2013**, *8*, 1253–1262. [CrossRef]

48. Patel, K.G.; Swartz, J.R. Surface functionalization of virus-like particles by direct conjugation using azide-alkyne click chemistry. *Bioconjug. Chem.* **2011**, *22*, 376–387. [CrossRef]

49. Smith, M.L.; Lindbo, J.A.; Dillard-Telm, S.; Brosio, P.M.; Lasnik, A.B.; McCormick, A.A.; Nguyen, L.V.; Palmer, K.E. Modified Tobacco mosaic virus particles as scaffolds for display of protein antigens for vaccine applications. *Virology* **2006**, *348*, 475–488. [CrossRef]

50. Koho, T.; Ihalainen, T.O.; Stark, M.; Uusi-Kerttula, H.; Wieneke, R.; Rahikainen, R.; Blazevic, V.; Marjomäki, V.; Tampé, R.; Kulomaa, M.S.; et al. His-tagged norovirus-like particles: A versatile platform for cellular delivery and surface display. *Eur. J. Pharm. Biopharm.* **2015**, *96*, 22–31. [CrossRef]

51. Brune, K.D.; Leneghan, D.B.; Brian, I.J.; Ishizuka, A.S.; Bachmann, M.F.; Draper, S.J.; Biswas, S.; Howarth, M.; Sapsford, K.E.; Gregorio, E.D.; et al. Plug-and-Display: Decoration of Virus-Like Particles via isopeptide bonds for modular immunization. *Sci. Rep.* **2016**, *6*, 1–13. [CrossRef] [PubMed]

52. Doucet, M.; El-Turabi, A.; Zabel, F.; Hunn, B.H.M.; Bengoa-Vergniory, N.; Cioroch, M.; Ramm, M.; Smith, A.M.; Gomes, A.C.; Cabral de Miranda, G.; et al. Preclinical development of a vaccine against oligomeric alpha-synuclein based on virus-like particles. *PLoS ONE* **2017**, *12*, 1–25. [CrossRef] [PubMed]

53. Cornuz, J.; Zwahlen, S.; Jungi, W.F.; Osterwalder, J.; Klingler, K.; van Melle, G.; Bangala, Y.; Guessous, I.; Müller, P.; Willers, J.; et al. A Vaccine against Nicotine for Smoking Cessation: A Randomized Controlled Trial. *PLoS ONE* **2008**, *3*, e2547. [CrossRef] [PubMed]

54. Stephanopoulos, N.; Francis, M.B. Choosing an effective protein bioconjugation strategy. *Nat. Chem. Biol.* **2011**, *7*, 876–884. [CrossRef] [PubMed]

55. Leneghan, D.B.; Miura, K.; Taylor, I.J.; Li, Y.; Jin, J.; Brune, K.D.; Bachmann, M.F.; Howarth, M.; Long, C.A.; Biswas, S. Nanoassembly routes stimulate conflicting antibody quantity and quality for transmission-blocking malaria vaccines. *Sci. Rep.* **2017**, *7*, 3811. [CrossRef] [PubMed]

56. Spohn, G.; Keller, I.; Beck, M.; Grest, P.; Jennings, G.T.; Bachmann, M.F. Active immunization with IL-1 displayed on virus-like particles protects from autoimmune arthritis. *Eur. J. Immunol.* **2008**, *38*, 877–887. [CrossRef]

57. Lu, Y.; Welsh, J.P.; Chan, W.; Swartz, J.R. Escherichia coli-based cell free production of flagellin and ordered flagellin display on virus-like particles. *Biotechnol. Bioeng.* **2013**, *110*, 2073–2085. [CrossRef]

58. Nwe, K.; Brechbiel, M.W. Growing applications of "click chemistry" for bioconjugation in contemporary biomedical research. *Cancer Biother. Radiopharm.* **2009**, *24*, 289–302. [CrossRef]

59. Akhras, S.; Toda, M.; Boller, K.; Himmelsbach, K.; Elgner, F.; Biehl, M.; Scheurer, S.; Gratz, M.; Vieths, S.; Hildt, E. Cell-permeable capsids as universal antigen carrier for the induction of an antigen-specific. *Sci. Rep.* **2017**, *7*, 1–16. [CrossRef]

60. Thrane, S.; Janitzek, C.M.; Agerbæk, M.Ø.; Ditlev, S.B.; Resende, M.; Nielsen, M.A.; Theander, T.G.; Salanti, A.; Sander, A.F.; Brabin, B.; et al. A Novel Virus-Like Particle Based Vaccine Platform Displaying the Placental Malaria Antigen VAR2CSA. *PLoS ONE* **2015**, *10*, 1–16. [CrossRef]

61. Zakeri, B.; Fierer, J.O.; Celik, E.; Chittock, E.C.; Schwarz-Linek, U.; Moy, V.T.; Howarth, M. Peptide tag forming a rapid covalent bond to a protein, through engineering a bacterial adhesin. *PNAS* **2012**, *109*, E690–E697. [CrossRef] [PubMed]

62. Li, L.; Fierer, J.O.; Rapoport, T.A.; Howarth, M. Structural Analysis and Optimization of the Covalent Association between SpyCatcher and a Peptide Tag. *J. Mol. Biol.* **2014**, *426*, 309–317. [CrossRef] [PubMed]

63. Veggiani, G.; Nakamura, T.; Brenner, M.D.; Gayet, R.V.; Yan, J.; Robinson, C.V.; Howarth, M. Programmable polyproteams built using twin peptide superglues. *Proc. Natl. Acad. Sci. USA.* **2016**, *113*, 1202–1207. [CrossRef] [PubMed]

64. Tan, L.L.; Hoon, S.S.; Wong, F.T. Kinetic Controlled Tag-Catcher Interactions for Directed Covalent Protein Assembly. *PLoS ONE* **2016**, *11*, 1–15. [CrossRef]

65. Fierer, J.O.; Veggiani, G.; Howarth, M. SpyLigase peptide-peptide ligation polymerizes affibodies to enhance magnetic cancer cell capture. *Proc. Natl. Acad. Sci. USA.* **2014**, *111*, E1176–E1181. [CrossRef]

66. Buldun, C.M.; Jean, J.X.; Bedford, M.R.; Howarth, M. SnoopLigase Catalyzes Peptide–Peptide Locking and Enables Solid-Phase Conjugate Isolation. *J. Am. Chem. Soc.* **2018**, *140*, 3008–3018. [CrossRef]

67. Khairil Anuar, I.N.A.; Banerjee, A.; Keeble, A.H.; Carella, A.; Nikov, G.I.; Howarth, M. Spy&Go purification of SpyTag-proteins using pseudo-SpyCatcher to access an oligomerization toolbox. *Nat. Commun.* **2019**, *10*, 1–13.

68. Pröschel, M.; Kraner, M.E.; Horn, A.H.C.; Schä Fer, L.; Sonnewald, U.; Sticht, H. Probing the potential of CnaB-type domains for the design of tag/catcher systems. *PLoS ONE* **2017**, *12*, 1–26. [CrossRef]

69. Wu, X.-L.; Liu, Y.; Liu, D.; Sun, F.; Zhang, W.-B. An Intrinsically Disordered Peptide-Peptide Stapler for Highly Efficient Protein Ligation Both *in Vivo* and *in Vitro*. *J. Am. Chem. Soc.* **2018**, *140*, 17474–17483. [CrossRef]

70. Keeble, A.H.; Turkki, P.; Stokes, S.; Khairil Anuar, I.N.A.; Rahikainen, R.; Hytönen, V.P.; Howarth, M. Approaching infinite affinity through engineering of peptide–protein interaction. *PNAS* **2019**, 1–11. [CrossRef]

71. Govasli, M.L.; Diaz, Y.; Puntervoll, P. Virus-like particle-display of the enterotoxigenic Escherichia coli heat-stable toxoid STh-A14T elicits neutralizing antibodies in mice. *Vaccine* **2019**, *37*, 6405–6414. [CrossRef] [PubMed]

72. Escolano, A.; Gristick, H.B.; Abernathy, M.E.; Merkenschlager, J.; Gautam, R.; Oliveira, T.Y.; Pai, J.; West, A.P.; Barnes, C.O.; Cohen, A.A.; et al. Immunization expands B cells specific to HIV-1 V3 glycan in mice and macaques. *Nature* **2019**, *570*, 468–473. [CrossRef] [PubMed]

73. E van den Worm, S.H.; Koning, R.I.; Warmenhoven, H.J.; Koerten, H.K.; van Duin, J. Cryo Electron Microscopy Reconstructions of the Leviviridae Unveil the Densest Icosahedral RNA Packing Possible. *J. Mol. Biol.* **2006**, *263*, 858–865. [CrossRef] [PubMed]

74. Shishovs, M.; Rumnieks, J.; Diebolder, C.; Jaudzems, K.; Andreas, L.B.; Stanek, J.; Kazaks, A.; Kotelovica, S.; Akopjana, I.; Pintacuda, G.; et al. Structure of AP205 Coat Protein Reveals Circular Permutation in ssRNA Bacteriophages. *J. Mol. Biol.* **2016**, *428*, 4267–4279. [CrossRef] [PubMed]

75. Pumpens, P.; Renhofa, R.; Dishlers, A.; Kozlovska, T.; Ose, V.; Pushko, P.; Tars, K.; Grens, E.; Bachmann, M.F. The True Story and Advantages of RNA Phage Capsids as Nanotools. *Intervirology* **2016**, *59*, 74–110. [CrossRef] [PubMed]

76. Bruun, T.U.J.; Andersson, A.M.C.; Draper, S.J.; Howarth, M. Engineering a Rugged Nanoscaffold to Enhance Plug-and-Display Vaccination. *ACS Nano* **2018**, *12*, 8855–8866. [CrossRef] [PubMed]

77. Marini, A.; Zhou, Y.; Li, Y.; Taylor, I.J.; Leneghan, D.B.; Jin, J.; Zaric, M.; Mekhaiel, D.; Long, C.; Miura, K.; et al. A universal Plug-and-Display vaccine carrier based on HBsAg VLP to maximize effective antibody response. *Front. Immunol.* **2019**, *10*, 1–12. [CrossRef]

78. Wang, C.; Zhu, W.; Wang, B.Z. Dual-linker gold nanoparticles as adjuvanting carriers for multivalent display of recombinant influenza hemagglutinin trimers and flagellin improve the immunological responses in vivo and in vitro. *Int. J. Nanomed.* **2017**, *12*, 4747–4762. [CrossRef]

79. Janitzek, C.M.; Peabody, J.; Thrane, S.H.R.; Carlsen, P.G.; Theander, T.; Salanti, A.; Chackerian, B.A.; Nielsen, M.; Sander, A.F. A proof-of-concept study for the design of a VLP-based combinatorial HPV and placental malaria vaccine. *Sci. Rep.* **2019**, *9*, 1–10. [CrossRef]

80. Bachmann, M.F.; Hoffmann Rohrer, U.; Kündig, T.M.; Bürki, K.; Hengartner, H.; Zinkernagel, R.M. The Influence of Antigen Organization on B Cell Responsiveness. *Science* **1993**, *262*, 1448–1451. [CrossRef]

81. Zhang, W.B.; Sun, F.; Tirrell, D.A.; Arnold, F.H. Controlling macromolecular topology with genetically encoded SpyTag-SpyCatcher chemistry. *J. Am. Chem. Soc.* **2013**, *135*, 13988–13997. [CrossRef] [PubMed]

82. Kasaraneni, N.; Chamoun-Emanuelli, A.M.; Wright, G.; Chen, Z. Retargeting lentiviruses via spyCatcher-spyTag chemistry for gene delivery into specific cell types. *MBio* **2017**, *8*, 1–12. [CrossRef]

83. Krammer, F. Strategies to induce broadly protective antibody responses to viral glycoproteins. *Expert Rev. Vaccines* **2017**, *16*, 503–513. [CrossRef] [PubMed]

84. Sattentau, Q.J.; Moore, J.P. Human immunodeficiency virus type 1 neutralization is determined by epitope exposure on the gp120 oligomer. *J. Exp. Med.* **1995**, *182*, 185–196. [CrossRef]

85. Sanders, R.W.; Moore, J.P. Native-like Env trimers as a platform for HIV-1 vaccine design. *Immunol. Rev.* **2017**, *275*, 161–182. [CrossRef] [PubMed]

86. He, L.; De Val, N.; Morris, C.D.; Vora, N.; Thinnes, T.C.; Kong, L.; Azadnia, P.; Sok, D.; Zhou, B.; Burton, D.R.; et al. Presenting native-like trimeric HIV-1 antigens with self-assembling nanoparticles. *Nat. Commun.* **2016**, *7*, 1–15. [CrossRef]

87. Zeltins, A. Construction and characterization of virus-like particles: A review. *Mol. Biotechnol.* **2013**, *53*, 92–107. [CrossRef]

88. Shirbaghaee, Z.; Bolhassani, A. Different applications of virus-like particles in biology and medicine: Vaccination and delivery systems. *Biopolymers* **2016**, *105*, 113–132. [CrossRef]

89. Mohsen, M.O.; Zha, L.; Cabral-miranda, G.; Bachmann, M.F. Major findings and recent advances in virus – like particle (VLP) -based vaccines. *Semin. Immunol.* **2017**, 1–10. [CrossRef]

90. Liekniņa, I.; Kalniņš, G.; Akopjana, I.; Bogans, J.; Šišovs, M.; Jansons, J.; Rūmnieks, J.; Tārs, K. Production and characterization of novel ssRNA bacteriophage virus-like particles from metagenomic sequencing data. *J. Nanobiotechnol.* **2019**, *17*, 61. [CrossRef]

91. Huang, X.; Wang, X.; Zhang, J.; Xia, N.; Zhao, Q. Escherichia coli-derived virus-like particles in vaccine development. *Npj Vaccines* **2017**, *2*, 1–9. [CrossRef] [PubMed]

92. Abdoli, A.; Soleimanjahi, H.; Fotouhi, F.; Teimoori, A.; Pour Beiranvand, S.; Kianmehr, Z. Human Papillomavirus Type16-L1 VLP Production in Insect Cells. *Iran. J. Basic Med. Sci.* **2013**, *16*, 891–895. [PubMed]

93. Biemelt, S.; Sonnewald, U.; Galmbacher, P.; Willmitzer, L.; Muller, M. Production of Human Papillomavirus Type 16 Virus-Like Particles in Transgenic Plants. *J. Virol.* **2003**, *77*, 9211–9220. [CrossRef] [PubMed]

94. Dennis, S.J.; O'Kennedy, M.M.; Rutkowska, D.; Tsekoa, T.; Lourens, C.W.; Hitzeroth, I.I.; Meyers, A.E.; Rybicki, E.P. Safety and immunogenicity of plant-produced African horse sickness virus-like particles in horses. *Vet. Res.* **2018**, *49*, 1–6. [CrossRef]

95. Chen, Q.; Lai, H. Plant-derived virus-like particles as vaccines. *Hum. Vaccines Immunother.* **2013**, *9*, 26–49. [CrossRef] [PubMed]

96. Rybicki, E.P. Plant molecular farming of virus-like nanoparticles as vaccines and reagents. *Wiley Interdiscip. Rev. Nanomed. Nanobiotechnol.* **2019**, *1587*, 1–22. [CrossRef]

97. Gomes, A.C.; Roesti, E.S.; El-Turabi, A.; Bachmann, M.F. Type of RNA packed in VLPs impacts IgG class switching—implications for an influenza vaccine design. *Vaccines* **2019**, *7*, 47. [CrossRef]

98. Lenz, P.; Day, P.M.; Pang, Y.-Y.S.; Frye, S.A.; Jensen, P.N.; Lowy, D.R.; Schiller, J.T. Papillomavirus-Like Particles Induce Acute Activation of Dendritic Cells. *J. Immunol.* **2001**, *166*, 5346–5355. [CrossRef]

99. Ruedl, C.; Storni, T.; Lechner, F.; Bächi, T.; Bachmann, M.F. Cross-presentation of virus-like particles by skin-derived CD8– dendritic cells: A dispensable role for TAP. *Eur. J. Immunol.* **2002**, *32*, 818–825. [CrossRef]

100. Gregson, A.L.; Oliveira, G.; Othoro, C.; Calvo-Calle, J.M.; Thorton, G.B.; Nardin, E.; Edelman, R. Phase I Trial of an Alhydrogel Adjuvanted Hepatitis B Core Virus-Like Particle Containing Epitopes of Plasmodium falciparum Circumsporozoite Protein. *PLoS ONE* **2008**, *3*, 1–9. [CrossRef]

101. Storni, T.; Ruedl, C.; Schwarz, K.; Schwendener, R.A.; Renner, W.A.; Bachmann, M.F. Nonmethylated CG Motifs Packaged into Virus-Like Particles Induce Protective Cytotoxic T Cell Responses in the Absence of Systemic Side Effects. *J. Immunol.* **2004**, *172*, 1777–1785. [CrossRef] [PubMed]

102. Bachmann, M.F.; Dyer, M.R. Therapeutic vaccination for chronic diseases: A new class of drugs in sight. *Nat. Rev. Drug Discov.* **2004**, *3*, 81–88. [CrossRef] [PubMed]

103. Gilman, S.; Koller, M.; Black, R.S.; Jenkins, L.; Griffith, S.G.; Fox, N.C.; Eisner, L.; Kirby, L.; Boada Rovira, M.; Forette, F.; et al. Clinical effects of Aβ immunization (AN1792) in patients with AD in an interrupted trial. *Neurology* **2005**, *64*, 1553–1562. [CrossRef] [PubMed]

104. Jegerlehner, A.; Wiesel, M.; Dietmeier, K.; Zabel, F.; Gatto, D.; Saudan, P.; Bachmann, M.F. Carrier induced epitopic suppression of antibody responses induced by virus-like particles is a dynamic phenomenon caused by carrier-specific antibodies. *Vaccine* **2010**, *28*, 5503–5512. [CrossRef]

105. De Filette, M.; Martens, W.; Smet, A.; Schotsaert, M.; Birkett, A.; Londoño-Arcila, P.; Fiers, W.; Saelens, X. Universal influenza A M2e-HBc vaccine protects against disease even in the presence of pre-existing anti-HBc antibodies. *Vaccine* **2008**, *26*, 6503–6507. [CrossRef]

106. Albanese, A.; Tang, P.S.; Chan, W.C.W. The Effect of Nanoparticle Size, Shape, and Surface Chemistry on Biological Systems. *Annu. Rev. Biomed. Eng.* **2012**, *14*, 1–16. [CrossRef]

107. Oyewumi, M.O.; Kumar, A.; Cui, Z. Nano-microparticles as immune adjuvants: Correlating particle sizes and the resultant immune responses. *Expert Rev. Vaccines* **2010**, *9*, 1095–1107. [CrossRef]

108. Li, X.; Sloat, B.R.; Yanasarn, N.; Cui, Z. Relationship between the size of nanoparticles and their adjuvant activity: Data from a study with an improved experimental design. *Eur. J. Pharm. Biopharm.* **2011**, *78*, 107–116. [CrossRef]

109. Hong, S.; Zhang, Z.; Liu, H.; Tang, H.; Hua, Z.; Correspondence, B.H. B Cells Are the Dominant Antigen-Presenting Cells that Activate Naive CD4 + T Cells upon Immunization with a Virus-Derived Nanoparticle Antigen. *Immunity* **2018**, *49*, 695–708. [CrossRef]

110. Wang, Z.; Gao, H.; Zhang, Y.; Liu, G.; Niu, G.; Chen, X. Functional ferritin nanoparticles for biomedical applications. *Front. Chem. Sci. Eng.* **2017**, *11*, 633–646. [CrossRef]

111. Dintzis, H.M.; Dintzis, R.Z.; Vogelstein, B. Molecular determinants of immunogenicity: The immunon model of immune response. *Proc. Natl. Acad. Sci. USA.* **1976**, *73*, 3671–3675. [CrossRef] [PubMed]

112. Chackerian, B.; Lenz, P.; Lowy, D.R.; Schiller, J.T. Determinants of Autoantibody Induction by Conjugated Papillomavirus Virus-Like Particles. *J. Immunol.* **2002**, *169*, 6120–6126. [CrossRef] [PubMed]

113. Chen, Z.; Wholey, W.-Y.; Hassani Najafabadi, A.; Moon, J.J.; Grigorova, I.; Chackerian, B.; Cheng, W. Self-Antigens Displayed on Liposomal Nanoparticles above a Threshold of Epitope Density Elicit Class-Switched Autoreactive Antibodies Independent of T Cell Help. *J. Immunol.* **2019**, *204*, 1–13. [CrossRef] [PubMed]

114. Dintzis, R.Z.; Vogelstein, B.; Dintzis, H.M. Specific cellular stimulation in the primary immune response: Experimental test of a quantized model. *Proc. Natl. Acad. Sci. USA.* **1982**, *79*, 884–888. [CrossRef] [PubMed]

115. Jegerlehner, A.; Storni, T.; Lipowsky, G.; Schmid, M.; Pumpens, P.; Bachmann, M.F. Regulation of IgG antibody responses by epitope density and CD21-mediated costimulation. *Eur. J. Immunol.* **2002**, *32*, 3305–3314. [CrossRef]

116. Brewer, M.G.; DiPiazza, A.; Acklin, J.; Feng, C.; Sant, A.J.; Dewhurst, S. Nanoparticles decorated with viral antigens are more immunogenic at low surface density. *Vaccine* **2017**, *35*, 774–781. [CrossRef]

117. Honorati, M.C.; Facchini, A. Immune response against HBsAg vaccine. *World J. Gastroenterol.* **1998**, *4*, 464–466. [CrossRef]

118. RTS, S.C.T.P. Efficacy and safety of RTS,S/AS01 malaria vaccine with or without a booster dose in infants and children in Africa: Final results of a phase 3, individually randomised, controlled trial. *Lancet* **2015**, *386*, 31–45.

119. Tokatlian, T.; Kulp, D.W.; Mutafyan, A.A.; Jones, C.A.; Menis, S.; Georgeson, E.; Kubitz, M.; Zhang, M.H.; Melo, M.B.; Silva, M.; et al. Enhancing Humoral Responses Against HIV Envelope Trimers via Nanoparticle Delivery with Stabilized Synthetic Liposomes. *Sci. Rep.* **2018**, *8*, 1–12. [CrossRef]

120. Wei, S.; Lei, Y.; Yang, J.; Wang, X.; Shu, F.; Wei, X.; Lin, F.; Li, B.; Cui, Y.; Zhang, H.; et al. Neutralization effects of antibody elicited by chimeric HBV S antigen viral-like particles presenting HCV neutralization epitopes. *Vaccine* **2018**, *36*, 2273–2281. [CrossRef]

121. Chackerian, B.; Peabody, D.S. Factors That Govern the Induction of Long-Lived Antibody Responses. *Viruses* **2020**, *12*, 74. [CrossRef] [PubMed]

122. Poland, G.A.; Jacobson, R.M. Prevention of Hepatitis B with the Hepatitis B Vaccine. *N. Engl. J. Med.* **2004**, *351*, 2832–2838. [CrossRef] [PubMed]

123. Barry, M.; Cooper, C. Review of hepatitis B surface antigen-1018 ISS adjuvant-containing vaccine safety and efficacy. *Expert Opin. Biol. Ther.* **2007**, *7*, 1731–1737. [CrossRef] [PubMed]

124. Kurtovic, L.; Agius, P.A.; Feng, G.; Drew, D.R.; Ubillos, I.; Sacarlal, J.; Aponte, J.J.; Fowkes, F.J.I.; Dobaño, C.; Beeson, J.G. Induction and decay of functional complement-fixing antibodies by the RTS,S malaria vaccine in children, and a negative impact of malaria exposure. *BMC Med.* **2019**, *17*, 1–14. [CrossRef]

125. Guinovart, C.; Aponte, J.J.; Sacarlal, J.; Aide, P.; Leach, A.; Bassat, Q.; Macete, E.; Dobaño, C.; Lievens, M.; Loucq, C.; et al. Insights into Long-Lasting Protection Induced by RTS,S/AS02A Malaria Vaccine: Further Results from a Phase IIb Trial in Mozambican Children. *PLoS ONE* **2009**, *4*, 1–8. [CrossRef]

126. David, M.P.; Van Herck, K.; Hardt, K.; Tibaldi, F.; Dubin, G.; Descamps, D.; Van Damme, P. Long-term persistence of anti-HPV-16 and -18 antibodies induced by vaccination with the AS04-adjuvanted cervical cancer vaccine: Modeling of sustained antibody responses. *Gynecol. Oncol.* **2009**, *115*, 1–6. [CrossRef]

127. Schiller, J.T.; Lowy, D.R. Raising expectations for subunit vaccine. *J. Infect. Dis.* **2015**, *211*, 1373–1375. [CrossRef]

 *viruses*

Review

Hepatitis B Virus (HBV) Subviral Particles as Protective Vaccines and Vaccine Platforms

Joan Kha-Tu Ho [1], Beena Jeevan-Raj [1] and Hans-Jürgen Netter [1,2,*]

[1] Victorian Infectious Diseases Reference Laboratory (VIDRL), Melbourne Health, The Peter Doherty Institute, Melbourne, Victoria 3000, Australia; Joan.Ho@mh.org.au (J.K.-T.H.); Beena.JeevanRaj@mh.org.au (B.J.-R.)
[2] Royal Melbourne Institute of Technology (RMIT) University, School of Science, Melbourne, Victoria 3001, Australia
* Correspondence: Hans.Netter@mh.org.au

Received: 19 December 2019; Accepted: 13 January 2020; Published: 21 January 2020

Abstract: Hepatitis B remains one of the major global health problems more than 40 years after the identification of human hepatitis B virus (HBV) as the causative agent. A critical turning point in combating this virus was the development of a preventative vaccine composed of the HBV surface (envelope) protein (HBsAg) to reduce the risk of new infections. The isolation of HBsAg sub-viral particles (SVPs) from the blood of asymptomatic HBV carriers as antigens for the first-generation vaccines, followed by the development of recombinant HBsAg SVPs produced in yeast as the antigenic components of the second-generation vaccines, represent landmark advancements in biotechnology and medicine. The ability of the HBsAg SVPs to accept and present foreign antigenic sequences provides the basis of a chimeric particulate delivery platform, and resulted in the development of a vaccine against malaria (RTS,S/AS01, Mosquirix™), and various preclinical vaccine candidates to overcome infectious diseases for which there are no effective vaccines. Biomedical modifications of the HBsAg subunits allowed the identification of strategies to enhance the HBsAg SVP immunogenicity to build potent vaccines for preventative and possibly therapeutic applications. The review provides an overview of the formation and assembly of the HBsAg SVPs and highlights the utilization of the particles in key effective vaccines.

Keywords: hepatitis B virus; surface (envelope) antigen; sub-viral particle; virus-like particle

1. Introduction

Hepatitis B is globally one of the most common infectious diseases in humans, which is associated with significant morbidity and mortality. Approximately 2 billion people worldwide have been infected with hepatitis B virus (HBV) and approximately 257 million people live with chronic HBV infections. An estimated 887,000 persons died in 2015 from acute or chronic consequences of hepatitis B [1–4]. The ability of the HBV structural proteins, including the hepatitis B surface (envelope) proteins (HBsAg) to assemble into non-infectious sub-viral particles (SVPs), allows the generation of highly organized particles displaying neutralizing epitopes that promote protective immune responses against the parent virus. The approval of the recombinant hepatitis B vaccine Recombivax HB (Merck Sharp and Dohme) in 1986, based on HBsAg SVPs and produced in the yeast *Saccharomyces cerevisiae*, was the first developed vaccine using recombinant DNA technology. The recombinant vaccine, together with the recombinant products, human insulin (licensed 1982), human growth hormone (licensed 1985), and alpha interferon (licensed 1986), demonstrated the capability of biotechnological approaches to generate innovative medicines [1]. The ability to accept foreign antigenic sequences into the SVP structure can provide the basis for the development of delivery platforms for targeted medically relevant sequences, as in the case of the RTS,S/AS01 (Mosquirix™) vaccine against malaria. The antigenic components of Mosquirix™ are chimeric SVPs containing HBsAg proteins fused to a *Plasmodium falciparum*-specific circumsporozoite (CS) polypeptide [5–7]. The design and generation of chimeric SVPs holds enormous potential in the treatment of infectious diseases, for which there are no effective vaccines [8–10].

2. Hepatitis B Virus, Classification, and Gene Products

HBV is a hepatocyte-tropic virus and is assigned to the family of hepatitis DNA viruses, *Hepadnaviridae* [11–13]. HBV is divided into 10 main genotypes, A–J, which differ by more than 8% at the nucleotide level [14,15]. The HBV genome has a size of approximately 3.2 kilobases (kb), and is represented by a relaxed circular, partially double-stranded DNA (rcDNA), which is delivered to the nucleus of the host cell and converted into a covalently closed circular DNA (cccDNA) molecule [12]. The cccDNA represents a non-integrated stable episome and forms the template for all viral RNA transcripts. In the absence of an origin of replication site required for DNA-dependent DNA amplification, one of the viral transcripts, the pre-genomic RNA (pgRNA), serves as the template for replication to generate rcDNA via reverse transcription [12]. HBV contains four open reading frames (C, P, S, and X) and encodes seven proteins (polymerase, X protein, HBcAg, HBeAg, HBsAgL, HBsAgM, and HBsAgS). The polymerase is essential for several steps in the replication pathway through its reverse transcriptase, RNaseH, and priming activities. The X protein supports efficient infection and replication in vivo [11,12,16]. The core protein (HBcAg) constitutes the subunit of the viral capsid and is essential for the formation of virions. The e-antigen (HBeAg) is derived from the pre-core protein by proteolytic processing and is not part of the viral capsid. It is involved in modulating the host immune response against HBV and represents an important serological marker [11–13]. The virus encodes for three related surface (envelope) proteins (HBsAg) that share a common S-domain. They are translated from different in-frame start codons and hence are distinguished by their N-terminal extensions. The small HBsAg (HBsAgS) comprises only the S-domain with a size of 226 amino acids (aa), the middle HBsAg protein (HBsAgM) has an N-terminal extension of 55 aa (pre-S2 domain), and the large HBsAg (HBsAgL) has an additional extension of 108 or 119 aa (preS1-domain) depending on the genotype [17] (Figure 1A,B). In addition to the classification by genotypes, HBV is distinguished by four main serotypes based on the reactivity against HBsAg. All genotypes have a common serotypic reactivity against a major antigenic site called the "a"-determinant, but further express two mutually exclusive allelic antigenic determinants "d" or "y" and "w" or "r" [18–20]. The antigenic determinants of HBsAg are located in an exposed loop region of the S-domain. HBsAg and antibodies against HBsAg (anti-HBs) are important serological markers. The loss of HBsAg and seroconversion to anti-HBs antibodies are a sign of immunity and recovery from acute or chronic hepatitis B [13].

Characteristic of a HBV infection is the generation of a large quantity of HBsAg SVPs and filaments devoid of capsid and of the viral genome. SVPs exceed the presence of infectious virions in host sera by a factor between 10^2 and 10^5 [17,21–24]. SVPs are predominately composed of HBsAgS, and their presence in the sera does not seem to interfere with HBV particle entry into hepatocytes, suggesting that SVPs represent decoys by binding to virus-neutralizing antibodies [25]. HBsAgS SVPs share important immunological determinants with the mature virus, and therefore, SVPs derived from patient serum or recombinant SVPs represent effective immunogens for the induction of a protective immune response [26–28]. Vaccinated individuals develop antibodies targeting the "a"-determinant region, which provides protection against the infection of all HBV serotypes [20,29]. The discovery and characterization of "empty" genome-free virions containing HBsAg and capsid is reviewed by Hu and Liu, 2017 [30].

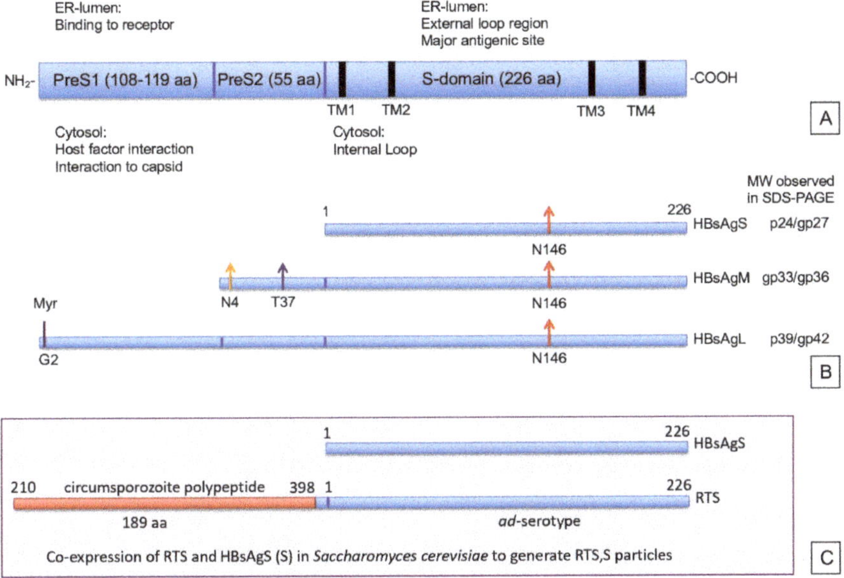

Figure 1. The surface (envelope) proteins (HBsAg) of hepatitis B virus (HBV): (**A**) The open reading frame encoding the complete hepatitis B surface antigen is depicted. The domain organization of preS1, preS2, and S, with the number of amino acids (aa) of the individual domains are specified. The four transmembrane regions (TM1–4) are indicated by the thick black lines. The function of the different domains in relation to their orientation towards the lumen of the endoplasmic reticulum (ER) or cytosol is indicated. (**B**) The individual HBsAg open reading frames for the small (HBsAgS), middle (HBsAgM), and large (HBsAgL) proteins, and their post-translational modifications are shown. The size of the HBsAgS protein is indicated by the number of amino acids. Arrows represent the utilized glycosylation sites. Red arrows mark asparagine 146 (N146) in the S-domain. Orange and purple arrows represent the N-4 and threonine 37 (T37) respectively, in the preS2 domain. The glycine residue at position 2 (G2) of the preS1 domain, indicated by a purple line, is myristolated. The observed molecular weights (MW) of the glycosylated (gp) and non-glycosylated proteins (p) separated under reducing conditions on a SDS-PAGE are indicated on the right. (**C**) Design of the RTS,S vaccine that is produced by co-expressing HBsAgS (aa 1–226) and the chimeric protein that is a fusion of the circumsporozoite polypeptide (210–398 aa) to the N-terminus of HBsAgS, including 4 aa from the preS2 domain.

3. Role of HBsAg in HBV, Filament, and SVP Formation

An essential step in the formation of virions, filaments, or SVPs is the cotranslational insertion of the HBsAgS protein into the membrane of the endoplasmic reticulum (ER) with a short luminal exposed N-terminal sequence, two transmembrane regions separated by a cytosolic loop, and a luminal domain, followed by a hydrophobic C-terminal region. The luminal domain corresponds to the external loop region of the S-domain, which contains multiple epitopes, including the immunodominant "a"-determinant region that is common to all HBV genotypes, and the allelic antigenic determinants "d/y" and "w/r" [17,31]. A conformational heparan sulfate binding site also overlaps with the "a"-determinant region and is essential to infectivity [32].

HBV virion formation depends on the presence of the viral capsid containing rcDNA and the HBsAg proteins for envelopment. The preS1 region of HBsAgL is essential for the assembly of infectious HBV particles by interacting with the capsid [17,33], possibly contributes to the binding to a proteoglycan attachment site [34,35], and is required for binding to the hepatocyte entry receptor, sodium taurochlorate cotransporting polypeptide (NTCP) [36,37]. Chaperones of the heat shock protein (hsp) 70 family facilitate the interaction of the HBsAgL preS1 domain with the capsid by retaining the preS1/preS2 sequence in the cytosolic (internal) orientation at the ER [38–40]. Virion secretion depends on host factors of the endosomal sorting complex required for transport (ESCRT) and sorting into late endosomal multivesicular bodies (MVBs), and finally, release of its intraluminal content at the hepatocyte surface [41–43]. The mature HBV virions are spherical, enveloped particles with a diameter of 42 nm with an inner capsid of 22 nm in diameter [11–13]. Due to the elevated presence of HBsAgL in virions and filaments compared to the SVPs, filaments seem to follow the secretion pathway taken by the virions (Figure 2) [44]. The filaments have a width of approximately 20 nm and are variable in length [12,23,45].

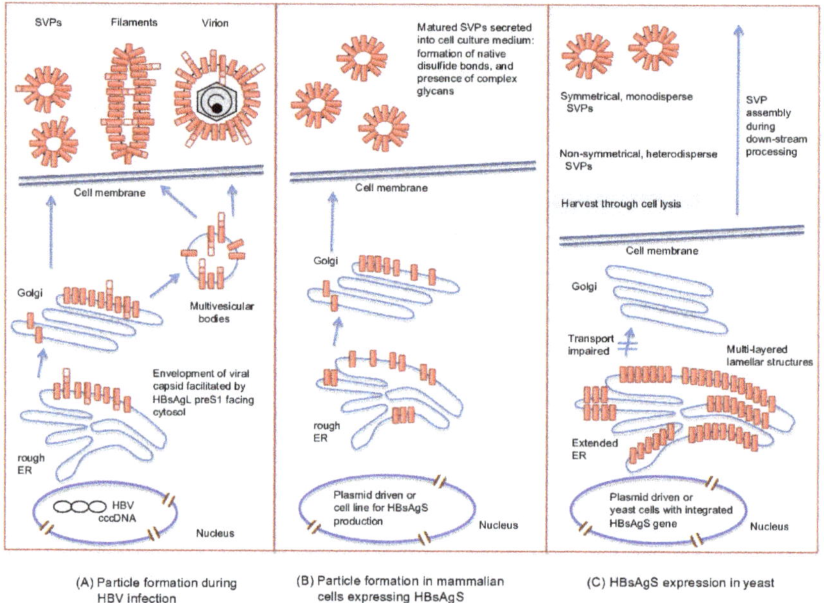

Figure 2. Illustration of HBsAg protein synthesis and assembly pathways during a natural infection (**A**), HBsAgS expression and assembly in mammalian cells in the absence of other viral gene products (**B**), or expressed in yeast (**C**). During a natural infection, virions and filaments are formed by budding from multivesicular bodies. The spherical subviral particles (SVPs) are produced and secreted through the endoplasmic reticulum (ER)-Golgi complex. The HBsAg subunits of the virions, filaments, and SVPs form intra- and intermolecular disulfide bonds, and are partially glycosylated (not shown) (**A**). Expression of the HBsAgS gene in mammalian cell lines leads to the formation of SVPs exclusively composed of HBsAgS permitting the formation of disulfide bonds and partial glycosylation (not shown) (**B**). HBsAgS protein expressed in yeast accumulates in the ER, causes an extended ER and forms multilayered lamellar structures. HBsAgS protein complexes are harvested from the cell lysate, and SVPs are assembled during down-stream processing (**C**). cccDNA: covalently closed circular DNA.

In contrast to the formation of infectious particles, the formation of SVP does not depend on the preS1 domain. HBsAgS proteins have the ability to assemble into secretion-competent SVPs and are composed solely of envelope proteins, lipids, and glycans. Contrary to virion secretion, SVP assembly follows the constitutive secretory pathway of the host cell for release [43,44]. The non-infectious SVPs are 17 to 25 nm in diameter. SVPs isolated from HBV carriers show spike-like features protruding from the surface similar to the surface projections observed on filaments and infectious virions (Dane particles) [24,45–48]. An alternative structure for recombinant SVPs expressed in transgenic mice has been proposed to possess an octahedral symmetry [49]. Mammalian cell lines expressing HBsAgS in the absence of any other viral component secrete SVPs, which are morphologically indistinguishable from serum-derived SVPs (Figure 2) [50–54]. Thus, particle morphogenesis substantially differs between SVPs and virions, but they have identical antigenic structures due to the S-domain, which is encoded by all HBV envelope proteins (Figure 1A,B).

3.1. Topology of HBsAgS

The correct folding of HBsAgS depends on two topogenic signal sequences, which determine the orientation of the S-domains in relation to the lipid layer. The insertion of the HBsAgS N-terminus into the ER membrane requires the presence of the topogenic N-terminal signal sequence 1 (transmembrane region 1, TM1, aa 8–22), which is not proteolytically cleaved by the host's peptidases, and allows the translocation of the N-terminus across the ER membrane. The second internal topogenic transmembrane sequence (TM2), which is located between aa 80 and 98, supports the translocation of flanking C-terminal sequences and serves as an anchor to hold the sequence in the membrane [17,55,56]. Both topogenic signal sequences are required for the correct folding of HBsAgS, resulting in the formation of a cytosolic loop (aa 23 to 79), and a loop reaching into the ER lumen (aa 99 to approximately 155), followed by a proposed amphipathic helix (aa 156 to 169) and the hydrophobic C-terminal region (aa 170 to 226) embedded in the ER membrane (Figures 1A and 3) [57]. The luminal loop region of the S-domain is located at the external surface of the mature SVPs and also infectious virions, and harbors the major HBsAg protein epitopes ("a"-determinant, and "d/y", "w/r" determinants) [13]. The hydrophobic C-terminal region (aa 170 to 226) possibly contains two additional transmembrane regions, as indicated by topological models [58,59]. The presence of transmembrane passages is supported by experimental data suggesting that the C-terminal sequences are exposed at internal and external surfaces. The sequence between residues 196 and 201 is important for packaging of hepatitis delta virus (HDV), a satellite of HBV, and therefore expected to be accessible to facilitate the interaction with the HDV ribonucleoprotein complex [60]. A second site between residue 178 and 186 is targeted by an anti-HBs monoclonal antibody, indicating surface exposure [61].

HBsAg proteins contain 14 cysteine (cys) residues located in the S-domain, which are highly conserved among different HBV genotypes; the preS1 and preS2 domains do not contain additional cysteine residues [12,17]. Eight cysteine residues are located in the external loop region, forming disulfide bonds which are important for the integrity of the major antigenic determinants [62,63], such as cys-107 in the external loop for retaining the "a"-determinant specific antigenicity [64]. The identification of HBsAgS oligomers and polymers by electrophoresis under non-reducing conditions suggests that the disulfide bonding is heterogeneous, consistent with the finding that only a fraction of HBsAgS subunits are exclusively linked by disulfide bonds formed between cys-121 and cys-147 (Figure 3) [62,63,65]. Reduction or absence of intermolecular disulfide bridges interferes with the native HBsAgS antigenicity but allows SVP formation [62,65,66].

The S-domain shared by the HBsAg proteins harbors an *N*-glycosylation site at position asparagine-146 (N146), which is partially utilized by the oligosaccharyltransferase [17,67], and hence, HBsAgS proteins are synthesized as unglycosylated p24 and *N*-glycosylated gp27 versions (Figure 1B). HBsAgS p24 and gp27 have identical transmembrane topologies and dimerize without preference for a specific pairing and form heterologous dimers with HBsAgM and HBsAgL [66]. Synthesis and secretion of HBsAgS SVPs do not depend on *N*-glycosylation at position N146 in the external loop region, in contrast to the formation and release of infectious HBV particles [67–70].

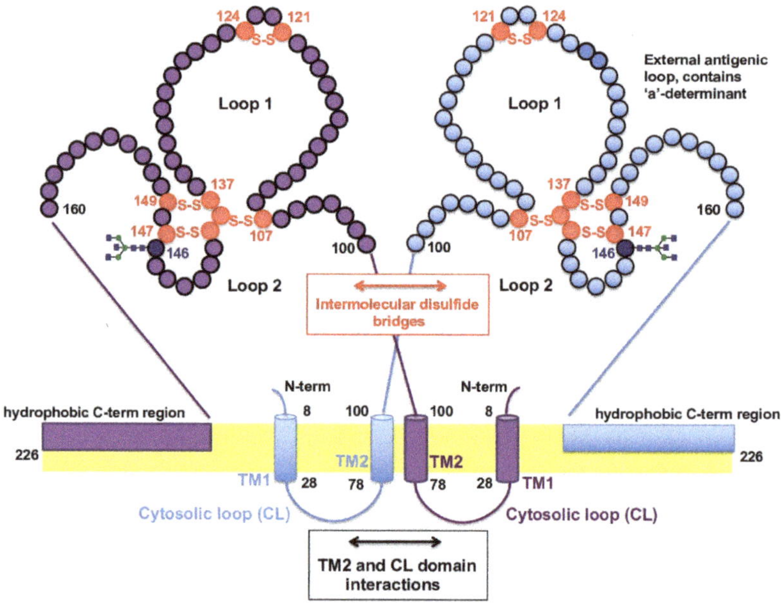

Figure 3. Proposed model for the formation of the HBsAgS homodimer based on Suffner et al. [57]. Dimerization of two HBsAgS monomers represented in purple and blue is facilitated by interactions of the transmembrane domains (TM2), cytosolic loops (CL) and intermolecular disulfide bridges (S-S) between cysteine residues (red circles) in the external loop region. The facultative *N*-glycosylation site at position N146 is indicated (dark blue). The orientation of the hydrophobic C-terminal region is illustrated in a simplified form.

3.2. Topology of HBsAgM

The middle HBsAg protein has the same transmembrane topology as HBsAgS, and hence is partly glycosylated at N146 in the S-domain. Translocation of the preS2 domain of HBsAgM is mediated by the topologic signal TM1 located in the S-domain [71]. The translocation event across the ER membrane into the lumen allows the *N*-glycosylation of the asparagine-4 (N4) residue in the preS2 domain, which is always glycosylated [72], resulting in glycoproteins with a molecular weight of 33 kDa and 36 kDa (Figure 1B). In addition to the *N*-glycosylation site in the preS2 domain, the preS2 domain can be partially *O*-glycosylated, depending on the genotype and the presence of the threonine (T37) (Figure 1B) [73]. Similar to HBsAgS, HBsAgM proteins assemble into SVPs and can be secreted independently from other viral proteins [74–77]. HBsAgM, however, is not essential for virion morphogenesis and infectivity [78].

3.3. Topology of HBsAgL

The large HBsAg protein adopts two distinguished transmembrane topologies facilitating an internal or external location of the preS1/S2 domain (Figure 1A) [17,79–81]. The internal orientation allows preS1/S2 to interact with the viral capsid, a critical step in viral morphogenesis. The preS2 region of the HBsAgL protein serves as a possible spacer to facilitate conformational changes of the preS1/S2 domain [17,81,82]. During the maturation process, the preS1/S2 domain is translocated to adopt an external orientation, which is essential for virus attachment to the host cell through a specific interaction with heparan sulfate proteoglycan [34,35] and binding to the entry receptor [36,37]. In addition, the preS1 domain is myristoylated at the N-terminus, which is required for efficient HBV entry into hepatocytes (Figure 1B) [83,84]. The potential *N*- and *O*-glycosylation sites in the HBsAgL preS1/S2 domain are not utilized due to the cytosolic orientation of the preS1/preS2 domain after translation. HBsAgL molecules have molecular weights of 39 kD (p39) or 42kD (gp42) depending on the glycosylation status at position N146 in the S-domain (Figure 1B). Expression of HBsAgL in mammalian cells in the absence of HBsAgS and HBsAgM does lead to particle formation, but not secretion, and it is retained in post-ER and pre-Golgi compartments [85]. HBsAgL causes a dose-dependent inhibition of particle release if co-expressed with HBsAgS [86,87].

Taken together, the S-domain is shared by the HBV envelope proteins and defines the backbone of the particle due to the presence of the topogenic transmembrane regions, glycans at position N146, cysteine residues to form inter- and intra-molecular disulfide bonds, and the external loop region. The correct folding of the external loop defines the HBsAg-specific antigenic determinants.

4. Biochemical Properties of SVPs

The proportion of the HBsAgL, -M, and -S proteins differ between Dane particles, filaments, and SVPs generated during a natural infection. It is estimated that the envelope of Dane particles contains HBsAgL, HBsAgM, and HBsAgS at a ratio of approximately 3:2:5, and filaments at a ratio of 1:1:4. SVPs contain less HBsAgM compared to filaments with trace amounts of HBsAgL [46,83,88]. HBsAgS SVPs have a molecular weight of 2–4×10^6, and are composed of protein (75% by weight), carbohydrates (in form of glycoproteins) and lipids (25% by weight) [89]. Approximately 100 HBsAgS proteins assemble with lipids into lipoprotein particles, fifty HBsAgS dimers were identified in SVPs purified from sera of transgenic mice [49,89,90]. Three different regions of the S-domain contribute to the oligomerization of the HBsAg proteins, the cytosolic loop, TM2, and the luminal loop (Figure 3) [57]. The SVPs are compact particles with a reported density of 1.21 g/mL in caesium chloride (CsCl) compared to a density of infectious virions between 1.24 and 1.26 g/mL [23,91,92]. The compact structure of the SVPs is due to the large number of intra- and inter-molecular disulfide bonds within and between the S-domains of the individual HBsAg subunits [62,65,90,93–95]. Kinetic studies demonstrated that disulfide-linked HBsAgS dimers are formed in the ER, then the immature particle precursors are transported to a post-ER, pre-Golgi compartment, which excludes the enzyme "protein disulfide isomerase" and allows the formation of HBsAgS oligomers [96]. Intracellular HBsAgS particles contain high-mannose oligosaccharide chains, and after secretion, SVPs contain complex oligosaccharide chains with terminal sialic acid *N*-acetylglucosamine residues representing glycosylation patterns conforming with the HBsAg movement from the ER through the Golgi cisternae [12,67,97].

Cryo-EM studies and biophysical analyses of the SVPs produced in cell culture or purified from sera of transgenic mice demonstrates a tight HBsAgS protein–lipid interaction. The lipid composition of HBsAgS SVPs purified from the plasma of several HBV carriers showed that phospholipids, in particular, phosphatidylcholine is a major lipid class; with palmitic, stearic, oleic, and linoleic acids being the major fatty acid components [89]. Consistently, SVPs produced in human hepatoma cell lines predominantly contain phospholipids, with phosphatidylcholine being the dominant component [98]. The tight protein–lipid interaction restricts lipid movement suggesting that the lipids are not aligned in a typical bilayer structure. HBsAg particles seem to contain the lipids in an unusual arrangement, with the lipids being closely intercalated with the proteins, located on the particle surface, and are hence likely arranged in a lipid monolayer [99,100]. HBsAg proteins contain a high content of alpha helices (45%–52%), which are lipid-associated, and provide an arrangement which allows the disposal of the loop regions in the particle interior or on its surface [100,101]. Assessing yeast (*Hansenula polymorpha*)-derived SVPs, the particles have an ordered and rigid lipid interface, possibly organized as a phospholipid monolayer, with a hydrophobic and fluid inner core. HBsAgS proteins penetrate into the lipid core, with parts of the protein protruding from the particle surface [89,102,103]. The lipids contribute to the antigenic activity of HBsAg particles [100], likely by stabilizing the proper helical structure of the HBsAg proteins and the conformation of their hydrophilic region, which contains the antigenic site. Removal of lipids decreases the helical content and reduces the antigenic activity of the particles [100]. S-domain antigenic structures seem to be strongly impacted by the lipid–protein interface, which defines the formation of alpha-helical structures and accommodates the arrangement of proper disulfide bonding patterns and the correct folding of HBsAg.

5. The First Vaccine Generation Against HBV: Vaccine Derived from Patient Plasma

The first vaccine against HBV was based on the unique approach to purify the HBsAg immunogen directly from the blood of asymptomatic HBV carriers [104]. The identification of the HBsAg as an important immune target was based on observations that anti-HBs human immunoglobulins conferred passive protection against hepatitis B [105,106]. Consistently, active immunization studies with HBsAg protected chimpanzees from hepatitis B [29,107,108], and clinical studies with recipients of blood transfusion indicated that patients who developed anti-HBs were less likely to develop hepatitis [109]. Patients on a renal dialysis unit and staff were less likely to acquire hepatitis if they had anti-HBs antibodies [110]. The first reported vaccination against hepatitis B was performed with a diluted, heat-inactivated HBsAg-positive serum in children, then the children challenged with infectious HBV, resulting in an incomplete but considerable protection [111]. For the generation of a proper vaccine, methodologies were developed to purify HBsAg using isopycnic banding and rate-zonal separation [26,27,112,113], affinity columns [114], adsorption onto colloidal silicate and desorption, differential polyethylene glycol (PEG) precipitations and gel filtration [27]. To minimize the risk of infections due to the presence of hepatitis B virions, the vaccine preparations were inactivated with formalin [112,114]. Plasma was from "ad" and "ay" donors, the SVPs mixed, then adsorbed onto aluminium hydroxide, or used in the absence of an adjuvant [26,27]. The pilot vaccines demonstrated that the plasma-derived SVPs induced anti-HBs antibody responses in different animal models [26]. Safety testing in chimpanzees, which are susceptible to HBV infection, did not provide any evidence of hepatitis in chimpanzees, which had received the inactivated plasma-derived SVP vaccine. Vaccinated chimpanzees were protected from a challenge with infectious HBV and did not show any indication of a hepatitis B infection [26,108].

The use of the first plasma-derived vaccines, Heptavax-B (Merck Sharp and Dohme, MSD) and Hevac-B (Institute Pasteur) provided good protection rates, and they were safe [114–117]. Depending on the purification and inactivation procedure, the composition of the plasma-derived HBsAg SVPs can vary, and they may or may not contain small quantities of HBsAgM, providing the preS2 domain in addition to HBsAgS. The Hevac-B vaccine contained 1%–2% HBsAgM. In contrast, the Heptavax-B vaccine did not contain preS2 proteins due to a treatment step with proteases, which however did not interfere with HBsAg-specific antigenicity [118–120]. Similar serum-derived vaccines were then produced from various manufacturers, such as Hepavax-B (Green Cross, Korea), Hepaccine-B (Cheil, Korea), and GCC VAC (Green Cross Corporation, Osaka) (Table 1). Limitations given by the supply of human plasma from chronically infected patients, and the risk associated with human-derived products due to contaminating proteins and the potential presence of other pathogens transmitted by blood, in particular a non-A non-B hepatitis virus, confronted the use of human plasma-derived HBsAg SVP vaccines [121–123]. Safety concerns about products from human sources, together with the advances in recombinant DNA technology and biotechnology, led to the development of recombinant hepatitis B vaccines.

Table 1. Key vaccines against HBV utilizing hepatitis B surface antigens.

Source	Cell type	Name	Antigen	Subtype	Manufacturer
Plasma-derived vaccines	-	Heptavax-B®	HBsAgS	ad	Merck [116]
	-	Hevac B®	HBsAgS, -M	ad *and* ay	Pasteur [120,124,125]
	-	GCC VAC	HBsAgS		Green Cross Operation, Osaka [125]
	-	Hepavax-B	HBsAgS		Korean Green Cross [126,127]
	-	Hepaccine B	HBsAgS		Cheil Foods & Chemicals Company [128]

Table 1. *Cont.*

Source	Cell type	Name	Antigen	Subtype	Manufacturer
Recombinant (yeast-derived)	*Saccharomyces cerevisiae*	Recombivax® HB HB-Vax II®	HBsAgS	adw	Merck [129]
	Saccharomyces cerevisiae	Engerix-B®	HBsAgS	adw	GlaxoSmithKline [129,130]
	Saccharomyces cerevisiae	TGP 943™	HBsAgS, -M	adr	Takeda Chemical Industries [131]
	Saccharomyces cerevisiae	Euvax B®	HBsAgS		LG Chemical Ltd, [130,132]
	Pichia pastoris	Shanvac B	HBsAgS	adw2	Shantha Biotechnics [133]
	Pichia pastoris	Heberbiovac-HB®	HBsAgS	adw2	Heber Biotech S.A., [130]
	Hansenula polymorpha	Heplisav-B®	HBsAgS	adw	Dynavax Technologies [134]
	Hansenula polymorpha	Hepavax-Gene™	HBsAgS	adr	Janssen Pharma [130,135]
Recombinant (mammalian cell-derived)	*Chinese hamster ovary cells*	Gen Hevac B®	HBsAgS, -M	ayw	Pasteur [120]
	Chinese hamster ovary cells	Sci-B-Vac/Bio-Hep-B™/Hepimmune™	HBsAgS, -M, -L	adw	VBI Vaccines [136]
	Mouse c127 clonal cell line	AG-3™ (Hepacare/Hepagene™)	HBsAgS, -M, -L	adw & ayw	Medeva [137]

6. The Second Vaccine Generation Against HBV: Yeast-Derived Recombinant HBsAgS SVPs

The identification of HBsAgS as the major HBV envelope protein of plasma-derived HBsAg SVPs and encoding the major antigenic determinants prompted the expression of HBsAgS in mammalian cell lines and yeast [13]. Yeast cell strains (*Saccharomyces cerevisiae, Pichia pastoris, Hansenula polymorpha*) were developed which express HBsAgS in high quantities. HBsAgS SVPs are isolated from yeast cell extracts with a sedimentation rate and buoyant density similar to particles from human samples or expressed in cell culture [138–140]. Yeast-derived SVPs have reduced antigenic reactivity compared to SVPs derived from human plasma but immunization studies established that yeast-derived SVPs induce anti-HBs antibody responses, which provide protection of immunized chimpanzees following a challenge with HBV. Importantly, the yeast-derived HBsAgS SVP vaccine of the "adw" subtype conferred protection against HBV subtypes "adr" and "ayw" [28]. Clinical trials using different age groups, healthy individuals, and special target populations confirmed that the yeast-derived vaccine is highly immunogenic, and generated qualitative and quantitative anti-HBs antibody responses with a protective efficacy similar to the plasma-derived vaccines [141]. The vaccines achieved 99% seroprotection rates in healthy children and adolescents but approximately 5%–7% of the adult population are non-responders and the rate can increase to 70% in elderly persons and in special risk groups [13,129,136,142]. Also, genetically determined resistance may contribute to non-responsiveness to HBsAgS SVP vaccines [143–145]. At the molecular level, yeast-derived HBsAg SVPs are not *N*-glycosylated at the N146 position within the S-domain, in contrast to SVPs produced in mammalian cell lines and harvested from the cell culture medium, or isolated from the blood of chronic hepatitis B patients. HBsAgS expressed in yeast generate SVPs [146], but experimental evidence indicates that the SVP are not formed within the yeast cell and generated during the down-stream purification procedures. HBsAgS expression in *Pichia pastoris* showed that HBsAgS assembles at the ER into multi-layered lamellar structures [147]. Monitoring SVP assembly during the purification procedure demonstrated that particulate structures are formed after eluting HBsAg bound to colloidal silica. Irregular SVP-like structures were visualized, and morphological changes observed after pH adjustment (colloidal silica eluate pH 10.8 to 8.0), ion-exchange, and size-exclusion chromatography. The monodispersity improved after potassium thiocyanate (KSCN) treatment, also the SVPs have a more-fine structured surface [148–150]. The correct disulfide bonding is the molecular basis for the formation of native epitopes as probed with an anti-HBs antibody [151]. Lipid-containing SVPs undergo KSCN-induced maturation by the formation of intra- and inter-molecular disulfide bonds to generate fully disulfide-bonded SVPs, resulting in a decreased conformational flexibility of HBsAgS in the matured particles [149,152]. Restricted conformational flexibility is possibly required for the formation of native HBsAg-specific antigenic structures, which allows epitope recognition by anti-HBs antibodies, and importantly in eliciting neutralizing antibodies [149]. The particle size (diameter) of yeast-derived SVPs is consistent with the data obtained from mammalian-cell-derived SVPs, but the size distribution can vary depending on the SVP maturation level [148,153]. Yeast-derived SVPs have a high content of alpha-helical structures, and the movement of the HBsAgS proteins is restricted due to the tight association with the lipid membrane [102,154]. Similarly, the lipid composition is characterized by high levels of phospholipids, in particular, phosphatidylcholine consistent with mammalian cell-derived SVPs [154–156]. No significant differences in the anti-HBs response induced by the plasma-derived or yeast-derived hepatitis B vaccines was observed [123,157].

The widely distributed, yeast cell-derived hepatitis B vaccines, Engerix®-B (GlaxoSmithKline) and Recombivax HB™ (Merck Sharp and Dohme) use aluminium hydroxide or aluminium hydroxyphosphate sulfate as adjuvanting substances, respectively. Aluminium-based adjuvants are widely used and activate the inflammasome pathway [158–160]. The adsorption of HBsAgS SVPs on aluminium hydroxide is mediated by binding of the phosphate groups of the HBsAgS SVPs phospholipids with hydroxyl groups of aluminium hydroxide through a ligand-exchange mechanism [161,162]. After vaccine administration, the SVPs are eluted from the aluminium adjuvant upon contact with the interstitial fluid [163,164]. Adsorption of HBsAgS SVPs derived from *Hansenula polymorpha* on an aluminium gel followed by a mild desorption step using competing phosphate anions demonstrated that the conformation of the HBsAgS protein is retained, and consistently, no significant changes of the lipid core and lipid membrane surface of the SVPs were identified [165]. To improve immunization outcomes in adults at risk of a hepatitis B infection, HBsAgS SVPs synthesized in *Hansenula polymorpha* are formulated with a Toll-like receptor 9 (TLR 9) agonist, cytidine-phosphate-guanosine oligodeoxynucleotide (CpG-ODN) 1018 as an adjuvant (Heplisav-B™, Dynavax Technologies). TLR9 is a pattern recognition receptor of the innate immune system, which induces the production of cytokines such as interleukin-12 and interferon-alpha to stimulate the adaptive immune response [134,166]. Heplisav-B™ induced earlier seroprotection rates allowing a two-dose regimen compared to three doses required for the Engerix-B vaccine, but it caused more injection-site reactions [134].

The ability of HBsAgS subunits expressed in yeast cells to form SVPs during down-stream procedures, and to reproduce native antigenic structures, allowed the development of a highly successful preventative vaccine. The advancement of the vaccine adjuvant technology allowed new HBsAgS SVP formulations and demonstrated strategies to enhance the magnitude of the anti-HBs immune response with immediate practical applications

7. Third Generation Vaccine Concepts against HBV

The development of third-generation HBsAg vaccines providing the S-domain in combination with preS1 and/or preS2 sequences was directed by the objective to enhance the protective efficacy of the human plasma-derived vaccines (which consisted predominantly of HBsAgS subunits), and the second-generation recombinant yeast-derived vaccines (which consisted exclusively of HBsAgS subunits). The third-generation vaccines attracted interest to improve the immunization outcomes in persons who do not respond to the conventional HBsAgS vaccines [13]. The importance of the preS1 domain for viral entry and assembly makes it a potential target for vaccine development, anti-preS1 antibodies protected chimpanzees from HBV infection [167]. PreS1/S2 sequences provide additional B-cell epitopes to generate protective antibody responses [167–169] and may also serve as a T cell immunogen to overcome the non-responsiveness to the S-domain [170–172]. Small quantities of HBsAgL present in the HBsAgS vaccine induced significant T-cell activation measured as in vitro proliferation specific for the preS domain [172]. In addition, preS2 peptide vaccines protect chimpanzees against a challenge with HBV [169], and the use of SVPs with preS1 and/or preS2 sequences generated anti-HBs immune responses in mouse strains, which are non-responsive to the standard yeast-derived HBsAgS SVP vaccine [173].

For the generation of third-generation vaccines, different yeast expression systems and mammalian cell lines were utilized to synthesize SVPs composed of HBsAgM and HBsAgL proteins in the presence or absence of HBsAgS. Utilizing *Saccharomyces cerevisiae*, the expression of a modified HBsAgM in the absence of HBsAgS allowed the formation of SVPs. The modified HBsAgM protein (P31c) contains a deletion of six amino acids to make it resistant to trypsin-like proteases in *S. cerevisiae*. The HBsAgM-P31c proteins assembled into SVPs with a diameter of approximately 20 nm and retained HBsAg antigenicity [174,175]. Immunization studies in BALB/c mice and guinea pigs demonstrated that anti-HBs antibody titres can be induced comparable to the plasma-derived Heptavax-B vaccine (MSD). Also, anti-preS2 antibodies were detected in animals immunized with the HBsAgM-P31c SVP vaccine. In contrast to the yeast-derived HBsAgS vaccine, the HBsAgM-P31c SVPs are glycosylated with N- and O-linked glycans located in the preS2 domain [174,175], which may facilitate interactions with lectin receptors expressed by antigen-presenting cells. In an independent study, HBsAgM SVPs induced anti-S and anti-preS2 antibodies in healthy young adults but the anti-S response was lower than in the patient group who received HBsAgS SVPs, and hence, the HBsAgM vaccine failed to achieve the objective of inducing an early and strong anti-S and anti-preS2 immune response [176]. The HBsAgM-P31c SVPs, however, were used to formulate a new vaccine (TGP-943, Takeda) and demonstrated a protective effect in the chimpanzee model and also generated protective levels of anti-preS2 antibodies in humans [177,178]. Clinical studies demonstrated that the vaccine TGP-943 induced both anti-S and anti-preS2 antibodies, approximately 50% of non-responders became positive for either or both anti-S and anti-preS2 [131].

For the development of a HBsAgL-based vaccine, HBsAgL expression during the exponential *S. cerevisiae* growth phase generated high levels of HBsAgL but did not assemble into the typical 20–25 nm SVPs, but generated a polydisperse population of small (2–3 nm) and large aggregates (15–50 nm) [179]. HBsAgL was glycosylated by *N*- and *O*-linked glycans in the preS1/S2 domain indicating that HBsAgL accessed the lumen of the ER of the yeast cell and caused morphological changes in the ER compartment [179,180]. The presence of *N*-linked and *O*-linked glycans in the preS1/S2 domain of yeast-derived HBsAgL proteins is in contrast to preS1/preS2 of HBsAgL isolated from human plasma due to the cytoplasmic exposure of the preS1/S2 domain during the orderly and regulated process of virion morphogenesis [83,88,179,180]. Particle formation using *S. cerevisiae* could be rescued providing an N-terminal signal sequence, which possibly allows a correct entry into the secretory pathway, and after purification from the yeast lysate, spheres and filaments with a diameter of 23 nm were obtained, the length of the filaments was in the range of 40 to 120 nm, visualized by negative staining electron microscopy. HBsAgL proteins expressed in the absence of the signal sequence did not form such an ordered structure [181]. The visualization of the particles by atomic force microscopy (AFM) demonstrated a heterogeneous population of rugged spherical forms between 50 and 500 nm in diameter [182]. Immunization of mice with the yeast-derived HBsAgL SVPs elicited anti-S, anti-preS2, and anti-preS1 antibodies, and the effective dose (ED_{50}) for anti-S and anti-preS2 antibodies were similar to those achieved with HBsAgM particles [182]. Using an alternative strategy, hybrid SVPs were generated in *S. cerevisiae* composed of HBsAgS and a modified HBsAgL (HBsAgL*). HBsAgL* contains a truncated preS1/S2 region with sequences relevant for the hepatocyte-binding site and immunologically important B- and T-helper epitopes but does not contain sites for proteolysis and the binding site for polymerized human serum albumin. The hybrid SVPs contained HBsAgS and HBsAgL* at a ratio of 75:25 [176]. The immunization of BALB/c mice with HBsAgL*/HBsAgS SVPs generated anti-S and anti-preS1 antibodies. The anti-S titers were similar to those found after immunization with Engerix B. Immunizations of African Green Monkeys (*Ceropithecus aethiops*) using HBsAgL*/HBsAgS SVPs induced anti-S, anti-preS2, and anti-preS1 antibodies [176]. Safety and immunogenicity studies in young, healthy adult persons, and in poor responders to hepatitis B vaccines demonstrated that the presence of the preS1/preS2 domain did not enhance the anti-S response compared to the control Engerix

B vaccine (GSK) [183,184], in spite of the preS1 sequence present as a strong T-cell immunogen [170]. The HBsAgL*/HBsAgS vaccine induced anti-preS1 antibodies in a young, healthy adult person cohort, and possibly provides additional neutralizing activity [183].

With the availability of mammalian cell culture technologies, recombinant hepatitis B vaccines composed of SVPs have been developed containing the HBsAgS and HBsAgM (GenHevacB, Sanofi Pasteur Vaccins) and the additional HBsAgL subunits (Sci-B-Vac, VBI Vaccines; Hepacare, Medeva Pharma) (Table 1). Chinese hamster ovary (CHO) cells (GenHevacB, Sci-B-Vac) [120,136] or murine cells (C1271) (Hepacare, Medeva Pharma) [137] were used to generate the vaccines. The GenHevac B vaccine is composed of HBsAgS/HBsAgM SVPs at a ratio 80:20, and compared in a clinical setting to the human plasma-derived Hevac B vaccine, both vaccines induced antibodies to the HBsAg in >90% of the participants (497 persons in the age range of 18–40 years). Compared to the plasma-derived vaccine, the recombinant vaccine produced early and high levels of anti-preS2 antibodies, which may provide an additional advantage in prevention of a HBV infection [120,185].

For Sci-B-Vac, the complete HBsAg gene encoding HBsAgS, -M, and -L, including native promoter, enhancer, and poly(A) signal, were used to establish a producer CHO cell line, which contains more than 100 HBsAg coding copies/cell [173]. Protein analysis of the secreted SVPs revealed the presence of all three HBsAg proteins and its glycosylated isoforms (HBsAgS p24 and gp27; HBsAgM gp33 and gp36, HBsAgL p39 and gp42). Sci-B-Vac induced anti-S and anti-preS1 antibody responses in BALB/c mice, and also in mouse strains which are resistant to immunizations with HBsAgS SVPs and/or HBsAgM [136,173]. Sci-B-Vac demonstrated an excellent safety record in clinical studies, which included healthy individuals, children, and neonates. In comparison with yeast-derived hepatitis B vaccines, more than 50% of vaccinees receiving Sci-B-Vac developed earlier seroprotection against HBV [136]. Sci-B-Vac performed superior to yeast-derived HBsAgS vaccines in specific patient risk groups and provided vaccine boosts in persons with no or low response to preceding immunizations with the conventional yeast-derived HBV vaccine. Specific risk groups including patients with renal failure, with overweight and immune-suppressed patients responded with higher seroprotection rates compared to conventional yeast derived vaccines. The Sci-B-Vac vaccine is widely used in Israel and licensed in various countries [136].

8. HBsAgS SVPs as Platforms for Medically Relevant Antigenic Sequences

SVPs display an array of antigenic sequences to the innate immune system facilitating the subsequent activation of the adaptive system [8,10,186,187]. The ability to accept foreign inserts into the SVP structure provides the basis for advanced delivery platforms for medically relevant sequences, such as malaria antigens. Chimeric SVPs can be constructed from viral capsid proteins, such as capsids from HBV, human papilloma virus, and Qβ phage that have been re-engineered to express foreign antigenic sequences at a high antigenic density [8,187]. Similarly, SVPs derived from the HBV envelope assemble into highly compact lipid-containing particles, and have been exploited as carrier platforms for foreign antigenic sequences by introducing N- or C-terminal extensions [188–191], N-terminal extensions in addition to substitutions of the HBsAgS N-terminal sequence [192], by replacing the HBsAgM preS2-domain [193,194], by insertions into the external loop region including replacing antigenic determinants [195–201], or by replacing HBsAgS-specific cytotoxic T lymphocyte (CTL) epitopes [202] (Table 2). The insertion of a poliovirus-specific epitope with a length of 11 amino acids into the external loop region of HBsAgS allowed the expression of the chimeric, assembly and secretion competent HBsPolioAg proteins in a mouse cell line [201,203]. The chimeric SVPs contained glycosylated and non-glycosylated HBsPolioAg subunits and formed particles with 22 nm in diameter, similar to wild-type HBsAgS proteins. HBsPolioAg SVPs were used in mouse immunization studies and induced anti-poliovirus peptide-specific antibodies with neutralizing activity and a low level of anti-HBs antibodies, possibly due to a partial loss of HBsAg-specific antigenicity. The co-expression of both wild-type HBsAgS proteins and HBsPolioAg generated hybrid SVPs composed of both proteins, which facilitated the induction of anti-HBs and anti-poliovirus epitope antibodies [203].

Similar studies inserting heterologous B-cell epitopes in an exposed site in the external loop region reduced HBsAg-specific antigenicity depending on the length of the insert, but the recombinant proteins retained the ability to induce anti-HBs antibodies [195,196,198,204]. Chimeric SVPs composed of subunits distinguished by the number of inserted epitope repeats from the *Plasmodium falciparum* circumsporozoite (CS) protein demonstrated that the CS epitope number influenced the activity of the anti-CS epitope antibodies. The effect of the epitope-specific density on the antibody quality may instruct chimeric SVP designs to optimize immunological outcomes and vaccine efficacy [195]. The RTS,S/AS01 vaccine (Mosquirix™) is the most advanced vaccine with a heterologous antigenic sequence arrayed on SVPs. The RTS,S vaccine is based on the fusion of a *Plasmodium falciparum* CS polypeptide of 189 aa with selected tandem repeats of B-cell and T-cell epitopes to the HBsAgS N-terminus (RTS) (Figure 1C). The CS-protein is expressed on the *Plasmodium* sporozoite surface, and essential for hepatocyte invasion and for establishing a productive infection, and therefore an important target for the development of a pre-erythrocyte vaccine [205]. The genes for HBsAgS and RTS are integrated into the genome of *S. cerevisiae* and co-expressed at a ratio RTS:HBsAgS of 1:4 to generate non-glycosylated mixed (hybrid) lipoprotein particles [190,191]. The RTS,S vaccine is well-tolerated, safe, and immunogenic, and is considered to be the first advanced vaccine against the pre-erythrocyte stage of the malaria parasite, and induces both anti-HBs and anti-CS protein antibodies. The anti-malaria RTS,S/AS01 vaccine in children of five months or older reduced clinical malaria episodes by 39% and life-threatening severe malaria episodes by 29%. The vaccine is licensed in three African countries [5,6,206]. Mosquirix™ (RTS,S/AS01) is adjuvanted with AS01, which is a liposome formulation and contains monophosphoryl lipid A (MPL) and the saponin QS-21. RTS,S in combination with AS01 resulted in higher anti-CS protein immune responses than AS02, which is an oil-in-water emulsion-based adjuvant [7,205–207]. To enhance the vaccine efficacy against malaria, a RTS,S-related vaccine (R21) with an increased proportion of CS-polypeptides was developed. CS-polypeptide-HBsAg fusion proteins were expressed in *Pichia pastoris*, and SVPs were obtained after caesium chloride density ultracentrifugation and gel filtration [208]. The R21 vaccine induces a sterile protection in mice against a challenge with transgenic sporozoites. The induction of anti-HBs antibodies is compromised, possibly because the high content of the CS-polypeptide blocks access to the HBsAg external loop region, which contains the antigenic "a"-determinant [208].

Table 2. Selection of chimeric HBsAg SVP platforms and vaccines.

Antigen	Target	Delivery Site	Expression System	Name	Study/Manufacturer
CS-HBsAg/HBsAgS	Malaria	N-terminal	*Saccharomyces cerevisiae*	RTS,S/AS01 Mosquirix™	GlaxoSmithKline [5,6,190]
CS-HBsAgS	Malaria	N-terminal	*Pichia pastoris*	R21	[208]
DENV-EDIII-HBsAgS/HBsAgS	Dengue virus	N-terminal	*Pichia pastoris*	DSV4	[188]
HBsAgS-gp120 (HIV-1)	HIV-1	C-terminal	CV-1 cell line	MR15, MR23	[189]
Env1-, Env2-HBsAgS	Hepatitis C virus	Substitution/N-terminal extension	CHO cells		[192,209]
HBsAgS-NANP repeats	Malaria	Insertion	HEK293F cell line	M-HBsAg-N4, -N9	[195]
HBsAgS-HCV env epitopes	Hepatitis C virus	Insertion	HEK293T cell line		[196,198]
HBsAgS-catalase epitope	*Helicobacter pylori*	Insertion	HuH-7	VLP-KatA	[197]
HBsAgS-VP1 capsid epitope	Poliovirus	Insertion	Mouse L cells	HBsPolioAg	[201,203]
HBsAgS-matrix CTL epitope	Influenza A virus	Insertion/Substitution	HEK293T cell line		[202]
Polyepitope-HBsAgS;	HIV	Substitution preS2 sequence	SW480 cells		[193]

CS—circumsporozoite polypeptide; DENV-EDIII—Dengue virus envelope domain III; env—envelope protein; HCV—hepatitis C virus; HIV—Human immunodeficiency virus; CTL—cytotoxic T lymphocyte. VLP—virus-like particle.

Preexisting immunity against vaccine vectors can impose a negative effect on the outcome of the vaccination [210–213]. HBsAgS SVPs are widely used in immunization programs to combat HBV, and therefore, the use of chimeric SVPs could pose a problem for recipients previously immunized against hepatitis B. Immunization studies in mice using chimeric SVPs with a foreign epitope inserted into the external loop of HBsAgS or fused to the N-terminus of HBsAgS demonstrated that pre-existing anti-HBs antibodies do not compromise the immunogenicity of the foreign antigenic sequence presented by the chimeric HBsAgS SVPs [214,215]. Consistently, clinical studies with human volunteers to assess the RTS,S malaria vaccine did not provide any evidence that a pre-existing anti-HBs status prevented an anti-CS-protein immune response [190]. In relation to a HBV chronic carrier status and the use of the RTS,S vaccine, there was no evidence that chronic HBV carriers (HBsAg positive) and HBsAg-negative individuals respond differently regarding an antibody response to the CS-protein [216].

9. Enhancement of Platform Immunogenicity through Biochemical Modifications

Based on the importance of SVPs as medical tools and platforms for the presentation of native viral antigenic sequences, it is critical to understand their immunogenicity in relation to antigen structure in order to enhance or to modulate their immunogenicity. Depending on the SVP type, targeted biochemical modifications of the SVP subunits may allow the generation of SVP variants with enhanced immunogenicity. HBsAgS SVPs are glycosylated lipoprotein particles and are stabilized by extensive intra- and inter-molecular disulfide bonds, which allows targeted modifications of the glycan content and level of disulphide bonding.

Changing disulfide bonding impacts on antigen processing and epitope selection by modifying the conformational flexibility [217]. The three-dimensional structure guides processing and presentation of T helper (Th) and CTL epitopes, and subtle changes in antigen structure can modulate T cell responses due to qualitative and quantitative differences in protein processing [217–220]. Distinct Th cell epitope profiles emerged from human immunodeficiency virus type 1 (HIV-1) gp120 molecules after destabilizing the three-dimensional structure as a consequence of deleted cysteine residues [221]. In an attempt to enhance immunogenicity, HBsAgS SVPs with a reduced level of disulfide bonds were generated. The biochemically modified SVPs showed a higher protease sensitivity, potentially due to introducing structural changes associated with enhanced cellular immunogenicity [65]. Altering SVP structure may represent an attractive strategy to modulate proteolytic sensitivity to influence antigen processing and promoting an enhanced immune response and/or a changed hierarchy of epitope presentation [218,222].

Manipulation of protein glycosylation represents an alternative strategy to promote antigen internalization and antigen presentation via MHC class I and class II molecules to enhance the adaptive immune responses [223,224]. The glycosylation status and glycan density of the immunogen impacts on its interaction with antigen-presenting cells and recognition by lectins [224]. Glycan-mediated interactions with immunocompetent cells impact on protein uptake and can enhance or modulate cell-mediated and humoral immune responses [225–229]. Contrarily, glycans can shield protein epitopes to evade recognition by antibodies and can block antigen processing [230]. Mannosylation provided an efficient strategy to improve uptake and processing of a SVP derived from the rabbit hemorrhagic disease virus [225]. Consistently, mannosylated solid lipid nanoparticles loaded with HBsAg induced stronger cellular responses than nanoparticles devoid of mannose [229]. Mutant HBsAgS subunits with additional N-glycosylation sites assembled into hyperglycosylated SVP. Antigenic fingerprints indicated that additional glycans do not extensively shield HBsAg-specific antigenic sites. Immunization studies demonstrated that the hyperglycosylated SVPs induced earlier and longer-lasting antibody responses than hypoglycosylated SVPs or wild type SVPs [231]. The ability of biochemically modified SVPs to promote immune responses possibly due to differences in their glycosylation-related interaction with cells of the innate immune system illustrates approaches for the design of immunogens with superior immunological characteristics.

10. Concluding Remarks

The development of preventative vaccines against hepatitis B resulted in remarkable advances in reducing HBV associated liver diseases. However, chronic hepatitis B is still difficult to control due to continuous viral replication driven by the episomal cccDNA present in the nuclei of infected hepatocytes. Novel strategies to generate vaccines based on structurally modified subunits to enhance immunogenicity and/or to modify antigen processing to change the hierarchy of epitope presentation may represent a pathway to overcome chronic viral infections or may complement a vaccine based on native proteins [65,218,224–226,231]. HBsAgS SVPs have been used as carrier platforms for various antigenic sequences to induce anti-foreign humoral and cellular immune responses [8,186]. One of the most advanced chimeric vaccines with a foreign antigenic sequence arrayed on a particulate carrier is based on the HBsAgS backbone fused to a *P. falciparum* CS-polypeptide [5,6]. For the design of next generation vaccines with therapeutic capabilities, formulations based on antigen combinations, such as mixtures of HBsAgS SVPs and SVPs composed of the HBV nucleocapsid antigen (HBcAg), may allow the induction of broad CD4 and CD8 T-cell responses suitable for therapeutic outcomes [232]. Alternatively, the assessment of synergistic effects between biochemically modified immunogens and adjuvant compounds possibly represent an avenue for the generation of optimized vaccines and delivery platforms, which may be suitable for therapeutic applications to overcome established chronic infections.

Author Contributions: H.-J.N. wrote the original draft version; J.K.-T.H. and B.J.-R. edited and reviewed the manuscript. All authors have read and agreed to the published version of the manuscript.

Funding: HJN was partly supported by a grant awarded by the Australian Centre for HIV and Viral Hepatitis Research (ACH^2).

Conflicts of Interest: The author H.-J.N. of this publication has an equity interest in, and serves as a consultant to ClearB Therapeutics. ClearB Therapeutics had no role in the design and writing of the review article.

References

1. Huzair, F.; Sturdy, S. Biotechnology and the transformation of vaccine innovation: The case of the hepatitis B vaccines 1968–2000. *Stud. Hist. Philos. Biol. Biomed. Sci.* **2017**, *64*, 11–21. [CrossRef]
2. Millman, I. The development of the hepatitis B vaccine. In *Hepatitis B. The Virus, the Disease, and the Vaccine*; Millman, I., Eisenstein, T.K., Blumberg, B.S., Eds.; Plenum Publishing Corp.: New York, NY, USA, 1984; pp. 137–147.
3. World Health Organization. *Hepatitis B. Key Facts*; World Health Organization: Geneva, Switzerland, 2019.
4. World Health Organization. *Prevention and Control of Viral Hepatitis Infection. Framework for Global Action*; World Health Organization: Geneva, Switzerland, 2012.
5. Adepoju, P. RTS, S malaria vaccine pilots in three African countries. *Lancet* **2019**, *393*, 1685. [CrossRef]
6. Schuerman, L. RTS, S malaria vaccine could provide major public health benefits. *Lancet* **2019**, *394*, 735–736. [CrossRef]
7. Cohen, J.; Nussenzweig, V.; Nussenzweig, R.; Vekemans, J.; Leach, A. From the circumsporozoite protein to the RTS, S/AS candidate vaccine. *Hum. Vaccines* **2010**, *6*, 90–96. [CrossRef] [PubMed]
8. Mohsen, M.O.; Zha, L.; Cabral-Miranda, G.; Bachmann, M.F. Major findings and recent advances in virus-like particle (VLP)-based vaccines. *Semin. Immunol.* **2017**, *34*, 123–132. [CrossRef]
9. López-Sagaseta, J.; Malito, E.; Rappuoli, R.; Bottomley, M.J. Self-assembling protein nanoparticles in the design of vaccines. *Comput. Struct. Biotechnol. J.* **2016**, *14*, 58–68. [CrossRef] [PubMed]
10. Tan, M.; Jiang, X. Recent advancements in combination subunit vaccine development. *Hum. Vaccines Immunother.* **2017**, *13*, 180–185. [CrossRef] [PubMed]
11. Yuen, M.-F.; Chen, D.-S.; Dusheiko, G.M.; Janssen, H.L.A.; Lau, D.T.Y.; Locarnini, S.A.; Peters, M.G.; Lai, C.-L. Hepatitis B virus infection. *Nat. Rev. Dis. Primers* **2018**, *4*, 18035. [CrossRef] [PubMed]
12. Seeger, C.; Mason, W.S. Molecular biology of hepatitis B virus infection. *Virology* **2015**, *479–480*, 672–686. [CrossRef]

13. Gerlich, W.H. Medical virology of hepatitis B: How it began and where we are now. *Virol. J.* **2013**, *10*, 239. [CrossRef]

14. Kramvis, A. Genotypes and genetic variability of hepatitis B virus. *Intervirology* **2014**, *57*, 141–150. [CrossRef] [PubMed]

15. Norder, H.; Couroucé, A.M.; Coursaget, P.; Echevarria, J.M.; Lee, S.D.; Mushahwar, I.K.; Robertson, B.H.; Locarnini, S.; Magnius, L.O. Genetic diversity of hepatitis B virus strains derived worldwide: Genotypes, subgenotypes and HBsAg subtypes. *Intervirology* **2004**, *47*, 289–309. [CrossRef] [PubMed]

16. Slagle, B.L.; Bouchard, M.J. Role of HBx in hepatitis B virus persistence and its therapeutic implications. *Curr. Opin. Virol.* **2018**, *30*, 32–38. [CrossRef]

17. Bruss, V. Hepatitis B virus morphogenesis. *World J. Gastroenterol.* **2007**, *13*, 65–73. [CrossRef] [PubMed]

18. Le Bouvier, G.L. The heterogeneity of Australia antigen. *J. Infect. Dis.* **1971**, *123*, 671–675. [CrossRef] [PubMed]

19. Bancroft, W.H.; Mundon, F.K.; Russell, P.K. Detection of additional antigenic determinants of hepatitis B antigen. *J. Immunol.* **1972**, *109*, 842–848.

20. Jazayeri, S.M.; Alavian, S.M.; Dindoost, P.; Thomas, H.C.; Karayiannis, P. Molecular variants of hepatitis B surface antigen (HBsAg). In *Viral Hepatitis*, 4th ed.; Thomas, H.C., Lok, A.S.F., Locarnini, S.A., Zuckerman, A.J., Eds.; John Wiley & Sons Ltd.: Hoboken, NJ, USA, 2014; pp. 107–126.

21. Désiré, N.; Ngo, Y.; Franetich, J.-F.; Dembele, L.; Mazier, D.; Vaillant, J.-C.; Poynard, T.; Thibault, V. Definition of an HBsAg to DNA international unit conversion factor by enrichment of circulating hepatitis B forms. *J. Viral Hepat.* **2015**, *22*, 718–726.

22. Brunetto, M.R. A new role for an old marker, HBsAg. *J. Hepatol.* **2010**, *52*, 475–477. [CrossRef]

23. Dreesman, G.R.; Hollinger, F.B.; Suriano, J.R.; Fujioka, R.S.; Brunschwig, J.P.; Melnick, J.L. Biophysical and biochemical heterogeneity of purified hepatitis B antigen. *J. Virol.* **1972**, *10*, 469–476. [CrossRef]

24. Dane, D.S.; Cameron, C.H.; Briggs, M. Virus-like particles in serum of patients with Australia-antigen associated hepatitis. *Lancet* **1970**, *1*, 695–698. [CrossRef]

25. Rydell, G.E.; Prakash, K.; Norder, H.; Lindh, M. Hepatitis B surface antigen on subviral particles reduces the neutralizing effect of anti-HBs antibodies on hepatitis B viral particles in vitro. *Virology* **2017**, *509*, 67–70. [CrossRef] [PubMed]

26. Hilleman, M.R.; Bertland, A.U.; Buynak, E.B.; Lampson, G.P.; McAleer, W.J.; McLean, A.A.; Roehm, R.R.; Tytell, A.A. Clinical and laboratory studies of HBsAg vaccine. In *Viral Hepatitis*; Vyas, G.N., Cohen, S.N., Schmid, R., Eds.; The Franklin Institute Press: Philadelphia, PA, USA, 1978; pp. 525–537.

27. Maupas, P.; Goudeau, A.; Coursaget, P.; Drucker, J.; Barin, F.; André, M. Immunization against hepatitis B in man: A pilot study of two years' duration. In *Viral Hepatitis*; Vyas, G.N., Cohen, S.N., Schmid, R., Eds.; The Franklin Institute Press: Philadelphia, PA, USA, 1978; pp. 539–556.

28. McAleer, W.J.; Buynak, E.B.; Maigetter, R.Z.; Wampler, D.E.; Miller, W.J.; Hilleman, M.R. Human hepatitis B vaccine from recombinant yeast. *Nature* **1984**, *307*, 178–180. [CrossRef] [PubMed]

29. Murphy, B.L.; Maynard, J.E.; Le Bouvier, G.L. Viral subtypes and cross-presentation in hepatitis B virus infections of chimpanzees. *Intervirology* **1974**, *3*, 378–381. [CrossRef] [PubMed]

30. Hu, J.; Liu, K. Complete and incomplete hepatitis B virus particles: Formation, function, and application. *Viruses* **2017**, *9*, 56. [CrossRef]

31. Selzer, L.; Zlotnick, A. Assembly and release of hepatitis B virus. *Cold Spring Harb. Perspect. Med.* **2015**, *5*, a021394. [CrossRef]

32. Sureau, C.; Salisse, J. A conformational heparan sulfate binding site essential to infectivity overlaps with the conserved hepatitis B virus a-determinant. *Hepatology* **2013**, *57*, 985–994. [CrossRef]

33. Bruss, V.; Vieluf, K. Functions of the internal pre-S domain of the large surface protein in hepatitis B virus particle morphogenesis. *J. Virol.* **1995**, *69*, 6652–6657. [CrossRef]

34. Schulze, A.; Gripon, P.; Urban, S. Hepatitis B virus infection initiates with a large surface protein-dependent binding to heparan sulfate proteoglycans. *Hepatology* **2007**, *46*, 1759–1768. [CrossRef]

35. Leistner, C.M.; Gruen-Bernhard, S.; Glebe, D. Role of glycosaminoglycans for binding and infection of hepatitis B virus. *Cell. Microbiol.* **2008**, *10*, 122–133. [CrossRef]

36. Neurath, A.R.; Kent, S.B.H.; Strick, N.; Parker, K. Identification and chemical synthesis of a host cell receptor binding site on hepatitis B virus. *Cell* **1986**, *46*, 429–436. [CrossRef]

37. Yan, H.; Zhong, G.; Xu, G.; He, W.; Jing, Z.; Gao, Z.; Huang, Y.; Qi, Y.; Peng, B.; Wang, H.; et al. Sodium taurocholate cotransporting polypeptide is a functional receptor for human hepatitis B and D virus. *eLife* **2012**, *1*, e00049. [CrossRef]

38. Lambert, C.; Prange, R. Chaperone action in the posttranslational topological reorientation of the hepatitis B virus large envelope protein: Implications for translocational regulation. *Proc. Natl. Acad. Sci. USA* **2003**, *100*, 5199–5204. [CrossRef] [PubMed]

39. Cho, D.Y.; Yang, G.H.; Ryu, C.J.; Hong, H.J. Molecular chaperone GRP78/BiP interacts with the large surface protein of hepatitis B virus in vitro and in vivo. *J. Virol.* **2003**, *77*, 2784–2788. [CrossRef] [PubMed]

40. Löffler-Mary, H.; Werr, M.; Prange, R. Sequence-specific repression of cotranslational translocation of the hepatitis B virus envelope proteins coincides with binding of heat shock protein Hsc70. *Virology* **1997**, *235*, 144–152. [CrossRef] [PubMed]

41. Stieler, J.T.; Prange, R. Involvement of ESCRT-II in hepatitis B virus morphogenesis. *PLoS ONE* **2014**, *9*, e91279. [CrossRef]

42. Lambert, C.; Döring, T.; Prange, R. Hepatitis B virus maturation is sensitive to functional inhibition of ESCRT-III, Vps4, and γ2-Adaptin. *J. Virol.* **2007**, *81*, 9050–9060. [CrossRef]

43. Watanabe, T.; Sorensen, E.M.; Naito, A.; Schott, M.; Kim, S.; Ahlquist, P. Involvement of host cellular multivesicular body functions in hepatitis B virus budding. *Proc. Natl. Acad. Sci. USA* **2007**, *104*, 10205–10210. [CrossRef]

44. Jiang, B.; Himmelsbach, K.; Ren, H.; Boller, K.; Hildt, E. Subviral hepatitis B virus filaments, like infectious viral particles, are released via multivesicular bodies. *J. Virol.* **2016**, *90*, 3330–3341. [CrossRef]

45. Chairez, R.; Hollinger, F.B.; Melnick, J.L.; Dreesman, G.R. Biophysical properties of purified morphologic forms of hepatitis B antigen. *Intervirology* **1974**, *3*, 129–140. [CrossRef]

46. Short, J.M.; Chen, S.; Roseman, A.M.; Butler, P.J.G.; Crowther, R.A. Structure of hepatitis B surface antigen from subviral tubes determined by electron cryomicroscopy. *J. Mol. Biol.* **2009**, *390*, 135–141. [CrossRef]

47. Dryden, K.A.; Wieland, S.F.; Whitten-Bauer, C.; Gerin, J.L.; Chisari, F.V.; Yeager, M. Native hepatitis B virions and capsids visualized by electron cryomicroscopy. *Mol. Cell* **2006**, *22*, 843–850. [CrossRef]

48. Cao, J.; Zhang, J.; Lu, Y.; Luo, S.; Zhang, J.; Zhu, P. Cryo-EM structure of native spherical subviral particles isolated from HBV carriers. *Virus Res.* **2019**, *259*, 90–96. [CrossRef]

49. Gilbert, R.J.C.; Beales, L.; Blond, D.; Simon, M.N.; Lin, B.Y.; Chisari, F.V.; Stuart, D.I.; Rowlands, D.J. Hepatitis B small surface antigen particles are octahedral. *Proc. Natl. Acad. Sci. USA* **2005**, *102*, 14783–14788. [CrossRef] [PubMed]

50. Crowley, C.W.; Liu, C.-C.; Levinson, A.D. Plasmid-directed synthesis of hepatitis B surface antigen in monkey cells. *Moll. Cell. Biol.* **1983**, *3*, 44–55. [CrossRef] [PubMed]

51. Liu, C.-C.; Yansura, D.; Levinson, A.D. Direct expression of hepatitis B surface antigen in monkey cells from an SV40 vector. *DNA* **1982**, *1*, 213–221. [CrossRef] [PubMed]

52. Moriarty, A.M.; Hoyer, B.H.; Shih, J.W.K.; Gerin, J.L.; Hamer, D.H. Expression of the hepatitis B virus surface antigen gene in cell culture by using a simian virus 40 vector. *Proc. Natl. Acad. Sci. USA* **1981**, *78*, 2606–2610. [CrossRef] [PubMed]

53. Dubois, M.F.; Pourcel, C.; Rousset, S.; Chany, C.; Tiollais, P. Excretion of hepatitis B surface antigen particles from mouse cells transformed with cloned viral DNA. *Proc. Natl. Acad. Sci. USA* **1980**, *77*, 4549–4553. [CrossRef] [PubMed]

54. Lee, Y.S.; Kim, B.K.; Choi, E.-C. Physiochemical properties of recombinant hepatitis B surface antigen expressed in mammalian cell (C127). *Arch. Pharm. Res.* **1998**, *21*, 521–526. [CrossRef]

55. Eble, B.E.; Lingappa, V.R.; Ganem, D. Hepatitis B surface antigen: An unusual secreted protein initially synthesized as a transmembrane polypeptide. *Mol. Cell. Biol.* **1986**, *6*, 1454–1463. [CrossRef]

56. Eble, B.E.; MacRae, D.R.; Lingappa, V.R.; Ganem, D. Multiple topogenic sequences determine the transmembrane orientation of hepatitis B surface antigen. *Mol. Cell. Biol.* **1987**, *7*, 3591–3601. [CrossRef]

57. Suffner, S.; Gerstenberg, N.; Patra, M.; Ruibal, P.; Orabi, A.; Schindler, M.; Bruss, V. Domains of the hepatitis B virus small surface protein S mediating oligomerization. *J. Virol.* **2018**, *92*, e02232-17. [CrossRef] [PubMed]

58. Stirk, H.J.; Thornton, J.M.; Howard, C.R. A topological model for hepatitis B surface antigen. *Intervirology* **1992**, *33*, 148–158. [CrossRef] [PubMed]

59. Berting, A.; Hahnen, J.; Kröger, M.; Gerlich, W.H. Computer-aided studies on the spatial structure of the small hepatitis B surface protein. *Intervirology* **1995**, *38*, 8–15. [CrossRef] [PubMed]

60. Komla-Soukha, I.; Sureau, C. A tryptophan-rich motif in the carboxyl terminus of the small envelope protein of hepatitis B virus is central to the assembly of hepatitis delta virus particles. *J. Virol.* **2006**, *80*, 4648–4655. [CrossRef] [PubMed]

61. Paulij, W.P.; de Wit, P.L.M.; Sünnen, C.M.G.; van Roosmalen, M.H.; Petersen-van Ettekoven, A.; Cooreman, M.P.; Heijtink, R.A. Localization of a unique hepatitis B virus epitope sheds new light on the structure of hepatitis B virus surface antigen. *J. Gen. Virol.* **1999**, *80*, 2121–2126. [CrossRef] [PubMed]

62. Mangold, C.M.T.; Unckell, F.; Werr, M.; Streeck, R.E. Analysis of intermolecular disulfide bonds and free sulfhydryl groups in hepatitis B surface antigen particles. *Arch. Virol.* **1997**, *142*, 2257–2267. [CrossRef] [PubMed]

63. Mangold, C.M.T.; Streeck, R.E. Mutational analysis of the cysteine residues in the hepatitis B virus small envelope protein. *J. Virol.* **1993**, *67*, 4588–4597. [CrossRef]

64. Mangold, C.M.T.; Unckell, F.; Werr, M.; Streeck, R.E. Secretion and antigenicity of hepatitis B virus small envelope proteins lacking cysteines in the major antigenic region. *Virology* **1995**, *211*, 535–543. [CrossRef]

65. Cheong, W.S.; Hyakumura, M.; Yuen, L.; Warner, N.; Locarnini, S.; Netter, H.J. Modulation of the immunogenicity of virus-like particles composed of mutant hepatitis B virus envelope subunits. *Antivir. Res.* **2012**, *93*, 209–218. [CrossRef]

66. Wunderlich, G.; Bruss, V. Characterization of early hepatitis B virus surface protein oligomers. *Arch. Virol.* **1996**, *141*, 1191–1205. [CrossRef]

67. Patzer, E.J.; Nakamura, G.R.; Yaffe, A. Intracellular transport and secretion of hepatitis B surface antigen in mammalian cells. *J. Virol.* **1984**, *51*, 346–353. [CrossRef]

68. Lu, X.; Mehta, A.; Dwek, R.; Butters, T.; Block, T. Evidence that N-linked glycosylation is necessary for hepatitis B virus secretion. *Virology* **1995**, *213*, 660–665. [CrossRef] [PubMed]

69. Ito, K.; Qin, Y.; Guarnieri, M.; Garcia, T.; Kwei, K.; Mizokami, M.; Zhang, J.; Li, J.; Wands, J.R.; Tong, S. Impairment of hepatitis B virus virion secretion by single-amino-acid substitutions in the small envelope protein and rescue by a novel glycosylation site. *J. Virol.* **2010**, *84*, 12850–12861. [CrossRef] [PubMed]

70. Julithe, R.; Abou-Jaoudé, G.; Sureau, C. Modification of the hepatitis B virus envelope protein glycosylation pattern interferes with secretion of viral particles, infectivity, and susceptibility to neutralizing antibodies. *J. Virol.* **2014**, *88*, 9049–9059. [CrossRef] [PubMed]

71. Eble, B.E.; Lingappa, V.R.; Ganem, D. The N-terminal (pre-S2) domain of a hepatitis B virus surface glycoprotein is translocated across membranes by downstream signal sequences. *J. Virol.* **1990**, *64*, 1414–1419. [CrossRef] [PubMed]

72. Stibbe, W.; Gerlich, W.H. Characterization of pre-s gene products in hepatitis B surface antigen. *Dev. Biol. Stand.* **1983**, *54*, 33–43.

73. Schmitt, S.; Glebe, D.; Tolle, T.K.; Lochnit, G.; Linder, D.; Geyer, R.; Gerlich, W.H. Structure of pre-S2 N- and O-linked glycans in surface proteins from different genotypes of hepatitis B virus. *J. Gen. Virol.* **2004**, *85*, 2045–2053. [CrossRef]

74. Cheng, K.-C.; Moss, B. Selective synthesis and secretion of particles composed of the hepatitis B virus middle surface protein directed by a recombinant vaccinia virus: Induction of antibodies to pre-S and S epitopes. *J. Virol.* **1987**, *61*, 1286–1290. [CrossRef]

75. McLachlan, A.; Milich, D.R.; Raney, A.K.; Riggs, M.G.; Hughes, J.L.; Sorge, J.; Chisari, F.V. Expression of hepatitis B virus surface and core antigens: Influences of pre-S and precore sequences. *J. Virol.* **1987**, *61*, 683–692. [CrossRef]

76. Molnar-Kimber, K.L.; Jarocki-Witek, V.; Dheer, S.K.; Vernon, S.K.; Conley, A.J.; Davis, A.R.; Hung, P.P. Distinctive properties of the hepatitis B virus envelope proteins. *J. Virol.* **1988**, *62*, 407–416. [CrossRef]

77. Sheu, S.Y.; Lo, S.J. Biogenesis of the hepatitis B viral middle (M) surface protein in a human hepatoma cell line: Demonstration of an alternative secretion pathway. *J. Gen. Virol.* **1994**, *75*, 3031–3039. [CrossRef] [PubMed]

78. Fernholz, D.; Galle, P.R.; Stemler, M.; Brunetto, M.; Bonino, F.; Will, H. Infectious hepatitis B virus variant defective in pre-S2 protein expression in a chronic carrier. *Virology* **1993**, *194*, 137–148. [CrossRef] [PubMed]

79. Ostapchuk, P.; Hearing, P.; Ganem, D. A dramatic shift in the transmembrane topology of a viral envelope glycoprotein accompanies hepatitis B viral morphogenesis. *EMBO J.* **1994**, *13*, 1048–1057. [CrossRef] [PubMed]

80. Prange, R.; Streeck, R.E. Novel transmembrane topology of the hepatitis B virus envelope proteins. *EMBO J.* **1995**, *14*, 247–256. [CrossRef]

81. Seitz, S.; Iancu, C.; Volz, T.; Mier, W.; Dandri, M.; Urban, S.; Bartenschlager, R. A slow maturation process renders hepatitis B virus infectious. *Cell Host Microbe* **2016**, *20*, 25–35. [CrossRef]

82. Ni, Y.; Sonnabend, J.; Seitz, S.; Urban, S. The pre-S2 domain of the hepatitis B virus is dispensable for infectivity but serves a spacer function for L-protein-connected virus assembly. *J. Virol.* **2010**, *84*, 3879–3888. [CrossRef]

83. Bruss, V.; Gerhardt, E.; Vieluf, K.; Wunderlich, G. Functions of the large hepatitis B virus surface protein in viral particle morphogenesis. *Intervirology* **1996**, *39*, 23–31. [CrossRef]

84. Gripon, P.; Le Seyec, J.; Rumin, S.; Guguen-Guillouzo, C. Myristylation of the hepatitis B virus large surface protein is essential for viral infectivity. *Virology* **1995**, *213*, 292–299. [CrossRef]

85. Xu, Z.; Bruss, V.; Yen, T.S.B. Formation of intracellular particles by hepatitis B virus large surface protein. *J. Virol.* **1997**, *71*, 5487–5494. [CrossRef]

86. Persing, D.H.; Varmus, H.E.; Ganem, D. Inhibition of secretion of hepatitis B surface antigen by a related presurface polypeptide. *Science* **1986**, *234*, 1388–1391. [CrossRef]

87. Ou, J.H.; Rutter, W.J. Regulation of secretion of the hepatitis B virus major surface antigen by the preS-1 protein. *J. Virol.* **1987**, *61*, 782–786. [CrossRef] [PubMed]

88. Heermann, K.H.; Goldmann, U.; Schwartz, W.; Seyffarth, T.; Baumgarten, H.; Gerlich, W.H. Large surface proteins of hepatitis B virus containing the pre-s sequence. *J. Virol.* **1984**, *52*, 396–402. [CrossRef] [PubMed]

89. Gavilanes, F.; Gonzalez-Ros, J.M.; Peterson, D.L. Structure of hepatitis B surface antigen. *J. Biol. Chem.* **1982**, *257*, 7770–7777. [PubMed]

90. Peterson, D.L. The structure of hepatitis B surface antigen and its antigenic sites. *BioEssays* **1987**, *6*, 258–262. [CrossRef]

91. Shih, C.; Li, L.S.; Roychoudhury, S.; Ho, M.H. In vitro propagation of human hepatitis B virus in a rat hepatoma cell line. *Proc. Natl. Acad. Sci. USA* **1989**, *86*, 6323–6327. [CrossRef]

92. Hruska, J.F.; Robinson, W.S. The proteins of hepatitis B Dane particle cores. *J. Med. Virol.* **1977**, *1*, 119–131. [CrossRef]

93. Sukeno, N.; Shirachi, R.; Yamaguchi, J.; Ishida, N. Reduction and reoxidation of Australia antigen: Loss and reconstitution of particle structure and antigenicity. *J. Virol.* **1972**, *9*, 182–183. [CrossRef]

94. Vyas, G.N.; Rao, K.R.; Ibrahim, A.B. Australia antigen (hepatitis B antigen): A conformational antigen dependent of disulfide bonds. *Science* **1972**, *178*, 1300–1301. [CrossRef]

95. Imai, M.; Gotoh, A.; Nishioka, K.; Kurashina, S.; Miyakawa, Y.; Mayumi, M. Antigenicity of reduced and alkylated Australia antigen. *J. Immunol.* **1974**, *112*, 416–419.

96. Huovila, A.-P.J.; Eder, A.M.; Fuller, S.D. Hepatitis B surface antigen assembles in a post-ER, pre-Golgi compartment. *J. Cell Biol.* **1992**, *118*, 1305–1320. [CrossRef]

97. Patzer, E.J.; Nakamura, G.R.; Simonsen, C.C.; Levinson, A.D.; Brands, R. Intracellular assembly and packaging of hepatitis B surface antigen particles occur in the endoplasmic reticulum. *J. Virol.* **1986**, *58*, 884–892. [CrossRef] [PubMed]

98. Satoh, O.; Umeda, M.; Imai, H.; Tunoo, H.; Inoue, K. Lipid composition of hepatitis B surface antigen particles and the particle-producing human hepatoma cell lines. *J. Lipid Res.* **1990**, *31*, 1293–1300. [PubMed]

99. Satoh, O.; Imai, H.; Yoneyama, T.; Miyamura, T.; Utsumi, H.; Inoue, K.; Umeda, M. Membrane structure of the hepatitis B virus surface antigen particle. *J. Biochem.* **2000**, *127*, 543–550. [CrossRef] [PubMed]

100. Gavilanes, F.; Gomez-Gutierrez, J.; Aracil, M.; Gonzales-Ros, J.M.; Ferragut, J.A.; Guerrero, E.; Peterson, D.L. Hepatitis B surface antigen. Role of lipids in maintaining the structural and antigenic properties of protein components. *Biochem. J.* **1990**, *265*, 857–864. [CrossRef]

101. Guerrero, E.; Gavilanes, F.; Peterson, D.L. Model for the protein arrangement in HBsAg particles based on physical and chemical studies. In *Viral Hepatitis and Liver Disease*; Zuckerman, A.J., Ed.; Alan, R. Liss Inc.: New York, NY, USA, 1988; pp. 606–613.

102. Greiner, V.J.; Egelé, C.; Oncul, S.; Ronzon, F.; Manin, C.; Klymchenko, A.; Mély, Y. Characterization of the lipid and protein organization in HBsAg viral particles by steady-state and time-resolved fluorescence spectroscopy. *Biochimie* **2010**, *92*, 994–1002. [CrossRef]

103. Milhiet, P.-E.; Dosset, P.; Godefroy, C.; Le Grimellec, C.; Guigner, J.M.; Larquet, E.; Ronzon, F.; Manin, C. Nanoscale topography of hepatitis B antigen particles by atomic force microscopy. *Biochimie* **2011**, *93*, 254–259. [CrossRef]

104. Blumberg, B.S. Australia antigen and the biology of hepatitis B. *Science* **1977**, *197*, 17–25. [CrossRef]

105. Courouce-Pauty, A.M.; Delons, S.; Soulier, J.P. Attempt to prevent hepatitis by using specific anti-HBs immunoglobulin. *Am. J. Med. Sci.* **1975**, *270*, 375–383.

106. Iwarson, S.; Kjellman, H.; Ahlmén, J.; Ljunggren, C.; Eriksson, E.; Selander, D.; Hermodsson, S. Hepatitis B immune serum globulin and standard gamma globulin in prevention of hepatitis B infection among hospital staff: A preliminary report. *Am. J. Med. Sci.* **1975**, *270*, 385–389. [CrossRef]

107. Hilleman, M.R.; Buynak, E.B.; Roehm, R.R.; Tytell, A.A.; Bertland, A.U.; Lampson, G.P. Purified and inactivated human hepatitis B vaccine: Progress report. *Am. J. Med. Sci.* **1975**, *270*, 401–404. [CrossRef]

108. Purcell, R.H.; Gerin, J.L. Hepatitis B subunit vaccine: A preliminary report of safety and efficacy tests in chimpanzees. *Am. J. Med. Sci.* **1975**, *270*, 395–399. [CrossRef] [PubMed]

109. Okochi, K.; Murakami, S.; Ninomiya, K.; Kaneko, M. Australia antigen, transfusion and hepatitis. *Vox Sang.* **1970**, *18*, 289–300. [CrossRef]

110. London, W.T.; Drew, J.S.; Lustbader, E.D.; Werner, B.G.; Blumberg, B.S. Host responses to hepatitis B infection in patients in a chronic hemodialysis unit. *Kidney Int.* **1977**, *12*, 51–58. [CrossRef] [PubMed]

111. Krugman, S.; Giles, J.P.; Hammond, J. Viral Hepatitis, type B (MS-2 strain). Studies on active immunization. *JAMA* **1971**, *217*, 41–45. [CrossRef]

112. Buynak, E.B.; Roehm, R.R.; Tytell, A.A.; Bertland, A.U., II; Lampson, G.P.; Hilleman, M.R. Vaccine against human hepatitis B. *J. Am. Med. Assoc.* **1976**, *235*, 2832–2834. [CrossRef]

113. Gerin, J.L.; Holland, P.V.; Purcell, R.H. Australia antigen: Large-scale purification from human serum and biochemical studies of its proteins. *J. Virol.* **1971**, *7*, 569–576. [CrossRef]

114. Maupas, P.; Goudeau, A.; Coursaget, P.; Drucker, J.; Bagros, P. Immunisation against hepatitis B in man. *Lancet* **1976**, *307*, 1367–1370. [CrossRef]

115. McMahon, B.J.; Helminiak, C.; Wainwright, R.B.; Bulkow, L.; Trimble, B.A.; Wainwright, K. Frequency of adverse reactions to hepatitis B vaccine in 43,618 persons. *Am. J. Med.* **1992**, *92*, 254–256. [CrossRef]

116. Szmuness, W.; Stevens, C.E.; Zang, E.A.; Harley, E.J.; Kellner, A. A controlled clinical trial of the efficacy of the hepatitis B vaccine (Heptavax B): A final report. *Hepatol* **1981**, *1*, 377–385. [CrossRef]

117. Shaw, F.E.; Graham, D.J.; Guess, H.A.; Milstien, J.B.; Johnson, J.M.; Schatz, G.C.; Hadler, S.C.; Kuritsky, J.N.; Hiner, E.E.; Bregman, D.J.; et al. Postmarketing surveillance for neurologic adverse events reported after hepatitis B vaccination. Experience of the first three years. *Am. J. Epidemiol.* **1988**, *127*, 337–352. [CrossRef]

118. Neurath, A.R.; Strick, N.; Kent, S.B.H.; Offensperger, W.; Wahl, S.; Christman, J.K.; Acs, G. Enzyme-linked immunoassay of pre-S gene-coded sequences in hepatitis B vaccines. *J. Virol. Method* **1985**, *12*, 185–192. [CrossRef]

119. Neurath, A.R.; Kent, S.B.H.; Strick, N.; Taylor, P.; Stevens, C.E. Hepatitis B virus contains pre-S gene-encoded domains. *Nature* **1985**, *315*, 154–156. [CrossRef] [PubMed]

120. Coursaget, P.; Bringer, L.; Sarr, G.; Bourdil, C.; Fritzell, B.; Blondeau, C.; Yvonnet, B.; Chiron, J.P.; Jeannée, E.; Guindo, S.; et al. Comparative immunogenicity in children of mammalian cell-derived recombinant hepatitis B vaccine and plasma-derived hepatitis B vaccine. *Vaccine* **1992**, *10*, 379–382. [CrossRef]

121. Zuckerman, A.J. Hepatitis-B vaccine. Safety criteria and non-B infection. *Lancet* **1976**, *307*, 1396–1397. [CrossRef]

122. Hilleman, M.R.; Ellis, R. Vaccines made from recombinant yeast cells. *Vaccine* **1986**, *4*, 75–76. [CrossRef]

123. Stephenne, J. Recombinant versus plasma-derived hepatitis B vaccines: Issues of safety, immunogenicity and cost-effectiveness. *Vaccine* **1988**, *6*, 299–303. [CrossRef]

124. Coursaget, P.; Adamowicz, P.; Bourdil, C.; Yvonnet, B.; Buisson, Y.; Barrès, J.L.; Saliou, P.; Chiron, J.P.; Diop Mar, I. Anti-preS2 antibodies in natural hepatitis B virus infection and after immunization. *Vaccine* **1988**, *6*, 357–361. [CrossRef]

125. Lau, J.Y.N.; Lai, C.L.; Wu, P.C.; Lin, H.J. Comparison of two plasma-derived hepatitis B vaccines: Long-term report of a prospective, randomized trial. *J. Gastroenterol. Hepatol.* **1989**, *4*, 331–337. [CrossRef]

126. Lee, H.S.; Kim, C.Y. Seroepidemiology of HBV and HCV in Korea: The decreasing prevalence rate of HBV infection after launching HB vaccination program. *Int. Hepatol. Commun.* **1996**, *5*, 53–61. [CrossRef]

127. Kalayanarooj, S.; Vaughn, D.W.; Ariyasriwatana, C.; Snitbhan, R. Protective antibody after a 'one dollar' hepatitis B vaccine. *Southeast Asian J. Trop. Med. Public Health* **1996**, *27*, 659–663.

128. Schoub, B.D.; Matai, U.; Singh, B.; Blackburn, N.K.; Levin, J.B. Universal immunization of infants with low doses of a low-cost, plasma-derived hepatitis B vaccine in South Africa. *Bull. World Health Organ.* **2002**, *80*, 277–281. [PubMed]

129. Coates, T.; Wilson, R.; Patrick, G.; André, F.; Watson, V. Hepatitis B vaccines: Assessment of the seroprotective efficacy of two recombinant DNA vaccines. *Clin. Ther.* **2001**, *23*, 392–403. [CrossRef]

130. Hernández-Bernal, F.; Aguilar-Betancourt, A.; Aljovin, V.; Arias, G.; Valenzuela, C.; de Alejo, K.P.; Hernández, K.; Oquendo, O.; Figueredo, N.; Figueroa, N.; et al. Comparison of four recombinant hepatitis B vaccines applied on an accelerated schedule in healthy adults. *Hum. Vaccine* **2011**, *7*, 1026–1036. [CrossRef] [PubMed]

131. Suzuki, H.; Iino, S.; Shiraki, K.; Akahane, Y.; Okamoto, H.; Domoto, K.; Mishiro, S. Safety and efficacy of a recombinant yeast-derived pre-S2 + S-containing hepatitis B vaccine (TGP-943): Phase 1, 2 and 3 clinical testing. *Vaccine* **1994**, *12*, 1090–1096. [CrossRef]

132. Teles, S.A.; Martins, R.M.B.; Lopes, C.L.R.; dos Santos Carneiro, M.A.; Souza, K.P.; Yoshida, C.F.T. Immunogenicity of a recombinant hepatitis B vaccine (Euvax-B) in haemodialysis patients and staff. *Eur.J. Epidemiol.* **2001**, *17*, 145–149. [CrossRef]

133. Abraham, P.; Mistry, F.P.; Bapat, M.R.; Sharma, G.; Reddy, G.R.; Prasad, K.S.N.; Ramanna, V. Evaluation of a new recombinant DNA hepatitis B vaccine (*Shanvac-B*). *Vaccine* **1999**, *17*, 1125–1129. [CrossRef]

134. Anonymous. A two-dose hepatitis B vaccine for adults (Heplisav-B). *Med. Lett. Drugs Ther.* **2018**, *60*, 17–18.

135. Zhu, F.; Deckx, H.; Roten, R.; Michiels, B.; Sarnecki, M. Comparative efficacy, safety and immunogenicity of Hepavax-Gene TF and Engerix-B recombinant hepatitis B vaccines in neonates in China. *Pediatr. Infect. Dis. J.* **2017**, *36*, 94–101. [CrossRef]

136. Shouval, D.; Roggendorf, H.; Roggendorf, M. Enhanced immune response to hepatitis B vaccination through immunization with a Pre-S1/Pre-S2/S vaccine. *Med. Microbiol. Immunol.* **2015**, *204*, 57–68. [CrossRef]

137. Young, M.D.; Schneider, D.L.; Zuckerman, A.J.; Du, W.; Dickson, B.; Maddrey, W.C. Adult hepatitis B vaccination using a novel triple antigen recombinant vaccine. *Hepatology* **2001**, *34*, 372–376. [CrossRef]

138. Valenzuela, P.; Medina, A.; Rutter, W.J.; Ammerer, G.; Hall, B.D. Synthesis and assembly of hepatitis B virus surface antigen particles in yeast. *Nature* **1982**, *298*, 347–350. [CrossRef] [PubMed]

139. Harford, N.; Cabezon, T.; Crabeel, M.; Simoen, E.; Rutgers, A.; De Wilde, M. Expression of hepatitis B surface antigen in yeast. *Dev. Biol. Stand.* **1983**, *54*, 125–130. [PubMed]

140. Miyanohara, A.; Toh-e, A.; Nozaki, C.; Hamada, F.; Ohtomo, N.; Matsubara, K. Expression of hepatitis B surface antigen gene in yeast. *Proc. Natl. Acad. Sci. USA* **1983**, *80*, 1–5. [CrossRef] [PubMed]

141. André, F.E. Overview of a 5-year clinical experience with a yeast-derived hepatitis B vaccine. *Vaccine* **1990**, *8* (Suppl. 1), S74–S78.

142. Wolters, B.; Junge, U.; Dziuba, S.; Roggendorf, M. Immunogenicity of combined hepatitis A and B vaccine in elderly persons. *Vaccine* **2003**, *21*, 3623–3628. [CrossRef]

143. Zhang, Z.; Wang, C.; Liu, Z.; Zou, G.; Li, J.; Lu, M. Host genetic determinants of hepatitis B virus infection. *Front. Genet.* **2019**, *10*, 696. [CrossRef]

144. Li, Z.K.; Nie, J.J.; Li, J.; Zhuang, H. The effec t of HLA on immunological response to hepatitis B vaccine in healthy people: A meta-analysis. *Vaccine* **2013**, *31*, 4355–4361. [CrossRef]

145. Milich, D.R.; Leroux-Roels, G.G. Immunogenetics of the response to HBsAg vaccination. *Autoimmun. Rev.* **2003**, *2*, 248–257. [CrossRef]

146. Cregg, J.M.; Tschopp, J.F.; Stillman, C.; Siegel, R.; Akong, M.; Craig, W.S.; Buckholz, R.G.; Madden, K.R.; Kellaris, P.A.; Davis, G.R.; et al. High-level expression and efficient assembly of hepatitis B surface antigen in the methylotrophic yeast, Pichia pastoris. *Biotechnology* **1987**, *5*, 479–485. [CrossRef]

147. Lünsdorf, H.; Gurramkonda, C.; Adnan, A.; Khanna, N.; Rinas, U. Virus-like particle production with yeast: Ultrastructural and immunocytochemical insights into *Pichia pastoris* producing high levels of the hepatitis B surface antigen. *Microb. Cell Factories* **2011**, *10*, 48. [CrossRef]

148. Zahid, M.; Lünsdorf, H.; Rinas, U. Assessing stability and assembly of the hepatitis B surface antigen into virus-like particles during down-stream processing. *Vaccine* **2015**, *33*, 3739–3745. [CrossRef] [PubMed]

149. Zhao, Q.; Wang, Y.; Freed, D.; Fu, T.-M.; Gimenez, J.A.; Sitrin, R.D.; Washabaugh, M.W. Maturation of recombinant hepatitis B virus surface antigen particles. *Hum. Vaccines* **2006**, *2*, 174–180. [CrossRef] [PubMed]

150. Gurramkonda, C.; Zahid, M.; Nemani, S.K.; Adnan, A.; Gudi, S.K.; Khanna, N.; Ebensen, T.; Lünsdorf, H.; Guzmán, C.A.; Rinas, U. Purification of hepatitis B surface antigen virus-like particles from recombinant *Pichia pastoris* and *in vivo* analysis of their immunogenic properties. *J. Chromatogr. B* **2013**, *940*, 104–111. [CrossRef] [PubMed]

151. Zhao, Q.; Towne, V.; Brown, M.; Wang, Y.; Abraham, D.; Oswald, C.B.; Gimenez, J.A.; Washabaugh, M.W.; Kennedy, R.; Sitrin, R.D. In-depth process understanding of RECOMBIVAX HB® maturation and potential epitope improvements with redox treatment: Multifaceted biochemical and immunochemical characterization. *Vaccine* **2011**, *29*, 7936–7941. [CrossRef] [PubMed]

152. Wampler, D.E.; Lehman, E.D.; Boger, J.; McAleer, W.J.; Scolnick, E.M. Multiple chemical forms of hepatitis B surface antigen produced in yeast. *Proc. Natl. Acad. Sci. USA* **1985**, *82*, 6830–6834. [CrossRef]

153. Yamaguchi, M.; Sugahara, K.; Shiosaki, K.; Mizokami, H.; Takeo, K. Fine structure of hepatitis B virus surface antigen produced by recombinant yeast: Comparison with HBsAg of human origin. *FEMS Microbiol. Lett.* **1998**, *165*, 363–367. [CrossRef]

154. Sonveaux, N.; Conrath, K.; Capiau, C.; Brasseur, R.; Goormaghtigh, E.; Ruysschaert, J.M. The topology of the S protein in the yeast-derived hepatitis B surface antigen particles. *J. Biol. Chem.* **1994**, *269*, 25637–25645.

155. Van der Meeren, P.; Van Criekinge, W.; Vanderdeelen, J.; Baert, L. Phospholipid composition of r-DNA hepatitis B surface antigens. *Int. J. Pharm.* **1994**, *106*, 89–92. [CrossRef]

156. Diminsky, D.; Schirmbeck, R.; Reimann, J.; Barenholz, Y. Comparison between hepatitis B surface antigen (HBsAg) particles derived from mammalian cells (CHO) and yeast cells (*Hansenula polymorpha*): Composition, structure and immunogenicity. *Vaccine* **1997**, *15*, 637–647. [CrossRef]

157. Brown, S.E.; Stanley, C.; Howard, C.R.; Zuckerman, A.J.; Steward, M.W. Antibody responses to recombinant and plasma derived hepatitis B vaccines. *Br. Med. J.* **1986**, *292*, 159–161. [CrossRef]

158. Franchi, L.; Núñez, G. The Nlrp3 inflammasome is critical for aluminium hydroxide-mediated IL-1β secretion but dispensable for adjuvant activity. *Eur. J. Immunol.* **2008**, *38*, 2085–2089. [CrossRef] [PubMed]

159. Li, H.; Willingham, S.B.; Ting, J.P.-Y.; Re, F. Cutting Edge: Inflammasome activation by alum and alum's adjuvant effect are mediated by NLRP3. *J. Immunol.* **2008**, *181*, 17–21. [CrossRef] [PubMed]

160. Kool, M.; Pétrilli, V.; De Smedt, T.; Rolaz, A.; Hammad, H.; van Nimwegen, M.; Bergen, I.M.; Castillo, R.; Lambrecht, B.N.; Tschopp, J. Cutting edge: Alum adjuvant stimulates inflammatory dendritic cells through activation of the NALP3 inflammasome. *J. Immunol.* **2008**, *181*, 3755–3759. [CrossRef] [PubMed]

161. HogenEsch, H. Mechanism of immunopotentiation and safety of aluminum adjuvants. *Front. Immunol.* **2013**, *3*, 406. [CrossRef] [PubMed]

162. Iyer, S.; Robinett, R.S.R.; HogenEsch, H.; Hem, S.L. Mechanism of adsorption of hepatitis B surface antigen by aluminum hydroxide adjuvant. *Vaccine* **2004**, *22*, 1475–1479. [CrossRef] [PubMed]

163. Hansen, B.; Belfast, M.; Soung, G.; Song, L.; Egan, P.M.; Capen, R.; HogenEsch, H.; Mancinelli, R.; Hem, S.L. Effect of the strength of adsorption of hepatitis B surface antigen to aluminum hydroxide adjuvant on the immune response. *Vaccine* **2009**, *27*, 888–892. [CrossRef] [PubMed]

164. Shi, Y.; HogenEsch, H.; Hem, S.L. Change in the degree of adsorption of proteins by aluminum-containing adjuvants following exposure to interstitial fluid: Freshly prepared and aged model vaccines. *Vaccine* **2002**, *20*, 80–85. [CrossRef]

165. Greiner, V.J.; Ronzon, F.; Larquet, E.; Desbat, B.; Estèves, C.; Bonvin, J.; Gréco, F.; Manin, C.; Klymchenko, A.S.; Mély, Y. The structure of HBsAg particles is not modified upon their adsorption on aluminium hydroxide gel. *Vaccine* **2012**, *30*, 5240–5245. [CrossRef]

166. Jackson, S.; Lentino, J.; Kopp, J.; Murray, L.; Ellison, W.; Rhee, M.; Shockey, G.; Akella, L.; Erby, K.; Heyward, W.L.; et al. Immunogenicity of a two-dose investigational hepatitis B vaccine, HBsAg-1018, using a toll-like receptor 9 agonist adjuvant compared with a licensed hepatitis B vaccine in adults. *Vaccine* **2018**, *36*, 668–674. [CrossRef]

167. Neurath, A.R.; Seto, B.; Strick, N. Antibodies to synthetic peptides from the preS1 region of the hepatitis B virus (HBV) envelope (env) protein are virus-neutralizing and protective. *Vaccine* **1989**, *7*, 234–236. [CrossRef]

168. Neurath, A.R.; Kent, S.B.H.; Parker, K.; Prince, A.M.; Strick, N.; Brotman, B.; Sproul, P. Anrtibodies to a synthetic peptide from the preS 120-145 region of the hepatitis B virus envelope are virus-neutralizing. *Vaccine* **1986**, *4*, 35–37. [CrossRef]

169. Itoh, Y.; Takai, E.; Ohnuma, H.; Kitajima, K.; Tsuda, F.; Machida, A.; Mishiro, S.; Nakamura, T.; Miyakawa, Y.; Mayumi, M. A synthetic peptide vaccine involving the product of the pre-S(2) region of hepatitis B virus DNA: Protective efficacy in chimpanzees. *Proc. Natl. Acad. Sci. USA* **1986**, *83*, 9174–9178. [CrossRef] [PubMed]

170. Ferrari, C.; Penna, A.; Bertoletti, A.; Cavalli, A.; Valli, A.; Schianchi, C.; Fiaccadori, F. The preS1 antigen of hepatitis B virus is highly immunogenic at the T cell level in man. *J. Clin. Investig.* **1989**, *84*, 1314–1319. [CrossRef] [PubMed]

171. Milich, D.R. T- and B-cell recognition of hepatitis B viral antigens. *Immunol. Today* **1988**, *9*, 380–386. [CrossRef]

172. Milich, D.R.; McLachlan, A.; Chisari, F.V.; Kent, S.B.H.; Thornton, G.B. Immune response to the pre-S(1) region of the hepatitis B surface antigen (HBsAg): A pre-S(1)-specific T cell response can bypass nonresponsiveness to the pre-S(2) and S regions of HBsAg. *J. Immunol.* **1986**, *137*, 315–322.

173. Shouval, D.; Ilan, Y.; Adler, R.; Deepen, R.; Panet, A.; Even-Chen, Z.; Gorecki, M.; Gerlich, W.H. Improved immunogenicity in mice of a mammalian cell-derived recombinant hepatitis B vaccine containing pre-S$_1$ and pre-S$_2$ antigens as compared with conventional yeast-derived vaccines. *Vaccine* **1994**, *12*, 1453–1459. [CrossRef]

174. Kobayashi, M.; Asano, T.; Utsunomiya, M.; Itoh, Y.; Fijisawa, Y.; Nishimura, O.; Kato, K.; Kakinuma, A. Recombinant hepatitis B virus surface antigen carrying the pre-S2 region derived from yeast: Purification and characterization. *J. Biotechnol.* **1988**, *8*, 1–22. [CrossRef]

175. Itoh, Y.; Fujisawa, Y. Synthesis in yeast of hepatitis B virus surface antigen modified P31 particles by gene modification. *Biochem. Biophys. Res. Commun.* **1986**, *141*, 942–948. [CrossRef]

176. De Wilde, M.; Rutgers, T.; Cabezon, T.; Hauser, P.; van Opstal, O.; Harford, N.; van Wijnendaele, F.; Desmons, P.; Comberbach, M.; Roelants, P.; et al. PreS-containing HBsAg particles from *Saccharomyces cerevisiae*: Production, antigenicity, and immunogenicity. In *Viral Hepatitis and Liver Disease*; Hollinger, F.B., Lemon, S.M., Margolis, H., Eds.; Williams & Wilkins: Baltimore, MD, USA, 1991; pp. 732–736.

177. Kuroda, S.; Fujisawa, Y.; Iino, S.; Akahane, Y.; Suzuki, H. Induction of protection level of anti-pre-S2 antibodies in humans immunized with a novel hepatitis B vaccine consisting of M (pre-S2+S) protein particles (a third generation vaccine). *Vaccine* **1991**, *9*, 163–169. [CrossRef]

178. Fujisawa, Y.; Kuroda, S.; Van Eerd, P.M.C.A.; Schellekens, H.; Kakinuma, A. Protective efficacy of a novel hepatitis B vaccine consisting of M (pre-S2+S) protein particles (a third generation vaccine). *Vaccine* **1990**, *8*, 192–198. [CrossRef]

179. Kniskern, P.J.; Hagopian, A.; Burke, P.; Dunn, N.; Emini, E.A.; Miller, W.J.; Yamazaki, S.; Ellis, R.W. A candidate vaccine for hepatitis B containing the complete viral surface protein. *Hepatology* **1988**, *8*, 82–87. [CrossRef] [PubMed]

180. Biemans, R.; Thines, D.; Rutgers, T.; de Wilde, M.; Cabezon, T. The large surface protein of hepatitis B virus is retained in the yeast endoplasmic reticulum and provokes its unique enlargement. *DNA Cell Biol.* **1991**, *10*, 191–200. [CrossRef] [PubMed]

181. Kuroda, S.; Otaka, S.; Miyazaki, T.; Nakao, M.; Fujisawa, Y. Hepatitis B virus envelope L protein particles. *J. Biol. Chem.* **1992**, *267*, 1953–1961. [PubMed]

182. Yamada, T.; Iwabuki, H.; Kanno, T.; Tanaka, H.; Kawai, T.; Fukuda, H.; Kondo, A.; Seno, M.; Tanizawa, K.; Kuroda, S. Physicochemical and immunological characterization of hepatitis B virus envelope particels exclusively consisting of the entire L (pre-S1+pre-S2+S) protein. *Vaccine* **2001**, *19*, 3154–3163. [CrossRef]

183. Leroux-Roels, G.; Desombere, I.; De Tollenaere, G.; Petit, M.A.; Desmons, P.; Hauser, P.; Delem, A.; De Grave, D.; Safary, A. Hepatitis B vaccine containing surface antigen and selected preS1 and preS2 sequences. 1. Safety and immunogenicity in young, healthy adults. *Vaccine* **1997**, *15*, 1724–1731. [CrossRef]

184. Leroux-Roels, G.; Desombere, I.; Cobbaut, L.; Petit, M.A.; Desmons, P.; Hauser, P.; Delem, A.; De Grave, D.; Safary, A. Hepatitis B vaccine containing surface antigen and selected preS1 and preS2 sequences. 2. Immunogenicity in poor responders to hepatitis B vaccines. *Vaccine* **1997**, *15*, 1732–1736. [CrossRef]

185. Tron, F.; Degos, F.; Bréchot, C.; Couroucé, A.M.; Goudeau, A.; Marie, F.N.; Adamowicz, P.; Saliou, P.; Laplanche, A.; Benhamou, J.P.; et al. Randomized dose range study of a recombinant hepatitis B vaccine produced in mammalian cells and containing the S and preS2 sequences. *J. Infect. Dis.* **1989**, *160*, 199–204. [CrossRef]

186. Yan, D.; Wei, Y.Q.; Guo, H.C.; Sun, S.Q. The application of virus-like particles as vaccines and biological vehicles. *Appl. Microbiol. Biotechnol.* **2015**, *99*, 10415–10432. [CrossRef]

187. Pushko, P.; Pumpens, P.; Grens, E. Development of virus-like particle technology from small highly symmetric to large complex virus-like particle structures. *Intervirology* **2013**, *56*, 141–165. [CrossRef]

188. Ramasamy, V.; Arora, U.; Shukla, R.; Poddar, A.; Shanmugam, R.K.; White, L.J.; Mattocks, M.M.; Raut, R.; Perween, A.; Tyagi, P.; et al. A tetravalent virus-like particle vaccine designed to display domain III of dengue envelope proteins induces multi-serotype neutralizing antibodies in mice and macaques which confer protection against antibody dependent enhancement in AG129 mice. *PLoS Negl. Trop. Dis.* **2018**, *12*, 0006191. [CrossRef]

189. Berkower, I.; Raymond, M.; Muller, J.; Spadaccini, A.; Aberdeen, A. Assembly, structure, and antigenic properties of virus-like particles rich in HIV-1 envelope gp120. *Virology* **2004**, *321*, 75–86. [CrossRef] [PubMed]

190. Gordon, D.M.; McGovern, T.W.; Krzych, U.; Cohen, J.C.; Schneider, I.; LaChance, R.; Heppner, D.G.; Yuan, G.; Hollingdale, M.; Slaoui, M.; et al. Safety, immunogenicity, and efficacy of a recombinantly produced *Plasmodium falciparum* circumsporozoite protein-hepatitis B surface antigen subunit vaccine. *J. Infect. Dis.* **1995**, *171*, 1576–1585. [CrossRef] [PubMed]

191. Rutgers, T.; Gordon, D.; Gathoye, A.M.; Hollingdale, M.; Hockmeyer, W.; Rosenberg, M.; De Wilde, M. Hepatitis B surface antigen as carrier matrix for the repetitive epitope of the circumsporozoite protein of *Plasmodium falciparum*. *Bio/Technology* **1988**, *6*, 1065–1070. [CrossRef]

192. Beaumont, E.; Patient, R.; Hourioux, C.; Dimier-Poisson, I.; Roingeard, P. Chimeric hepatitis B virus/hepatitis C virus envelope proteins elicit broadly neutralizing antibodies and constitute a potential bivalent prophylactic vaccine. *Hepatology* **2013**, *57*, 1303–1313. [CrossRef] [PubMed]

193. Cervantes Gonzalez, M.; Kostrzak, A.; Guetard, D.; Pniewski, T.; Sala, M. HIV-1 derived peptides fused to HBsAg affects its immunogenicity. *Virus Res.* **2009**, *146*, 107–114. [CrossRef] [PubMed]

194. Schlienger, K.; Mancini, M.; Rivière, Y.; Dormont, D.; Tiollais, P.; Michel, M.L. Human immunodeficiency virus type 1 major nautralizing determinant exposed on hepatitis B surface antigen particles is highly immunogenic in primates. *J. Virol.* **1992**, *66*, 2570–2576. [CrossRef]

195. Kingston, N.J.; Kurtovic, L.; Walsh, R.; Joe, C.; Lovrecz, G.; Locarnini, S.; Beeson, J.G.; Netter, H.J. Hepatitis B virus-like particles expressing *Plasmodium falciparum* epitopes induce complement-fixing antibodies against the circumsporozoite protein. *Vaccine* **2019**, *37*, 1674–1684. [CrossRef]

196. Wei, S.; Lei, Y.; Yang, J.; Wang, X.; Shu, F.; Wei, X.; Lin, F.; Li, B.; Cui, Y.; Zhang, H.; et al. Neutralization effects of antibody elicited by chimeric HBV S antigen viral-like particles presenting HCV neutralization epitopes. *Vaccine* **2018**, *36*, 2273–2281. [CrossRef]

197. Kotiw, M.; Johnson, M.; Pandey, M.; Fry, S.; Hazell, S.L.; Netter, H.J.; Good, M.F.; Olive, C. Immunological response to parenteral vaccination with recombinant hepatitis B virus surface antigen virus-like particles expressing *Helicobacter pylori* KatA epitopes in a murine *H. pylori* challenge model. *Clin. Vaccine Immunol.* **2012**, *19*, 268–276. [CrossRef]

198. Vietheer, P.T.K.; Boo, I.; Drummer, H.E.; Netter, H.J. Immunizations with chimeric hepatitis B virus-like particles to induce potential anti-hepatitis C virus neutralizing antibodies. *Antivir. Ther.* **2007**, *12*, 477–487.

199. Pumpens, P.; Razanskas, R.; Pushko, P.; Renhof, R.; Gusars, I.; Skrastina, D.; Ose, V.; Borisova, G.; Sominskaya, I.; Petrovskis, I.; et al. Evaluation of HBs, HBc, and frCP virus-like particles for expression of human papillomavirus 16 E7 oncoprotein epitopes. *Intervirology* **2002**, *45*, 24–32. [CrossRef]

200. von Brunn, A.; Früh, K.; Müller, H.M.; Zentgraf, H.W.; Bujard, H. Epitopes of the human malaria parasite P. *falciparum* carried on the surface of HBsAg particles elicit an immune response against the parasite. *Vaccine* **1991**, *9*, 477–484. [CrossRef]

201. Delpeyroux, F.; Chenciner, N.; Lim, A.; Malpièce, Y.; Blondel, B.; Crainic, R.; van der Werf, S.; Streeck, R.E. A poliovirus neutralization epitope expressed on hybrid hepatitis B surface antigen particles. *Science* **1986**, *233*, 472–475. [CrossRef] [PubMed]

202. Cheong, W.S.; Reiseger, J.; Turner, S.J.; Boyd, R.; Netter, H.J. Chimeric virus-like particles for the delivery of an inserted influenza A-specific CTL epitope. *Antivir. Res.* **2009**, *81*, 113–122. [CrossRef] [PubMed]

203. Delpeyroux, F.; Peillon, N.; Blondel, B.; Crainic, R.; Streeck, R.E. Presentation and immunogenicity of the hepatitis B surface antigen and a poliovirus neutralization antigen on mixed empty envelope particles. *J. Virol.* **1988**, *62*, 1836–1839. [CrossRef]

204. Netter, H.J.; Macnaughton, T.B.; Woo, W.P.; Tindle, R.; Gowans, E.J. Antigenicity and immunogenicity of novel chimeric hepatitis B surface antigen particles with exposed hepatitis C virus epitopes. *J. Virol.* **2001**, *75*, 2130–2141. [CrossRef]

205. Casares, S.; Brumeanu, T.D.; Richie, T.L. The RTS, S malaria vaccine. *Vaccine* **2010**, *28*, 4880–4894. [CrossRef]

206. RTS,S Clinical Trials Partnership. Efficacy and safety of RTS,S/AS01 malaria vaccine with or without a booster dose in infants and children in Africa: Final results of a phase 3, individually randomised, controlled trial. *Lancet* **2015**, *386*, 31–45. [CrossRef]

207. Kester, K.E.; Cummings, J.F.; Ofori-Anyinam, O.; Ockenhouse, C.F.; Krzych, U.; Moris, P.; Schwenk, R.; Nielsen, R.A.; Debebe, Z.; Pinelis, E.; et al. Randomized, double-blind, phase 2a trial of falciparum malaria vaccines RTS,S/AS01B and RTS,S/AS02A in malaria-naive adults: Safety, efficacy, and immunologic associates of protection. *J. Infect. Dis.* **2009**, *200*, 337–346. [CrossRef]

208. Collins, K.A.; Snaith, R.; Cottingham, M.G.; Gilbert, S.C.; Hill, A.V.S. Enhancing protective immunity to malaria with a highly immunogenic virus-like particle vaccine. *Sci. Rep.* **2017**, *7*, 46621. [CrossRef]

209. Patient, R.; Hourioux, C.; Vaudin, P.; Pagès, J.C.; Roingeard, P. Chimeric hepatitis B and C viruses envelope proteins can form subviral particles: Implications for the design of new vaccine strategies. *New Biotechnol.* **2009**, *25*, 226–234. [CrossRef] [PubMed]

210. McCluskie, M.J.; Evans, D.M.; Zhang, N.; Benoit, M.; McElhiney, S.P.; Unnithan, M.; DeMarco, S.C.; Clay, B.; Huber, C.; Deora, A.; et al. The effect of pre-existing anti-carrier immunity on subsequent responses to CRM197 or Qb-VLP conjugate vaccines. *Immunopharmacol. Immunotoxicol.* **2016**, *38*, 184–196. [CrossRef] [PubMed]

211. Saxena, M.; Van, T.T.H.; Baird, F.J.; Coloe, P.J.; Smooker, P.M. Pre-existing immunity against vaccine vectors—Friend or foe? *Microbiology* **2013**, *159*, 1–11. [CrossRef] [PubMed]

212. Jegerlehner, A.; Wiesel, M.; Dietmeier, K.; Zabel, F.; Gatto, D.; Saudan, P.; Bachmann, M.F. Carrier induced epitopic suppression of antibody responses induced by virus-like particles is a dynamic phenomenon caused by carrier-specific antibodies. *Vaccine* **2010**, *28*, 5503–5512. [CrossRef]

213. Schutze, M.P.; Leclerc, C.; Jolivet, M.; Audibert, F.; Chedid, L. Carrier-induced epitopic suppression, a major issue for future synthetic vaccines. *J. Immunol.* **1985**, *135*, 2319–2322.

214. Beaumont, E.; Roingeard, P. Chimeric hepatitis B virus (HBV)/hepatitis C virus (HCV) subviral envelope particles induce efficient anti-HCV antibody production in animals pre-immunized with HBV vaccine. *Vaccine* **2015**, *33*, 973–976. [CrossRef]

215. Netter, H.J.; Woo, W.P.; Tindle, R.; Macfarlan, R.I.; Gowans, E.J. Immunogenicity of recombinant HBsAg/HCV particles in mice pre-immunised with hepatitis B virus-specific vaccine. *Vaccine* **2003**, *21*, 2692–2697. [CrossRef]

216. Bojang, K.A.; Milligan, P.J.M.; Pinder, M.; Vigneron, L.; Alloueche, A.; Kester, K.E.; Ballou, W.R.; Conway, D.J.; Reece, W.H.H.; Gothard, P.; et al. Efficacy of RTS,S/AS02 malaria vaccine against *Plasmodium falciparum* infection in semi-immune adult men in The Gambia: A randomised trial. *Lancet* **2001**, *358*, 1927–1934. [CrossRef]

217. Li, P.; Haque, M.A.; Blum, J.S. Role of disulfide bonds in regulating antigen processing and epitope selection. *J. Immunol.* **2002**, *169*, 2444–2450. [CrossRef]

218. Carmicle, S.; Steede, N.K.; Landry, S.J. Antigen three-dimensional structure guides the processing and presentation of helper T-cell epitopes. *Mol. Immunol.* **2007**, *44*, 1159–1168. [CrossRef] [PubMed]

219. Prato, S.; Fleming, J.; Schmidt, C.W.; Corradin, G.; Lopez, J.A. Cross-presentation of a human malaria CTL epitope is conformation dependent. *Mol. Immunol.* **2006**, *43*, 2031–2036. [CrossRef] [PubMed]

220. Dai, G.; Carmicle, S.; Steede, N.K.; Landry, S.J. Structural basis for helper T-cell and antibody epitope immunodominance in bacteriophage T4 Hsp10. *J. Biol. Chem.* **2002**, *277*, 161–168. [CrossRef] [PubMed]

221. Mirano-Bascos, D.; Steede, N.K.; Robinson, J.E.; Landry, S.J. Influence of disulfide-stabilized structure on the specificity of helper T-cell and antibody responses to HIV envelope glycoprotein gp120. *J. Virol.* **2010**, *84*, 3303–3311. [CrossRef] [PubMed]

222. Landry, S.J. Three-dimensional structure determines the pattern of CD4+ T-cell epitope dominance in influenza virus hemagglutinin. *J. Virol.* **2008**, *82*, 1238–1248. [CrossRef]

223. Wolfert, M.A.; Boons, G.J. Adaptive immune activation: Glycosylation does matter. *Nat. Chem. Biol.* **2013**, *9*, 776–784. [CrossRef]

224. Dam, T.K.; Brewer, C.F. Lectins as pattern recognition molecules: The effects of epitope density in innate immunity. *Glycobiology* **2010**, *20*, 270–279. [CrossRef]

225. Al-Barwani, F.; Young, S.L.; Baird, M.A.; Larsen, D.S.; Ward, V.K. Mannosylation of virus-like particles enhances internalization by antigen-presenting cells. *PLoS ONE* **2014**, *9*, e104523. [CrossRef]

226. Freire, T.; Zhang, X.; Dériaud, E.; Ganneau, C.; Vichier-Guerre, S.; Azria, E.; Launay, O.; Lo-Man, R.; Bay, S.; Leclerc, C. Glycosidic Tn-based vaccines targeting dermal dendritic cells favor germinal center B-cell development and potent antibody response in the absence of adjuvant. *Blood* **2010**, *116*, 3526–3536. [CrossRef]

227. Sheng, K.C.; Kalkanidis, M.; Pouniotis, D.S.; Esparon, S.; Tang, C.K.; Apostolopoulos, V.; Pietersz, G.A. Delivery of antigen using a novel mannosylated dendrimer potentiates immunogenicity in vitro and in vivo. *Eur. J. Immunol.* **2008**, *38*, 424–436. [CrossRef]

228. Doe, B.; Steimer, K.S.; Walker, C.M. Induction of HIV-1 envelope (gp120)-specific cytotoxic T lymphocyte responses in mice by recombinant CHO cell-derived gp120 is enhanced by enzymatic removal of N-linked glycans. *Eur. J. Immunol.* **1994**, *24*, 2369–2376. [CrossRef]

229. Mishra, H.; Mishra, D.; Mishra, P.K.; Nahar, M.; Dubey, V.; Jain, N.K. Evaluation of solid lipid nanoparticles as carriers for delivery of hepatitis B surface antigen for vaccination using subcutaneous route. *J. Pharm. Pharm. Sci.* **2010**, *13*, 495–509. [CrossRef] [PubMed]

230. Vigerust, D.J.; Shepherd, V.L. Virus glycosylation: Role in virulence and immune interactions. *Trends Microbiol.* **2007**, *15*, 211–218. [CrossRef] [PubMed]

231. Hyakumura, M.; Walsh, R.; Thaysen-Andersen, M.; Kingston, N.J.; La, M.; Lu, L.; Lovrecz, G.; Packer, N.H.; Locarnini, S.; Netter, H.J. Modification of asparagine-linked glycan density for the design of hepatitis B virus virus-like particles with enhanced immunogenicity. *J. Virol.* **2015**, *89*, 11312–11322. [CrossRef] [PubMed]

232. Lopez, M.; Rodriguez, E.N.; Lobaina, Y.; Musacchio, A.; Falcon, V.; Guillen, G.; Aguilar, J.C. Characterization of the size distribution and aggregation of virus-like nanoparticles used as active ingredients of the HeberNasvac therapeutic vaccine against chronic hepatitis B. *Adv. Nat. Sci. Nanosci. Nanotechnol.* **2017**, *8*, 025009. [CrossRef]

Review

Prophylactic Hepatitis E Vaccines: Antigenic Analysis and Serological Evaluation

Yike Li [1], Xiaofen Huang [1], Zhigang Zhang [1], Shaowei Li [1,2], Jun Zhang [1], Ningshao Xia [1,2] and Qinjian Zhao [1,*]

1 State Key Laboratory of Molecular Vaccinology and Molecular Diagnostics, National Institute of Diagnostics and Vaccine Development in Infectious Diseases, School of Public Health, Xiamen University, Xiamen 361102, Fujian, China; yike_li@163.com (Y.L.); xiaofen_huang@stu.xmu.edu.cn (X.H.); zhigang_zhang@stu.xmu.edu.cn (Z.Z.); shaowei@xmu.edu.cn (S.L.); zhangj@xmu.edu.cn (J.Z.); nsxia@xmu.edu.cn (N.X.)

2 State Key Laboratory of Molecular Vaccinology and Molecular Diagnostics, National Institute of Diagnostics and Vaccine Development in Infectious Diseases, School of Life Sciences, Xiamen University, Xiamen 361102, Fujian, China

* Correspondence: qinjian_zhao@xmu.edu.cn; Tel.: +86-59-2218-0936

Received: 10 December 2019; Accepted: 13 January 2020; Published: 16 January 2020

Abstract: Hepatitis E virus (HEV) infection causes sporadic outbreaks of acute hepatitis worldwide. HEV was previously considered to be restricted to resource-limited countries with poor sanitary conditions, but increasing evidence implies that HEV is also a public health problem in developed countries and regions. Fortunately, several vaccine candidates based on virus-like particles (VLPs) have progressed into the clinical development stage, and one of them has been approved in China. This review provides an overview of the current HEV vaccine pipeline and future development with the emphasis on defining the critical quality attributes for the well-characterized vaccines. The presence of clinically relevant epitopes on the VLP surface is critical for eliciting functional antibodies against HEV infection, which is the key to the mechanism of action of the prophylactic vaccines against viral infections. Therefore, the epitope-specific immunochemical assays based on monoclonal antibodies (mAbs) for HEV vaccine antigen are critical methods in the toolbox for epitope characterization and for in vitro potency assessment. Moreover, serological evaluation methods after immunization are also discussed as biomarkers for clinical performance. The vaccine efficacy surrogate assays are critical in the preclinical and clinical stages of VLP-based vaccine development.

Keywords: antigenic analysis; epitope characterization; hepatitis E vaccine; serological evaluation; virion-like epitopes; well-characterized vaccines

---

1. Introduction

Hepatitis E virus (HEV) belongs to the genus *Orthohepevirus* within the family *Hepeviridae*. HEV was first identified in the 1970s, and its genome was successfully sequenced a decade later [1]. One third of the population in the world may be infected with HEV during their lifetime [2]. HEV infection usually causes acute, self-limiting hepatitis and may cause chronic hepatitis among immunocompromised individuals and solid organ transplant recipients. Moreover, HEV infection poses a threat to pregnant women, with a mortality of 10–50% [3]. A study on the global burden of the disease estimated that approximately 20.1 million people were infected with HEV, leading annually to 3.4 million symptomatic cases, 70,000 deaths and 3000 stillbirths [4]. The burden of HEV-related disease is considerable in developing countries, especially in Africa and Asia [5]. Although there is no evidence of outbreaks of hepatitis E in developed countries, sporadic locally acquired cases of HEV infections have been reported in France [6], Germany [7], Switzerland [8], Australia [9], Japan [10] and so on. To date, eight genotypes of HEV have been isolated, and at least five genotypes (genotypes 1–4 and 7) can cause human infection [11–13].

HEV contains a single-stranded, positive-sense RNA with a size of approximately 7.2 kb. The genome of HEV comprises three open reading frames (ORFs) [14]. ORF1 encodes a non-structural protein that is responsible for viral RNA replication [15]. Recently, a protein encoded by the newly discovered ORF4 (within ORF1) was shown to be able to stimulate the replication of genotype-1 HEV [16]. ORF3 encodes a phosphoprotein, which is involved in virus release from host cells [14]. HEV was recently found to be a quasi-enveloped virus. It exists as non-enveloped virions in faeces and urine for transmission, whereas its form could be predominantly enveloped in serum for evading neutralizing antibodies [17–19]. The existence of two different forms of virions might be related to the function of ORF3 [20]. Notably, the most well-studied ORF, ORF2, encodes the sole capsid protein (pORF2), which has the ability to self-assemble into viral capsids to package the viral RNA after each replication cycle. Therefore, the capsid protein pORF2 is a rational target for vaccine design [21].

Based on pORF2, different truncations using recombinant DNA technology were performed to yield various proteins with different assembly forms [22]. Some of the truncated proteins form into virus-like particles (VLPs) or subviral particles, presenting virion-like epitopes on the particle surface [23]. In this review, we use the term VLP in a loose term regardless of the size of the VLPs. Some VLPs could be more virion-like, including the size and the array of the epitope, while others could be smaller in size and less regular in particle formation, which are essential for effectively stimulating the immune response. Among these truncated pORF2 molecules, three vaccine candidates have been studied in clinical trials. One vaccine, with a trade name of Hecolin®, was licensed in China in 2011. However, it was not prequalified by the World Health Organization (WHO), which was necessary for introduction into countries where the disease burden was considerable [24]. The WHO issued a recommendation in 2018 to provide guidance to national regulatory authorities and manufacturers on the manufacturing process and on nonclinical and clinical aspects to assure the quality, safety and efficacy of recombinant hepatitis E vaccines [3]. Needless to say, quality assurance is of utmost importance for a licensed vaccine for widespread use. For the VLP-based vaccines, the presence of virion-like epitopes on the surface of VLPs is the structural basis to elicit protection against the specific pathogen [25]. The process comparability, analytical comparability, and the link of a change to the clinical outcome were considered as essential issues for a VLP-based vaccine. One A-vax case study discussed the risk assessment and control strategy, which supports comparability studies of vaccines [26]. Various methods have been established as a "toolbox" to evaluate the vaccine antigen for quality assurance during bioprocessing and for stability during storage and transportation. Among them, a large panel of monoclonal antibodies (mAbs) against the HEV capsid protein was developed, and assays based on these mAbs are used during bioprocessing and in lot release and stability testing [25]. In addition, immunization with a vaccine could induce a response against viral B cell epitopes in vaccine recipients [27]. The functional polyclonal antibodies in serum elicited by immunization can be analysed by a series of methods. In this review, we focus on epitope-specific antigenic analyses of hepatitis E vaccine antigen and evaluation assays of vaccine-elicited functional antibodies for its overall neutralization activity or against well-characterized epitopes.

2. The Design of Hepatitis E Vaccines

2.1. Molecular Structure of Different Truncated Versions of pORF2

HEV ORF2 encodes a viral capsid protein containing 660 amino acids (aa). A recent study reported that a secreted form of pORF2 was observed in serum samples from both HEV-infected rhesus macaques and humans. Two different forms, pORF2^C (capsid) and pORF2^S (secreted), are two different translation products of the same viral ORF2 gene [28]. Compared to ORF2^C, ORF2^S contains an additional 15 aa. This 15-aa peptide segment could represent a signal sequence that drives ORF2^S secretion [28]. The prolonged existence of ORF2^S in the blood raises the possibility of decoying, leading to partial or full depletion of neutralizing antibodies in the sera of convalescent patients. Thus, ORF2^S antigen in HEV-infected patient serum could reduce the protective efficiency after vaccination. Further characterization of ORF2^S, such as the complexing forms and kinetics in blood is of clinical importance [28].

To test the immunogenicity of the capsid protein, a series of truncated forms of pORF2 were prepared in different laboratories [22]. They include the nearly full-length p595 (aa 14–608) and p495 (aa 112–606), as well as much smaller particulate forms, such as p239 (aa 368–606) and p179 (aa 439–617), and the even shorter proteins, such as E2 (aa 394–606) and E2s (aa 459–606), that are amenable for crystallization (Figure 1A). The N- and C-terminally truncated version of pORF2, i.e., aa 14–608 (p595), can form T = 3 icosahedral VLPs that are highly analogous to native virions [29]. Additional highly virion-like VLPs were observed with p495, with a further truncation from p595 containing aa 112–606. P495 self-assembled into well-formed T = 1 icosahedral VLPs [30]. Through analysis of pORF2 using cryo-EM and X-ray crystallographic studies, three functional domains (S, P1 and P2) were identified [29]. The P2 domain, also named the E2s domain, harbours the major neutralizing epitopes on the viral capsid. The E2s domain, which has available high-resolution structural data, forms a tight homodimer with key interacting residues identified at the dimeric interface [31]. The E2s domain may play a key role in the interaction of HEV with host cells because antibodies directed towards this region tend to be neutralizing and functional. Moreover, the structure of the E2s domain in a complex with a potent neutralizing mAb, 8C11 [32], was determined. The epitopes recognized by the mAb 8C11 were identified to comprise three different peptide segments, namely, Asp496-Thr499, Val510-Leu514, and Asn573-Arg578. Among these, Arg512 was found to be the most crucial residue for 8C11 interaction with and neutralization of HEV. In addition, the complex crystal structure of 8G12 [33], another highly efficient neutralizing mAb, with E2s was also determined. Several important residues (Glu549, Lys554 and Gly591) were revealed that played an important role in 8G12 neutralization.

A 66-aa extension from the N terminus of E2s and could stabilize E2 (aa 394–606) to form hexamers in solution [34]. Further extended with another 26-aa extension on E2, p239 (aa 368–606) could self-assemble into a particulate form or VLPs with a 20–30 nm diameter [35]. Although there is a certain degree of particle size and particle regularity, the immunogenicity of the p239-based antigen was quite impressive. Notably, in experiments conducted in parallel, the immunogenicity of p239-based VLPs was shown to be approximately 240 times stronger than that of E2 in mice [35]. The enhanced immunogenicity is likely due to the multiplicity of arrayed virion-like epitopes on the VLP surface. More recently, p179 (aa 439–617) was found to self-assemble into VLPs with a diameter of approximately 20 nm [36]. This VLP form of the antigen also shares the major neutralization epitopes with p239. This antigen, along with the previously mentioned p239 and p495, was tested in clinical trials with the goal of human vaccine development (Figure 1B).

2.2. Hepatitis E Vaccines

While more than fourteen HEV vaccine candidates have been studied [37], only three candidates have progressed to clinical trials (Figure 1B). Aluminium-based adjuvants were used in the formulation of all three vaccine candidates. The p495-based vaccine was the first vaccine candidate evaluated in clinical trials with the support of GlaxoSmithKline. It was produced in insect cells with a baculovirus expression system. Desirable safety and efficacy in a phase II clinical trial study were demonstrated for the p495-based vaccine. The vaccine efficacy was 95.5% after three immunizations. In addition, the reports of any adverse event were similar in vaccine group and placebo group [38]. However, this project did not progress further after the phase II clinical trial possibly due to the lack of commercial value. Another vaccine antigen based on p239 was expressed in an *Escherichia coli* (*E. coli*) expression system. The p239-based vaccine was developed and licensed, with safety and efficacy demonstrated in a large-scale phase III clinical trial. The efficacy was 100% over 12 months in preventing hepatitis E among participants receiving all three doses of the immunization [39]. Moreover, a follow-up study showed that immunization with the p239-based vaccine could provide long-term (up to 4.5 years) protection against hepatitis E, with an efficacy of 86.8% [40]. A post-licensure study showed that the p239-based vaccine was immunogenic and well tolerated in the elderly population (>65 years old), setting the stage for expanded recommendation of the vaccine to the aged populations in whom HEV infection could be more harmful [41]. More recently, a p179 (expressed in *E. coli*)-based vaccine candidate was tested in a phase I clinical trial. This study showed safety and good tolerance for the 16- to 65-year-old population [36] (Figure 1B), and a phase II clinical trial is ongoing. The recreation of the neutralizing epitopes on the truncated pORF2 forms as vaccine antigens is the key for assuring the elicitation of functional antibodies. The existence of these virion-like epitopes can be characterized using a series of immunochemical techniques.

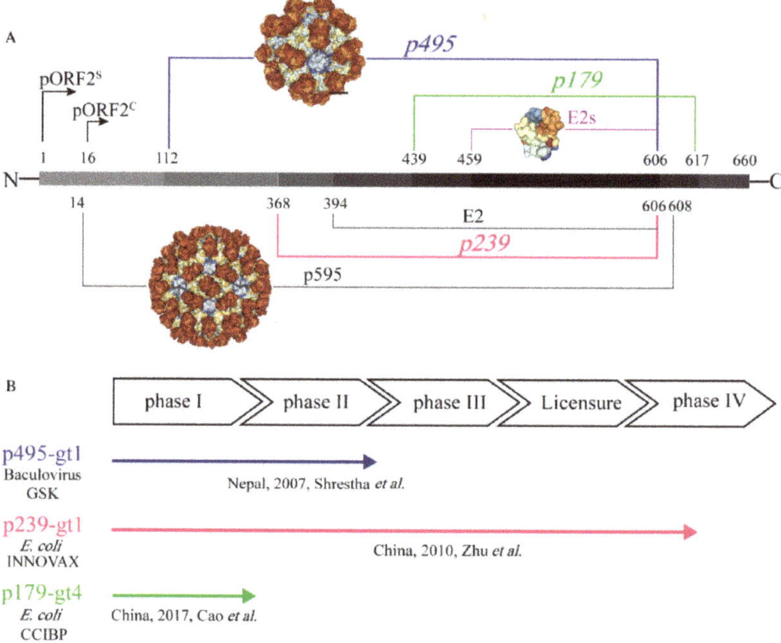

Figure 1. Presentation of different truncated versions of hepatitis E virus (HEV) pORF2. (**A**) shows the molecular structure of truncated pORF2, and (**B**) shows three existing HEV vaccines, which have been studied in clinical trials. HEV pORF2 consists of 660 amino acids. HEV p595 (aa 14–608) can form a virus-like particle (VLP) that is similar to the native virion. The structure of p595 was demonstrated by cryo-EM. HEV p495 (aa 112–608) can form a VLP, and the structure has been determined by X-ray. HEV p495 was used as a vaccine antigen manufactured by GSK, which showed good safety and efficacy in a phase II clinical trial. HEV p239 (aa 368–606), named Hecolin®, has been licensed in China. The HEV p179 (aa 439−617)-based vaccine, which was manufactured by Changchun Institute of Biological Products Co., Ltd. (CCIBP), was safe and well tolerated in a phase I clinical trial. E2 was a useful candidate for diagnostic reagents and was able to form hexamers in solution. The structure of E2s (aa 459–606), the shortest version to form a dimer harbouring the major neutralizing epitopes, was determined at a high resolution.

3. Analysis of the Native-Like Epitopes on Recombinant Antigens

A recombinant VLP-based antigen is an ideal candidate for use in vaccine formulation due to its high immunogenicity and desirable safety performance. Comprehensive analysis of a vaccine antigen is important for the quality assurance of a vaccine. Therefore, it is essential to develop a series of methods to evaluate the vaccine antigen during processing, formulation and storage/transportation. Various methods, such as biophysical, biochemical, immunochemical and in vivo potency assays, were established and applied to two licensed recombinant VLP-based vaccines: hepatitis B vaccine and human papillomavirus vaccine [42]. For the hepatitis E vaccine, the toolbox was composed of various analytical approaches, especially specific epitope-based immunochemical assays, which will be discussed in the following section. Among all the critical quality attributes of prophylactic vaccines, the presence and the integrity of the virion-like epitopes on the recombinant antigens are critically important for eliciting functional antibodies and conferring protection against viral infection [25].

3.1. Epitope-Specific Analysis of the HEV Vaccine Antigen by mAbs

The presence of clinically relevant epitopes on the VLP surface is normally the structural basis for eliciting functional antibodies against viral proteins against the viral capsid. For HEV, the E2s domain, an important part of pORF2, was shown to harbour the major neutralizing epitopes. To study the clinically relevant epitopes, recombinant p239 protein was used as an immunogen for the preparation of a large panel (96) of murine monoclonal antibodies (mAbs) [43]. Among them, approximately 50% of the mAbs were chosen for further analysis due to their significant reactivity with p239. Overall, 20% of the mAbs recognized linear epitopes (such as 3A11, 16D7, 12A10, *etc.*), and 30% of the mAbs recognized conformational epitopes (such as 8G12, 8C11, 9F7, *etc.*) on the E2s domain. There are 23 conformation-dependent mAbs that were selected for further study due to their ability to capture authentic HEV virions in vitro. Six distinct conformation-dependent epitopes (C1-C6) were identified in the E2s domain by clustering analyses [43]. As a representative conformational and neutralizing antibody, 8G12 could react with p239 dimer and capture HEV virions. The epitopes recognized by 8G12 were located around the E2s dimerization interface, and the key epitope residues were Glu549 and Gly59 in the E2s domain [33]. Another neutralizing antibody, 8C11, recognized C5 epitopes, which are located in the groove zone of the E2s domain, away from the dimeric interface [32]. Due to the neutralizing activity of these two well-characterized mAbs, 8C11 and 8G12, their targets likely represent two distinct functional epitopes on the HEV viral capsid. These epitopes are likely the clinically relevant epitopes. Thus, based on 8C11, 8G12 and other mAbs, various immunochemical assays were developed. These assays, along with a battery of other physico-chemical assays, have been used to evaluate the critical quality attributes of the vaccine products.

Surface plasma resonance (SPR)-based BIAcore is a one-site binding assay probing the antibody-antigen interaction on a sensorchip in real time. The assay is label-free, with no need to label the antibody or the antigen. Due to the high throughput and the degree of automation, the antigenicity for multiple batches of p239 was measured using a panel of 5 mAbs. They included mAbs that recognize linear epitopes (12A10 and 3A11) and mAbs that recognize conformational epitopes (8G12, 8C11 and 12F12). Good lot-to-lot consistency and desirable stability of the HEV p239 antigen was demonstrated [25]. Another one-site binding assay is a solution competition enzyme-linked immunosorbent assay (ELISA), where one mAb is used for interrogating the antigen immune reactivity in solution. As an example, the solution interaction of recombinant HBsAg and the mAb 5F11 was monitored for quality analysis of the antigen [44]. This assay could detect subtle differences in antigen epitopes in solution, avoiding potential conformational changes during the surface adsorption process [45]. Using this solution competition ELISA, comparable antigenicity among different lots of p239 was demonstrated with multiple antibodies, namely, 8C11, 8G12, 9F7, 12A10, etc. [25].

Vaccine formulation in general contains particulate-form adjuvants, making antigenicity more difficult. Recently, a new in situ antigenicity analysis method, high content analysis (HCA), which is capable of antigenicity analysis of aluminium-adsorbed antigen without the need for dissolution, has been developed [46]. Using fluorescence-labelled 8G12, the real-time stability of multiple lots of p239-based vaccines was demonstrated after long-term storage. HCA-based in situ analysis is a fluorescence-based measurement. Thus, it enabled the simultaneous detection of two or more differently labelled mAbs. These could be multiple mAbs against the same antigen or against different antigens in the vaccine formulation. Zhang et al. [47] reported the antigenic analysis of VLP-based antigens adsorbed on adjuvants without dissolution using two distinctly fluorescence-labelled mAbs. HEV VLP-based vaccines or HEV antigens in combination vaccines could be analysed using technology with multiple detection antibodies. In addition, based on two-site binding, a highly sensitive and robust sandwich ELISA with the mAb 3A11 as the capture antibody and the mAb 8C11 as the detection antibody was developed for routine antigenicity testing. Comparable antigenicity of p239 pre- or post-dissolution treatment was determined, showing no change in epitopes after proper dissolution treatment with adjuvants [48]. Meanwhile, multiple lots of hepatitis E vaccines retained antigenicity after 36 months of storage [46]. As a robust technical assay, the sandwich ELISA has the potential to become a product release assay since this assay format is amenable for set up in a manufacturing setting.

3.2. The Application of an In Vitro Relative Potency Assay (IVRP)

The potency assay is the most critical for vaccine characterization, comparability evaluation and lot-release testing. Currently, potency assays for vaccines can be roughly divided into in vivo animal-based potency assays and in vitro relative potency assays [49]. Traditional evaluation of vaccine potency is an in vivo animal-based potency assay, which is the closest mimic of a human response to a vaccine. It is widely used in the preclinical development stage of a vaccine. However, in vivo animal-based potency assays are time-consuming (4–6 weeks) and exhibit poor precision and high relative standard deviations. In addition, based on the "3R" principle (reduction, replacement and refinement of animals), it is highly desirable to establish an alternative in vitro potency assay to assess the binding activity of the antigen to functional antibodies during bioprocessing [26,50]. Therefore, additional alternative in vitro assays could be implemented to minimize animal use. With good reproducibility and robustness, the sandwich ELISA has the potential to be a candidate in vitro relative potency assay for product release [51]. Notably, Sandwich ELISA was chosen as in vitro alternatives to evaluate the potency of human rabies vaccines and was accepted for an international collaborative study [52]. In addition, it has been reported that IVRP has a good correlation with animal-based potency assays for licensed human papillomavirus and hepatitis B virus vaccines [53,54]. For the hepatitis E vaccine, the correlation between animal-based efficacy and IVRP needs to be evaluated in the future. In general, a mouse potency assay, as indicated by an ED_{50} value, is a potency assessment for a vaccine formulation to cause seroconversion in the sera of the test animals. Either binding antibodies or neutralizing antibodies could be evaluated via such a seroconversion analysis. This fact is also true for clinical serological testing using either binding titres or neutralization titres. While the latter is more desirable, its limited throughput would normally exclude its use in support of clinical sample testing.

4. The Serological Evaluation

Like most prophylactic VLP-based vaccines, the hepatitis E vaccine elicits a strong humoral response. Effective presentation with orderly arrayed epitopes on the antigen surface and high local epitope density could account for the effective B cell response. The generation of neutralizing antibodies against clinically relevant epitopes is the underlying mechanism of efficacious prophylactic vaccines [49]. In general, the titre of neutralizing antibodies elicited by vaccination correlates with the efficacy of the specific vaccine. In addition, the p239-based hepatitis E vaccine could also stimulate a cell-mediated immune response. Khateri et al. [55] demonstrated via IFN-γ ELISPOT assay that p239 vaccination was able to induce cellular immunity. However, due to limited research, the contribution of cellular immunity to protection against HEV infection is not clear. In the following section, a series of quantitative methods for evaluation of the B cell response elicited by the hepatitis E vaccine are discussed.

4.1. The Binding Antibody Analysis

While the clinical end point for vaccine efficacy in a clinical trial could be a disease or a pathological marker, the seroconversion induced by vaccines is always a quantitative index to be measured, owing to the vaccination with a test vaccine. The binding titre in serum samples in an ELISA is always a straightforward method for serological analysis [56]. In ELISAs for measuring titres, 3 recombinant proteins and VLPs were used as coating antigens. For the hepatitis E vaccine, the coating antigens are generally the same or similar proteins as the vaccine antigens derived from ORF2 proteins. For the p239-based vaccine, anti-HEV IgG in serum samples from vaccinated humans and rhesus monkeys was analysed via E2-coated ELISA [57] (Table 1). For another vaccine candidate based on p495, the total immunoglobulin in serum samples derived from cynomolgus monkeys after immunization was evaluated in a p495-coated ELISA. Anti-HEV antibodies in monkeys' serum samples after receiving two doses of p495-based vaccine were elicited with titres of $1:10^4$, and monkeys developed neither hepatitis nor viremia when challenged with virus [58].

Table 1. Different antigens used in the ELISA methods for supporting the vaccine clinical trials or preclinical development.

Vaccine Antigen.	Antibody Type	Species	Coating Antigen	Reference
p495	total immunoglobulin	rhesus monkey human	p495 (112–606)	Tsarev et al., 1997 [58] Shrestha et al., 2007 [38]
p239	IgG	rhesus monkey human	E2 (394–606) *	Li et al., 2005 [35] Zhu et al., 2010 [39]

* E2 and p239 share most of the critical epitopes.

Since different vaccine developers use different coating antigens, standardization of the serological assay would facilitate cross-lab and cross-product comparison if the same coating antigen could be used in the assay. Wen et al. [59] investigated the immunogenicity differences between hepatitis E vaccines based on p239 and p179. HEV p166 (aa 452–617) was used as a coating antigen to evaluate anti-HEV IgG in the serum of mice and humans immunized with p179 or p239 [59]. However, p166 was composed of aa 452–617, which is much shorter than p239 and may exclude some important functional epitopes. With this caveat in mind, a p495-based VLP antigen was more virion-like and contained all functional epitopes of p179 and p239 (Figure 1A), making it a better coating antigen. Similarly, efforts are being made to standardize the human papillomavirus serological assays using a more native- or virion-like VLPs as coating antigen in the ELISAs. Although the cross-lab comparability of ELISAs for human papillomavirus−16 and −18 has been formed, international standards are still required for the additional types in Gardasil®9 [56]. In most cases, correlations were observed between the binding titres and neutralizing titres. This result supports the use of binding titre measurement to support the clinical trials while using a subset of the clinical samples to establish such a correlation.

4.2. The Neutralizing Antibody Analysis

An evaluation method of virus neutralizing efficiency of serum samples is essential to assess the vaccine-elicited immune response against the virus. However, inefficient propagation of HEV in cell models has posed a challenge to the evaluation of neutralizing antibodies in sera. The propagation and production of HEV were attempted in primary hepatocytes. Based on this cell system, an in vitro neutralization assay was developed to evaluate the anti-HEV antibodies. However, the difficulty in culturing the cells has limited its application [60]. To date, at least two cell culture systems for HEV have been developed. They are the culture system in PLC/PRF/5 and A549 cells established by

Tanaka et al. [61] and the culture system in HepG2/C3A cells established by Shukla et al. [62] (Table 2). These cell culture systems allow the virus to propagate enough authentic HEV virions to support the neutralizing analysis [63]. Based on these HEV cell culture systems, polyclonal antibodies in serum samples or mAbs derived from immunized animals showed neutralizing ability against different viruses (Table 2). In one study, the HEV genotype 3 strain Kernow was shown to have high replication efficiency, using HepG2/C3A as a cell substrate. The infectivity of HEV was inhibited efficiently by the serum samples from immunized individuals [64] (Table 2). Apart from polyclonal antibodies in serum, several mAbs showed neutralizing ability. The neutralizing activity of two represent mAbs (8G12 and H6225) was studied with different cell systems. Gu et al. [33] reported that 8G12 was able to inhibit the replication of HEV based on Huh7 cells. H6225 was able to neutralize a genotype 3 HEV strain (JE03-1760F) in PLC/PRF/5 cells [65] (Table 2). Although many improvements were made, there was still a lack of an efficient and reliable cell-culture system for an HEV virus neutralization assay. Therefore, finding a surrogate of native virions may be a good choice to develop cell-based assays as HEV serological assays for vaccine evaluation.

Table 2. The neutralization analysis of monoclonal or polyclonal antibodies after vaccination or infection.

Host Cell	Virus/Genotype	Type of Sample	Reference
Native virus			
HepG2/C3A cells	Kernow (gt3)	serum (rhesus macaque & human)	Liu et al., 2019 [64] *
Huh7 cells	stool and bile-derived HEV (gt1/4)	mAb 8G12	Gu et al., 2015 [33]
PLC/PRF/5 cells	JE03-1760F (gt3)	mAb H6225	Takahashi et al., 2008 [65]
Primary hepatocytes	Burma (gt1)	purified IgGs (cynomolgus monkey)	Tam et al., 1997 [66]
PLC/PRF/5 cells	F23, SAR-55 (gt1)	serum (cynomolgus monkey & human)	Meng et al., 1997 [67]
HepG2/C3A cells	Sar55, Mex14, Meng (gt1)	serum (rhesus macaque)	Emerson et al., 2006 [68]
PLC/PRF/5 and A549 cells	JE03-1760F (gt3)	serum (human)	Tanaka et al., 2007 [61]
PLC/PRF/5 and A549 cells	serum-derived HEV (gt3)	serum (human)	Takahashi et al., 2010 [20]
Virus surrogate			
HepG2 cells	recombinant protein 239	serum (rhesus macaque)	Cai et al., 2016 [69]

"*" Using similar cell lines and virus strain, a robust HEV infection and replication system was reported recently (Todt et al. Robust hepatitis E virus infection and transcriptional response in human hepatocytes. *Proc Natl Acad Sci* 2020) with high virus titres obtained consistently. It is conceivable that the use of this system should facilitate the development of a virus neutralization assay.

The vaccine antigen p239 was used as a surrogate for native HEV virions in a cell-based functional assay for antibodies. He et al. [70] used p239 to simulate native HEV for virus attachment onto hepatocytes. P239 could attach to and enter the cells of four susceptible cell lines, i.e., HepG2, Huh7, PLC/PRF5 and A549 [71]. When neutralizing mAbs were used, the cell attachment of p239 was effectively blocked. More recently, a "neutralizing-like" blocking assay based on HepG2 cells was developed by Cai et al. [69] for antibody functionality assessment (Table 2). The assay was based on biotin-conjugated p239 and staining with allophycocyanin-conjugated streptavidin to amplify the fluorescence signal. Using this assay, the p239-blocking activity of serum samples from HEV-infected and vaccinated macaques was quantitatively evaluated [69].

With a well-characterized murine mAb (8G12), a competitive ELISA was developed for functional antibody evaluation of serum samples. 8G12 could efficiently block the binding of polyclonal antibodies in immunized human and rhesus macaque serum to vaccine antigen. 8G12 was applied to develop a competitive ELISA assay to detect 8G12-like antibodies in mouse and human serum samples. The 8G12-like antibody was predominant among the vaccine-induced anti-HEV antibodies in both human and mouse sera. Therefore, 8G12-like antibodies might be a promising surrogate for neutralizing antibodies and have the potential to be used as an indicator of the neutralization ability of the hepatitis E vaccine [72]. In another study, the mAb 8C11 was also used in a competitive ELISA to evaluate 8C11-like antibodies in serum samples from immunized mice [73]. In a similar format but with another virus, Palivizumab-like antibodies against respiratory syncytial virus were tested via a competitive ELISA to reflect the neutralization antibodies after vaccination [74]. Similarly, in the case of human papillomavirus, the neutralizing antibody level of each human papillomavirus serotype elicited by vaccination was assessed by a multiplex competitive Luminex immunoassay with functional and type-specific mAbs (such as H16.V5 for type 16) as a specific probe for each type [75].

Recently, some modifications of HEV were performed at the genetic level to quantitatively monitor the infection and replication of HEV. Swiss scientists claimed that HEV genomes harbouring a haemagglutinin epitope tag or a small luciferase were found to be fully functional. This approach could enable the efficient production of infectious viruses with specific tags for ease of virus quantitation [76]. Based on HepG2/C3A cells infected with the HEV genotype 3 strain, virus replication and infection were monitored efficiently. Since these viruses can be easily engineered, different genotypes of HEV strains should be prepared in parallel, facilitating evaluation of the cross-genotype virus neutralization efficiency of various serum samples.

5. Can Current Vaccines Protect against Hetero-Genotypes?

To date, three VLP-based hepatitis E vaccines (p495-, p239-, and p179-based vaccine) have been tested in human volunteers and showed desirable safety and efficacy. Each vaccine antigen was derived from one given virus genotype. However, there are at least five genotypes (genotypes 1–4 and 7) that are capable of infecting humans. In addition, various HEV isolates have been identified from diverse animal species. Recently, rabbit HEV of genotype 3ra was found to be capable of infecting humans in Switzerland [8]. With different circulating strains of HEV in different continents, a question is always asked—is the current vaccine containing one antigen from a given genotype capable of protecting against other different genotypes?

In clinical trials conducted in China, where the majority of HEV isolated from patients is genotype 4, the efficacy of a p239-based vaccine was evaluated by a large-scale randomized double-blind phase III trial. The vaccine antigen derived from genotype 1 could elicit a robust antibody response in most patients, showing excellent cross-genotype protection against HEV [39] (Table 3). While the clinical efficacy data against genotypes 1, 2 and 3 are not available, the cross-genotype protection of the hepatitis E vaccine was also determined in preclinical tests. The p239-based vaccine was efficacious in preventing infection in rhesus monkeys challenged with genotype 1 or genotype 4 virus [35]. For the p495-based vaccine, the cross-genotype protection among non-human primates was also determined among HEV genotypes 1, 2 and 3 [77]. Furthermore, the p179-based vaccine derived from genotype 4 showed complete cross-protection against genotype 1 and genotype 4 virus in rhesus monkeys [36].

Table 3. The protection of the hepatitis E vaccine among different virus genotypes.

Vaccination	Subject	Genotype 1	Genotype 2	Genotype 3	Genotype 4	Reference
Animal- or human-based analysis						
p495-gt1	human	√	-	-	-	Shrestha et al., 2007 [38]
p239-gt1	human	-	-	-	√	Zhu et al., 2010 [39]
p495-gt1	rhesus monkey	√ a	√ b	√ c	-	Purcell et al., 2003 [77]
p495-gt1	rhesus monkey	√ a	√ b	-	-	Tsarev et al., 1997 [58]
p239-gt1	rhesus monkey	√	-	-	√	Li et al., 2005 [35]
p179-gt4	rhesus monkey	√	-	-	√	Cao et al., 2017 [36]
Cell-based analysis						
p239	HepG2/C3A	-	-	√ d	-	Liu et al., 2019 [64]
8G12	Huh7 cells	√	-	-	√	Gu et al., 2015 [33]

"-" means no data; "√" means the cross-genotype protection; "a" SAR-55 (GenBank accession no. AF444002); "b" Mex-14 (GenBank accession no. M74506); "c" US-2 (GenBank accession no. AF060669.1); "d" Kernow (GenBank accession no. JQ679013).

In addition, a series of cell-based analyses also showed cross-genotype protection against HEV. The infection of Kernow (genotype 3) was blocked by vaccinated human serum (genotype 1) based on HepG2 cells, suggesting that the genotype 1 HEV vaccine was able to protect against genotype 3 virus infection [64]. In addition, the mAb 8G12 could inhibit genotype 1 and genotype 4 HEV infections in the Huh7 cell model [33] (Table 3). At the biochemistry level, according to the crystal structure of different forms of truncated pORF2, the protrusion domains (E2s domain) from different genotypes show highly similar structural features. The crystal structures of the immune complexes of 8G12 and the E2s domains (genotypes 1 and 4) were very similar. Thus, 8G12 could efficiently bind to the E2s molecules derived from all 4 different types [33]. All lines of accumulating evidence from serological, biochemical and structural studies indicate that the protrusion domain of the viral capsid plays an important role in protection. These functional epitopes in association with these protrusions are immuno-dominant and are highly conserved across different HEV genotypes. It is conceivable that vaccination with an antigen derived from a given genotype could confer protection from infection by all the HEV genotypes.

6. Conclusions and Prospects

Prophylactic vaccination is an effective method to protect against HEV infection and control HEV infection-associated diseases. The ordered arrangement of the clinically relevant epitopes is the structural basis of a VLP-based hepatitis E vaccine. A multifaceted and comprehensive toolbox composed of orthogonal methods was established to ensure vaccine quality, safety and efficacy. Similar analytical assays, built with the same concept, could also be used for other VLP-based vaccine antigens. Serological evaluation after vaccination should be standardized to allow cross-laboratory comparison when multiple developers are producing vaccines against the same indication.

Outbreaks of HEV infections in low-resource regions continue to pose challenges to public health. Outbreaks were reported in at least 30 countries in Africa, Asia, and North America in recent years. During the past decade (2009–2019), six outbreaks of hepatitis E, each causing at least 1000 infections, were reported in African countries, including South Sudan, Namibia, Uganda, Nigeria, Niger and Chad. The most recent and ongoing (since 2017) HEV outbreak in Namibia has caused at least 4669 infections and 41 deaths [78]. In addition to the outbreaks in developing regions, the high rate of asymptomatic HEV infections worldwide has raised concern of infection via blood donation in recent years. For instance, HEV IgG antibody seroprevalence in blood donors was found to be 19.8% in Denmark, owing to the better awareness and improved assays [79]. Although most patients remain asymptomatic after accepting infected blood product, those immune-suppressed individuals take a considerable risk of developing chronic HEV infection. Use of the licensed vaccine should be further promoted to enhance coverage, particularly for the outbreak-prone regions or among high-risk populations, such as pregnant women or immuno-suppressed or immune-compromised individuals. Recently, the protective effect of the hepatitis E vaccine for pregnant women was evaluated in Bangladesh (NCT02759991) with support from the Norwegian Institute of Public Health, showing no safety concerns for the participants related to immunization [80]. Furthermore, another safety clinical trial of the hepatitis E vaccine was initiated in May 2019 in a healthy US adult population (NCT03827395) with the support of the National Institute of Allergy and Infectious Disease [81]. With efforts from multiple fronts towards the control of this vaccine-preventable disease, the effect of this public health tool should be maximized in the future.

Funding: This work was funded by National Natural Science Foundation of China, grant numbers 31670939, 31730029 and 81993149041.

Conflicts of Interest: The authors declare no conflict of interest.

References

1. Khuroo, M.S. Study of an epidemic of non-A, non-B hepatitis. Possibility of another human hepatitis virus distinct from post-transfusion non-A, non-B type. *Am. J. Med.* **1980**, *68*, 818–824. [CrossRef]
2. Webb, G.W.; Dalton, H.R. Hepatitis E: An underestimated emerging threat. *Adv. Infect. Dis.* **2019**, *6*, 2049936119837162. [CrossRef]

3. World Health Organization. Recommendations to Assure the Quality, Safety and Efficacy of Recombinant Hepatitis E Vaccines. Available online: https://www.who.int/biologicals/BS.2018.2348.Recommendations_HEP_E_vaccines.pdf (accessed on 10 December 2019).

4. World Health Organization. Global Hepatitis Report. 2017. Available online: https://www.who.int/hepatitis/publications/global-hepatitis-report2017/en/ (accessed on 8 January 2020).

5. Kamar, N.; Bendall, R.; Legrand-Abravanel, F.; Xia, N.S.; Ijaz, S.; Izopet, J.; Dalton, H.R. Hepatitis E. *Lancet* **2012**, *379*, 2477–2488. [CrossRef]

6. Gallian, P.; Pouchol, E.; Djoudi, R.; Lhomme, S.; Mouna, L.; Gross, S.; Bierling, P.; Assal, A.; Kamar, N.; Mallet, V.; et al. Transfusion-Transmitted Hepatitis E Virus Infection in France. *Transfus. Med. Rev.* **2019**, *33*, 146–153. [CrossRef]

7. Vollmer, T.; Diekmann, J.; Eberhardt, M.; Knabbe, C.; Dreier, J. Hepatitis E in blood donors: Investigation of the natural course of asymptomatic infection, Germany, 2011. *Euro. Surveill.* **2016**, *21*. [CrossRef]

8. Sahli, R.; Fraga, M.; Semela, D.; Moradpour, D.; Gouttenoire, J. Rabbit HEV in immunosuppressed patients with hepatitis E acquired in Switzerland. *J. Hepatol.* **2019**, *70*, 1023–1025. [CrossRef]

9. Yapa, C.M.; Furlong, C.; Rosewell, A.; Ward, K.A.; Adamson, S.; Shadbolt, C.; Kok, J.; Tracy, S.L.; Bowden, S.; Smedley, E.J.; et al. First reported outbreak of locally acquired hepatitis E virus infection in Australia. *Med. J. Aust.* **2016**, *204*, 274. [CrossRef]

10. Nanmoku, K.; Owada, Y.; Oshiro, Y.; Kurosawa, A.; Kubo, T.; Shinzato, T.; Shimizu, T.; Kimura, T.; Sakuma, Y.; Ishikawa, N.; et al. Prevalence and characteristics of hepatitis E virus infection in kidney transplant recipients: A single-center experience in Japan. *Transpl. Infect. Dis.* **2019**, *21*, e13033. [CrossRef]

11. Donnelly, M.C.; Scobie, L.; Crossan, C.L.; Dalton, H.; Hayes, P.C.; Simpson, K.J. Review article: Hepatitis E-a concise review of virology, epidemiology, clinical presentation and therapy. *Aliment. Pharm.* **2017**, *46*, 126–141. [CrossRef]

12. Lee, G.H.; Tan, B.H.; Teo, E.C.; Lim, S.G.; Dan, Y.Y.; Wee, A.; Aw, P.P.; Zhu, Y.; Hibberd, M.L.; Tan, C.K.; et al. Chronic Infection With Camelid Hepatitis E Virus in a Liver Transplant Recipient Who Regularly Consumes Camel Meat and Milk. *Gastroenterology* **2016**, *150*, 355–357. [CrossRef]

13. Woo, P.C.; Lau, S.K.; Teng, J.L.; Cao, K.Y.; Wernery, U.; Schountz, T.; Chiu, T.H.; Tsang, A.K.; Wong, P.C.; Wong, E.Y.; et al. New Hepatitis E Virus Genotype in Bactrian Camels, Xinjiang, China, 2013. *Emerg. Infect. Dis.* **2016**, *22*, 2219–2221. [CrossRef] [PubMed]

14. Debing, Y.; Moradpour, D.; Neyts, J.; Gouttenoire, J. Update on hepatitis E virology: Implications for clinical practice. *J. Hepatol.* **2016**, *65*, 200–212. [CrossRef] [PubMed]

15. Nan, Y.; Yu, Y.; Ma, Z.; Khattar, S.K.; Fredericksen, B.; Zhang, Y.J. Hepatitis E virus inhibits type I interferon induction by ORF1 products. *J. Virol.* **2014**, *88*, 11924–11932. [CrossRef]

16. Nair, V.P.; Anang, S.; Subramani, C.; Madhvi, A.; Bakshi, K.; Srivastava, A.; Nayak, B.; Surjit, M. Endoplasmic Reticulum Stress Induced Synthesis of a Novel Viral Factor Mediates Efficient Replication of Genotype-1 Hepatitis E Virus. *PLOS Pathog.* **2016**, *12*, e1005521. [CrossRef]

17. Kamar, N.; Izopet, J.; Pavio, N.; Aggarwal, R.; Labrique, A.; Wedemeyer, H.; Dalton, H.R. Hepatitis E virus infection. *Nat. Rev. Dis. Primers* **2017**, *3*, 17086. [CrossRef]

18. Feng, Z.; Hirai-Yuki, A.; McKnight, K.L.; Lemon, S.M. Naked Viruses That Aren't Always Naked: Quasi-Enveloped Agents of Acute Hepatitis. *Annu. Rev. Virol.* **2014**, *1*, 539–560. [CrossRef]

19. Marion, O.; Lhomme, S.; Nayrac, M.; Dubois, M.; Pucelle, M.; Requena, M.; Migueres, M.; Abravanel, F.; Peron, J.M.; Carrere, N.; et al. Hepatitis E virus replication in human intestinal cells. *Gut* **2019**. [CrossRef]

20. Takahashi, M.; Tanaka, T.; Takahashi, H.; Hoshino, Y.; Nagashima, S.; Mizuo, H.; Yazaki, Y.; Takagi, T.; Azuma, M.; Kusano, E.; et al. Hepatitis E Virus (HEV) strains in serum samples can replicate efficiently in cultured cells despite the coexistence of HEV antibodies: Characterization of HEV virions in blood circulation. *J. Clin. Microbiol.* **2010**, *48*, 1112–1125. [CrossRef]

21. Mori, Y.; Matsuura, Y. Structure of hepatitis E viral particle. *Virus Res.* **2011**, *161*, 59–64. [CrossRef]

22. Li, S.W.; Zhao, Q.; Wu, T.; Chen, S.; Zhang, J.; Xia, N.S. The development of a recombinant hepatitis E vaccine HEV 239. *Hum. Vaccin. Immunother.* **2015**, *11*, 908–914. [CrossRef]

23. Zhang, X.; Xin, L.; Li, S.; Fang, M.; Zhang, J.; Xia, N.; Zhao, Q. Lessons learned from successful human vaccines: Delineating key epitopes by dissecting the capsid proteins. *Hum. Vaccin. Immunother.* **2015**, *11*, 1277–1292. [CrossRef]

24. Innis, B.L.; Lynch, J.A. Immunization against Hepatitis E. *Cold Spring Harb. Perspect. Med.* **2018**, *8*, a032573. [CrossRef]

25. Wang, X.; Li, M.; Lin, Z.; Pan, H.; Tang, Z.; Zheng, Z.; Li, S.; Zhang, J.; Xia, N.; Zhao, Q. Multifaceted characterization of recombinant protein-based vaccines: An immunochemical toolbox for epitope-specific analyses of the hepatitis E vaccine. *Vaccine* **2018**, *36*, 7650–7658. [CrossRef]

26. Group, C.-V.W. A-VAX: Applying Quality by Design to Vaccines. 2012. Available online: https://www.dcvmn.org/IMG/pdf/a-vax-applying-qbd-to-vaccines_2012.pdf. (accessed on 10 December 2019).

27. Huang, X.; Wang, X.; Zhang, J.; Xia, N.; Zhao, Q. Escherichia coli-derived virus-like particles in vaccine development. *Npj. Vaccines* **2017**, *2*, 3. [CrossRef]

28. Yin, X.; Ying, D.; Lhomme, S.; Tang, Z.; Walker, C.M.; Xia, N.; Zheng, Z.; Feng, Z. Origin, antigenicity, and function of a secreted form of ORF2 in hepatitis E virus infection. *Proc. Natl. Acad. Sci. USA* **2018**, *115*, 4773–4778. [CrossRef]

29. Yamashita, T.; Mori, Y.; Miyazaki, N.; Cheng, R.H.; Yoshimura, M.; Unno, H.; Shima, R.; Moriishi, K.; Tsukihara, T.; Li, T.C.; et al. Biological and immunological characteristics of hepatitis E virus-like particles based on the crystal structure. *Proc. Natl. Acad. Sci. USA* **2009**, *106*, 12986–12991. [CrossRef]

30. Huang, C.C.; Nguyen, D.; Fernandez, J.; Yun, K.Y.; Fry, K.E.; Bradley, D.W.; Tam, A.W.; Reyes, G.R. Molecular cloning and sequencing of the Mexico isolate of hepatitis E virus (HEV). *Virology* **1992**, *191*, 550–558. [CrossRef]

31. Li, S.; Tang, X.; Seetharaman, J.; Yang, C.; Gu, Y.; Zhang, J.; Du, H.; Shih, J.W.; Hew, C.L.; Sivaraman, J.; et al. Dimerization of hepatitis E virus capsid protein E2s domain is essential for virus-host interaction. *PLOS Pathog.* **2009**, *5*, e1000537. [CrossRef]

32. Tang, X.; Yang, C.; Gu, Y.; Song, C.; Zhang, X.; Wang, Y.; Zhang, J.; Hew, C.L.; Li, S.; Xia, N.; et al. Structural basis for the neutralization and genotype specificity of hepatitis E virus. *Proc. Natl. Acad. Sci. USA* **2011**, *108*, 10266–10271. [CrossRef]

33. Gu, Y.; Tang, X.; Zhang, X.; Song, C.; Zheng, M.; Wang, K.; Zhang, J.; Ng, M.H.; Hew, C.L.; Li, S.; et al. Structural basis for the neutralization of hepatitis E virus by a cross-genotype antibody. *Cell Res.* **2015**, *25*, 604–620. [CrossRef]

34. Zhang, J.Z.; Ng, M.H.; Xia, N.S.; Lau, S.H.; Che, X.Y.; Chau, T.N.; Lai, S.T.; Im, S.W. Conformational antigenic determinants generated by interactions between a bacterially expressed recombinant peptide of the hepatitis E virus structural protein. *J. Med. Virol.* **2001**, *64*, 125–132. [CrossRef]

35. Li, S.W.; Zhang, J.; Li, Y.M.; Ou, S.H.; Huang, G.Y.; He, Z.Q.; Ge, S.X.; Xian, Y.L.; Pang, S.Q.; Ng, M.H.; et al. A bacterially expressed particulate hepatitis E vaccine: Antigenicity, immunogenicity and protectivity on primates. *Vaccine* **2005**, *23*, 2893–2901. [CrossRef]

36. Cao, Y.F.; Tao, H.; Hu, Y.M.; Shi, C.B.; Wu, X.; Liang, Q.; Chi, C.P.; Li, L.; Liang, Z.L.; Meng, J.H.; et al. A phase 1 randomized open-label clinical study to evaluate the safety and tolerability of a novel recombinant hepatitis E vaccine. *Vaccine* **2017**, *35*, 5073–5080. [CrossRef]

37. Cao, Y.; Bing, Z.; Guan, S.; Zhang, Z.; Wang, X. Development of new hepatitis E vaccines. *Hum. Vaccin. Immunother.* **2018**, *14*, 2254–2262. [CrossRef]

38. Shrestha, M.P.; Scott, R.M.; Joshi, D.M.; Mammen, M.P., Jr.; Thapa, G.B.; Thapa, N.; Myint, K.S.; Fourneau, M.; Kuschner, R.A.; Shrestha, S.K.; et al. Safety and efficacy of a recombinant hepatitis E vaccine. *N. Engl. J. Med.* **2007**, *356*, 895–903. [CrossRef]

39. Zhu, F.C.; Zhang, J.; Zhang, X.F.; Zhou, C.; Wang, Z.Z.; Huang, S.J.; Wang, H.; Yang, C.L.; Jiang, H.M.; Cai, J.P.; et al. Efficacy and safety of a recombinant hepatitis E vaccine in healthy adults: A large-scale, randomised, double-blind placebo-controlled, phase 3 trial. *Lancet* **2010**, *376*, 895–902. [CrossRef]

40. Zhang, J.; Zhang, X.F.; Huang, S.J.; Wu, T.; Hu, Y.M.; Wang, Z.Z.; Wang, H.; Jiang, H.M.; Wang, Y.J.; Yan, Q.; et al. Long-term efficacy of a hepatitis E vaccine. *N. Engl. J. Med.* **2015**, *372*, 914–922. [CrossRef]

41. Yu, X.Y.; Chen, Z.P.; Wang, S.Y.; Pan, H.R.; Wang, Z.F.; Zhang, Q.F.; Shen, L.Z.; Zheng, X.P.; Yan, C.F.; Lu, M.; et al. Safety and immunogenicity of hepatitis E vaccine in elderly people older than 65years. *Vaccine* **2019**, *37*, 4581–4586. [CrossRef]

42. Zhang, X.; Wei, M.; Pan, H.; Lin, Z.; Wang, K.; Weng, Z.; Zhu, Y.; Xin, L.; Zhang, J.; Li, S.; et al. Robust manufacturing and comprehensive characterization of recombinant hepatitis E virus-like particles in Hecolin((R)). *Vaccine* **2014**, *32*, 4039–4050. [CrossRef]

43. Zhao, M.; Li, X.J.; Tang, Z.M.; Yang, F.; Wang, S.L.; Cai, W.; Zhang, K.; Xia, N.S.; Zheng, Z.Z. A Comprehensive Study of Neutralizing Antigenic Sites on the Hepatitis E Virus (HEV) Capsid by Constructing, Clustering, and Characterizing a Tool Box. *J. Biol. Chem.* **2015**, *290*, 19910–19922. [CrossRef]

44. Weng, Z.; Zhao, Q. Utilizing ELISA to monitor protein-protein interaction. *Methods Mol. Biol.* **2015**, *1278*, 341–352. [CrossRef] [PubMed]

45. Cao, L.; Wang, X.; Fang, M.; Xia, N.; Zhao, Q. Detection of subtle differences in analogous viral capsid proteins by allowing unrestricted specific interaction in solution competition ELISA. *J. Virol. Methods* **2016**, *236*, 1–4. [CrossRef] [PubMed]

46. Yin, X.; Wang, X.; Zhang, Z.; Li, Y.; Lin, Z.; Pan, H.; Gu, Y.; Li, S.; Zhang, J.; Xia, N.; et al. Demonstration of real-time and accelerated stability of hepatitis E vaccine with a combination of different physicochemical and immunochemical methods. *J. Pharm. Biomed.* **2020**, *177*. [CrossRef] [PubMed]

47. Zhang, Z.; Zhang, T.; Cao, L.; Wang, X.; Cao, J.; Huang, X.; Cai, Y.; Lin, Z.; Pan, H.; Yuan, Q.; et al. Simultaneous in situ visualization and quantitation of dual antigens adsorbed on adjuvants using high content analysis. *Nanomed. (Lond)* **2019**, *14*, 2535–2548. [CrossRef] [PubMed]

48. Zhang, Y.; Li, M.; Yang, F.; Li, Y.; Zheng, Z.; Zhang, X.; Lin, Q.; Wang, Y.; Li, S.; Xia, N.; et al. Comparable quality attributes of hepatitis E vaccine antigen with and without adjuvant adsorption-dissolution treatment. *Hum. Vaccin. Immunother.* **2015**, *11*, 1129–1139. [CrossRef]

49. Zhao, Q.; Li, S.; Yu, H.; Xia, N.; Modis, Y. Virus-like particle-based human vaccines: Quality assessment based on structural and functional properties. *Trends Biotechnol.* **2013**, *31*, 654–663. [CrossRef]

50. Sitrin, R.D.; Zhao, Q.; Potter, C.S.; Carragher, B.; Washabaugh, M.W. Recombinant Virus-like Particle Protein Vaccines. In *Vaccine Analysis: Strategies*; Nunnally, B., Turula, V., Sitrin, R., Eds.; Springer: Berlin, Germany, 2015.

51. Wang, X.; An, Z.; Luo, W.; Xia, N.; Zhao, Q. Molecular and functional analysis of monoclonal antibodies in support of biologics development. *Protein Cell* **2018**, *9*, 74–85. [CrossRef]

52. Morgeaux, S.; Poirier, B.; Ragan, C.I.; Wilkinson, D.; Arabin, U.; Guinet-Morlot, F.; Levis, R.; Meyer, H.; Riou, P.; Shaid, S.; et al. Replacement of in vivo human rabies vaccine potency testing by in vitro glycoprotein quantification using ELISA - Results of an international collaborative study. *Vaccine* **2017**, *35*, 966–971. [CrossRef]

53. Zhao, Q.; Towne, V.; Brown, M.; Wang, Y.; Abraham, D.; Oswald, C.B.; Gimenez, J.A.; Washabaugh, M.W.; Kennedy, R.; Sitrin, R.D. In-depth process understanding of RECOMBIVAX HB(R) maturation and potential epitope improvements with redox treatment: Multifaceted biochemical and immunochemical characterization. *Vaccine* **2011**, *29*, 7936–7941. [CrossRef]

54. Mostafa, M.M.; Al-Ghobashy, M.A.; Fathalla, F.A.; Salem, M.Y. Optimization and validation of ELISA immunoassay for the evaluation of in-vitro relative potency of a four-component human papillomavirus vaccine products. *Biologicals* **2016**, *44*, 596–599. [CrossRef]

55. Khateri, M.; Abdoli, A.; Motevalli, F.; Fotouhi, F.; Bolhassani, A.; Arashkia, A.; Jazaeri, E.O.; Shahbazi, S.; Mehrbod, P.; Naziri, H.; et al. Evaluation of autophagy induction on HEV 239 vaccine immune response in a mouse model. *Iubmb. Life* **2018**, *70*, 207–214. [CrossRef] [PubMed]

56. Pinto, L.A.; Dillner, J.; Beddows, S.; Unger, E.R. Immunogenicity of HPV prophylactic vaccines: Serology assays and their use in HPV vaccine evaluation and development. *Vaccine* **2018**, *36*, 4792–4799. [CrossRef] [PubMed]

57. Zhang, J.; Ge, S.X.; Huang, G.Y.; Li, S.W.; He, Z.Q.; Wang, Y.B.; Zheng, Y.J.; Gu, Y.; Ng, M.H.; Xia, N.S. Evaluation of antibody-based and nucleic acid-based assays for diagnosis of hepatitis E virus infection in a rhesus monkey model. *J. Med. Virol.* **2003**, *71*, 518–526. [CrossRef] [PubMed]

58. Tsarev, S.A.; Tsareva, T.S.; Emerson, S.U.; Govindarajan, S.; Shapiro, M.; Gerin, J.L.; Purcell, R.H. Recombinant vaccine against hepatitis E: Dose response and protection against heterologous challenge. *Vaccine* **1997**, *15*, 1834–1838. [CrossRef]

59. Wen, J.; Behloul, N.; Dai, X.; Dong, C.; Liang, J.; Zhang, M.; Shi, C.; Meng, J. Immunogenicity difference between two hepatitis E vaccines derived from genotype 1 and 4. *Antivir. Res.* **2016**, *128*, 36–42. [CrossRef]

60. Tam, A.W.; White, R.; Reed, E.; Short, M.; Zhang, Y.; Fuerst, T.R.; Lanford, R.E. In vitro propagation and production of hepatitis E virus from in vivo-infected primary macaque hepatocytes. *Virology* **1996**, *215*, 1–9. [CrossRef]

61. Tanaka, T.; Takahashi, M.; Kusano, E.; Okamoto, H. Development and evaluation of an efficient cell-culture system for Hepatitis E virus. *J. Gen. Virol.* **2007**, *88*, 903–911. [CrossRef]

62. Shukla, P.; Nguyen, H.T.; Torian, U.; Engle, R.E.; Faulk, K.; Dalton, H.R.; Bendall, R.P.; Keane, F.E.; Purcell, R.H.; Emerson, S.U. Cross-species infections of cultured cells by hepatitis E virus and discovery of an infectious virus-host recombinant. *Proc. Natl. Acad. Sci. USA* **2011**, *108*, 2438–2443. [CrossRef]

63. Wang, Y.; Zhao, C.; Qi, Y.; Geng, Y. Hepatitis E Virus. *Adv. Exp. Med. Biol.* **2016**, *948*, 1–16. [CrossRef]

64. Liu, C.; Cai, W.; Yin, X.; Tang, Z.; Wen, G.; Ambardekar, C.; Li, X.; Ying, D.; Feng, Z.; Zheng, Z.; et al. An Optimized High-Throughput Neutralization Assay for Hepatitis E Virus (HEV) Involving Detection of Secreted Porf2. *Viruses* **2019**, *11*. [CrossRef]

65. Takahashi, M.; Hoshino, Y.; Tanaka, T.; Takahashi, H.; Nishizawa, T.; Okamoto, H. Production of monoclonal antibodies against hepatitis E virus capsid protein and evaluation of their neutralizing activity in a cell culture system. *Arch. Virol.* **2008**, *153*, 657–666. [CrossRef] [PubMed]

66. Tam, A.W.; White, R.; Yarbough, P.O.; Murphy, B.J.; McAtee, C.P.; Lanford, R.E.; Fuerst, T.R. In vitro infection and replication of hepatitis E virus in primary cynomolgus macaque hepatocytes. *Virology* **1997**, *238*, 94–102. [CrossRef]

67. Meng, J.; Dubreuil, P.; Pillot, J. A new PCR-based seroneutralization assay in cell culture for diagnosis of hepatitis E. *J. Clin. Microbiol.* **1997**, *35*, 1373–1377. [CrossRef]

68. Emerson, S.U.; Clemente-Casares, P.; Moiduddin, N.; Arankalle, V.A.; Torian, U.; Purcell, R.H. Putative neutralization epitopes and broad cross-genotype neutralization of Hepatitis E virus confirmed by a quantitative cell-culture assay. *J. Gen. Virol.* **2006**, *87*, 697–704. [CrossRef]

69. Cai, W.; Tang, Z.M.; Wen, G.P.; Wang, S.L.; Ji, W.F.; Yang, M.; Ying, D.; Zheng, Z.Z.; Xia, N.S. A high-throughput neutralizing assay for antibodies and sera against hepatitis E virus. *Sci. Rep.* **2016**, *6*, 25141. [CrossRef]

70. He, S.; Miao, J.; Zheng, Z.; Wu, T.; Xie, M.; Tang, M.; Zhang, J.; Ng, M.H.; Xia, N. Putative receptor-binding sites of hepatitis E virus. *J. Gen. Virol.* **2008**, *89*, 245–249. [CrossRef]

71. Zheng, Z.Z.; Miao, J.; Zhao, M.; Tang, M.; Yeo, A.E.; Yu, H.; Zhang, J.; Xia, N.S. Role of heat-shock protein 90 in hepatitis E virus capsid trafficking. *J. Gen. Virol.* **2010**, *91*, 1728–1736. [CrossRef]

72. Wu, X.; Chen, P.; Lin, H.; Su, Y.; Hao, X.; Cao, Y.; Li, L.; Zhu, F.; Liang, Z. Dynamics of 8G12 competitive antibody in "prime-boost" vaccination of Hepatitis E vaccine. *Hum. Vaccin. Immunother.* **2017**, *13*, 1–6. [CrossRef]

73. Chen, S.; Huang, X.; Li, Y.; Wang, X.; Pan, H.; Lin, Z.; Zheng, Q.; Li, S.; Zhang, J.; Xia, N.; et al. Altered antigenicity and immunogenicity of human papillomavirus virus-like particles in the presence of thimerosal. *Eur. J. Pharm. Biopharm.* **2019**, *141*, 221–231. [CrossRef]

74. Smith, G.; Raghunandan, R.; Wu, Y.; Liu, Y.; Massare, M.; Nathan, M.; Zhou, B.; Lu, H.; Boddapati, S.; Li, J.; et al. Respiratory syncytial virus fusion glycoprotein expressed in insect cells form protein nanoparticles that induce protective immunity in cotton rats. *PLoS ONE* **2012**, *7*, e50852. [CrossRef]

75. Sudenga, S.L.; Torres, B.N.; Botha, M.H.; Zeier, M.; Abrahamsen, M.E.; Glashoff, R.H.; Engelbrecht, S.; Schim Van der Loeff, M.F.; Van der Laan, L.E.; Kipping, S.; et al. HPV serostatus pre- and post-vaccination in a randomized phase II preparedness trial among young Western Cape, South African women: The evri trial. *Papillomavirus Res.* **2017**, *3*, 50–56. [CrossRef] [PubMed]

76. Szkolnicka, D.; Pollan, A.; Da Silva, N.; Oechslin, N.; Gouttenoire, J.; Moradpour, D. Recombinant Hepatitis E Viruses Harboring Tags in the ORF1 Protein. *J. Virol.* **2019**, *93*. [CrossRef] [PubMed]

77. Purcell, R.H.; Nguyen, H.; Shapiro, M.; Engle, R.E.; Govindarajan, S.; Blackwelder, W.C.; Wong, D.C.; Prieels, J.P.; Emerson, S.U. Pre-clinical immunogenicity and efficacy trial of a recombinant hepatitis E vaccine. *Vaccine* **2003**, *21*, 2607–2615. [CrossRef]

78. World Health Organization; Regional Office for Africa. Weekly Bulletin on Outbreak and other Emergencies: Week 14: 01–07 April 2019. Available online: https://apps.who.int/iris/handle/10665/1642?locale=en (accessed on 10 December 2019).

79. Holm, D.K.; Moessner, B.K.; Engle, R.E.; Zaaijer, H.L.; Georgsen, J.; Purcell, R.H.; Christensen, P.B. Declining prevalence of hepatitis E antibodies among Danish blood donors. *Transfusion* **2015**, *55*, 1662–1667. [CrossRef]

80. Effectiveness Trial to Evaluate Protection of Pregnant Women by Hepatitis E Vaccine in Bangladesh. Available online: https://clinicaltrials.gov/ct2/show/NCT02759991?term=NCT02759991&draw=2&rank=1 (accessed on 9 January 2020).

81. Safety Study of Hepatitis E Vaccine (HEV239). Available online: https://clinicaltrials.gov/ct2/show/NCT03827395?term=NCT03827395&draw=2&rank=1 (accessed on 9 January 2020).

Review

Virus-Like Particles and Nanoparticles for Vaccine Development against HCMV

Michela Perotti [1,2] **and Laurent Perez** [1,3,*]

[1] Faculty of Biomedical Sciences, Institute for Research in Biomedicine, Università della Svizzera Italiana, 6500 Bellinzona, Switzerland; michela.perotti@irb.usi.ch

[2] Institute of Microbiology, ETH Zürich, 8093 Zürich, Switzerland

[3] European Virus Bioinformatics Center, 07743 Jena, Germany

* Correspondence: laurent.perez@irb.usi.ch

Received: 22 November 2019; Accepted: 25 December 2019; Published: 28 December 2019

Abstract: Human cytomegalovirus (HCMV) infects more than 70% of the human population worldwide. HCMV is responsible for high morbidity and mortality in immunocompromised patients and remains the leading viral cause of congenital birth defects. Despite considerable efforts in vaccine and therapeutic development, HCMV infection still represents an unmet clinical need and a life-threatening disease in immunocompromised individuals and newborns. Immune repertoire interrogation of HCMV seropositive patients allowed the identification of several potential antigens for vaccine design. However, recent HCMV vaccine clinical trials did not lead to a satisfactory outcome in term of efficacy. Therefore, combining antigens with orthogonal technologies to further increase the induction of neutralizing antibodies could improve the likelihood of a vaccine to reach protective efficacy in humans. Indeed, presentation of multiple copies of an antigen in a repetitive array is known to drive a more robust humoral immune response than its soluble counterpart. Virus-like particles (VLPs) and nanoparticles (NPs) are powerful platforms for multivalent antigen presentation. Several self-assembling proteins have been successfully used as scaffolds to present complex glycoprotein antigens on their surface. In this review, we describe some key aspects of the immune response to HCMV and discuss the scaffolds that were successfully used to increase vaccine efficacy against viruses with unmet medical need.

Keywords: HCMV; cytomegalovirus; VLP; nanoparticle; vaccine; immune response

---

1. Introduction

Human cytomegalovirus (HCMV) is a ubiquitously distributed member of the *Herpesviridae* family belonging to the *Betaherpesvirinae* sub-family; it infects 70% of the human adult population worldwide [1,2] and primary infection is usually asymptomatic in immunocompetent individuals. Nevertheless, the virus establishes a lifelong latency of the host after a primary infection and viral reactivation is common, also it does not necessarily imply clinical symptoms [1,3]. However, infection or viral reactivation in immunodeficient individuals such as AIDS patients, solid organ (SOT) or hematopoietic stem cells (HSCT) transplant patients causes morbidity and mortality [4,5]. Furthermore, HCMV is the most common viral cause of congenital birth defects affecting 0.7% of the newborns with permanent sequelae such as sensorineural hearing loss, growth restriction, and cognitive disabilities [6]. Current antiviral therapies and transfusion with hyper-immune globulins to control viremia are not efficient [7]. Therefore, given the severity and importance of these diseases, as well as the associated socioeconomic cost, the need for an HCMV vaccine has been assigned in the highest priority category by the Institute of Medicine and it represents the second highest priority target after HIV by the Centers for Disease Control [8,9]. Over the last decades, considerable efforts were deployed to develop a vaccine capable of preventing HCMV infection, congenital transmission and viral spreading after SOT or HSCT from seropositive donors to seronegative recipients [10]. The potential vaccine included live virus, attenuated or vectored viral vaccines expressing HCMV immunogens as well as purified recombinant protein vaccines that have been evaluated in clinical trials [11,12]. The abundant virion envelope glycoprotein B (gB) was shown to elicit vigorous T cell and antibody responses and represents the basis of most vaccines developed so far [13]. In a recent phase II clinical trial the recombinant gB vaccine formulated in MF59 adjuvant (gB/MF59), an oil-in-water emulsion, generated antibody titers comparable to natural infection. However, the vaccine demonstrated only modest efficacy in preventing primary HCMV infections in seronegative women and in the reduction of viremia in transplant recipients [14,15]. Surprisingly, a recent study demonstrated that seronegative patients vaccinated with the gB/MF59 vaccine developed a faster humoral response against gB after solid organ transplantation from seropositive donors [16]. Still, it is of the utmost importance to evaluate possible strategies for HCMV next generation vaccines. In this review, we will summarize the current findings on the adaptive immune response to HCMV and provide an update on the new methodologies available to boost the immune response against infectious diseases using virus-like particles (VLPs) and nanoparticles (NPs).

2. Immune Response against HCMV

The establishment of a long-lasting immunity in response to a primary HCMV infection is required to control subsequent HCMV reactivation and prevent uncontrolled viral replication or serious HCMV diseases [17,18]. HCMV infection activates both humoral and cellular immune responses. It is commonly believed that cellular immunity controls most of HCMV replication. Nonetheless, HCMV-specific antibodies have been associated with the prevention and protection from reinfection as well as the congenital transmission of HCMV [6,19,20]. Despite this, it is clear that the humoral response on its own is not able to clear the virus and prevent viral reactivation [21,22]. Nevertheless, the two arms of the adaptive immune response appear to be necessary for protection and resolution of HCMV infection [23].

2.1. Cellular Immune Response against HCMV

The primary cellular response to HCMV infection arises from natural killer (NK) cells, which trigger the release of inflammatory cytokines and cause lysis or apoptosis of infected cells [24]. Secondly, T cells restrict HCMV viral replication and viral spreading, although they are not able to definitively clear the virus [25–27]. HCMV-specific T cells represent 10% of both the CD4$^+$ and CD8$^+$ memory compartments in the peripheral blood of CMV-seropositive individuals [28]. CD8$^+$ T cells release perforin and granzyme B to promote lysis of infected cells that present CMV peptides on MHC-I [29,30], while CD4$^+$ specific T cells present a typical Th1 cytokine secretion signature (IFN-γ and TNF-α) [31]. Immune repertoire interrogation using 13,687 spanning peptides to cover 213 predicted CMV-proteins [32] revealed that human T cells recognize at least 151 HCMV proteins. Among these, CD4$^+$ T cells recognize 125 proteins, while CD8$^+$ T cells recognize 107 proteins, and 81 proteins are recognized by both cell types. Although the HCMV-specific T cell response is very broad, the vast majority of the results points toward antigens that are highly conserved among different HCMV strains, such as the 65 kDa phosphoprotein (pp65) and the immediate early 1 (IE1) proteins, which are recognized by more than 50% of seropositive individuals [28]. CD8$^+$ T cells were proven to have a protective role in both HSCT and SOT recipients [33–35] and a function in preventing congenital HCMV infection [36]. However, one of the paradoxes of the cellular immunity against HCMV comes from the observation that HCMV does induce a very large T cell response, meanwhile the virus also possesses sophisticated and numerous immune evasion mechanisms [37]. Among those, the capacity of the virus to enter a latency state has been the subject of intense investigation. Indeed, while primary infection of immunocompetent individuals by HCMV is rarely the source of disease, the opposite is true for individuals with an immune system that is still immature (fetus) or compromised (HIV/AIDS and transplantations). Consequently, several studies sought to identify the antigens that are relevant for the latency phase. A vaccine able to boost the immune system against the latter would help clearing the latent reservoir of the virus. It is known that HCMV establishes latency in the myeloid cell lineage and viral genome has been detected in peripheral blood monocytes and in CD34$^+$ progenitors in the bone marrow [38–40]. The viral gene expression program in these cells is restricted and extremely different from the one established during lytic infections. HCMV latent transcriptome is an area of extensive research, and at present, the exact roles of many latency-associated genes remain unclear. Nevertheless, several studies identified a number of viral genes that are specifically associated with latency [41,42]. Notably, RNAs from the major IE region (UL122–123 CLTs) [43], Latency Unique Natural Antigen (LUNA) [44], UL138 [41], UL111a (encoding a viral IL-10 homologue (LAcmvIL-10)) [45], UL144 [46] and US28 [47] were identified [22]. Moreover, viral reactivation is also associated with the expression of the major immediate early proteins (IE72 and IE86) [48]. T-cell repertoire interrogation identified four viral proteins expressed during latency (LUNA, UL138, US28, and LAcmvIL-10). Interestingly, these proteins can be recognized by CD4$^+$ T cells and T-cell responses specific for all four proteins are detectable in healthy HCMV-positive donors [49]. Accordingly, future vaccine candidates should boost cellular immunity against some of these proteins representing key targets for viral clearance.

2.2. Humoral Immune Response against HCMV

Upon HCMV infection, a large quantity of the antibodies generated target the most abundant viral proteins. These include proteins from the tegument (i.e., pp65), the immediate early proteins (i.e., IE1), and envelope glycoproteins such as gB, gM/gN, and the gH/gL complexes (trimer and pentamer). Antibodies targeting antigens that are not exposed on the surface of the virion are unlikely to generate an efficient protection against the virus [10]. Nonetheless, antibodies targeting the envelope glycoproteins are able to neutralize viral infection in vitro and correlate with a decrease of viral transmission in primary infected pregnant women [50]. It is known that antibodies targeting the pentamer are the most potent neutralizers at least in vitro [51,52], while antibodies targeting gHgL and gB are a thousand times less potent [51]. This result is still not fully understood, since the gHgL and gB proteins form the core machinery for viral membrane fusion [53] and monoclonal antibodies (mAbs) targeting the fusion machinery of other herpesviruses, such as Epstein–Barr virus, are potent neutralizers [54]. Immune repertoire interrogation of HCMV seropositive donors allowed the identification of multiple mAbs targeting the pULs subunits. We recently demonstrated that these mAbs neutralize the virus by blocking the molecular recognition of the pentamer to its cellular host receptor Neuropilin 2 [55]. In contrast, mAbs targeting gH are supposed to prevent the activation of the fusion machinery [56]. These mAbs were assigned to only two antigenic regions of gH, indeed 3G16 and MSL-109 mAbs bind toward the C-terminus of gH, while the 13H11 mAb binds close to the gL interaction site [57].

Recently, two studies reported the results of the recombinant gB vaccine formulated in MF59 adjuvant, which demonstrated partial efficacy in reducing viraemia after SOT and preventing primary infection in women and adolescents [14,15]. Sera analysis from gB-vaccinated individuals allowed the characterization of at least five gB antigenic domains (AD-1 to AD-5). The AD-1 is a domain of 80 amino acids between amino acid 560 and 640 whereas AD-2 consists of two discontinuous binding sites: Site I, located between amino acids 68 and 77, is highly conserved among strains and is considered as a target of neutralizing responses, while Site II, located between amino acids 50 and 54, is less conserved. AD-3 is a linear epitope corresponding to amino acids 798–805 in the intraluminal region of gB. AD-4, like AD-2, is a discontinuous domain corresponding to amino acids 121–132 and 344–438. AD-5 matches to amino acids between 133 and 343 [58–60]. Interestingly, Baraniak et. al. demonstrated that vaccination only boosted AD-2 responses in the 50% of CMV+ individuals with a preexisting response and did not induce a new AD-2 response in those who lacked AD-2 antibodies following natural infection [15,61]. In parallel, Nelson et al. observed an increase of anti-gB IgG titer induced by the same vaccine in a cohort of CMV-post-partum women [14]. However, the authors also subsequently observed that 76% of the vaccine-induced IgG response recognized AD-3. Considering that AD-3 is located in the intraluminal part of the virus, antibodies against this region will be most probably non-protective [12].

The hypothesis that AD-3 is diverting the immune response away from antigenic sites that more likely induce protective antibody responses (AD-2), was recently proposed [62]. The next HCMV vaccines will have to contain multiple antigens that should induce a strong neutralizing antibody response protecting from primary infections and an efficient T-cell response to prevent viral reactivation and to clear the latent virus.

3. HCMV Antigens as Vaccine Candidates

HCMV is the largest virus infecting humans, with a genome containing a linear double-stranded DNA of about 230 kb, from which at least 170 open reading frames (ORFs) have been identified [63,64]. The virus contains over 30 structural proteins and encodes at least 20 membrane glycoproteins expressed on the virion envelope, which are involved in immune evasion, cell attachment, and viral entry. HCMV exhibits an extremely broad cellular tropism being able to infect different cell types ranging from epithelial, endothelial, fibroblasts, myeloid hematopoietic precursors, monocytes/macrophages to smooth muscle cells, stromal cells, neurons, astrocytes, hepatocytes and glial cells [65]. Identification of the host cellular receptors has been a subject of debate during the last decade. It is now accepted that the two main host cell receptors for HCMV involve the platelet-derived growth factor receptor α (PDGFRα) [66–68] and Neuropilin 2 (Nrp2) [55]. In addition, other cellular host factors were described as facilitators of viral entry such as the olfactory receptor family member OR14I1 [69], the adipocyte plasma membrane-associated protein (APMAP) [70], CD46 [71], and CD147 [72]. However, the biological relevance of this multitude of receptors remains to be understood. In contrast, the viral ligands responsible for the broad cellular tropism were clearly identified and correspond to two viral envelope complexes: the gHgLgO (trimer) and the gHgLpUL128pUL130pUL131A (pentamer) (Figure 1) [73]. The trimer is a heterocomplex, in which the heterodimer consisting of gH (UL75) and gL (UL115) is disulfide-linked to glycoprotein O (gO), a heavily N-glycosylated polypeptide encoded by UL74 [74–76]. The trimer complex binds PDGFRα and is required for entry in all cell types [77], representing an attractive target for vaccine design. However, the gO sequence heterogeneity and the large glycan shield present on this subunit are impairing the generation of potent neutralizing mAbs. Along with others, we speculate that these issues will probably slow down possible efforts to use the trimer as a vaccine candidate [78]. The pentamer complex is composed of the gHgL heterodimer bound to three additional glycoproteins encoded by pUL128, pUL130, and pUL131A. The pentamer complex is required for viral entry in epithelial, endothelial, and myeloid cells and binds Nrp2 [55,65,79]. It represents the main target of HCMV neutralizing antibodies, eliciting a protective response targeting the pULs subunits, which is several orders of magnitude more potent than any other response against HCMV glycoprotein [51,52,80]. However, anti-pULs mAbs are not able to block infection in all cell types, in contrast to mAbs directed against gH and gB subunits [81]. Indeed, membrane fusion and cell entry in all cell types is dictated by the core fusion machinery composed of gHgL and the glycoprotein B (gB) which are thus considered as primary targets for vaccine development [2]. gB is a class III viral fusion protein that forms homotrimers (Figure 1) and was initially reported to bind cell–surface proteins such as PDGFRα and EGFR. However, it is more likely that gB functions as a viral fusogen that is triggered when viral and host membranes are in close proximity [82]. In addition to the HCMV membrane glycoproteins, other viral antigens have been considered as potential vaccine targets. The phosphoprotein 65 (pp65) encoded by pUL83 and the immediate-early protein 1 (IE1) encoded by pUL123, both eliciting a robust T cell response, have undergone evaluation as subunit or vectored vaccines in human [83,84]. In conclusion, the aforementioned vaccine candidates failed to confer an appropriate protection from infection, viral reinfection or reactivation. We believe that the development of a protective vaccine should include antigens from the viral envelope to trigger the humoral response (mainly pentamer and newly designed gB) and antigens from the tegument or proteins that are specifically associated with latency (i.e., US28) to prompt the cellular response able to control viral reactivation.

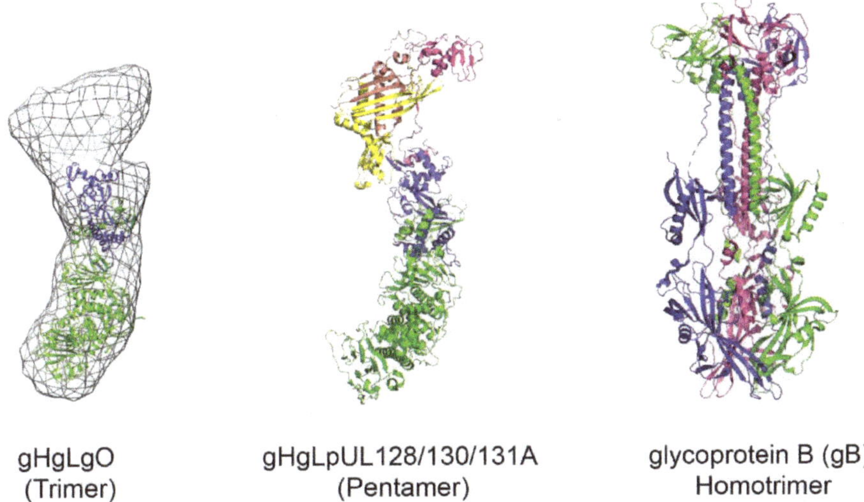

Figure 1. Structural representation of HCMV glycoproteins. Shown is the trimer, composed of the gHgL heterodimer and gO. Representation prepared with PDB: 5voc and EMD-3391. The pentamer is composed of the gHgL heterodimer and the three additional subunits pUL128, pUL130, and pUL131A. Representation prepared with PDB: 5voc. The gB homotrimer in its postfusion conformation is shown based on PDB: 5cxf. The figure was prepared using PyMOL software (The PyMOL Molecular Graphics System, Version 4.5 Schrödinger, LLC).

4. Virus-Like Particles and Nanoparticles for Antigen Display

Virus-like particles (VLPs) and nanoparticles (NPs) are protein structures that resemble wild type viruses but do not have a viral genome nor infectious ability, creating in principle safer vaccine candidates. Both VLPs and NPs are composed of self-assembling proteins displaying the epitope of interest at a high density on their surface. Antigen presentation as a repetitive array is known to drive stronger humoral immune responses when compared to the single soluble antigen [85]. This effect is thought to derive from some key properties of the VLPs/NPs presentation such as the increased antigen density that creates avidity leading to a stronger B cell activation through antigen-driven cross-linking of B cell receptors (BCRs) [86]. Moreover, VLPs/NPs are also supposed to enhance antigen uptake by antigen presenting cells (APCs) which subsequently support the adaptive arms of the immune system [87]. Finally, VLPs/NPs have also been proven, in some cases, to generate an epitope-focusing effect in which antigenic sites promoting the generation of potent neutralizing mAbs were preferentially targeted [88,89]. Over the last decades, several platforms for VLPs/NPs design have been generated, including the usage of viral core proteins, covalent links of individual folded proteins through site-specific ligations and de novo protein nanostructure formation via non-covalent intramolecular and intermolecular interactions. VLPs/NPs are currently recognized as one of the most promising and extensively studied molecular carriers for new vaccine development. Here we will present the current VLP HCMV vaccines in development and provide some examples that can be relevant for novel vaccine design.

4.1. Developed HCMV VLP Vaccine

Enveloped virus-like particles (eVLPs) are VLP structures wrapped with a lipid envelope obtained by transfection of the Moloney murine leukemia virus (MLV) gag protein in human or hamster cell lines (i.e., HEK or CHO cells). The MLV gag expression induces burgeoning of particles from membranes of transfected cells. Variation Biotechnologies Incorporated (VBI) laboratories generated an eVLP vaccine expressing an HCMV gB full-length or a chimeric gB protein, where the extracellular domain (ECD) of gB is membrane-anchored using the transmembrane and cytoplasmic domains of the vesicular stomatitis virus (VSV) G protein [90]. The chimeric gB-VSV-G protein is thought to maintain gB in a prefusion conformation. Both vaccines were used to immunize mice and induce a neutralizing antibody response 10-fold higher compared to their soluble recombinant protein counterpart [90]. The vaccine was entered in phase I clinical studies enrolling HCMV-seronegative individuals for evaluation of safety and immunogenicity in early 2016 [91]. An additional eVLP vaccine candidate, targeting both the gB and pp65 antigens, has also been manufactured by VBI. However, we believe that a combination of gB, pentamer, and pp65 or US28 would be more appealing. Finally, another eVLP HCMV vaccine candidate was developed by Redvax GmbH, a Swiss biopharmaceutical company. Unlike VBI, that used mammalian cells, they used the rePAX baculovirus co-expression technology minimizing the purification issues due to the relative large size of VLPs for the efficient generation of a VLP vaccine candidate containing HCMV pp65 and gB [92]. Redvax GmbH was later acquired by Pfizer and the vaccine was able to elicit high-titer neutralizing antibodies and good T cell responses after immunization. Nonetheless, no protection was demonstrated from viremia upon challenge.

4.2. Hepatitis B Core Antigen Nanoparticle

The core protein of the hepatitis B virus (HBc) self-assembles into highly immunogenic virus-like structures with icosahedral symmetry. The hepatitis B core nanoparticle (HBc) is assembled from dimers of 183 or 185 amino acids and occurs in two size classes with a T = 3 or T = 4 symmetry. Depending on the symmetry, HBc is formed by 90 or 120 dimeric subunits with a respective diameter of 24 or 31 nm [93]. The antigen of interest needs to be inserted between alpha helices 3 and 4 of the major immune-dominant region (MIR) domain (amino acids 78–82) (Figure 2A) of the HBc protein. This insertion allows the antigen to be exposed on spike structures at the surface of the assembled nanoparticle. HBc protomers first dimerize and then multimerize (Figure 2B,C). Several developments led to the success of this platform, such as splitting of the HBc protein between the α3 and α4 helices or fusing two monomers together using a flexible glycine serine linker that allowed the introduction of larger protein domains [94,95]. However, one limitation of this technology is the need for both the N- and C-termini of the inserted antigen to fit with the geometry of the core acceptor site, a requirement that is not always successfully met by natively folded proteins.

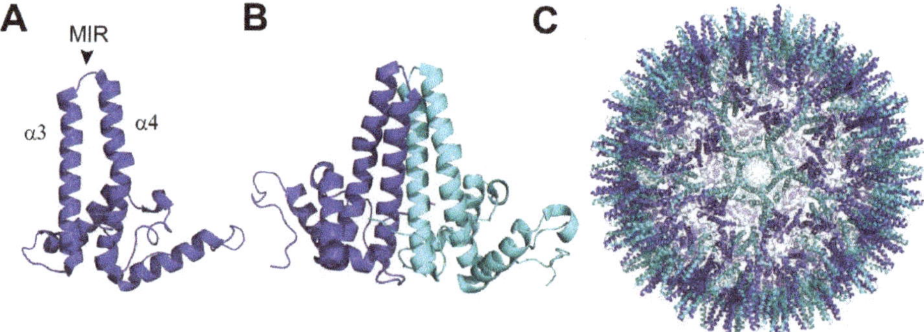

Figure 2. Hepatitis B core antigen and relative nanoparticle. (**A**) Polypeptide fold and structure of HBcAg monomer. The antigen of interest needs to be inserted in the MIR loop connecting alpha helices 3 and 4. (**B**) Two HBcAg monomers spontaneously interact to form a dimer, which acts as an intermediate in the nanoparticle formation. (**C**) Assembled nanoparticle containing either 90 or 120 dimers. The figure was prepared with PDB: 6htx using Pymol software (The PyMOL Molecular Graphics System, Version 4.5 Schrödinger, LLC).

In summary, the HBc is a powerful platform to generate nanoparticles and it is particularly useful in the case of small antigens. Indeed, it is unlikely that complex envelope glycoproteins such as gB or the gHgL complexes could be correctly displayed on HBc particles. Nevertheless, HBc could be very efficient to generate strong T cell responses [96]. For example, the viral-encoded G protein-coupled receptor US28 was recently shown to aid in the establishment and the maintenance of viral latency. US28 modulates host–cell proteins to suppress viral processes associated with lytic replication and thereby represent an important target promoting latent infection [97]. HBc particles displaying US28 peptides recognized by T cells could provide an important advancement to boost the existing immune responses necessary to control reactivation of latently infected individuals [98].

4.3. Ferritin-Based Nanoparticle

Ferritin is an intracellular globular protein that stores iron and releases it in a controlled fashion [99]. The protein is expressed in many tissues, mainly as a cytosolic protein, but it can be also found at low concentration in the serum where it functions as an iron carrier [100]. Ferritin is a four-helix bundle protein consisting of two couples of anti-parallel α-helices connected by a long loop and a short C-terminal α-helix (Figure 3A). The monomer can dimerize and further self-assemble to form a nanocage of 24 protein subunits, which has an internal and external diameter of 8 and 12 nm, respectively [101]. Ferritin nanoparticle (ferritin-NP) presents the advantage of being resistant to thermal and chemical stress [102]. Moreover, antigen display is achieved through genetic fusion at the N-terminus of each ferritin protomer. Ferritin-NP self-arranges as an octahedron composed of eight trimeric units (Figure 3B), each with a 3-fold symmetry axis that allows the correct presentation of trimeric antigens at the nanoparticle surface (Figure 3C).

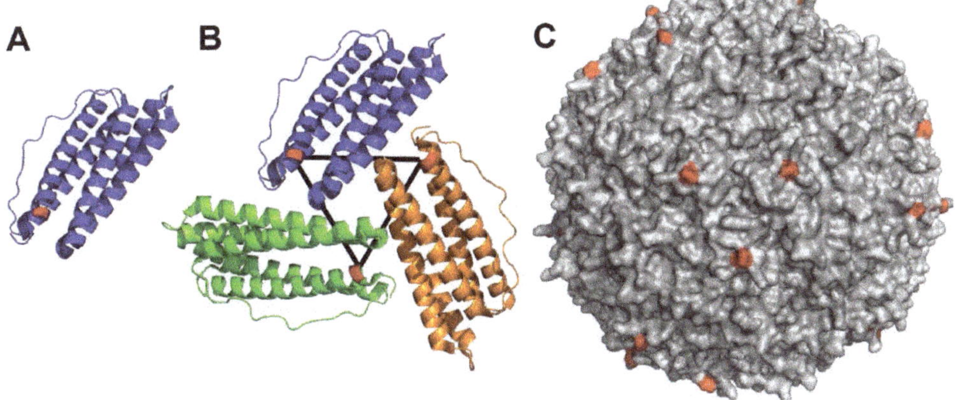

Figure 3. Structure of ferritin nanoparticle. (**A**) The antigen of interest is fused to the N-terminus of the ferritin polypeptide (shown in red). (**B**) Ferritin assembles to form symmetry units with threefold axes. (**C**) Fully assembled nanoparticle with surface-exposed N-termini colored in red. The figure was prepared using PDB: 3bve using PyMOL software (The PyMOL Molecular Graphics System, Version 4.5 Schrödinger, LLC).

A recent development and one of the major advantages of this carrier is the possibility to generate multivalent displays of different antigenic proteins. For example, ferritin-NP was first used to present HIV-1 and influenza antigens on the same particle [103]. Furthermore, Kanekiyo and coworkers at the Vaccine Research Center in the United States. further developed and exploited the ferritin-NP system using a ferritin hybrid protein [89] to generate a universal vaccine against influenza. The team co-displayed the receptor binding domain (RBD) of the last 90 years H1N1 influenza viruses [104] and showed that immunization with this mosaic RBD-NP elicited a broad antibody response covering all the known H1N1 strains. Multiple displays on ferritin-NP have opened up new possibilities for novel vaccines against different pathogens or against different antigens of a given pathogen.

In the case of HCMV, we can speculate that a mosaic ferritin-NP displaying both pentamer and gB ECDs should generate a potent humoral immune reaction. The antibodies generated targeting the gHgL and gB subunits will block infection of connective tissues and stroma (containing fibroblasts). On the other hand, antibodies directed against the pUL subunits will block infection of epithelial tissues.

4.4. Qβ Nanoparticle

The bacteriophage Qβ is a member of the *Leviviridae* family, forming 30 nm icosahedral nanoparticles of 180 copies from the 14 kDa coat protein [105]. The coat protein monomer (Figure 4A) forms Qβ dimers (Figure 4B) that further assemble into pentamers and hexamers by disulfide and hydrogen bond interactions to form a Qβ nanoparticle under physiological conditions (Figure 4C). This nanoparticle is highly stable and is used to display ligands and proteins on its exterior surface [106,107]. Co-expression of Qβ proteins displaying different antigens of interest or engineered Qβ proteins displaying functionalized groups for subsequent direct antigen conjugation can be easily produced in bacterial cells. Of note, the genetic fusion occurs at the C-terminus of the protein, a property that might impair the correct folding of some antigens of interest. Genes encoding the viral Qβ protein have been also successfully expressed in yeast cells but the possibility to generate a Qβ nanoparticle in prokaryotic hosts, maintaining therefore a low production cost, will be of great interest for the generation of a personalized T cell vaccine targeting the proteins associated to latency. Regarding mammalian cell expression, there is currently no evidence that the system is also compatible with it, which is a prerequisite for the production of most of the HCMV glycoprotein antigens.

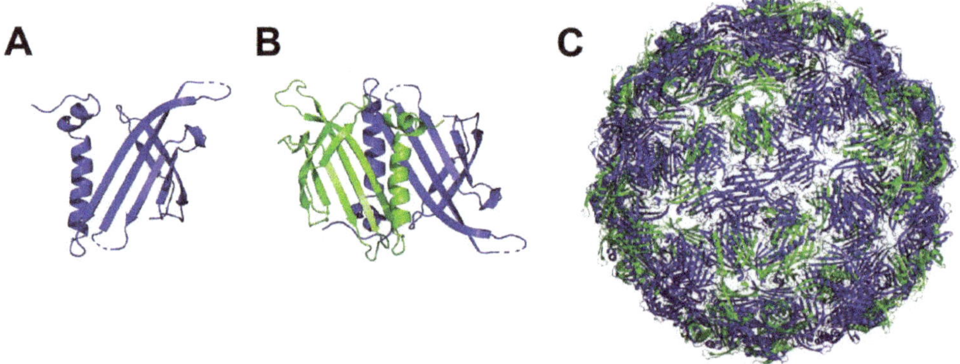

Figure 4. Structure and assembly of Qβ nanoparticle. Qβ protein is depicted as a monomer (**A**) and noncovalent dimer (**B**). (**C**) The fully self-assembled Qβ nanoparticle consists of 20 hexamers and 12 pentamers. The figure was prepared using 1qbe PDB structure and PyMOL software (The PyMOL Molecular Graphics System, Version 4.5 Schrödinger, LLC).

4.5. De Novo Design-Based Nanoparticle

Structure-based design of nanoparticle immunogens has been limited by the restricted number of scaffolds available and the fact that their physico-chemical properties are fixed (i.e., ferritin, encapsulins). Moreover, most of the self-assembling scaffolds spontaneously do so upon expression in the transfected host cell, a property that might lead to poor production yield. These constraints have pushed the exploration of new structural and functional geometries (i.e., icosahedron, dodecahedron) in the nanoparticle immunogen design field. Recent computational methods for designing novel self-assembling proteins with atomic-level accuracy offer the possibility to design self-assembling proteins with customized structures, offering new opportunities for structure-based vaccine design [108–110]. For instance, the HIV gp41 ectodomain trimer was displayed onto the I3–01 particle [108,111]. The gp41 antigen exposed on NPs, stimulated mAb-expressing B cells more effectively than the soluble trimers [111]. Two additional publications highlighted the potential of this de novo self-assembling nanoparticle for antigen display. The I5350 nanoparticle (Figure 5) was used to generate an enhanced vaccine against the respiratory syncytial virus (RSV). Self-assembling protein nanoparticles displaying 20 copies of the stabilized version of the RSV fusion glycoprotein trimer (DS-Cav1) induced a neutralizing antibody response ~10-fold higher than soluble trimeric DS-Cav1 [112]. In the second case, native-like HIV-1 envelope trimer antigens were displayed in a multivalent fashion on the I5350 nanoparticle scaffold and immunization studies revealed a more effective priming compared to the soluble SOSIP trimers [113]. SOSIP corresponds to a stabilized trimer with mutations that cross-link the cleaved gp120 and gp41. In contrast, the natural HIV trimer is comprised of three copies of a non-covalently linked gp120/gp41 heterodimer arising from cleavage of the viral gp160 precursor protein. It can be speculated that de novo design-based nanoparticles will have an increasing impact on next generation vaccines since they are customizable at will, supported by a multitude of expression platforms and able to generate multivalent display as well as encapsulation of small proteins or compounds [114,115].

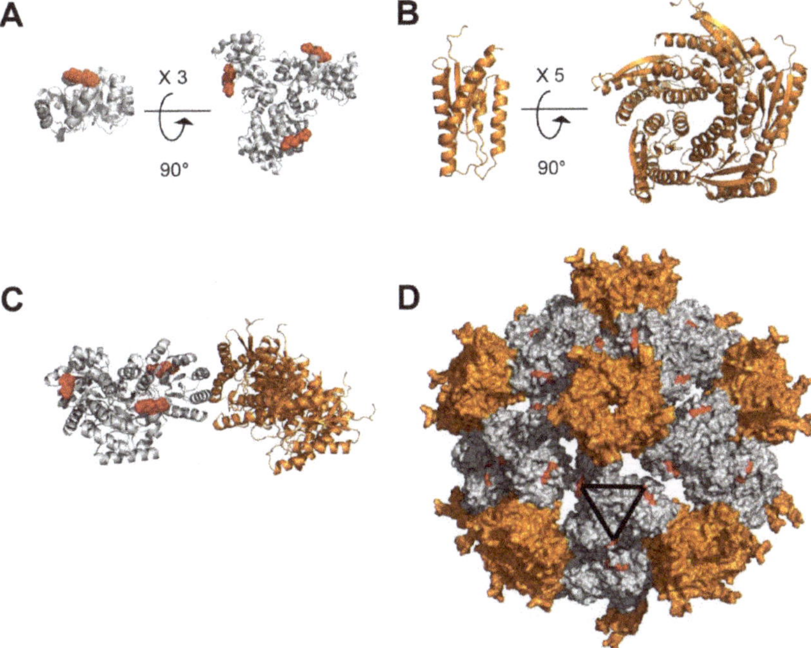

Figure 5. Self-assembly of the I5350 nanoparticle. (**A**) Polypeptide I53–50A.1NT1; shown in red is the N-terminus projecting outward, where antigens can be displayed. I53–50A.1NT1 naturally trimerizes forming a 3-fold symmetry axis. (**B**) The I53–50B.4PT1 protein (orange) assembles into a pentamer. (**C**) I53–50A.1NT1 trimer and I53–50B.4PT1 pentamer self-assemble as a nanoparticle. (**D**) Assembled nanoparticle formed by 20 trimers and 12 pentamers, the N-termini of I53–50A.1NT1 is shown in red, and the black triangle represents the 3-fold symmetry axis of each pair of trimers. The figure was prepared using PyMOL software (The PyMOL Molecular Graphics System, Version 4.5 Schrödinger, LLC).

Ideally, I5350 particles should present on their surface the pentamer and gB to generate a humoral response against the gH, gL, pULs, and gB molecules. Moreover, as I5350 nanoparticles are hollow, the in vitro assembly step with I53–50B.4PT1 could be performed in the presence of T-cell stimulating peptides (i.e., pp65, IE1 and US28) that would be encapsulated inside the particles. The resulting vaccine candidates should in this way stimulate both arms of the immune system, offering a maximal protection.

4.6. SpyCatcher and SpyTag for Nanoparticle Formation

The SpyTag/SpyCatcher system (referred as Spy-System) is based on the internal isopeptide bond of the CnaB2 domain of FbaB protein from *Streptococcus pyogenes*. An internal isopeptide bond forms spontaneously in this domain between the reactive amine of lysine K31 and the side chain carboxyl of aspartate D117. This reaction is catalyzed by the spatially adjacent glutamate E77. The resulting isopeptide bond confers high stability to the CnaB2 domain. The CnaB2 domain is cleaved into two subcomponents, the large and incomplete immunoglobulin-like domain, named SpyCatcher, composed of 138 amino acids, and the short beta strand, named SpyTag, composed of 13 amino acids [116]. While the SpyCatcher contains the reactive lysine (K31) and catalytic glutamate (E77), the SpyTag includes the reactive aspartate (D117). The two components can still recognize each other with high affinity and the isopeptide is formed within minutes. Despite the fact that this system is not able to generate nanoparticles on its own, the two components of the Spy-System can be genetically fused to a number of viral or designed self-assembling proteins leading to the generation of a stable nanoparticle displaying a large variety of antigens. Moreover, the Spy-System offers a great flexibility since components can be inserted at the N- or C-termini of the target proteins. Its simplicity is of particular interest to quickly screen if a nanoparticle formulation will be of benefit for vaccine design, before generating genetic fusion on the nanoparticle of interest. The first nanoparticle assembly with Spy-System was performed by genetically fusing the SpyCatcher to the N-terminus of the viral coat protein (CP3) of the RNA bacteriophage AP205 [117]. AP205 VLPs have been then further investigated as vaccine candidates against a number of targets, including malaria and cancer [117–119].

5. Conclusions and Future Directions

We described a large variety of systems to generate both VLPs and NPs. Self-assembling scaffolds have been used to present complex glycoprotein antigens and for vaccination studies against influenza [104], HIV [113,120–123], Epstein–Barr virus [89], and RSV [112]. In all cases, antigen immunogenicity was increased by multivalent presentation and in some cases, an epitope focusing effect could be observed. The rise of computational design to generate novel self-assembling proteins has now paved the way for new possibilities to display antigens without limitation in term of quantity or oligomeric state of the antigen. HCMV is still the principal viral cause of congenital malformations. About one quarter of infants infected by HCMV in utero will develop severe sequelae including microcephaly, sensorineural hearing and vision loss, or cognitive delay [1,6]. The recent identification of key HCMV antigenic targets for both humoral and cellular immune responses and the capacity to display and encapsulate proteins or nucleic acids (RNAs) on VLPs/NPs is likely to generate improved vaccine candidates able to redirect the immune response against specific targets.

Author Contributions: M.P. and L.P. conceptualized and wrote the review. All authors have read and agreed to the published version of the manuscript.

Funding: This work was supported by the Novartis Foundation for Medical–Biological Research, Novartis grant (application #19B116 to LP).

Acknowledgments: The authors acknowledge Mathilde Foglierini and Jessica Marcandalli for critical reading and helpful comments on the manuscript.

Conflicts of Interest: The authors declare no conflict of interest. The funders had no role in the design of the study; in the collection, analyses, or interpretation of data; in the writing of the manuscript; or in the decision to publish the results.

References

1. Kenneson, A.; Cannon, M.J. Review and meta-analysis of the epidemiology of congenital cytomegalovirus (CMV) infection. *Rev. Med. Virol.* **2007**, *17*, 253–276. [CrossRef] [PubMed]
2. Plotkin, S.A.; Boppana, S.B. Vaccination against the human cytomegalovirus. *Vaccine* **2018**, *37*, 7437–7442. [CrossRef] [PubMed]
3. Britt, W. Manifestations of human cytomegalovirus infection: Proposed mechanisms of acute and chronic disease. *Curr. Top. Microbiol. Immunol.* **2008**, *325*, 417–470. [CrossRef] [PubMed]
4. Gerna, G.; Zavattoni, M.; Baldanti, F.; Furione, M.; Chezzi, L.; Revello, M.G.; Percivalle, E. Circulating cytomegalic endothelial cells are associated with high human cytomegalovirus (HCMV) load in AIDS patients with late-stage disseminated HCMV disease. *J. Med. Virol.* **1998**, *55*, 64–74. [CrossRef]
5. Lilleri, D.; Gerna, G. Strategies to control human cytomegalovirus infection in adult hematopoietic stem cell transplant recipients. *Immunotherapy* **2016**, *8*, 1135–1149. [CrossRef] [PubMed]
6. Boppana, S.B.; Britt, W.J. Antiviral antibody responses and intrauterine transmission after primary maternal cytomegalovirus infection. *J. Infect. Dis.* **1995**, *171*, 1115–1121. [CrossRef]
7. Spinillo, A.; Gerna, G. Hyperimmune globulin to prevent congenital CMV infection. *N. Engl. J. Med.* **2014**, *370*, 2544–2545. [CrossRef]
8. James, S.H.; Kimberlin, D.W. Advances in the prevention and treatment of congenital cytomegalovirus infection. *Curr. Opin. Pediatrics* **2016**, *28*, 81–85. [CrossRef]
9. Marsico, C.; Kimberlin, D.W. Congenital Cytomegalovirus infection: Advances and challenges in diagnosis, prevention and treatment. *Ital. J. Pediatrics* **2017**, *43*, 38. [CrossRef]
10. Gomes, A.C.; Griffiths, P.D.; Reeves, M.B. The Humoral Immune Response Against the gB Vaccine: Lessons Learnt from Protection in Solid Organ Transplantation. *Vaccines* **2019**, *7*, 67. [CrossRef]
11. Schleiss, M.R. Recombinant cytomegalovirus glycoprotein B vaccine: Rethinking the immunological basis of protection. *Proc. Natl. Acad. Sci. USA* **2018**, *115*, 6110–6112. [CrossRef] [PubMed]
12. Nelson, C.S.; Herold, B.C.; Permar, S.R. A new era in cytomegalovirus vaccinology: Considerations for rational design of next-generation vaccines to prevent congenital cytomegalovirus infection. *Npj Vaccines* **2018**, *3*, 38. [CrossRef] [PubMed]
13. Foglierini, M.; Marcandalli, J.; Perez, L. HCMV Envelope Glycoprotein Diversity Demystified. *Front. Microbiol.* **2019**, *10*, 1005. [CrossRef]
14. Nelson, C.S.; Huffman, T.; Jenks, J.A.; Cisneros de la Rosa, E.; Xie, G.; Vandergrift, N.; Pass, R.F.; Pollara, J.; Permar, S.R. HCMV glycoprotein B subunit vaccine efficacy mediated by nonneutralizing antibody effector functions. *Proc. Natl. Acad. Sci. USA* **2018**, *115*, 6267–6272. [CrossRef]
15. Baraniak, I.; Kropff, B.; Ambrose, L.; McIntosh, M.; McLean, G.R.; Pichon, S.; Atkinson, C.; Milne, R.S.B.; Mach, M.; Griffiths, P.D.; et al. Protection from cytomegalovirus viremia following glycoprotein B vaccination is not dependent on neutralizing antibodies. *Proc. Natl. Acad. Sci. USA* **2018**, *115*, 6273–6278. [CrossRef] [PubMed]
16. Baraniak, I.; Gomes, A.C.; Sodi, I.; Langstone, T.; Rothwell, E.; Atkinson, C.; Pichon, S.; Piras-Douce, F.; Griffiths, P.D.; Reeves, M.B. Seronegative patients vaccinated with cytomegalovirus gB-MF59 vaccine have evidence of neutralising antibody responses against gB early post-transplantation. *EBioMedicine* **2019**, *50*, 45–54. [CrossRef] [PubMed]
17. Gerna, G.; Lilleri, D.; Fornara, C.; Bruno, F.; Gabanti, E.; Cane, I.; Furione, M.; Revello, M.G. Differential kinetics of human cytomegalovirus load and antibody responses in primary infection of the immunocompetent and immunocompromised host. *J. Gen. Virol.* **2015**, *96*, 360–369. [CrossRef]
18. Smith, C.J.; Quinn, M.; Snyder, C.M. CMV-Specific CD8 T Cell Differentiation and Localization: Implications for Adoptive Therapies. *Front. Immunol.* **2016**, *7*, 352. [CrossRef]
19. Bialas, K.M.; Westreich, D.; Cisneros de la Rosa, E.; Nelson, C.S.; Kauvar, L.M.; Fu, T.M.; Permar, S.R. Maternal Antibody Responses and Nonprimary Congenital Cytomegalovirus Infection of HIV-1-Exposed Infants. *J. Infect. Dis.* **2016**, *214*, 1916–1923. [CrossRef]
20. Itell, H.L.; Nelson, C.S.; Martinez, D.R.; Permar, S.R. Maternal immune correlates of protection against placental transmission of cytomegalovirus. *Placenta* **2017**, *60* (Suppl. 1), S73–S79. [CrossRef]

21. Stern, L.; Withers, B.; Avdic, S.; Gottlieb, D.; Abendroth, A.; Blyth, E.; Slobedman, B. Human Cytomegalovirus Latency and Reactivation in Allogeneic Hematopoietic Stem Cell Transplant Recipients. *Front. Microbiol.* **2019**, *10*, 1186. [CrossRef] [PubMed]

22. Collins-McMillen, D.; Buehler, J.; Peppenelli, M.; Goodrum, F. Molecular Determinants and the Regulation of Human Cytomegalovirus Latency and Reactivation. *Viruses* **2018**, *10*, 444. [CrossRef] [PubMed]

23. Krishna, B.A.; Wills, M.R.; Sinclair, J.H. Advances in the treatment of cytomegalovirus. *Br. Med. Bull.* **2019**, *131*, 5–17. [CrossRef] [PubMed]

24. Wilkinson, G.W.; Tomasec, P.; Stanton, R.J.; Armstrong, M.; Prod'homme, V.; Aicheler, R.; McSharry, B.P.; Rickards, C.R.; Cochrane, D.; Llewellyn-Lacey, S.; et al. Modulation of natural killer cells by human cytomegalovirus. *J. Clin. Virol.* **2008**, *41*, 206–212. [CrossRef]

25. Rist, M.; Cooper, L.; Elkington, R.; Walker, S.; Fazou, C.; Tellam, J.; Crough, T.; Khanna, R. Ex vivo expansion of human cytomegalovirus-specific cytotoxic T cells by recombinant polyepitope: Implications for HCMV immunotherapy. *Eur. J. Immunol.* **2005**, *35*, 996–1007. [CrossRef]

26. Rohrlich, P.S.; Cardinaud, S.; Lule, J.; Montero-Julian, F.A.; Prodhomme, V.; Firat, H.; Davignon, J.L.; Perret, E.; Monseaux, S.; Necker, A.; et al. Use of a lentiviral vector encoding a HCMV-chimeric IE1-pp65 protein for epitope identification in HLA-Transgenic mice and for ex vivo stimulation and expansion of CD8(+) cytotoxic T cells from human peripheral blood cells. *Hum. Immunol.* **2004**, *65*, 514–522. [CrossRef]

27. Boppana, S.B.; Britt, W.J. Recognition of human cytomegalovirus gene products by HCMV-specific cytotoxic T cells. *Virology* **1996**, *222*, 293–296. [CrossRef]

28. Sylwester, A.W.; Mitchell, B.L.; Edgar, J.B.; Taormina, C.; Pelte, C.; Ruchti, F.; Sleath, P.R.; Grabstein, K.H.; Hosken, N.A.; Kern, F.; et al. Broadly targeted human cytomegalovirus-specific CD4+ and CD8+ T cells dominate the memory compartments of exposed subjects. *J. Exp. Med.* **2005**, *202*, 673–685. [CrossRef]

29. Hertoghs, K.M.; Moerland, P.D.; van Stijn, A.; Remmerswaal, E.B.; Yong, S.L.; van de Berg, P.J.; van Ham, S.M.; Baas, F.; ten Berge, I.J.; van Lier, R.A. Molecular profiling of cytomegalovirus-induced human CD8+ T cell differentiation. *J. Clin. Invest.* **2010**, *120*, 4077–4090. [CrossRef]

30. Gillespie, G.M.; Wills, M.R.; Appay, V.; O'Callaghan, C.; Murphy, M.; Smith, N.; Sissons, P.; Rowland-Jones, S.; Bell, J.I.; Moss, P.A. Functional heterogeneity and high frequencies of cytomegalovirus-specific CD8(+) T lymphocytes in healthy seropositive donors. *J. Virol.* **2000**, *74*, 8140–8150. [CrossRef]

31. van Leeuwen, E.M.; Remmerswaal, E.B.; Vossen, M.T.; Rowshani, A.T.; Wertheim-van Dillen, P.M.; van Lier, R.A.; ten Berge, I.J. Emergence of a CD4+CD28- granzyme B+, cytomegalovirus-specific T cell subset after recovery of primary cytomegalovirus infection. *J. Immunol.* **2004**, *173*, 1834–1841. [CrossRef]

32. Kern, F.; Faulhaber, N.; Frommel, C.; Khatamzas, E.; Prosch, S.; Schonemann, C.; Kretzschmar, I.; Volkmer-Engert, R.; Volk, H.D.; Reinke, P. Analysis of CD8 T cell reactivity to cytomegalovirus using protein-spanning pools of overlapping pentadecapeptides. *Eur. J. Immunol.* **2000**, *30*, 1676–1682. [CrossRef]

33. Lilleri, D.; Zelini, P.; Fornara, C.; Zavaglio, F.; Rampino, T.; Perez, L.; Gabanti, E.; Gerna, G. Human cytomegalovirus (HCMV)-specific T cell but not neutralizing or IgG binding antibody responses to glycoprotein complexes gB, gHgLgO, and pUL128L correlate with protection against high HCMV viral load reactivation in solid-organ transplant recipients. *J. Med. Virol.* **2018**, *90*, 1620–1628. [CrossRef] [PubMed]

34. Reusser, P.; Gambertoglio, J.G.; Lilleby, K.; Meyers, J.D. Phase I-II trial of foscarnet for prevention of cytomegalovirus infection in autologous and allogeneic marrow transplant recipients. *J. Infect. Dis.* **1992**, *166*, 473–479. [CrossRef] [PubMed]

35. Greenberg, P.D.; Reusser, P.; Goodrich, J.M.; Riddell, S.R. Development of a treatment regimen for human cytomegalovirus (CMV) infection in bone marrow transplantation recipients by adoptive transfer of donor-derived CMV-specific T cell clones expanded in vitro. *Ann. N. Y. Acad. Sci.* **1991**, *636*, 184–195. [CrossRef] [PubMed]

36. Fornara, C.; Cassaniti, I.; Zavattoni, M.; Furione, M.; Adzasehoun, K.M.G.; De Silvestri, A.; Comolli, G.; Baldanti, F. Human Cytomegalovirus-Specific Memory CD4+ T-Cell Response and Its Correlation With Virus Transmission to the Fetus in Pregnant Women With Primary Infection. *Clin. Infect. Dis.* **2017**, *65*, 1659–1665. [CrossRef]

37. Reddehase, M.J. Antigens and immunoevasins: Opponents in cytomegalovirus immune surveillance. *Nat. Rev. Immunol.* **2002**, *2*, 831–844. [CrossRef]

38. Taylor-Wiedeman, J.; Sissons, J.G.; Borysiewicz, L.K.; Sinclair, J.H. Monocytes are a major site of persistence of human cytomegalovirus in peripheral blood mononuclear cells. *J. Gen. Virol.* **1991**, *72 Pt 9*, 2059–2064. [CrossRef]

39. Mendelson, M.; Monard, S.; Sissons, P.; Sinclair, J. Detection of endogenous human cytomegalovirus in CD34+ bone marrow progenitors. *J. Gen. Virol.* **1996**, *77 Pt 12*, 3099–3102. [CrossRef]

40. Hargett, D.; Shenk, T.E. Experimental human cytomegalovirus latency in CD14+ monocytes. *Proc. Natl Acad. Sci. USA* **2010**, *107*, 20039–20044. [CrossRef]

41. Goodrum, F.; Reeves, M.; Sinclair, J.; High, K.; Shenk, T. Human cytomegalovirus sequences expressed in latently infected individuals promote a latent infection in vitro. *Blood* **2007**, *110*, 937–945. [CrossRef] [PubMed]

42. Reeves, M.B.; Sinclair, J.H. Analysis of latent viral gene expression in natural and experimental latency models of human cytomegalovirus and its correlation with histone modifications at a latent promoter. *J. Gen. Virol.* **2010**, *91*, 599–604. [CrossRef] [PubMed]

43. Kondo, K.; Kaneshima, H.; Mocarski, E.S. Human cytomegalovirus latent infection of granulocyte-macrophage progenitors. *Proc. Natl Acad. Sci. USA* **1994**, *91*, 11879–11883. [CrossRef] [PubMed]

44. Bego, M.; Maciejewski, J.; Khaiboullina, S.; Pari, G.; St Jeor, S. Characterization of an antisense transcript spanning the UL81-82 locus of human cytomegalovirus. *J. Virol.* **2005**, *79*, 11022–11034. [CrossRef]

45. Avdic, S.; Cao, J.Z.; Cheung, A.K.; Abendroth, A.; Slobedman, B. Viral interleukin-10 expressed by human cytomegalovirus during the latent phase of infection modulates latently infected myeloid cell differentiation. *J. Virol.* **2011**, *85*, 7465–7471. [CrossRef]

46. Poole, E.; Walther, A.; Raven, K.; Benedict, C.A.; Mason, G.M.; Sinclair, J. The myeloid transcription factor GATA-2 regulates the viral UL144 gene during human cytomegalovirus latency in an isolate-specific manner. *J. Virol.* **2013**, *87*, 4261–4271. [CrossRef]

47. Beisser, P.S.; Laurent, L.; Virelizier, J.L.; Michelson, S. Human cytomegalovirus chemokine receptor gene US28 is transcribed in latently infected THP-1 monocytes. *J. Virol.* **2001**, *75*, 5949–5957. [CrossRef]

48. Gatherer, D.; Seirafian, S.; Cunningham, C.; Holton, M.; Dargan, D.J.; Baluchova, K.; Hector, R.D.; Galbraith, J.; Herzyk, P.; Wilkinson, G.W.; et al. High-resolution human cytomegalovirus transcriptome. *Proc. Natl. Acad. Sci. USA* **2011**, *108*, 19755–19760. [CrossRef]

49. Mason, G.M.; Jackson, S.; Okecha, G.; Poole, E.; Sissons, J.G.; Sinclair, J.; Wills, M.R. Human cytomegalovirus latency-associated proteins elicit immune-suppressive IL-10 producing CD4(+) T cells. *PLoS Pathog.* **2013**, *9*, e1003635. [CrossRef]

50. Lilleri, D.; Kabanova, A.; Revello, M.G.; Percivalle, E.; Sarasini, A.; Genini, E.; Sallusto, F.; Lanzavecchia, A.; Corti, D.; Gerna, G. Fetal human cytomegalovirus transmission correlates with delayed maternal antibodies to gH/gL/pUL128-130-131 complex during primary infection. *PLoS ONE* **2013**, *8*, e59863. [CrossRef]

51. Kabanova, A.; Perez, L.; Lilleri, D.; Marcandalli, J.; Agatic, G.; Becattini, S.; Preite, S.; Fuschillo, D.; Percivalle, E.; Sallusto, F.; et al. Antibody-driven design of a human cytomegalovirus gHgLpUL128L subunit vaccine that selectively elicits potent neutralizing antibodies. *Proc. Natl. Acad. Sci. USA* **2014**, *111*, 17965–17970. [CrossRef] [PubMed]

52. Macagno, A.; Bernasconi, N.L.; Vanzetta, F.; Dander, E.; Sarasini, A.; Revello, M.G.; Gerna, G.; Sallusto, F.; Lanzavecchia, A. Isolation of human monoclonal antibodies that potently neutralize human cytomegalovirus infection by targeting different epitopes on the gH/gL/UL128-131A complex. *J. Virol.* **2010**, *84*, 1005–1013. [CrossRef] [PubMed]

53. Vanarsdall, A.L.; Howard, P.W.; Wisner, T.W.; Johnson, D.C. Human Cytomegalovirus gH/gL Forms a Stable Complex with the Fusion Protein gB in Virions. *PLoS Pathog.* **2016**, *12*, e1005564. [CrossRef] [PubMed]

54. Bu, W.; Joyce, M.G.; Nguyen, H.; Banh, D.V.; Aguilar, F.; Tariq, Z.; Yap, M.L.; Tsujimura, Y.; Gillespie, R.A.; Tsybovsky, Y.; et al. Immunization with Components of the Viral Fusion Apparatus Elicits Antibodies That Neutralize Epstein-Barr Virus in B Cells and Epithelial Cells. *Immunity* **2019**. [CrossRef] [PubMed]

55. Martinez-Martin, N.; Marcandalli, J.; Huang, C.S.; Arthur, C.P.; Perotti, M.; Foglierini, M.; Ho, H.; Dosey, A.M.; Shriver, S.; Payandeh, J.; et al. An Unbiased Screen for Human Cytomegalovirus Identifies Neuropilin-2 as a Central Viral Receptor. *Cell* **2018**, *174*, 1158–1171.e1119. [CrossRef]

56. Malito, E.; Chandramouli, S.; Carfi, A. From recognition to execution-the HCMV Pentamer from receptor binding to fusion triggering. *Curr. Opin. Virol.* **2018**, *31*, 43–51. [CrossRef]

57. Ciferri, C.; Chandramouli, S.; Leitner, A.; Donnarumma, D.; Cianfrocco, M.A.; Gerrein, R.; Friedrich, K.; Aggarwal, Y.; Palladino, G.; Aebersold, R.; et al. Antigenic Characterization of the HCMV gH/gL/gO and Pentamer Cell Entry Complexes Reveals Binding Sites for Potently Neutralizing Human Antibodies. *PLoS Pathog.* **2015**, *11*, e1005230. [CrossRef]

58. Wiegers, A.K.; Sticht, H.; Winkler, T.H.; Britt, W.J.; Mach, M. Identification of a neutralizing epitope within antigenic domain 5 of glycoprotein B of human cytomegalovirus. *J. Virol.* **2015**, *89*, 361–372. [CrossRef]

59. Spindler, N.; Diestel, U.; Stump, J.D.; Wiegers, A.K.; Winkler, T.H.; Sticht, H.; Mach, M.; Muller, Y.A. Structural basis for the recognition of human cytomegalovirus glycoprotein B by a neutralizing human antibody. *PLoS Pathog.* **2014**, *10*, e1004377. [CrossRef]

60. Potzsch, S.; Spindler, N.; Wiegers, A.K.; Fisch, T.; Rucker, P.; Sticht, H.; Grieb, N.; Baroti, T.; Weisel, F.; Stamminger, T.; et al. B cell repertoire analysis identifies new antigenic domains on glycoprotein B of human cytomegalovirus which are target of neutralizing antibodies. *PLoS Pathog.* **2011**, *7*, e1002172. [CrossRef]

61. Baraniak, I.; Kropff, B.; McLean, G.R.; Pichon, S.; Piras-Douce, F.; Milne, R.S.B.; Smith, C.; Mach, M.; Griffiths, P.D.; Reeves, M.B. Epitope-Specific Humoral Responses to Human Cytomegalovirus Glycoprotein-B Vaccine With MF59: Anti-AD2 Levels Correlate With Protection From Viremia. *J. Infect. Dis.* **2018**, *217*, 1907–1917. [CrossRef] [PubMed]

62. Baraniak, I.; Kern, F.; Holenya, P.; Griffiths, P.; Reeves, M. Original Antigenic Sin Shapes the Immunological Repertoire Evoked by Human Cytomegalovirus Glycoprotein B/MF59 Vaccine in Seropositive Recipients. *J. Infect. Dis.* **2019**, *220*, 228–232. [CrossRef] [PubMed]

63. Varnum, S.M.; Streblow, D.N.; Monroe, M.E.; Smith, P.; Auberry, K.J.; Pasa-Tolic, L.; Wang, D.; Camp, D.G., 2nd; Rodland, K.; Wiley, S.; et al. Identification of proteins in human cytomegalovirus (HCMV) particles: The HCMV proteome. *J. Virol.* **2004**, *78*, 10960–10966. [CrossRef] [PubMed]

64. Stern-Ginossar, N.; Weisburd, B.; Michalski, A.; Le, V.T.; Hein, M.Y.; Huang, S.X.; Ma, M.; Shen, B.; Qian, S.B.; Hengel, H.; et al. Decoding human cytomegalovirus. *Science* **2012**, *338*, 1088–1093. [CrossRef] [PubMed]

65. Nguyen, C.C.; Kamil, J.P. Pathogen at the Gates: Human Cytomegalovirus Entry and Cell Tropism. *Viruses* **2018**, *10*, 704. [CrossRef] [PubMed]

66. Wu, K.; Oberstein, A.; Wang, W.; Shenk, T. Role of PDGF receptor-alpha during human cytomegalovirus entry into fibroblasts. *Proc. Natl. Acad. Sci. USA* **2018**, *115*, E9889–E9898. [CrossRef] [PubMed]

67. Wu, Y.; Prager, A.; Boos, S.; Resch, M.; Brizic, I.; Mach, M.; Wildner, S.; Scrivano, L.; Adler, B. Human cytomegalovirus glycoprotein complex gH/gL/gO uses PDGFR-alpha as a key for entry. *PLoS Pathog.* **2017**, *13*, e1006281. [CrossRef]

68. Kabanova, A.; Marcandalli, J.; Zhou, T.; Bianchi, S.; Baxa, U.; Tsybovsky, Y.; Lilleri, D.; Silacci-Fregni, C.; Foglierini, M.; Fernandez-Rodriguez, B.M.; et al. Platelet-derived growth factor-alpha receptor is the cellular receptor for human cytomegalovirus gHgLgO trimer. *Nat. Microbiol.* **2016**, *1*, 16082. [CrossRef]

69. E, X.; Meraner, P.; Lu, P.; Perreira, J.M.; Aker, A.M.; McDougall, W.M.; Zhuge, R.; Chan, G.C.; Gerstein, R.M.; Caposio, P.; et al. OR14I1 is a receptor for the human cytomegalovirus pentameric complex and defines viral epithelial cell tropism. *Proc. Natl. Acad. Sci. USA* **2019**, *116*, 7043–7052. [CrossRef]

70. Ye, X.; Gui, X.; Freed, D.C.; Ku, Z.; Li, L.; Chen, Y.; Xiong, W.; Fan, X.; Su, H.; He, X.; et al. Identification of adipocyte plasma membrane-associated protein as a novel modulator of human cytomegalovirus infection. *PLoS Pathog.* **2019**, *15*, e1007914. [CrossRef]

71. Stein, K.R.; Gardner, T.J.; Hernandez, R.E.; Kraus, T.A.; Duty, J.A.; Ubarretxena-Belandia, I.; Moran, T.M.; Tortorella, D. CD46 facilitates entry and dissemination of human cytomegalovirus. *Nat. Commun.* **2019**, *10*, 2699. [CrossRef] [PubMed]

72. Vanarsdall, A.L.; Pritchard, S.R.; Wisner, T.W.; Liu, J.; Jardetzky, T.S.; Johnson, D.C. CD147 Promotes Entry of Pentamer-Expressing Human Cytomegalovirus into Epithelial and Endothelial Cells. *MBio* **2018**, *9*. [CrossRef] [PubMed]

73. Wang, D.; Shenk, T. Human cytomegalovirus virion protein complex required for epithelial and endothelial cell tropism. *Proc. Natl. Acad. Sci. USA* **2005**, *102*, 18153–18158. [CrossRef] [PubMed]

74. Huber, M.T.; Compton, T. The human cytomegalovirus UL74 gene encodes the third component of the glycoprotein H-glycoprotein L-containing envelope complex. *J. Virol.* **1998**, *72*, 8191–8197.

75. Wille, P.T.; Knoche, A.J.; Nelson, J.A.; Jarvis, M.A.; Johnson, D.C. A human cytomegalovirus gO-null mutant fails to incorporate gH/gL into the virion envelope and is unable to enter fibroblasts and epithelial and endothelial cells. *J. Virol.* **2010**, *84*, 2585–2596. [CrossRef]

76. Ciferri, C.; Chandramouli, S.; Donnarumma, D.; Nikitin, P.A.; Cianfrocco, M.A.; Gerrein, R.; Feire, A.L.; Barnett, S.W.; Lilja, A.E.; Rappuoli, R.; et al. Structural and biochemical studies of HCMV gH/gL/gO and Pentamer reveal mutually exclusive cell entry complexes. *Proc. Natl. Acad. Sci. USA* **2015**, *112*, 1767–1772. [CrossRef]

77. Zhou, M.; Lanchy, J.M.; Ryckman, B.J. Human Cytomegalovirus gH/gL/gO Promotes the Fusion Step of Entry into All Cell Types, whereas gH/gL/UL128-131 Broadens Virus Tropism through a Distinct Mechanism. *J. Virol.* **2015**, *89*, 8999–9009. [CrossRef]

78. Bagdonaite, I.; Norden, R.; Joshi, H.J.; King, S.L.; Vakhrushev, S.Y.; Olofsson, S.; Wandall, H.H. Global Mapping of O-Glycosylation of Varicella Zoster Virus, Human Cytomegalovirus, and Epstein-Barr Virus. *J. Biol. Chem.* **2016**, *291*, 12014–12028. [CrossRef]

79. Nogalski, M.T.; Chan, G.C.; Stevenson, E.V.; Collins-McMillen, D.K.; Yurochko, A.D. The HCMV gH/gL/UL128-131 complex triggers the specific cellular activation required for efficient viral internalization into target monocytes. *PLoS Pathog.* **2013**, *9*, e1003463. [CrossRef]

80. Fouts, A.E.; Chan, P.; Stephan, J.P.; Vandlen, R.; Feierbach, B. Antibodies against the gH/gL/UL128/UL130/UL131 complex comprise the majority of the anti-cytomegalovirus (anti-CMV) neutralizing antibody response in CMV hyperimmune globulin. *J. Virol.* **2012**, *86*, 7444–7447. [CrossRef]

81. Zydek, M.; Petitt, M.; Fang-Hoover, J.; Adler, B.; Kauvar, L.M.; Pereira, L.; Tabata, T. HCMV infection of human trophoblast progenitor cells of the placenta is neutralized by a human monoclonal antibody to glycoprotein B and not by antibodies to the pentamer complex. *Viruses* **2014**, *6*, 1346–1364. [CrossRef] [PubMed]

82. Wille, P.T.; Wisner, T.W.; Ryckman, B.; Johnson, D.C. Human cytomegalovirus (HCMV) glycoprotein gB promotes virus entry in trans acting as the viral fusion protein rather than as a receptor-binding protein. *MBio* **2013**, *4*, e00332-13. [CrossRef] [PubMed]

83. Selinsky, C.; Luke, C.; Wloch, M.; Geall, A.; Hermanson, G.; Kaslow, D.; Evans, T. A DNA-based vaccine for the prevention of human cytomegalovirus-associated diseases. *Hum. Vaccin.* **2005**, *1*, 16–23. [CrossRef] [PubMed]

84. Yue, Y.; Kaur, A.; Zhou, S.S.; Barry, P.A. Characterization and immunological analysis of the rhesus cytomegalovirus homologue (Rh112) of the human cytomegalovirus UL83 lower matrix phosphoprotein (pp65). *J. Gen. Virol.* **2006**, *87*, 777–787. [CrossRef] [PubMed]

85. Bachmann, M.F.; Jennings, G.T. Vaccine delivery: A matter of size, geometry, kinetics and molecular patterns. *Nat. Reviews. Immunol.* **2010**, *10*, 787–796. [CrossRef]

86. Lopez-Sagaseta, J.; Malito, E.; Rappuoli, R.; Bottomley, M.J. Self-assembling protein nanoparticles in the design of vaccines. *Comput Struct Biotechnol J.* **2016**, *14*, 58–68. [CrossRef]

87. Manolova, V.; Flace, A.; Bauer, M.; Schwarz, K.; Saudan, P.; Bachmann, M.F. Nanoparticles target distinct dendritic cell populations according to their size. *Eur. J. Immunol.* **2008**, *38*, 1404–1413. [CrossRef]

88. Duan, H.; Chen, X.; Boyington, J.C.; Cheng, C.; Zhang, Y.; Jafari, A.J.; Stephens, T.; Tsybovsky, Y.; Kalyuzhniy, O.; Zhao, P.; et al. Glycan Masking Focuses Immune Responses to the HIV-1 CD4-Binding Site and Enhances Elicitation of VRC01-Class Precursor Antibodies. *Immunity* **2018**, *49*, 301–311.e305. [CrossRef]

89. Kanekiyo, M.; Bu, W.; Joyce, M.G.; Meng, G.; Whittle, J.R.; Baxa, U.; Yamamoto, T.; Narpala, S.; Todd, J.P.; Rao, S.S.; et al. Rational Design of an Epstein-Barr Virus Vaccine Targeting the Receptor-Binding Site. *Cell* **2015**, *162*, 1090–1100. [CrossRef]

90. Kirchmeier, M.; Fluckiger, A.C.; Soare, C.; Bozic, J.; Ontsouka, B.; Ahmed, T.; Diress, A.; Pereira, L.; Schodel, F.; Plotkin, S.; et al. Enveloped virus-like particle expression of human cytomegalovirus glycoprotein B antigen induces antibodies with potent and broad neutralizing activity. *Clin. Vaccine Immunol. Cvi.* **2014**, *21*, 174–180. [CrossRef]

91. Cui, X.; Snapper, C.M. Development of novel vaccines against human cytomegalovirus. *Hum. Vaccines Immunother.* **2019**, *15*, 2673–2683. [CrossRef] [PubMed]

92. Vicente, T.; Burri, S.; Wellnitz, S.; Walsh, K.; Rothe, S.; Liderfelt, J. Fully aseptic single-use cross flow filtration system for clarification and concentration of cytomegalovirus-like particles. *Eng. Life Sci.* **2014**, *14*, 318–326. [CrossRef]

93. Crowther, R.A.; Kiselev, N.A.; Bottcher, B.; Berriman, J.A.; Borisova, G.P.; Ose, V.; Pumpens, P. Three-dimensional structure of hepatitis B virus core particles determined by electron cryomicroscopy. *Cell* **1994**, *77*, 943–950. [CrossRef]

94. Roseman, A.M.; Berriman, J.A.; Wynne, S.A.; Butler, P.J.; Crowther, R.A. A structural model for maturation of the hepatitis B virus core. *Proc. Natl. Acad. Sci. USA* **2005**, *102*, 15821–15826. [CrossRef] [PubMed]

95. Walker, A.; Skamel, C.; Vorreiter, J.; Nassal, M. Internal core protein cleavage leaves the hepatitis B virus capsid intact and enhances its capacity for surface display of heterologous whole chain proteins. *J. Biol. Chem.* **2008**, *283*, 33508–33515. [CrossRef] [PubMed]

96. Schwarz, K.; Meijerink, E.; Speiser, D.E.; Tissot, A.C.; Cielens, I.; Renhof, R.; Dishlers, A.; Pumpens, P.; Bachmann, M.F. Efficient homologous prime-boost strategies for T cell vaccination based on virus-like particles. *Eur. J. Immunol.* **2005**, *35*, 816–821. [CrossRef]

97. Krishna, B.A.; Humby, M.S.; Miller, W.E.; O'Connor, C.M. Human cytomegalovirus G protein-coupled receptor US28 promotes latency by attenuating c-fos. *Proc. Natl. Acad. Sci. USA* **2019**, *116*, 1755–1764. [CrossRef]

98. Krishna, B.A.; Miller, W.E.; O'Connor, C.M. US28: HCMV's Swiss Army Knife. *Viruses* **2018**, *10*, 445. [CrossRef]

99. Watt, G.D.; Jacobs, D.; Frankel, R.B. Redox reactivity of bacterial and mammalian ferritin: Is reductant entry into the ferritin interior a necessary step for iron release? *Proc. Natl. Acad. Sci. USA* **1988**, *85*, 7457–7461. [CrossRef]

100. Theil, E.C. Ferritin: Structure, gene regulation, and cellular function in animals, plants, and microorganisms. *Annu. Rev. Biochem.* **1987**, *56*, 289–315. [CrossRef]

101. Cho, K.J.; Shin, H.J.; Lee, J.H.; Kim, K.J.; Park, S.S.; Lee, Y.; Lee, C.; Park, S.S.; Kim, K.H. The crystal structure of ferritin from Helicobacter pylori reveals unusual conformational changes for iron uptake. *J. Mol. Biol.* **2009**, *390*, 83–98. [CrossRef] [PubMed]

102. Pulsipher, K.W.; Villegas, J.A.; Roose, B.W.; Hicks, T.L.; Yoon, J.; Saven, J.G.; Dmochowski, I.J. Thermophilic Ferritin 24mer Assembly and Nanoparticle Encapsulation Modulated by Interdimer Electrostatic Repulsion. *Biochemistry* **2017**, *56*, 3596–3606. [CrossRef] [PubMed]

103. Georgiev, I.S.; Joyce, M.G.; Chen, R.E.; Leung, K.; McKee, K.; Druz, A.; Van Galen, J.G.; Kanekiyo, M.; Tsybovsky, Y.; Yang, E.S.; et al. Two-Component Ferritin Nanoparticles for Multimerization of Diverse Trimeric Antigens. *Acs. Infect. Dis.* **2018**, *4*, 788–796. [CrossRef] [PubMed]

104. Kanekiyo, M.; Joyce, M.G.; Gillespie, R.A.; Gallagher, J.R.; Andrews, S.F.; Yassine, H.M.; Wheatley, A.K.; Fisher, B.E.; Ambrozak, D.R.; Creanga, A.; et al. Mosaic nanoparticle display of diverse influenza virus hemagglutinins elicits broad B cell responses. *Nat. Immunol.* **2019**, *20*, 362–372. [CrossRef] [PubMed]

105. Golmohammadi, R.; Fridborg, K.; Bundule, M.; Valegard, K.; Liljas, L. The crystal structure of bacteriophage Q beta at 3.5 A resolution. *Structure* **1996**, *4*, 543–554. [CrossRef]

106. Cornuz, J.; Zwahlen, S.; Jungi, W.F.; Osterwalder, J.; Klingler, K.; van Melle, G.; Bangala, Y.; Guessous, I.; Muller, P.; Willers, J.; et al. A vaccine against nicotine for smoking cessation: A randomized controlled trial. *PLoS ONE* **2008**, *3*, e2547. [CrossRef]

107. Vasiljeva, I.; Kozlovska, T.; Cielens, I.; Strelnikova, A.; Kazaks, A.; Ose, V.; Pumpens, P. Mosaic Qbeta coats as a new presentation model. *Febs. Lett.* **1998**, *431*, 7–11. [CrossRef]

108. Hsia, Y.; Bale, J.B.; Gonen, S.; Shi, D.; Sheffler, W.; Fong, K.K.; Nattermann, U.; Xu, C.; Huang, P.S.; Ravichandran, R.; et al. Corrigendum: Design of a hyperstable 60-subunit protein icosahedron. *Nature* **2016**, *540*, 150. [CrossRef]

109. King, N.P.; Bale, J.B.; Sheffler, W.; McNamara, D.E.; Gonen, S.; Gonen, T.; Yeates, T.O.; Baker, D. Accurate design of co-assembling multi-component protein nanomaterials. *Nature* **2014**, *510*, 103–108. [CrossRef]

110. Bale, J.B.; Gonen, S.; Liu, Y.; Sheffler, W.; Ellis, D.; Thomas, C.; Cascio, D.; Yeates, T.O.; Gonen, T.; King, N.P.; et al. Accurate design of megadalton-scale two-component icosahedral protein complexes. *Science* **2016**, *353*, 389–394. [CrossRef]

111. He, L.; Kumar, S.; Allen, J.D.; Huang, D.; Lin, X.; Mann, C.J.; Saye-Francisco, K.L.; Copps, J.; Sarkar, A.; Blizard, G.S.; et al. HIV-1 vaccine design through minimizing envelope metastability. *Sci. Adv.* **2018**, *4*, eaau6769. [CrossRef]

112. Marcandalli, J.; Fiala, B.; Ols, S.; Perotti, M.; de van der Schueren, W.; Snijder, J.; Hodge, E.; Benhaim, M.; Ravichandran, R.; Carter, L.; et al. Induction of Potent Neutralizing Antibody Responses by a Designed Protein Nanoparticle Vaccine for Respiratory Syncytial Virus. *Cell* **2019**, *176*, 1420–1431.e1417. [CrossRef] [PubMed]

113. Brouwer, P.J.M.; Antanasijevic, A.; Berndsen, Z.; Yasmeen, A.; Fiala, B.; Bijl, T.P.L.; Bontjer, I.; Bale, J.B.; Sheffler, W.; Allen, J.D.; et al. Enhancing and shaping the immunogenicity of native-like HIV-1 envelope trimers with a two-component protein nanoparticle. *Nat. Commun.* **2019**, *10*, 4272. [CrossRef] [PubMed]

114. Boyken, S.E.; Benhaim, M.A.; Busch, F.; Jia, M.; Bick, M.J.; Choi, H.; Klima, J.C.; Chen, Z.; Walkey, C.; Mileant, A.; et al. De novo design of tunable, pH-driven conformational changes. *Science* **2019**, *364*, 658–664. [CrossRef] [PubMed]

115. Votteler, J.; Ogohara, C.; Yi, S.; Hsia, Y.; Nattermann, U.; Belnap, D.M.; King, N.P.; Sundquist, W.I. Designed proteins induce the formation of nanocage-containing extracellular vesicles. *Nature* **2016**, *540*, 292–295. [CrossRef]

116. Zakeri, B.; Fierer, J.O.; Celik, E.; Chittock, E.C.; Schwarz-Linek, U.; Moy, V.T.; Howarth, M. Peptide tag forming a rapid covalent bond to a protein, through engineering a bacterial adhesin. *Proc. Natl. Acad. Sci. USA* **2012**, *109*, E690–E697. [CrossRef]

117. Brune, K.D.; Leneghan, D.B.; Brian, I.J.; Ishizuka, A.S.; Bachmann, M.F.; Draper, S.J.; Biswas, S.; Howarth, M. Plug-and-Display: Decoration of Virus-Like Particles via isopeptide bonds for modular immunization. *Sci. Rep.* **2016**, *6*, 19234. [CrossRef]

118. Palladini, A.; Thrane, S.; Janitzek, C.M.; Pihl, J.; Clemmensen, S.B.; de Jongh, W.A.; Clausen, T.M.; Nicoletti, G.; Landuzzi, L.; Penichet, M.L.; et al. Virus-like particle display of HER2 induces potent anti-cancer responses. *Oncoimmunology* **2018**, *7*, e1408749. [CrossRef]

119. Thrane, S.; Janitzek, C.M.; Matondo, S.; Resende, M.; Gustavsson, T.; de Jongh, W.A.; Clemmensen, S.; Roeffen, W.; van de Vegte-Bolmer, M.; van Gemert, G.J.; et al. Bacterial superglue enables easy development of efficient virus-like particle based vaccines. *J. Nanobiotechnology* **2016**, *14*, 30. [CrossRef]

120. Abbott, R.K.; Lee, J.H.; Menis, S.; Skog, P.; Rossi, M.; Ota, T.; Kulp, D.W.; Bhullar, D.; Kalyuzhniy, O.; Havenar-Daughton, C.; et al. Precursor Frequency and Affinity Determine B Cell Competitive Fitness in Germinal Centers, Tested with Germline-Targeting HIV Vaccine Immunogens. *Immunity* **2018**, *48*, 133–146.e6. [CrossRef]

121. He, L.; de Val, N.; Morris, C.D.; Vora, N.; Thinnes, T.C.; Kong, L.; Azadnia, P.; Sok, D.; Zhou, B.; Burton, D.R.; et al. Presenting native-like trimeric HIV-1 antigens with self-assembling nanoparticles. *Nat. Commun.* **2016**, *7*, 12041. [CrossRef] [PubMed]

122. McGuire, A.T.; Gray, M.D.; Dosenovic, P.; Gitlin, A.D.; Freund, N.T.; Petersen, J.; Correnti, C.; Johnsen, W.; Kegel, R.; Stuart, A.B.; et al. Specifically modified Env immunogens activate B-cell precursors of broadly neutralizing HIV-1 antibodies in transgenic mice. *Nat. Commun.* **2016**, *7*, 10618. [CrossRef] [PubMed]

123. Sliepen, K.; Ozorowski, G.; Burger, J.A.; van Montfort, T.; Stunnenberg, M.; LaBranche, C.; Montefiori, D.C.; Moore, J.P.; Ward, A.B.; Sanders, R.W. Presenting native-like HIV-1 envelope trimers on ferritin nanoparticles improves their immunogenicity. *Retrovirology* **2015**, *12*, 82. [CrossRef] [PubMed]

 *viruses*

Review

Virus-like Particle-Based L2 Vaccines against HPVs: Where Are We Today?

Rashi Yadav [1], Lukai Zhai [1,2] and Ebenezer Tumban [1,*

[1]   Department of Biological Sciences, Michigan Technological University, Houghton, MI 49931, USA; rashiy@mtu.edu (R.Y.); LZhai@mtu.edu (L.Z.)
[2]   Current address: Van Andel Research Institute, Grand Rapids, MI 49503, USA
*   Correspondence: etumban@mtu.edu; Tel.: +1-906-487-2256; Fax: +1-906-487-3167

Received: 25 November 2019; Accepted: 18 December 2019; Published: 23 December 2019

Abstract: Human papillomaviruses (HPVs) are the most common sexually transmitted infections worldwide. Ninety percent of infected individuals clear the infection within two years; however, in the remaining 10% of infected individuals, the infection(s) persists and ultimately leads to cancers (anogenital cancers and head and neck cancers) and genital warts. Fortunately, three prophylactic vaccines have been approved to protect against HPV infections. The most recent HPV vaccine, Gardasil-9 (a nonavalent vaccine), protects against seven HPV types associated with ~90% of cervical cancer and against two HPV types associated with ~90% genital warts with little cross-protection against non-vaccine HPV types. The current vaccines are based on virus-like particles (VLPs) derived from the major capsid protein, L1. The L1 protein is not conserved among HPV types. The minor capsid protein, L2, on the other hand, is highly conserved among HPV types and has been an alternative target antigen, for over two decades, to develop a broadly protective HPV vaccine. The L2 protein, unlike the L1, cannot form VLPs and as such, it is less immunogenic. This review summarizes current studies aimed at developing HPV L2 vaccines by multivalently displaying L2 peptides on VLPs derived from bacteriophages and eukaryotic viruses. Recent data show that a monovalent HPV L1 VLP as well as bivalent MS2 VLPs displaying HPV L2 peptides (representing amino acids 17–36 and/or consensus amino acids 69–86) elicit robust broadly protective antibodies against diverse HPV types (6/11/16/18/26/31/33/34/35/39/43/44/45/51/52/53/56/58/59/66/68/73) associated with cancers and genital warts. Thus, VLP-based L2 vaccines look promising and may be favorable, in the near future, over current L1-based HPV vaccines and should be explored further.

Keywords: HPVs; vaccines; virus-like particles (VLPs); minor capsid protein (L2)

---

1. Background

Virus-like particles (VLPs) are empty viral shells derived from the expression, in a suitable host cell, of viral structural proteins such as capsid or coat proteins. Over-expression of these proteins allows them to spontaneously self-assemble into VLPs (Figure 1). VLPs can also be derived from envelope proteins following over-expression of envelope proteins together with other structural proteins such as the pre-membrane proteins, capsid proteins, or over-expression of the envelope protein with Gag protein [1]. VLPs derived from the latter are known as enveloped VLPs. Thus, VLPs are empty shells which consist of one or more types of multimeric coat or envelope proteins arranged geometrically into dense, repetitive (multivalent) arrays [2–4]. They are morphologically and structurally similar to viruses from which the coat proteins are derived, except for the fact that they lack the viral genome. VLPs are highly immunogenic, even at small doses of antigens (as low as 500 ng) [5]. The following features have been credited for making VLPs highly immunogenic:

(i)   Coat proteins that form the capsid are multivalently displayed to the immune system. Multivalent display enhances cross-linking of B-cell receptors on naïve B cells, leading to a stronger B-cell activation, proliferation, and differentiation, secretion of high-affinity antibodies, and the generation of long-lived memory B cells [6–9].

(ii)   VLPs have virally encoded T-helper cell epitopes, which enhance T-cell activation. Presentation of these epitopes by antigen-presenting cells (APCs) in association with major histocompatibility complex class II to T-helper cells leads to the activation of T-helper cells. In addition to this, co-stimulatory molecules from APCs help to activate the T-helper cells. Activated T-helper cells then secret cytokines, leading to the activation of B-cells, T-cells, and macrophages [6].

(iii)   Most VLPs are between 25 and 100 nm. This size range is very important for the following reasons. Firstly, nanoparticles between 10 and 200 nm, unlike those >200 nm, have been shown to efficiently enter the lymphatic system by direct diffusion through cell junctions. This allows VLPs to be efficiently exposed to immune cells. It is worth mentioning that VLPs can also be taken up and transported by APCs to the lymphatic system. Secondly, the small size of VLPs enables them to be transported to the lymphoid organs within a short period of time and to be efficiently taken up by APCs for presentation to T-helper cells [9].

(iv)   Some VLPs (especially bacteriophage VLPs) contain single-stranded (ss)RNA, which serves as endogenous adjuvant. Bacteriophage coat proteins (from MS2, PP7, etc.) have the potential to encapsudate ssRNA that codes for its capsid/coat protein during VLP assembly [10,11]. This ssRNA serves as endogenous adjuvant by directly activating immunostimulatory molecules such as toll-like receptors (TLR7 and TLR8), which in turn activate innate immune responses [9,12–14].

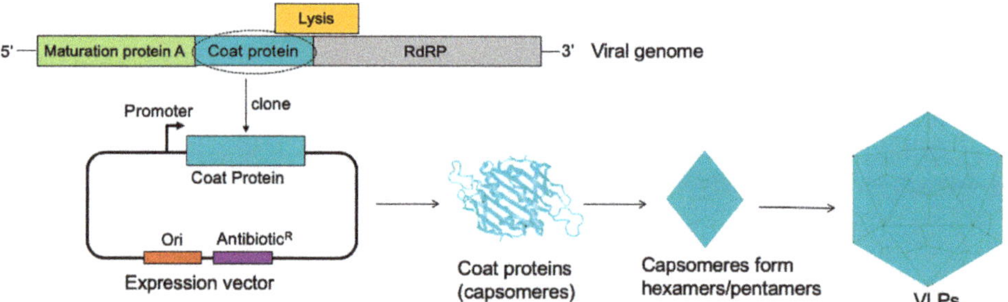

Figure 1. The assembly of bacteriophage MS2 coat proteins into VLPs (virus-like particles). The genome of MS2 bacteriophage (top image) with the coat protein (light blue). Cloning of the coat protein into a bacterial expression vector and expression of the protein in a suitable bacterial host gives rise to coat proteins (capsomers). The capsomeres form pentamers and hexamers; twelve pentamers and 20 hexamers spontaneously self-assemble to form a VLP (composed of 180 capsomeres). RdRP stands for RNA dependent RNA polymerase.

The aforementioned features, in addition to the fact that VLPs are noninfectious and are safe, have made VLPs attractive biological agents for vaccine design and development. A number of VLP-based prophylactic vaccines have been approved by the Food and Drug Administration to protect against human papillomaviruses (HPVs) and hepatitis B virus (HBV) infections. Two VLP-based HBV vaccines are credited for reducing incidences of HBV-related hepatocellular carcinomas and mortalities, worldwide, with protection lasting for up to 30 years in some individuals [2,15]. On the other hand, three VLP-based HPV vaccines (Gadarsil-9, Cervarix, and Gardasil-4 (discontinued in the US)) have been approved within the last decade to protect against HPV infections.

2. HPV Vaccines

Approximately 42 HPV types can be transmitted sexually via anogenital to anogenital sex or anogenital to oral sex [16]. Out of these HPVs, ~19 types called high-risk types (oncogenic types; types 16, 18, 26, 31, 33, 35, 39, 45, 51–53, 56, 58, 59, 66, 68, 70, 73, and 82) are associated with cancers [17,18]. The remaining types, known as low-risk HPV types (types 6, 11, 40–44, 54, 61, 72, 81, etc.), are associated with genital warts and recurrent respiratory papillomatosis. VLP-based HPV vaccines have recently been shown to prevent cases of cervical intraepithelial neoplasias, with protection levels lasting for at least 10 years [19,20]. Moreover, recent studies show that a single dose of the HPV VLP-based vaccine can lead to long-lasting protection from HPV infection [21]. These vaccines are based on VLPs derived from over-expression of the capsid proteins in yeast (Gardasil vaccines) or in insect cells (Cervarix vaccine). HPV capsid is composed of two capsid proteins, the major capsid protein (L1) and the minor capsid protein (L2) (Figure 2). The L1 protein forms pentamers, and 72 copies of the pentamer assemble to form a capsid [22]. The L2 protein is suggested to be present as canyons at the vertices of pentamers and it is only transiently exposed following binding of the virion to heparan sulfate proteoglycans (HSPG) on the basement membrane [23,24]. The exact number of L2 protein on a virion is debatable. Studies suggest that about 12–72 copies of L2 proteins are present on a virion [25,26]. It is worth mentioning that L1 can assemble into the capsid without L2. L2 enhances encapsidation of a double-stranded circular DNA genome into the capsid, thus forming a virion [27]. The L2 protein also has other functions in the life cycle of the virus. Binding of a virion to HSPG promotes conformational change that exposes L2 on the capsid, and thus L2 enhances binding of the virion to epithelial cells [23,28]. It also promotes egress of the virion from the endosome [29] and trafficking of the viral genome towards the nucleus for replication [30,31].

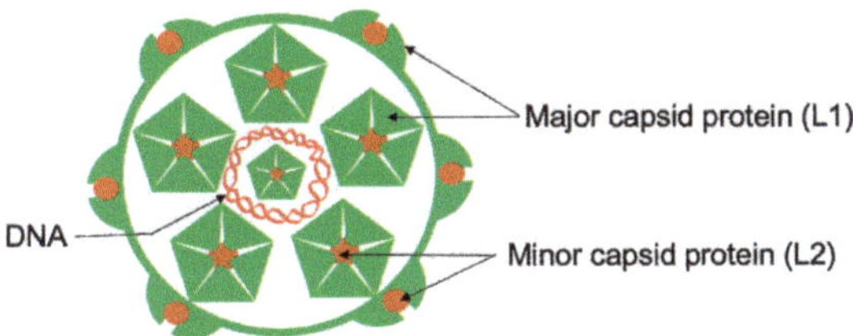

Figure 2. A schematic of HPV (human papillomavirus virion). A double-stranded DNA genome (red) is surrounded by the capsid, which is composed of two proteins: the major capsid protein (L1, shown in light green) and the minor capsid protein (L2, shown in brown color). The L1 protein forms pentamers, and L2 protein is inserted on vertices of the pentamers. Seventy-two copies of the pentamers and about 12–72 copies of L2 protein assemble to form a virion.

Although current L1-based HPV vaccines are highly immunogenic, they protect mostly against the HPV types included in the vaccines [32–36]; the L1 protein is not conserved among HPV types. For example, the most recent HPV vaccine (Gardasil-9, an upgrade of Gardasil-4) offers protection against seven HPV types (HPV16, 18, 31, 45, 33, 52, 58) associated with ~90% of cervical cancer and two HPV types (HPV6 and 11) associated with ~90% genital warts [17]. Thus, complete protection from HPV-associated cancers/warts may require the addition, to current HPV vaccines, of VLPs from HPV types that are not currently included in the vaccines. With this in consideration and given the fact that L1-based vaccines offer little cross-protection against other HPV types, the L2 protein has been explored within the last two decades to develop next-generation HPV vaccines. As shown in Figure 3, the L2 protein is conserved among different HPV types. This thus suggests that vaccines targeting L2 are going to be broadly protective against different HPV types. In fact, results from preclinical studies targeting L2 protein look very promising [37–41].

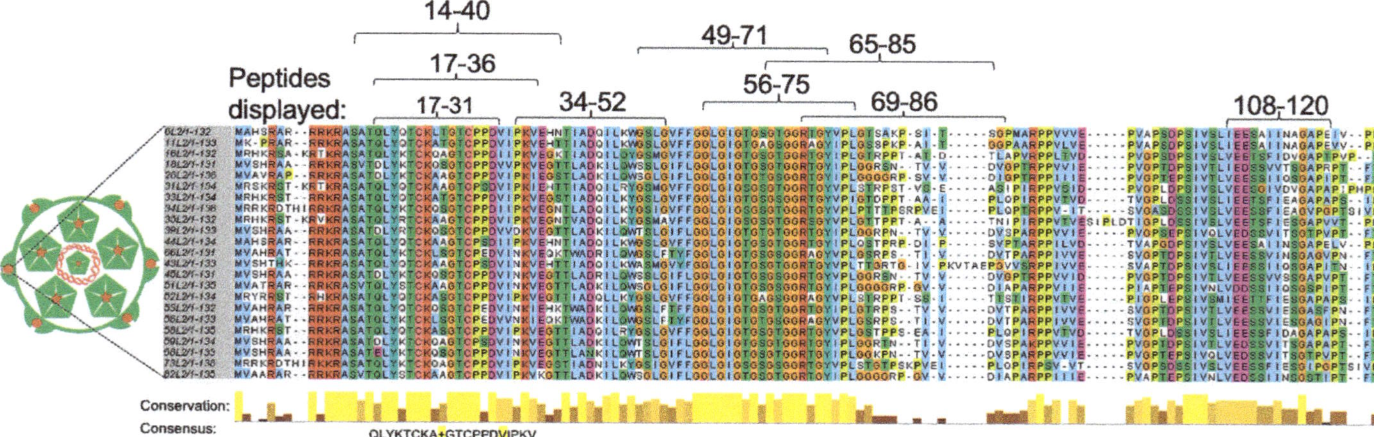

Figure 3. Sequence alignment of L2 (aa 1–135) from different HPV types. Sequence alignment was done using sequences from both high-risk (HPVs 16, 18, 26, 31, 33, 34, 35, 39, 45, 51–53, 56, 58, 59, 66, 68, 73, and 82) and low-risk HPV types (HPVs 6, 11, 43, and 44). Alignment was done using Jalview software. Residues that are highly conserved among different HPV types are highlighted, below the alignment, in gold bars. A consensus sequence of amino acid 17–36 from different HPV types is shown below the bars. Amino acid residues (in the consensus sequence) that differ from HPV16 L2 are highlighted in yellow background. The numbers above sequence alignment represent peptides (amino acid residues) that have been displayed on different VLP platforms (see Tables 1 and 2 for details).

3. L2 Protein

Immunization with the N-terminus of L2 protein (out of the context of the L1 protein) elicits antibodies that protect and cross-protect against heterologous HPV types. For example, unlike antibodies elicited against L1 VLPs, antibodies elicited against an L2 peptide representing amino acids 1–88 from HPV type 18 or even from L2 of bovine papillomavirus (BPV) cross-protect, albeit at low titers, against diverse mucosal (HPV6, 11, 16, 18, 31) and cutaneous HPV types (HPV5) [37–41]. Given these data, efforts to develop next-generation HPV vaccines have focused on eliciting protective antibodies against L2 protein, especially against the first 130 amino acids of L2 peptide. Unfortunately, the L2 protein cannot form VLPs like L1 protein. Thus, studies to develop an HPV L2 vaccine have relied on using protein/peptide antigens. As mentioned earlier, L2 peptide antibody titers are very low. This could be explained by the fact that peptide antigens are very unstable in serum and are rapidly degraded following immunization [42,43]. As such, different strategies have been explored in preclinical studies to enhance protective antibody responses against L2 protein. For example, immunizations with an L2 peptide conjugated to thioredoxin [44,45], immunizations with concatemers of L2 protein fused to a self-adjuvanting protein (flagellin) [46,47] as well as immunizations with a concatemer of L2 proteins derived from different HPV types [48,49] have been used. More recently, an L2 polypeptide and a heptamerizing coiled-coil polypeptide OVX313 fused to a nanoparticle derived from a thermostable thioredoxin has been used [50–52]. These approaches have enhanced the immunogenicity of L2 peptides, especially immunization with the nanoparticle thioredoxin-L2-OVX313 candidate vaccine. The nanoparticle candidate vaccine offers protection against more than 14 HPV types [50]. Although the aforementioned strategies enhanced immune responses, the responses were observed mostly when large doses (up to 25 µg) of antigens, coupled with large amount of adjuvant with multiple immunizations regimens were used. With this in consideration, we and others have focused on immunizing with L2 peptides displayed multivalently on the surface of VLPs.

4. Multivalent Display of HPV L2 on VLPs

Not only can VLPs be used to develop vaccines against the virus from which the structural proteins are derived from, they can also be used as platforms to display heterologous antigens from other viruses [53], bacteria [54], pathophysiological diseases like cholesterol [55], and even tumor-associated antigens [56]. The goal of a chimeric VLP is to induce antibodies against a heterologous antigen displayed on the platform, but not against the platform. This technology has been exploited to develop VLP-based L2 vaccines against HPV, targeting some of the L2 peptides or epitopes shown in Figure 3. VLPs from viruses that infect bacteria (bacteriophages) or from viruses that infect eukaryotic cells have both been explored as platforms to display L2 peptides (Tables 1 and 2).

4.1. The Display of L2 Peptides on Bacteriophage VLP Platforms

Different HPV L2 peptides (shown in Figure 3) have been displayed, by genetic insertion or by chemical conjugation, on VLPs derived from viruses (MS2, PP7, Qβ, AP205) that infect bacteria. Genetic insertion of L2 allows for DNA sequences that code for L2 peptides to be inserted by polymerase chain reaction on the single-chain dimer (two fused copies) of the coat proteins of bacteriophages MS2, PP7 or to be inserted on the coat protein monomer of bacteriophage AP205 in expression vectors (Table 1). Expression of the chimeric coat proteins in a suitable bacterial host allows 90 copies of the coat protein dimers (for MS2 and PP7) and 180 copies of coat protein monomers (AP205) to spontaneously assemble into VLPs, thus displaying 90 copies and 180 copies of L2 on the VLPs, respectively [5,11,57–60]. Chemical conjugation, on the other hand, allows chemically synthesized L2 peptides to be displayed mostly on Qβ VLPs as follows. Qβ VLPs are first expressed and purified from bacterial cells. Synthesized L2 peptides, carrying a terminal cysteine residue, are then cross-linked to lysine residues on the VLPs using a bi-functional cross-linker, succinimidyl 6-((beta-maleimidopropionamido)hexanoate) [61]. This allows for at least 360 copies of L2 to be displayed on Qβ VLPs.

Table 1. Candidate bacteriophage VLP-based L2 HPV vaccines.

Source of VLPs	HPV L2 Epitope Displayed (aa)	Length of Epitope	Position of Insertion on VLPs	Dose Immunized with	Neutralized or Protected against HPV Pseudoviruses (PsVs) *	Ref [#].
MS2	HPV16 L2 (aa 17–31)	15	N-terminus	Mice immunized with two doses of VLPs (5 µg/dose)	PsVs (genital): 5/**6**/**16**/33/**35**/39/**45**/51/53/58	[5]
	HPV16 L2 (aa 20–31)	12	''	''	Not tested	
	HPV16 L2 (aa 14–40)	26	''	''	''	
	HPV16 L2/31L2 (aa 20–31 + aa 17-31)	27	N-terminus	Mice immunized with two doses of a mixture of VLPs displaying 16L2/31L2 and VLPs displaying consensus epitope (10 µg/dose)	PsVs (genital): **16/18/31/33/45**/58	[57]
	Consensus (aa 69–86)	18		Mice orally immunized with three doses of a mixture of VLPs displaying 16/31L2 and VLPs displaying consensus epitope (100 µg/dose)	PsVs (genital): 11/**16/53/56**	[62]
					PsVs (oral): **16/35/39/52**/58	
	HPV16 L2 (aa 17–31)	15	16L2 on N-terminus of one coat protein and 31L2 on the N-terminus of another coat protein	Mice immunized with two doses of hybrid VLPs (5 µg/dose)	PsVs (neutralized): **16/31/45/18/58**	[59]
	HPV31 L2 (aa 17–31)	15				
PP7	HPVs: 1L2, 5L2, 6L2, 11L2, 16L2, 18L2, 45L2, 58L2 (aa 17–31)	15 from each L2 on each VLP	AB-loop (internal insertion)	Mice immunized with three doses of a mixture of VLPs displaying all eight L2 peptides (10 µg/dose)	PsVs (genital): 6/**16/18/31/45/52**/58 PsVs (cutaneous): **5**	[58]
PP7	HPV16 L2 (aa 17–31)	15	16L2 on AB-loop of one coat protein and 18L2 on the AB-loop of another coat protein	Mice immunized with two doses of hybrid VLPs (5 µg/dose)	PsVs (genital): 6	[59]
	HPV18 L2 (aa 17–31)	15				
	HPV16 L2 (aa 35–50)	16	Chemical conjugation on surface	Mice immunized with three doses each VLP (5 µg/dose)	PsVs (genital): 16 (no protection)	[61]
	HPV16 L2 (aa 51–65)	15			PsVs (genital): 16	
Qβ	HPV16 L2 (aa 34–52)	19	Chemical conjugation on surface	Mice immunized with three doses of each VLP (5 µg/dose)	PsVs (genital): 16 (no protection)	[61]
	HPV16 L2 (aa 49–71)	23			PsVs (genital): 16	
	Consensus (aa 65–85)	18			PsVs (genital): **16**	
	HPV16 L2 (aa 108–120)	13			PsVs (genital): 6/**16**/31/45/**58**	
AP205	A concatemer of HPV: 16L2, 18L2, 31L2, 35L2, 52L2 (aa 17–38)	110	C-terminus	Mice immunized with two doses of VLPs (5 µg/dose)	PsVs (genital): 16, 31, 35, 45, **52**	[60]

* Bolded PsVs indicate that protection/neutralization was both significant and complete, while PsVs not bolded (in black) indicate that although the protection/neutralization was significant, it was not complete. PsVs not bolded (in gray background) indicate that there was minimal protection/neutralization or no protection at all (not significant protection); [#] Ref: Reference.

As mentioned above, peptides representing different L2 epitopes have been displayed on bacteriophage VLPs. These peptides, which range in size from 12 amino acids (aa) to 110 aa, can be grouped into three categories based on the source of peptides:

(i)   L2 peptides derived from different individual HPV types, especially HPV16 and HPV18. They correspond to aa 17–31, 20–29, 14–40, 34–52, 49–71, 56–75, 108–120, and so forth. [5,11,58,61].

(ii) L2 peptides derived from a concatemer of aa 17–31 and aa 20–31 from HPV16/HPV31 and a concatemer of aa 17–38 from five HPV types (HPV16, 18, 31, 35, 52) [57,60].

(iii) Peptides derived from a consensus sequence of aa 65–85 or aa 69–86 following the alignment of cancer-causing and wart-causing HPV types (Figure 3 and Table 1) [57,61].

Of all the above HPV L2 peptides that have been displayed on bacteriophage VLPs, only bacteriophage VLPs displaying peptides representing aa 17–31 from HPV5, 6, 16, 18 [5,58,59], bacteriophage VLPs displaying consensus aa 65–85 or 69–86 [57,61], and bacteriophage VLPs displaying peptides from the L2 concatemers (16L2/31L2 and 16L2/18L2/31L2/35L2/52L2) [57,60,62] elicited broadly neutralizing/protective antibodies against diverse HPV types (Table 1). For instance, immunization with a mixture of eight PP7 VLPs each displaying L2 peptide (aa 17–31) from eight different HPV types offered broader and better protection against HPV pseudoviruses 5/6/16/18/31/45/52/58 (Table 1) [58]. Each L2 peptide was inserted on the AB-loop of PP7 coat protein. Similarly, immunization with a mixture of two MS2–L2 VLPs (mixed MS2-L2 VLPs; one VLP displaying a concatemer peptide, aa 17–31 from 16L2 and aa 20–31 from 31L2 and another VLP displaying a consensus peptide from aa 69–86) offered robust broad protection against genital and oral infections with HPV pseudoviruses 11/16/18/31/33/35/39/45/52/53/56/58 (Table 1) [57,62]; the L2 peptides were inserted on the N-terminus of MS2 coat protein. Mixed MS2-L2 VLPs have the potential to protect against all eleven HPV types (tested so far) associated with ~95% of cervical cancer and against ~99% of HPV-associated head and neck cancers. Additionally, they can protect against one HPV type (HPV11, tested so far) associated with 36% of genital warts and ~32% of recurrent respiratory papillomatosis [57,62]. More HPV types need to be tested to assess the spectrum of protection. Thus, mixed MS2-L2 is an excellent next-generation candidate vaccine against HPV.

4.2. The Display of L2 Peptides on Eukaryotic VLP Platforms

L2 peptides have also been displayed on VLPs derived from eukaryotic viruses such as HPV, bovine papillomavirus type 1 (BPV1), adenovirus, adeno-associated virus, hepatitis B virus, and potato virus A (Table 2). All L2 peptides displayed on eukaryotic VLPs have been done by genetic insertion as described above. Inserted L2 peptides ranged from aa 17–36 (the most commonly inserted), aa 18–38, aa 56–75, aa 96–115, aa 108–120, aa 414–426, and so forth, and were derived mostly from HPV16. Inserted peptides were also derived from HPV31, HPV33, and HPV58 to a small extent (Table 2). The insertions have been made at different locations on the coat proteins. For example, on helix 4 loop (H4 helix, between aa 414/433 or between aa 430/433) on L1 coat protein of HPV16 (Table 2). Also, insertions have also been made between aa 136/177 (DE loop of L1 of HPV16), between aa 134/135 (DE loop of L1 of HPV18 and BPV1), and on the C-terminus of L1 coat protein of HPV18. Furthermore, insertions have also been made on some viruses/virus-associated particles. For example, insertions have been made on VP3 of Adeno-associated virus 2 virus particles at aa positions 587 and 453. Moreover, insertions have also been made on adenovirus types 5 and 35 (Table 2).

Table 2. Candidate eukaryotic VLP-based L2 HPV vaccines.

Source of VLPs	HPV L2 Peptide Displayed (aa)	Length of Peptide	Position of Insertion on VLPs	Dose Immunized with	Neutralized or Protected against HPV Pseudoviruses (PsVs) *	Ref #.
HPV16 L1	HPV16 L2 (aa 414–426)	13	Helix 4 loop (aa 414–426 replaced with L2)	Mice immunized with three doses of VLPs (100 µg/dose)	Assays not conducted; however, the peptide has been shown, in previous studies [63], to neutralize HPV 16 and HPV 6	[64]
HPV16 L1	HPV16 L2 (aa 18–38)	21	Helix 4 loop (between amino acids 430 and 433)	Rabbits immunized with five doses of VLPs (concentration not provided)	PsVs (neutralized): **16**/18/**31**/52/58	[65]
	HPV16 L2 (aa 56–75)	20			PsVs (neutralized): **16**/18/**31**/**52**/58	
	HPV16 L2 (aa 96–115)	20			PsVs (neutralized): **16**/18/**31**/52/58	
BPV1	HPV16 L2 (aa 69–81)	13	Between amino acids 133 and 134 (DE loop)	Rabbits immunized with three doses of VLPs (50 µg/dose)	PsVs (neutralized): PsV11/PsV16	[66]
	HPV16 L2 (aa 108–120)	13			PsVs (neutralized): PsV11/PsV16	
BPV1	HPV16 L2 (aa 17–36)	20	Between amino acids 133 and 134	Rabbits immunized with four doses of VLPs (25–50 µg/dose)	PsVs (neutralized): **5**/**16**/18/**45**/58	[67]
HPV16 L1	HPV16 L2 (aa 17–36)	20	Between amino acids 136 and 137 (DE loop)	Rabbits immunized with four doses of VLPs (25–50 µg/dose)	PsVs (neutralized): 5/6/11/**16**/18/**31**/45/**52**/58/70	[67,68]
					Passive transferred sera protected mice against PsVs (genital): **6**/**16**/**18**/**26**/**31**/**33**/**34**/**35**/**39**/**43**/**44**/**45**/**51**/**52**/**53**/**56**/**58**/**59**/**66**/**68**/**73**	[69]
				Mice immunized with two doses of a mixture of 16L1 VLPs displaying 16L2 and HPV18 L1 VLPs (10 µg/dose)	PsV (genital): **58**	[68]
HPV18 L1	HPV33 L2 (aa 17–36)	20	Between amino acids 134 and 135 (DE loop)	Rabbits immunized with three doses of each VLP (20 µg/dose or 100 µg/dose)	QsV $ (neutralized): 5/6/**11**/**18**/16/31/33/45/52/**58** QsVs (cutaneous): **11**/**18**/**58**	[70]
	HPV58 L2 (aa 56–75)	20	C-terminus		QsV (neutralized): 31/45	
HPV18 L1	HPV33 L2 (aa 17–36)	20	HPV33 L2 inserted on DE loop and HPV58 L2 inserted on C-terminus	Rabbits immunized with three doses of a mixture of 18L1 VLPs displaying 33L2 on DE loop and 58L2 on C- terminus mixed human dose of 16L1/18L1 VLPs (Cervarix vaccine). Two µg/dose of chimeric 18L1-L2 and 1/10th per dose of Cervarix was used	QsVs (cutaneous): 6/**11**/**16**/**18**/**31**/**35**/**39**/**45**/59	[70]
	HPV58 L2 (aa 56–75)	20		Mice immunized with two of a mixture of 18L1 VLPs displaying 33L2 on DE loop and 58L2 on C- terminus mixed human dose of 16L1/18L1 VLPs (Cervarix vaccine). Two µg/dose of chimeric 18L1-L2 and 1/10th per dose of Cervarix was used	PsVs (genital): **11**/**16**/**35**/**45**/**58**/**59**	
HPV18 L1	HPV45 L2 (aa 16–35)	20	Between amino acids 134 and 135	Rabbits immunized with five doses of VLPs (50 µg/dose)	Passive transferred sera protected mice against PsVs (genital): **16**/**18**/**39**/**45**/**59**/**68**	[71]

Table 2. *Cont.*

Source of VLPs	HPV L2 Peptide Displayed (aa)	Length of Peptide	Position of Insertion on VLPs	Dose Immunized with	Neutralized or Protected against HPV Pseudoviruses (PsVs) *	Ref #.
HPV16 L1	HPV31 L2 (aa 17–38)	22	Helix 4 loop (between amino acids 430 and 433)	Mice immunized with four doses of VLPs (10 µg/dose)	PsVs (neutralized): 2/5/6/11/**16**/18/27/**31**/33/35/39/45/52/57/58/**59**/68	[72]
Adeno-associated virus 2 VP3 virus particles	HPV16 L2 (aa 17–36)	20	HPV16 L2 inserted at position 587 and HPV31 L2 inserted at position 453 of VP3	Mice immunized with three doses of particles (1 × 10^{11} per dose)	PsVs (neutralized): **16**/18/**31**/45/52/58	[73]
	HPV31 L2 (aa 17–36)	20		Rabbits immunized with four doses of particles (2 × 10^{12} per dose)	PsVs (neutralized): **16**/18/**31**/45/52/58 Passive transferred sera protected mice against PsVs (genital): **16**	
Adeno-associated virus 2 VP3 virus	HPV16 L2 (aa 17–36)	20	HPV16 L2 inserted at position 587 and HPV31 L2 inserted at position 453 of VP3	Mice immunized with three doses of particles (1 × 10^{12} per dose)	PsVs (neutralized): **16**/18/31/45/58	[74]
	HPV31 L2 (aa 17–36)	20		Rabbits immunized with three doses of particles (1 × 10^{19} per dose)	QsVs (cutaneous): **16/18/31/35/39/45/58/59**	
Adenovirus type 5	HPV16 L2 (12–41)	30	Hexon protein hypervariable regions 1 and 5	Mice immunized with three doses of particles (1 × 10^{10} per dose)	PsVs (genital): **16**/56 PsVs (neutralized): 16/73	[75]
Adenovirus type 35	Concatemers of HPV: 6L2, 11L2, 16L2, 18L2, 31L2, 33L2, 45L2, 52L2, 58L2 (aa 17–36)	60, 80, 100	C-terminus of pIX	Mice immunized with two doses of particles (1 × 10^{10} per dose)	PsVs (neutralized): **16/18/31/59**	[76]
Hepatitis B core (HBc) VLPs	HPV16 L2 (aa 14–122)	108	Inserted between a heterodimer of HBc	Mice immunized with three doses of VLPs (5µg/dose)	PsVs (neutralized): **16**	[77]
Potato virus A VLPs	HPV16 L2 (aa 108–120)	13	L2 inserted on the N-terminus and E7 inserted on the C-terminus	Immunization studies not conducted	No studies	[78]
	HPV16 E7 (aa 44–60)	17				

* Bolded PsVs indicate that protection/neutralization was both significant and complete, while PsVs not bolded (in black) indicate that although the protection/neutralization was significant, it was not complete. PsVs not bolded (in gray background) indicate that there was minimal protection/neutralization or no protection at all (not significant protection); # Ref: Reference; $ QsV: quasivirus (composed of HPV capsid and cottontail rabbit papillomavirus genome).

Of all these VLP insertions, HPV16 L1 VLPs displaying on the DE loop, L2 peptides representing aa 17–36 from HPV16 elicited antibodies that efficiently neutralized/protected against diverse HPV pseudovirus types (6/16/18/26/31/33/34/35/39/43/44/45/51/52/53/56/58/59/66/68/73) [69]. This is followed by HPV18 L1 VLPs displaying, on the DE loop and on the C-terminus, L2 peptides representing aa 17–36 from HPV33 and L2 peptides representing aa 56–75 from HPV58, respectively (Table 2). The VLPs protected rabbits from developing papillomas following challenge with diverse quasiviruses (6/11/16/18/31/35/39/45/59) [70]; the VLPs also protected mice from genital infection with HPV pseudoviruses (11/16/35/45/58/59). It is worth highlighting that HPV16 L1 VLPs displaying, between aa 130/433, L2 peptides representing aa 17–38 from HPV31 elicited antibodies that neutralized (to some degree) diverse HPV pseudovirus types (2/5/6/11/16/18/27/31/33/35/39/45/52/57/58/59/68) [72] (Table 2). Among the L2 insertions displayed on virus/virus-associated particles, Adeno-associated virus 2 VP3 virus particles displaying L2 peptides representing aa 17–36 from both HPV16 and HPV31 protected rabbits. The recombinant particles protected rabbits from developing papillomas following infection with eight different quasiviruses one year after immunization. Moreover, the rabbits did not develop papillomas for 10 weeks (the length of the study).

In summary, the type of peptide and the location of peptide insertion on eukaryotic VLPs make a difference in the neutralization potential. L2 peptide (aa 17–36) inserted on the DE loop elicits efficient broadly protective antibodies against diverse HPV types.

5. Expert Commentary and Perspectives for the Future

Current HPV vaccines based on L1 VLPs are highly immunogenic and offer robust protection against the HPV types included in the vaccines. However, they offer little cross-protection against non-vaccine HPV types [32–36]. Thus, complete protection against all HPV-associated cancers/warts requires the addition of VLPs from other HPV types not included in the vaccines, especially for Cervarix vaccine, which protects mostly against HPV16 and HPV18 (both HPV types are associated with ~70% of cervical cancer). From a production stand-point, the development and addition of more VLPs to current HPV vaccines may be costly. Also, although antigenic competition has not been reported for Gardasil-9 (with VLPs from nine HPV types), it is not clear whether the addition of more VLP types to the vaccine will lead to the immunodominance of certain VLP types. As an alternative to L1 vaccines, the L2 protein looks like an attractive target for a next-generation vaccine against HPV. Nevertheless, as mentioned above, L2 does not form VLPs and as such, it is less immunogenic. Fortunately, the display of peptides from L2 protein on VLPs enhances the immunogenicity of the peptides as well as its potential to protect against diverse HPV types. Although L2 peptides can be displayed by chemical conjugation or by genetic insertions, display by genetic insertion from a production standpoint seems to be the most cost-effective strategy. Constructs with genetically inserted L2 peptides can easily be expressed cheaply, at a large scale, in a host cell compared with chemical conjugation, whereby the peptides have to be synthesized commercially. Thus, chemical conjugation approach can be cost-prohibitive.

Among all the L2 peptides that have been displayed on VLPs, peptides representing amino acids 17–31 or 17–36 from HPV16 (and consensus sequence amino acid 69–86 to an extent) elicit antibodies with the broadest and the most efficient level of protection against diverse HPV types, regardless of whether a bacteriophage or eukaryotic VLP-platform was used to display the peptides. Within bacteriophage VLP platforms, MS2 VLPs displaying HPV16 L2 peptide (amino acids 17–31) elicit better cross-protection, while within eukaryotic VLP-platforms, HPV16 L1 VLPs displaying the HPV16 L2 peptide (amino acids 17–36) elicit better cross-protection. It is worth mentioning that the location of L2 insertion on the coat protein affects the immunogenicity of the peptide as well as the level of cross-protection. For bacteriophage VLPs, the insertion of amino acids 17–31 on the N-terminus seems to elicit better cross-protective antibodies against diverse HPV types compared to the same peptide inserted on the AB-loop ([5] and our unpublished data). While for HPV16L1 VLPs, the insertion of amino acids 17–36 on the DE loop seems to elicit better cross-protection against diverse HPV types [69] compared with insertions of same peptide on H4 helix (Table 2) [65,72].

As already highlighted, peptides representing amino acids 17–31 or 17–36 from HPV16 (compared to the same peptide from other HPV types) elicit robust broadly protective antibodies against diverse HPV types (6/11/16/18/26/31/33/34/35/39/43/44/45/51/52/53/56/58/59/66/68/73) [57,62,67–69]. This is probably due to the fact that amino acid 17–36 from HPV16 is 90% identical to a consensus sequence of amino acid 17–36 (Figure 3). Thus, immunization with L2 peptide (amino acid 17–36) from HPV16 is almost like immunizing with a consensus peptide from this region. It is unclear why the insertion of this epitope on the N-terminus of MS2 VLPs as well as on the DE loop of HPV16L1 VLPs elicits robust cross-protective antibodies against diverse HPV types but marginal cross-protection when the epitope is inserted on H4 helix of HPV16L1 VLPs (Table 2). This could be explained by the location of the insertion and whether the whole peptide is readily displayed on the surface of the VLP and to the immune system. Modelled cryo-electron microscopy images of MS2 [79] and HPV16 L1 [80] seem to show that the N-termini of MS2 coat proteins and the DE loops of HPV16 L1 coat proteins (where the peptides are inserted) are readily exposed on the surface of their icosahedral structures compared with the H4 helix of HPV16 L1 (Figure 4). In addition to this, insertions at these different locations can affect the assembly of the chimeric coat proteins into VLPs and ultimately, their immunogenicity. Transmission electron microscopy images of chimeric MS2 VLPs and HPV16 L1 VLPs with L2 peptide amino acid 17–36 insertions at the N-terminus and DE loop, respectively, show VLPs with conformations similar to their respective wild-type VLPs [5,67]. However, insertion of the same peptide on the H4 helix of HPV16 L1 coat protein gives rise to VLPs, some of which have irregular conformations and difference sizes [65]. Thus, the location of the insertion and/or the display of L2 peptide on the surface of a VLP platform is important in eliciting robust broadly protective antibodies against diverse HPV types. In summary, the N-terminus of MS2 and the DE loop of HPV16 L1 seem to be optimal locations to display L2 peptides, especially peptide 17–36.

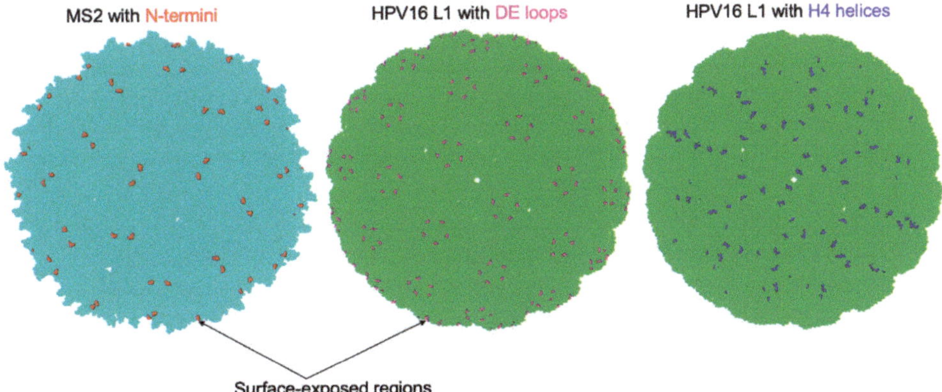

Figure 4. Model cryo-electron microscopy images of icosahedral structures of MS2 (light blue) and HPV16 L1 (light green) derived from protein data bank (PDB) with PDB identification numbers 2WBH and 5KEP, respectively. The N-termini on MS2 coat proteins where L2 peptides have been inserted are shown in red (left image). The DE loops on HPV16 L1 coat proteins where L2 peptides have been inserted are shown in magenta (middle image). The H4 helices on HPV16 L1 coat proteins where L2 peptides have been inserted are shown in blue (right image).

Overall, VLP-based L2 vaccines against diverse HPV infections look promising and may be favorable, in the near future, over current L1-based HPV vaccines. HPV16 L1-16L2 (aa 17–31) VLPs and mixed MS2-L2 VLPs ((MS2–16L2/31L2 and MS2–consL2 (69–86)) can be formulated as a monovalent vaccine [67,69] or bivalent vaccine [57,62], respectively, compared with current HPV vaccines (especially the nonavalent Gardasil-9). HPV16 L1 VLPs displaying peptide 17–36 on the DE loop neutralize/protect against sexually transmitted HPVs associated with cancers/warts as well as against non-sexually transmitted HPV types (HPV 3, HPV5) associated with cutaneous warts [67,69]. On the other hand, mixed MS2-L2 VLPs have been shown to protect, so far, against sexually transmitted HPVs associated with cancers/warts [57,62]. Mixed MS2-L2 VLPs have also been assessed against oral HPV infection, and the results look promising.

Although aforementioned preclinical protective studies look promising, there are still some questions that remain to be answered with both candidate vaccines. For example, we still do not know if immune responses elicited by HPV16 L1-16L2 (aa 17–31) VLPs or mixed MS2-L2 VLPs will last 3–8 years, as has been shown with a single dose of Gardasil-4 [21]. Recent preclinical studies with four doses (25–50 µg/dose) of HPV16 L1-16L2 (aa 17–31) VLPs show that the antibodies are protective 1 year after immunization [69]. Ten micrograms/dose (two doses total) of MS2 VLPs displaying peptide 17–36 has been shown in preclinical studies to be protective 18 months after immunization [81]. It is yet to be seen how long immune responses elicited by the mixed MS2-L2 will last. Thus, studies are required to assess the longevity of protection at different doses for these VLP-based L2 candidate vaccines.

Additionally, studies are required to assess the thermostability of candidate vaccines. Gardasil-9 has recently been shown to be thermostable at temperatures of up to 40 °C for three months [14,82]. Insertion of peptides on VLPs has been shown to decrease the thermostability of chimeric VLPs [83]. Thus, it would be nice to know if these candidate VLP-based L2 vaccines are thermostable given the fact that refrigeration and temperature-monitoring facilities are underdeveloped in developing countries, which have the highest burden of HPV infection and are in dire need of the vaccines. If the candidate vaccines are not thermostable, their thermostability can be enhanced by spray drying or spray freeze-drying. We have recently shown that spray freeze-dried mixed MS2-L2 VLPs stored at room temperature for two months are thermostable and protective [62].

Studies are also required to assess the effect of pre-existing L1 antibodies (from natural infection) to the immunogenicity of HPV16 L1-16L2(aa 17–31) VLPs. Studies have shown that pre-existing antibodies to some platforms can reduce the immunogenicity of the platforms and consequently, the immunogenicity of peptide displayed on the platforms [84–89]. This may not be an issue with mixed MS2-L2 VLPs given the fact that antibodies against MS2 (including PP7 or Qβ bacteriophages) have never been detected in humans. We have not come across any data documenting the presence of antibodies against these viruses in humans. Taken together, VLPs derived from bacteriophages (MS2) or eukaryotic viruses (HPV16 L1) displaying L2 peptides have the potential to serve as next-generation vaccines against HPVs and should be explored further.

Author Contributions: R.Y., L.Z., and E.T. reviewed the literature, generated the figures/tables, and wrote the review paper. All authors have read and agreed to the published version of the manuscript.

Funding: This work was supported by a grant (1R15 DE025812-01A1) from the US National Institute of Dental & Craniofacial Research of the National Institutes of Health.

Conflicts of Interest: Ebenezer Tumban is a co-inventor of an L2-bacteriophage VLP-related patent application licensed to Agilvax Biotech. Interactions with Agilvax are managed by the University of New Mexico in accordance with its conflict of interest policies. The content is solely the responsibility of the authors and does not necessarily represent the official views of the National Institutes of Health. The funding agency had no role in analyses or interpretation of data nor in the writing of the review; or the decision to publish the review paper.

References

1. Dai, S.; Wang, H.; Deng, F. Advances and challenges in enveloped virus-like particle (VLP)-based vaccines. *J. Immunol. Sci.* **2018**, *2*, 36–41.

2. Chroboczek, J.; Szurgot, I.; Szolajska, E. Virus-like particles as vaccine. *Acta Biochim. Pol.* **2014**, *61*, 531–539. [CrossRef] [PubMed]

3. Schwarz, B.; Uchida, M.; Douglas, T. Biomedical and Catalytic Opportunities of Virus-Like Particles in Nanotechnology. *Adv. Virus Res.* **2017**, *97*, 1–60. [PubMed]

4. Zeltins, A. Construction and characterization of virus-like particles: A review. *Mol. Biotechnol.* **2013**, *53*, 92–107. [CrossRef]

5. Tumban, E.; Peabody, J.; Tyler, M.; Peabody, D.S.; Chackerian, B. VLPs displaying a single L2 epitope induce broadly cross-neutralizing antibodies against human papillomavirus. *PLoS ONE* **2012**, *7*, e49751. [CrossRef]

6. Chackerian, B.; Durfee, M.R.; Schiller, J.T. Virus-like display of a neo-self antigen reverses B cell anergy in a B cell receptor transgenic mouse model. *J. Immunol.* **2008**, *180*, 5816–5825. [CrossRef]

7. Yuseff, M.I.; Pierobon, P.; Reversat, A.; Lennon-Dumenil, A.M. How B cells capture, process and present antigens: A crucial role for cell polarity. *Nat. Rev. Immunol.* **2013**, *13*, 475–486. [CrossRef]

8. Zabel, F.; Kundig, T.M.; Bachmann, M.F. Virus-induced humoral immunity: On how B cell responses are initiated. *Curr. Opin. Virol.* **2013**, *3*, 357–362. [CrossRef]

9. Bachmann, M.F.; Jennings, G.T. Vaccine delivery: A matter of size, geometry, kinetics and molecular patterns. *Nat. Rev. Immunol.* **2010**, *10*, 787–796. [CrossRef]

10. Peabody, D.S.; Manifold-Wheeler, B.; Medford, A.; Jordan, S.K.; do Carmo Caldeira, J.; Chackerian, B. Immunogenic display of diverse peptides on virus-like particles of RNA phage MS2. *J. Mol. Biol.* **2008**, *380*, 252–263. [CrossRef]

11. Caldeira Jdo, C.; Medford, A.; Kines, R.C.; Lino, C.A.; Schiller, J.T.; Chackerian, B.; Peabody, D.S. Immunogenic display of diverse peptides, including a broadly cross-type neutralizing human papillomavirus L2 epitope, on virus-like particles of the RNA bacteriophage PP7. *Vaccine* **2010**, *28*, 4384–4393. [CrossRef] [PubMed]

12. Ibanez, L.I.; Roose, K.; De Filette, M.; Schotsaert, M.; De Sloovere, J.; Roels, S.; Pollard, C.; Schepens, B.; Grooten, J.; Fiers, W.; et al. M2e-displaying virus-like particles with associated RNA promote T helper 1 type adaptive immunity against influenza A. *PLoS ONE* **2013**, *8*, e59081. [CrossRef] [PubMed]

13. Manolova, V.; Flace, A.; Bauer, M.; Schwarz, K.; Saudan, P.; Bachmann, M.F. Nanoparticles target distinct dendritic cell populations according to their size. *Eur. J. Immunol.* **2008**, *38*, 1404–1413. [CrossRef] [PubMed]

14. Tumban, E.; Peabody, J.; Peabody, D.S.; Chackerian, B. A universal virus-like particle-based vaccine for human papillomavirus: Longevity of protection and role of endogenous and exogenous adjuvants. *Vaccine* **2013**, *31*, 4647–4654. [CrossRef]

15. Gara, N.; Abdalla, A.; Rivera, E.; Zhao, X.; Werner, J.M.; Liang, T.J.; Hoofnagle, J.H.; Rehermann, B.; Ghany, M.G. Durability of antibody response against hepatitis B virus in healthcare workers vaccinated as adults. *Clin. Infect. Dis. Off. Publ. Infect. Dis. Soc. Am.* **2015**, *60*, 505–513. [CrossRef]

16. U.S. Department of Health and Human Services. Human Papillomavirus (HPV). Available online: https://www.hhs.gov/opa/reproductive-health/fact-sheets/sexually-transmitted-diseases/hpv/index.htm (accessed on 17 September 2019).

17. Zhai, L.; Tumban, E. Gardasil-9: A global survey of projected efficacy. *Antivir. Res.* **2016**, *130*, 101–109. [CrossRef]

18. Tumban, E. A Current Update on Human Papillomavirus-Associated Head and Neck Cancers. *Viruses* **2019**, *11*, 922. [CrossRef]

19. Harper, D.M.; DeMars, L.R. HPV vaccines—A review of the first decade. *Gynecol. Oncol.* **2017**, *146*, 196–204. [CrossRef]

20. Haghshenas, M.R.; Mousavi, T.; Kheradmand, M.; Afshari, M.; Moosazadeh, M. Efficacy of Human Papillomavirus L1 Protein Vaccines (Cervarix and Gardasil) in Reducing the Risk of Cervical Intraepithelial Neoplasia: A Meta-analysis. *Int. J. Prev. Med.* **2017**, *8*, 44.

21. Gilca, V.; Sauvageau, C.; Panicker, G.; De Serres, G.; Ouakki, M.; Unger, E.R. Antibody persistence after a single dose of quadrivalent HPV vaccine and the effect of a dose of nonavalent vaccine given 3–8 years later—An exploratory study. *Hum. Vaccin. Immunother.* **2018**, *155*, 503–507. [CrossRef]

22. Kirnbauer, R.; Booy, F.; Cheng, N.; Lowy, D.R.; Schiller, J.T. Papillomavirus L1 major capsid protein self-assembles into virus-like particles that are highly immunogenic. *Proc. Natl. Acad. Sci. USA* **1992**, *89*, 12180–12184. [CrossRef] [PubMed]

23. Kines, R.C.; Thompson, C.D.; Lowy, D.R.; Schiller, J.T.; Day, P.M. The initial steps leading to papillomavirus infection occur on the basement membrane prior to cell surface binding. *Proc. Natl. Acad. Sci. USA* **2009**, *106*, 20458–20463. [CrossRef] [PubMed]

24. Bywaters, S.M.; Brendle, S.A.; Tossi, K.P.; Biryukov, J.; Meyers, C.; Christensen, N.D. Antibody Competition Reveals Surface Location of HPV L2 Minor Capsid Protein Residues 17–36. *Viruses* **2017**, *9*, 336. [CrossRef]

25. Buck, C.B.; Cheng, N.; Thompson, C.D.; Lowy, D.R.; Steven, A.C.; Schiller, J.T.; Trus, B.L. Arrangement of L2 within the papillomavirus capsid. *J. Virol.* **2008**, *82*, 5190–5197. [CrossRef] [PubMed]

26. Volpers, C.; Schirmacher, P.; Streeck, R.E.; Sapp, M. Assembly of the major and the minor capsid protein of human papillomavirus type 33 into virus-like particles and tubular structures in insect cells. *Virology* **1994**, *200*, 504–512. [CrossRef] [PubMed]

27. Holmgren, S.C.; Patterson, N.A.; Ozbun, M.A.; Lambert, P.F. The minor capsid protein L2 contributes to two steps in the human papillomavirus type 31 life cycle. *J. Virol.* **2005**, *79*, 3938–3948. [CrossRef] [PubMed]

28. Wang, J.W.; Roden, R.B. L2, the minor capsid protein of papillomavirus. *Virology* **2013**, *445*, 175–186. [CrossRef]

29. Kamper, N.; Day, P.M.; Nowak, T.; Selinka, H.C.; Florin, L.; Bolscher, J.; Hilbig, L.; Schiller, J.T.; Sapp, M. A membrane-destabilizing peptide in capsid protein L2 is required for egress of papillomavirus genomes from endosomes. *J. Virol.* **2006**, *80*, 759–768. [CrossRef]

30. Yang, R.; Yutzy, W.H., 4th; Viscidi, R.P.; Roden, R.B. Interaction of L2 with beta-actin directs intracellular transport of papillomavirus and infection. *J. Biol. Chem.* **2003**, *278*, 12546–12553. [CrossRef]

31. Campos, S.K. Subcellular Trafficking of the Papillomavirus Genome during Initial Infection: The Remarkable Abilities of Minor Capsid Protein L2. *Viruses* **2017**, *9*, 370. [CrossRef]

32. Brown, D.R.; Kjaer, S.K.; Sigurdsson, K.; Iversen, O.E.; Hernandez-Avila, M.; Wheeler, C.M.; Perez, G.; Koutsky, L.A.; Tay, E.H.; Garcia, P.; et al. The impact of quadrivalent human papillomavirus (HPV; types 6, 11, 16, and 18) L1 virus-like particle vaccine on infection and disease due to oncogenic nonvaccine HPV types in generally HPV-naive women aged 16-26 years. *J. Infect. Dis.* **2009**, *199*, 926–935. [CrossRef] [PubMed]

33. Joura, E.A.; Giuliano, A.R.; Iversen, O.E.; Bouchard, C.; Mao, C.; Mehlsen, J.; Moreira, E.D., Jr.; Ngan, Y.; Petersen, L.K.; Lazcano-Ponce, E.; et al. A 9-valent HPV vaccine against infection and intraepithelial neoplasia in women. *N. Engl. J. Med.* **2015**, *372*, 711–723. [CrossRef] [PubMed]

34. Smith, J.F.; Brownlow, M.; Brown, M.; Kowalski, R.; Esser, M.T.; Ruiz, W.; Barr, E.; Brown, D.R.; Bryan, J.T. Antibodies from women immunized with Gardasil cross-neutralize HPV 45 pseudovirions. *Hum. Vaccin.* **2007**, *3*, 109–115. [CrossRef] [PubMed]

35. Toft, L.; Storgaard, M.; Muller, M.; Sehr, P.; Bonde, J.; Tolstrup, M.; Ostergaard, L.; Sogaard, O.S. Comparison of the immunogenicity and reactogenicity of Cervarix and Gardasil human papillomavirus vaccines in HIV-infected adults: A randomized, double-blind clinical trial. *J. Infect. Dis.* **2014**, *209*, 1165–1173. [CrossRef]

36. Wheeler, C.M.; Kjaer, S.K.; Sigurdsson, K.; Iversen, O.E.; Hernandez-Avila, M.; Perez, G.; Brown, D.R.; Koutsky, L.A.; Tay, E.H.; Garcia, P.; et al. The impact of quadrivalent human papillomavirus (HPV; types 6, 11, 16, and 18) L1 virus-like particle vaccine on infection and disease due to oncogenic nonvaccine HPV types in sexually active women aged 16-26 years. *J. Infect. Dis.* **2009**, *199*, 936–944. [CrossRef]

37. Alphs, H.H.; Gambhira, R.; Karanam, B.; Roberts, J.N.; Jagu, S.; Schiller, J.T.; Zeng, W.; Jackson, D.C.; Roden, R.B. Protection against heterologous human papillomavirus challenge by a synthetic lipopeptide vaccine containing a broadly cross-neutralizing epitope of L2. *Proc. Natl. Acad. Sci. USA* **2008**, *105*, 5850–5855. [CrossRef]

38. Christensen, N.D.; Kreider, J.W.; Kan, N.C.; DiAngelo, S.L. The open reading frame L2 of cottontail rabbit papillomavirus contains antibody-inducing neutralizing epitopes. *Virology* **1991**, *181*, 572–579. [CrossRef]

39. Gambhira, R.; Jagu, S.; Karanam, B.; Gravitt, P.E.; Culp, T.D.; Christensen, N.D.; Roden, R.B. Protection of rabbits against challenge with rabbit papillomaviruses by immunization with the N terminus of human papillomavirus type 16 minor capsid antigen L2. *J. Virol.* **2007**, *81*, 11585–11592. [CrossRef]

40. Pastrana, D.V.; Gambhira, R.; Buck, C.B.; Pang, Y.Y.; Thompson, C.D.; Culp, T.D.; Christensen, N.D.; Lowy, D.R.; Schiller, J.T.; Roden, R.B. Cross-neutralization of cutaneous and mucosal Papillomavirus types with anti-sera to the amino terminus of L2. *Virology* **2005**, *337*, 365–372. [CrossRef]

41. Roden, R.B.; Yutzy, W.H., 4th; Fallon, R.; Inglis, S.; Lowy, D.R.; Schiller, J.T. Minor capsid protein of human genital papillomaviruses contains subdominant, cross-neutralizing epitopes. *Virology* **2000**, *270*, 254–257. [CrossRef]

42. Slingluff, C.L., Jr. The present and future of peptide vaccines for cancer: Single or multiple, long or short, alone or in combination? *Cancer J.* **2011**, *17*, 343–350. [CrossRef] [PubMed]

43. Li, W.; Joshi, M.D.; Singhania, S.; Ramsey, K.H.; Murthy, A.K. Peptide Vaccine: Progress and Challenges. *Vaccines* **2014**, *2*, 515–536. [CrossRef] [PubMed]

44. Rubio, I.; Bolchi, A.; Moretto, N.; Canali, E.; Gissmann, L.; Tommasino, M.; Muller, M.; Ottonello, S. Potent anti-HPV immune responses induced by tandem repeats of the HPV16 L2 (20–38) peptide displayed on bacterial thioredoxin. *Vaccine* **2009**, *27*, 1949–1956. [CrossRef] [PubMed]

45. Seitz, H.; Canali, E.; Ribeiro-Muller, L.; Palfi, A.; Bolchi, A.; Tommasino, M.; Ottonello, S.; Muller, M. A three component mix of thioredoxin-L2 antigens elicits broadly neutralizing responses against oncogenic human papillomaviruses. *Vaccine* **2014**, *32*, 2610–2617. [CrossRef] [PubMed]
46. Kalnin, K.; Chivukula, S.; Tibbitts, T.; Yan, Y.; Stegalkina, S.; Shen, L.; Cieszynski, J.; Costa, V.; Sabharwal, R.; Anderson, S.F.; et al. Incorporation of RG1 epitope concatemers into a self-adjuvanting Flagellin-L2 vaccine broaden durable protection against cutaneous challenge with diverse human papillomavirus genotypes. *Vaccine* **2017**, *35*, 4942–4951. [CrossRef]
47. Zhang, T.; Chen, X.; Liu, H.; Bao, Q.; Wang, Z.; Liao, G.; Xu, X. A rationally designed flagellin-L2 fusion protein induced serum and mucosal neutralizing antibodies against multiple HPV types. *Vaccine* **2019**, *37*, 4022–4030. [CrossRef] [PubMed]
48. Jagu, S.; Karanam, B.; Gambhira, R.; Chivukula, S.V.; Chaganti, R.J.; Lowy, D.R.; Schiller, J.T.; Roden, R.B. Concatenated multitype L2 fusion proteins as candidate prophylactic pan-human papillomavirus vaccines. *J. Natl. Cancer Inst.* **2009**, *101*, 782–792. [CrossRef]
49. Jagu, S.; Kwak, K.; Garcea, R.L.; Roden, R.B. Vaccination with multimeric L2 fusion protein and L1 VLP or capsomeres to broaden protection against HPV infection. *Vaccine* **2010**, *28*, 4478–4486. [CrossRef]
50. Pouyanfard, S.; Spagnoli, G.; Bulli, L.; Balz, K.; Yang, F.; Odenwald, C.; Seitz, H.; Mariz, F.C.; Bolchi, A.; Ottonello, S.; et al. Minor Capsid Protein L2 Polytope Induces Broad Protection against Oncogenic and Mucosal Human Papillomaviruses. *J. Virol.* **2018**, *92*, e01930-17. [CrossRef]
51. Seitz, H.; Ribeiro-Muller, L.; Canali, E.; Bolchi, A.; Tommasino, M.; Ottonello, S.; Muller, M. Robust In Vitro and In Vivo Neutralization against Multiple High-Risk HPV Types Induced by a Thermostable Thioredoxin-L2 Vaccine. *Cancer Prev. Res. (Phila.)* **2015**, *8*, 932–941. [CrossRef]
52. Spagnoli, G.; Pouyanfard, S.; Cavazzini, D.; Canali, E.; Maggi, S.; Tommasino, M.; Bolchi, A.; Muller, M.; Ottonello, S. Broadly neutralizing antiviral responses induced by a single-molecule HPV vaccine based on thermostable thioredoxin-L2 multiepitope nanoparticles. *Sci. Rep.* **2017**, *7*, 18000. [CrossRef] [PubMed]
53. Ramasamy, V.; Arora, U.; Shukla, R.; Poddar, A.; Shanmugam, R.K.; White, L.J.; Mattocks, M.M.; Raut, R.; Perween, A.; Tyagi, P.; et al. A tetravalent virus-like particle vaccine designed to display domain III of dengue envelope proteins induces multi-serotype neutralizing antibodies in mice and macaques which confer protection against antibody dependent enhancement in AG129 mice. *PLoS Negl. Trop. Dis.* **2018**, *12*, e0006191. [CrossRef] [PubMed]
54. Daly, S.M.; Joyner, J.A.; Triplett, K.D.; Elmore, B.O.; Pokhrel, S.; Frietze, K.M.; Peabody, D.S.; Chackerian, B.; Hall, P.R. VLP-based vaccine induces immune control of Staphylococcus aureus virulence regulation. *Sci. Rep.* **2017**, *7*, 637. [CrossRef] [PubMed]
55. Crossey, E.; Amar, M.J.A.; Sampson, M.; Peabody, J.; Schiller, J.T.; Chackerian, B.; Remaley, A.T. A cholesterol-lowering VLP vaccine that targets PCSK9. *Vaccine* **2015**, *33*, 5747–5755. [CrossRef] [PubMed]
56. Ong, H.K.; Tan, W.S.; Ho, K.L. Virus like particles as a platform for cancer vaccine development. *PeerJ* **2017**, *5*, e4053. [CrossRef] [PubMed]
57. Zhai, L.; Peabody, J.; Pang, Y.S.; Schiller, J.; Chackerian, B.; Tumban, E. A novel candidate HPV vaccine: MS2 phage VLP displaying a tandem HPV L2 peptide offers similar protection in mice to Gardasil-9. *Antivir. Res.* **2017**, *147*, 116–123. [CrossRef]
58. Tumban, E.; Peabody, J.; Peabody, D.S.; Chackerian, B. A pan-HPV vaccine based on bacteriophage PP7 VLPs displaying broadly cross-neutralizing epitopes from the HPV minor capsid protein, L2. *PLoS ONE* **2011**, *6*, e23310. [CrossRef]
59. Tyler, M.; Tumban, E.; Peabody, D.S.; Chackerian, B. The use of hybrid virus-like particles to enhance the immunogenicity of a broadly protective HPV vaccine. *Biotechnol. Bioeng.* **2014**, *111*, 2398–2406. [CrossRef]
60. Janitzek, C.M.; Peabody, J.; Thrane, S.; Carlsen, P.H.; Theander, T.G.; Salanti, A.; Chackerian, B.; Nielsen, M.A.; Sander, A.F. A proof-of-concept study for the design of a VLP-based combinatorial HPV and placental malaria vaccine. *Sci. Rep.* **2019**, *9*, 5260. [CrossRef]
61. Tyler, M.; Tumban, E.; Dziduszko, A.; Ozbun, M.A.; Peabody, D.S.; Chackerian, B. Immunization with a consensus epitope from human papillomavirus L2 induces antibodies that are broadly neutralizing. *Vaccine* **2014**, *32*, 4267–4274. [CrossRef]
62. Zhai, L.; Yadav, R.; Kunda, N.K.; Anderson, D.; Bruckner, E.; Miller, E.K.; Basu, R.; Muttil, P.; Tumban, E. Oral immunization with bacteriophage MS2-L2 VLPs protects against oral and genital infection with multiple HPV types associated with head & neck cancers and cervical cancer. *Antivir. Res.* **2019**, *166*, 56–65. [PubMed]

63. Kawana, K.; Kawana, Y.; Yoshikawa, H.; Taketani, Y.; Yoshiike, K.; Kanda, T. Nasal immunization of mice with peptide having a cross-neutralization epitope on minor capsid protein L2 of human papillomavirus type 16 elicit systemic and mucosal antibodies. *Vaccine* **2001**, *19*, 1496–1502. [CrossRef]

64. Varsani, A.; Williamson, A.L.; de Villiers, D.; Becker, I.; Christensen, N.D.; Rybicki, E.P. Chimeric human papillomavirus type 16 (HPV-16) L1 particles presenting the common neutralizing epitope for the L2 minor capsid protein of HPV-6 and HPV-16. *J. Virol.* **2003**, *77*, 8386–8393. [CrossRef] [PubMed]

65. Kondo, K.; Ochi, H.; Matsumoto, T.; Yoshikawa, H.; Kanda, T. Modification of human papillomavirus-like particle vaccine by insertion of the cross-reactive L2-epitopes. *J. Med. Virol.* **2008**, *80*, 841–846. [CrossRef] [PubMed]

66. Slupetzky, K.; Gambhira, R.; Culp, T.D.; Shafti-Keramat, S.; Schellenbacher, C.; Christensen, N.D.; Roden, R.B.; Kirnbauer, R. A papillomavirus-like particle (VLP) vaccine displaying HPV16 L2 epitopes induces cross-neutralizing antibodies to HPV11. *Vaccine* **2007**, *25*, 2001–2010. [CrossRef] [PubMed]

67. Schellenbacher, C.; Roden, R.; Kirnbauer, R. Chimeric L1-L2 virus-like particles as potential broad-spectrum human papillomavirus vaccines. *J. Virol.* **2009**, *83*, 10085–10095. [CrossRef] [PubMed]

68. Schellenbacher, C.; Huber, B.; Skoll, M.; Shafti-Keramat, S.; Roden, R.B.S.; Kirnbauer, R. Incorporation of RG1 epitope into HPV16L1-VLP does not compromise L1-specific immunity. *Vaccine* **2019**, *37*, 3529–3534. [CrossRef] [PubMed]

69. Schellenbacher, C.; Kwak, K.; Fink, D.; Shafti-Keramat, S.; Huber, B.; Jindra, C.; Faust, H.; Dillner, J.; Roden, R.B.S.; Kirnbauer, R. Efficacy of RG1-VLP vaccination against infections with genital and cutaneous human papillomaviruses. *J. Investig. Dermatol.* **2013**, *133*, 2706–2713. [CrossRef]

70. Boxus, M.; Fochesato, M.; Miseur, A.; Mertens, E.; Dendouga, N.; Brendle, S.; Balogh, K.K.; Christensen, N.D.; Giannini, S.L. Broad Cross-Protection Is Induced in Preclinical Models by a Human Papillomavirus Vaccine Composed of L1/L2 Chimeric Virus-Like Particles. *J. Virol.* **2016**, *90*, 6314–6325. [CrossRef]

71. Huber, B.; Schellenbacher, C.; Jindra, C.; Fink, D.; Shafti-Keramat, S.; Kirnbauer, R. A chimeric 18L1-45RG1 virus-like particle vaccine cross-protects against oncogenic alpha-7 human papillomavirus types. *PLoS ONE* **2015**, *10*, e0120152. [CrossRef]

72. Chen, X.; Zhang, T.; Liu, H.; Hao, Y.; Liao, G.; Xu, X. Displaying 31RG-1 peptide on the surface of HPV16 L1 by use of a human papillomavirus chimeric virus-like particle induces cross-neutralizing antibody responses in mice. *Hum. Vaccin. Immunother.* **2018**, *14*, 2025–2033. [CrossRef] [PubMed]

73. Nieto, K.; Weghofer, M.; Sehr, P.; Ritter, M.; Sedlmeier, S.; Karanam, B.; Seitz, H.; Muller, M.; Kellner, M.; Horer, M.; et al. Development of AAVLP(HPV16/31L2) particles as broadly protective HPV vaccine candidate. *PLoS ONE* **2012**, *7*, e39741. [CrossRef] [PubMed]

74. Jagu, S.; Karanam, B.; Wang, J.W.; Zayed, H.; Weghofer, M.; Brendle, S.A.; Balogh, K.K.; Tossi, K.P.; Roden, R.B.S.; Christensen, N.D. Durable immunity to oncogenic human papillomaviruses elicited by adjuvanted recombinant Adeno-associated virus-like particle immunogen displaying L2 17-36 epitopes. *Vaccine* **2015**, *33*, 5553–5563. [CrossRef] [PubMed]

75. Wu, W.H.; Alkutkar, T.; Karanam, B.; Roden, R.B.; Ketner, G.; Ibeanu, O.A. Capsid display of a conserved human papillomavirus L2 peptide in the adenovirus 5 hexon protein: A candidate prophylactic hpv vaccine approach. *Virol. J.* **2015**, *12*, 140. [CrossRef]

76. Vujadinovic, M.; Khan, S.; Oosterhuis, K.; Uil, T.G.; Wunderlich, K.; Damman, S.; Boedhoe, S.; Verwilligen, A.; Knibbe, J.; Serroyen, J.; et al. Adenovirus based HPV L2 vaccine induces broad cross-reactive humoral immune responses. *Vaccine* **2018**, *36*, 4462–4470. [CrossRef]

77. Diamos, A.G.; Larios, D.; Brown, L.; Kilbourne, J.; Kim, H.S.; Saxena, D.; Palmer, K.E.; Mason, H.S. Vaccine synergy with virus-like particle and immune complex platforms for delivery of human papillomavirus L2 antigen. *Vaccine* **2019**, *37*, 137–144. [CrossRef]

78. Cerovska, N.; Hoffmeisterova, H.; Pecenkova, T.; Moravec, T.; Synkova, H.; Plchova, H.; Veleminsky, J. Transient expression of HPV16 E7 peptide (aa 44–60) and HPV16 L2 peptide (aa 108–120) on chimeric potyvirus-like particles using Potato virus X-based vector. *Protein Expr. Purif.* **2008**, *58*, 154–161. [CrossRef]

79. Plevka, P.; Tars, K.; Liljas, L. Structure and stability of icosahedral particles of a covalent coat protein dimer of bacteriophage MS2. *Protein Sci.* **2009**, *18*, 1653–1661. [CrossRef]

80. Guan, J.; Bywaters, S.M.; Brendle, S.A.; Ashley, R.E.; Makhov, A.M.; Conway, J.F.; Christensen, N.D.; Hafenstein, S. Cryoelectron Microscopy Maps of Human Papillomavirus 16 Reveal L2 Densities and Heparin Binding Site. *Structure* **2017**, *25*, 253–263. [CrossRef]

81. Tumban, E.; Muttil, P.; Escobar, C.A.; Peabody, J.; Wafula, D.; Peabody, D.S.; Chackerian, B. Preclinical refinements of a broadly protective VLP-based HPV vaccine targeting the minor capsid protein, L2. *Vaccine* **2015**, *33*, 3346–3353. [CrossRef]

82. Kunda, N.K.; Peabody, J.; Zhai, L.; Price, D.N.; Chackerian, B.; Tumban, E.; Muttil, P. Evaluation of the thermal stability and the protective efficacy of spray-dried HPV vaccine, Gardasil(R) 9. *Hum. Vaccin. Immunother.* **2019**, *15*, 1995–2002. [CrossRef] [PubMed]

83. Caldeira Jdo, C.; Peabody, D.S. Thermal stability of RNA phage virus-like particles displaying foreign peptides. *J. Nanobiotechnol.* **2011**, *9*, 22. [CrossRef] [PubMed]

84. Zak, D.E.; Andersen-Nissen, E.; Peterson, E.R.; Sato, A.; Hamilton, M.K.; Borgerding, J.; Krishnamurty, A.T.; Chang, J.T.; Adams, D.J.; Hensley, T.R.; et al. Merck Ad5/HIV induces broad innate immune activation that predicts CD8(+) T-cell responses but is attenuated by preexisting Ad5 immunity. *Proc. Natl. Acad. Sci. USA* **2012**, *109*, E3503–E3512. [CrossRef] [PubMed]

85. Knuchel, M.C.; Marty, R.R.; Morin, T.N.; Ilter, O.; Zuniga, A.; Naim, H.Y. Relevance of a pre-existing measles immunity prior immunization with a recombinant measles virus vector. *Hum. Vaccin. Immunother.* **2013**, *9*, 599–606. [CrossRef]

86. Zarnitsyna, V.I.; Lavine, J.; Ellebedy, A.; Ahmed, R.; Antia, R. Multi-epitope Models Explain How Pre-existing Antibodies Affect the Generation of Broadly Protective Responses to Influenza. *PLoS Pathog.* **2016**, *12*, e1005692. [CrossRef]

87. Voysey, M.; Kelly, D.F.; Fanshawe, T.R.; Sadarangani, M.; O'Brien, K.L.; Perera, R.; Pollard, A.J. The Influence of Maternally Derived Antibody and Infant Age at Vaccination on Infant Vaccine Responses: An Individual Participant Meta-analysis. *JAMA Pediatr.* **2017**, *171*, 637–646. [CrossRef]

88. Masat, E.; Pavani, G.; Mingozzi, F. Humoral immunity to AAV vectors in gene therapy: Challenges and potential solutions. *Discov. Med.* **2013**, *15*, 379–389.

89. McCluskie, M.J.; Evans, D.M.; Zhang, N.; Benoit, M.; McElhiney, S.P.; Unnithan, M.; DeMarco, S.C.; Clay, B.; Huber, C.; Deora, A.; et al. The effect of preexisting anti-carrier immunity on subsequent responses to CRM197 or Qb-VLP conjugate vaccines. *Immunopharmacol. Immunotoxicol.* **2016**, *38*, 184–196. [CrossRef]

MDPI

St. Alban-Anlage 66

4052 Basel

Switzerland

Tel. +41 61 683 77 34

Fax +41 61 302 89 18

www.mdpi.com

Viruses Editorial Office

E-mail: viruses@mdpi.com

www.mdpi.com/journal/viruses

www.ingramcontent.com/pod-product-compliance
Lightning Source LLC
LaVergne TN
LVHW070135170726
843515LV00007B/1798